Principles of Colloid and Surface Chemistry

UNDERGRADUATE CHEMISTRY

A Series of Textbooks

edited by
J. J. Lagowski
Department of Chemistry
The University of Texas at Austin

Volume 1: Modern Inorganic Chemistry, *J. J. Lagowski*

Volume 2: Modern Chemical Analysis and Instrumentation, *Harold F. Walton and Jorge Reyes*

Volume 3: Problems in Chemistry, Revised and Expanded, *Henry O. Daley, Jr. and Robert F. O'Malley*

Volume 4: Principles of Colloid and Surface Chemistry, *Paul C. Hiemenz*

Principles of Colloid and Surface Chemistry

Paul C. Hiemenz

Chemistry Department
California State Polytechnic University, Pomona
Pomona, California

MARCEL DEKKER, INC. New York and Basel

Library of Congress Cataloging in Publication Data

Hiemenz, Paul C 1936-
Principles of colloid and surface chemistry

(Undergraduate chemistry; v. 4)
Includes bibliographical references and index.
1. Colloids. 2. Surface chemistry. I. Title.
QD549.H53 541'.345 76-55600
ISBN 0-8247-6573-7

MARCEL DEKKER, INC.
270 Madison Avenue, New York, New York 10016

Current printing (last digit):
10 9 8 7 6 5 4 3 2 1

PRINTED IN THE UNITED STATES OF AMERICA

Preface

Colloid and surface chemistry occupy a paradoxical position among the topics of physical chemistry. These are areas which have tranditionally been considered part of physical chemistry and are currently enjoying more widespread application than ever due to their relevance to environmental and biological problems. At the same time, however, colloid and surface chemistry have virtually disappeared from physical chemistry courses. These topics are largely absent from the contemporary general chemistry course as well. It is possible, therefore, that a student could complete a degree in chemistry without even being able to identify what colloid and surface chemistry are about.

The primary objective of this book is to bridge the gap between today's typical physical chemistry course and the literature of colloid and surface chemistry. The reader is assumed to have completed a course in physical chemistry, but no prior knowledge of the topics under consideration is assumed. The book is, therefore, introductory as far as the topic subjects are concerned, although familiarity with numerous other aspects of physical chemistry is required background.

Since physical chemistry is the point of departure for this presentation, the undergraduate chemistry major is the model reader toward whom the book is addressed. This in no way implies that these are the only students who will study the material contained herein. Students majoring in engineering, biology, physics, materials science, and so on, at both the undergraduate and graduate levels will find aspects of this subject highly useful. The interdisciplinary nature of colloid and surface chemistry is another aspect of these subjects that contributes to their relevance in today's curricula.

This is primarily a textbook, written with student backgrounds, needs, and objectives in mind. There are several ways in which this fact manifests itself in the organization of this book. First, no attempt has been made to review the literature or to describe research frontiers in colloid and surface chemistry. A large literature exits which does these things admirably. Our purpose is to provide the beginner with enough background to make intelligible the journals and monographs which present these topics. References have been limited to monographs, textbooks, and reviews which are especially comprehensive and/or accessible. Second, where derivations are presented, this is done in sufficient detail so that the reader should find them self-explanatory. In areas in which undergraduate chemistry majors have minimal backgrounds or have chronic difficulties—for example, fluid mechanics, classical electromagnetic theory, and electrostatics—the presentations begin at the level of general physics, which may be the student's only prior contact with these topics.

Third, an effort has been made to facilitate calculations by paying special attention to dimensional considerations. The cgs–esu system of units has been used throughout, even though this is gradually being phased out of most books. The reason for keeping these units is the stated objective of relating the student's experiences to the existing literature of colloid and surface chemistry. At present, the cgs–esu system is still the common denominator between the two. A fairly detailed list of conversions between cgs and SI units is included in Appendix C. Finally, a few problems are included in each chapter. These provide an opportunity to apply the concepts of the chapter and indicate the kinds of applications these ideas find.

Not all who use the book will have the time or interest to cover it entirely. In the author's course, about two-thirds of the material is discussed in a one-quarter course. With the same level of coverage, the entire book could be completed in a semester. To cover the amount of material involved, very little time is devoted to derivations except to answer questions. Lecture time is devoted instead to outlining highlights of the material and presenting supplementary examples.

The underlying unity which connects the various topics discussed here is seen most clearly when the book is studied in its entirety and in the order presented. Time limitations and special interests often interfere with this ideal. Those who choose to rearrange the sequence of topics should note the subthemes that unify certain blocks of chapters. Chapters 1 through 5 are primarily concerned with particle characterization, especially with respect to molecular weight; Chapters 6 through 8, with surface tension/free energy and adsorption; and Chapters 9 through 11, with flocculation and the electrical double layer. Subjects of special interest to students of the biological sciences are given in Chapters 2 to 5, 7, and 11.

Colloid chemistry and surface chemistry each span virtually the entire field of chemistry. The former may be visualized as a chemistry whose "atoms" are considerably larger than actual atoms; the latter, as a two-dimensional chemistry. The point is that each encompasses all the usual subdivisions of chemistry: reaction chemistry, analytical chemistry, physical chemistry, and so on. The various subdivisions of physical chemistry are also repesented: thermodynamics, structure elucidation, rate processes, and so on. As a consequence, these traditional categories could be used as the basis for organization in a book of this sort. For example, "The Thermodynamics of Surfaces" would be a logical chapter heading according to such a plan of organization. In this book, however, no such chapter exists (although not only chapters but entire volumes on this topic exist elsewhere). The reason goes back to the premise stated earlier: These days most undergraduates know more about thermodynamics than about surfaces, and this is probably true regardless of their thermodynamic literacy/illiteracy! Accordingly, this book discusses surfaces: flat and curved, rigid and mobile, pure substances and solutions, condensed phases and gases. Thermodynamic arguments are presented—along with arguments derived from other sources—in developing an overview of surface chemistry (with the emphasis on "surface"). A more systematic, formal presentation of surface thermodynamics (with the emphasis on "thermodynamics") would be a likely sequel to the study of this book for those who desire still more insight into that aspect of two-dimensional chemistry. Similarly, other topics could be organized differently as well. Only time will tell whether the plan followed in this book succeeds in convincing students that chemistry they have learned in other courses is also applicable to the "in between" dimensions of colloids and the two dimensions of surface chemistry.

The notion that molecules at a surface are in a two-dimensional state of matter is reminiscent of E. A. Abbott's science fiction classic, *Flatland*.* Perusal of this little book for quotations suitable for Chapters 6, 7, and 8 revealed other parallels also: the color revolt and light scattering, "Attend to your Configuration" and the shape of polymer molecules, and so on. Eventually, the objective of beginning each chapter with a quote from *Flatland* replaced the requirement that the passage cited have some actual connection with the contents of the chapter. As it ends up, the quotes are merely for fun: Perhaps those who are not captivated by colloids and surfaces will at least enjoy this glimpse of *Flatland*.

Finally, it is a pleasure to acknowledge those whose contributions helped bring this book into existence. I am grateful to Maurits and Marcel Dekker for the confidence they showed and the encouragement they gave throughout the entire project. I wish to thank Phyllis Bartosh, Felecia Granderson, Jennifer Woodruff, and, especially, Mickie McConnell and Lynda Parzick for making my sloppy manuscript presentable. My appreciation also goes to Bob Marvos, George Phillips, and, especially, Dottie Holmquist for their work on the figures, which are such an important part of any textbook. I also wish to thank Michael Goett for helping with proofreading and indexing. Finally, due to the diligence of the class on whom this material was tested in manuscript form, the book has 395 fewer errors than when it started. For the errors that remain, and I hope they are few in number and minor in magnitude, I am responsible. Reports from readers of errors and/or obscurities will be very much appreciated.

Paul C. Hiemenz

* E. A. Abbott, *Flatland* (6th ed.), Dover, New York, 1952. Used with permission.

Contents

Principles of Colloid and Surface Chemistry

Colloid and Surface Chemistry: Their Scope and Variables

1

Next . . . come the Nobility, of whom there are several degrees, beginning at Six-Sided Figures, . . . and from thence rising in the number of their sides till they receive the honorable title of Polygonal . . . Finally when the number of sides becomes so numerous, and the sides themselves so small, that the figure cannot be distinguished from a circle, he is included in . . . the highest class of all. [From Abbott's *Flatland*]

1.1 INTRODUCTION

"Yesterday, I couldn't define colloid chemistry; today, I'm doing it." This variation of an old quip could apply to many a recent chemistry graduate upon entering employment in the "real world." Two facts underlie this situation. First, colloid and surface chemistry, although traditional parts of physical chemistry, have largely disappeared from introductory physical chemistry courses. Second, in research, technology, and manufacture, countless problems are encountered which fall squarely within the purview of colloid and surface chemistry. Later in this section, we shall enumerate some examples which illustrate this statement.

The paradoxical situation just described means that it is entirely possible for a chemistry student to have completed a course in physical chemistry and still not have any clear idea of what colloid and surface chemistry are about. A book like this one is therefore in the curious position of being simultaneously "advanced" and "introductory." Our discussions are often advanced in the sense of building on topics from physical chemistry. At the same time, we shall have to describe the phenomena under consideration pretty much from scratch, since they are largely unfamiliar. In keeping with this, this chapter is concerned primarily with a broad description of the scope of colloid and surface chemistry and the kinds of variables with which they deal. In subsequent chapters, different specific phenomena are developed in detail.

Our first tasks are to define what we mean by colloid chemistry and how this is related to surface chemistry. For our purposes, any particle which has some linear dimension between 10^{-7} cm (10 Å) and 10^{-4} cm (1 μm or 1 μ)* is considered a

* Appendix C contains a fairly detailed list of conversion factors between cgs and SI units.

colloid. For us, linear dimensions rather than particle weights or the number of atoms in a particle will define the colloidal size range. However, other definitions may be encountered elsewhere. It should be emphasized that these limits are rather arbitrary. Smaller particles are considered within other branches of chemistry and larger ones are considered within sciences other than chemistry. The preceding statement may be expanded still further. Colloid chemistry is interdisciplinary in many respects; its field of interest overlaps physics, biology, materials science, and several other disciplines. It is the particle dimension—not the chemical composition (organic or inorganic), source of the sample (biological or mineralogical), or physical state (one phase or two)—that consigns it to our attention. With this in mind, it is evident that colloid chemistry is the science of both large molecules and finely subdivided multiphase systems.

It is in systems of more than one phase that colloid and surface chemistry meet. The word "surface" is thus used in the chemical sense of a phase boundary, rather than in a strictly geometrical sense. Geometrically, a surface has area, but not thickness. Chemically, however, it is a region in which the properties vary from those of one phase to those of the adjoining phase. This transition occurs over distances of molecular dimensions at least. For us, therefore, a surface has a thickness which we may imagine as shrinking to zero when we desire a purely geometric description.

It is self-evident that the more finely subdivided a given weight of material is the higher the surface area will be for that weight of sample. In the following section we discuss this in considerable detail since it is the basis for combining a discussion of surface and colloid chemistry in a single book.

In subsequent sections of this chapter, we shall discuss further the distinction between macromolecular colloids and multiphase dispersions (Sect. 1.3), the use of the term "stability" in colloid chemistry (Sect. 1.4), the size and shape of colloidal particles (Sect. 1.6), states of aggregation among particles (Sect. 1.7), and the distribution of particle sizes that is typical of virtually all colloidal preparations (Sect. 1.8). The fact that particles in the colloidal size range are not all identical in size also requires a preliminary discussion of statistics which is the subject matter of several sections at the end of this chapter.

One of the basic premises underlying the selection of topics included in this book is that areas of similarity between diverse fields should be stressed. This is not to say, of course, that differences are unimportant. Rather, it seems more valuable to point out to the beginner that useful methods and insights are frequently part of the well-established procedures of other disciplines which deal with related phenomena. Too provincial a viewpoint, especially at the beginning, is apt to isolate the worker from many potentially valuable sources of information. In the long run, this seems like a greater loss than the loss of time that occurs when a worker concludes too hastily that a technique which works well in one system should work equally well in another system where the particles are larger or smaller by several orders of magnitude. Errors of this last sort are generally discovered quickly enough!

Any attempt to enumerate the areas in which surface and colloid chemical concepts find applicability is bound to be incomplete and quite variable with time because of changing technology. Nonetheless, we shall conclude this section with a partial listing of such applications. If there is any difficulty in doing this, it is because of the abundance rather than scarcity of such examples.

Some areas of science and technology in which particles in the colloidal size range are regularly encountered are the following:

1. In analytical chemistry: adsorption indicators, ion exchange, nephelometry, precipitate filterability, chromatography, and decolorization.
2. In physical chemistry: nucleation; superheating, supercooling, and supersaturation; and liquid crystals.
3. In biochemistry and molecular biology: electrophoresis; osmotic and Donnan equilibria and other membrane phenomena; viruses, nucleic acids, and proteins; and hematology.
4. In chemical manufacturing: catalysis, soaps and detergents, paints, adhesives, ink; paper and paper coating; pigments; thickening agents; and lubricants.
5. In environmental science: aerosols, fog and smog, foams, water purification and sewage treatment; cloud seeding; and clean rooms.
6. In materials science: powder metallurgy, alloys, ceramics, cement, fibers, and plastics of all sorts.
7. In petroleum science, geology, and soil science: oil retrieval, emulsification, soil porosity, flotation, and ore enrichment.
8. In household and consumer products: milk and dairy products, beer, waterproofing, cosmetics, and encapsulated products.

It is evident from this partial list how many materials or phenomena of current scientific or everyday interest touch on colloid and surface chemistry to some extent. Many of these areas, of course, have enormous technological and/or theoretical facets which are totally outside our perspective. Nevertheless, all share a common interest in small particles and/or large molecules.

1.2 THE IMPORTANCE OF THE SURFACE FOR SMALL PARTICLES

The contemporary science student is probably aware that the concept of the atom is traceable to early Greek philosophers, notably Democritus. More than likely, however, few have bothered to follow through the hypothetical subdivision process that led to the original concept of an atom. The time has come to remedy this situation since the colloidal size range lies between microscopic chucks of material and individual atoms.

Consider a spherical particle of some unspecified material in which the sphere has a radius of exactly 1.0 cm. What we propose to do is to "reapportion" this fixed quantity of material by subdividing it, first, into an array of spheres, each with a radius half that of the original sphere. In a second subdivision, the radius of each of these spheres will be cut in half again. In the third "cut" the radii will be halved again, and so on. The results of such an exercise are summarized in Table 1.1.

The first line in Table 1.1 shows the volume and area of the original sphere whose radius is 1.0 cm. The second line in the table defines the symbols we use to signify

TABLE 1.1 *The radius, area, and volume per particle, number of particles, and total area for any array of spheres after n "cuts" where a cut is defined to be the reapportionment of material into particles whose radius is half the starting value*

Cut number	Radius (cm)	Number of spheres	Volume per sphere (cm^3)	Area per sphere (cm^2)	Total area (cm^2)
Original	1	1	4.19	1.26×10^1	1.26×10^1
Original, symbol	R_0	N_0	V_0	A_0	$A_{T,0}$
1	5×10^{-1}	8	5.24×10^{-1}	3.14	2.51×10^1
2	2.5×10^{-1}	6.4×10^1	6.55×10^{-2}	7.86×10^{-1}	5.03×10^1
3	1.25×10^{-1}	5.12×10^2	8.18×10^{-3}	1.96×10^{-1}	1.01×10^2
4	6.25×10^{-2}	4.10×10^3	1.02×10^{-3}	4.91×10^{-1}	2.01×10^2
5	3.13×10^{-2}	3.28×10^4	1.28×10^{-4}	1.23×10^{-2}	4.02×10^2
⋮	⋮	⋮	⋮	⋮	⋮
n	$(\frac{1}{2})^n R_0$	$8^n N_0$	$(\frac{1}{8})^n V_0$	$(\frac{1}{4})^n A_0$	$2^n A_{T,0}$
⋮	⋮	⋮	⋮	⋮	⋮
13.29	10^{-4}	10^{12}	4.2×10^{-12}	1.26×10^{-7}	1.26×10^5
16.61	10^{-5}	10^{15}	4.2×10^{-15}	1.26×10^{-9}	1.26×10^6
19.93	10^{-6}	10^{18}	4.2×10^{-18}	1.26×10^{-11}	1.26×10^7
23.25	10^{-7}	10^{21}	4.2×10^{-21}	1.26×10^{-13}	1.26×10^8
26.58	10^{-8}	10^{24}	4.2×10^{-24}	1.26×10^{-15}	1.26×10^9

these and other quantities in the exercise. The next entries in the table show the number of particles, the volume per particle, the area per particle, and the total area of all the particles after five successive reductions by a factor of $\frac{1}{2}$ in the radius. The arithmetical relationships between these quantities should be noted. The volume of a particle varies with the cube of its radius R. Therefore, halving the radius decreases the volume by a factor of 8. Since the total amount of material is unchanged, the number of spheres must be increased by a factor of 8 by this cut. The area varies with R^2; therefore, halving the radius decreases the area per sphere by a factor of 4. Since the number of particles is increased eightfold by the same cut, however, the total area of the array is increased by a factor of 2. These generalizations are summarized in the next line of entries in Table 1.1 which gives general formulas relating each quantity after n halvings to its initial value. In the final entries of the table, the various quantities have been evaluated for successive R values which differ by an order of magnitude.

These results are purely geometrical and, therefore, are independent of any characteristic of the material—almost. It is implied in the calculations reported in Table 1.1 that the density of the material, whatever it may be, remains the same throughout the subdivision process. By assuming that mass and volume bear a constant proportion to one another, the number of particles increases by the same factor that describes the decrease in particle volume. We shall return to this assumption of constant density presently.

We know from other chemical studies that atomic dimensions are of the order of 10^{-8} cm; therefore, from a chemical point of view, the calculations of Table 1.1 have definitely run into trouble by the cut shown as the final entry in the table. To pursue this point a little further, assume the material from which these spheres are made is water. Taking the density of water to be exactly 1.0 g cm^{-3}, it is easy to convert the particle volumes in Table 1.1 into the number of water molecules per sphere after each cut. This information is shown in Table 1.2 for the same cuts that were used in Table 1.1. It is evident that by the time we reach spheres of 1 Å radius we have begun chopping up water molecules. As a matter of fact, calculating by this procedure, we reach Avogadro's number of spheres after 26.33 halvings of all spheres, starting from one sphere of $R = 1.0$ cm. If we use the formulas from Table 1.1 to evaluate the radius of the spheres we have at this stage—that is, the radius of the water molecules themselves—we find this to be 1.18 Å. The radius of a water molecule from the van der Waals b value is about 1.45 Å, so a substantial discrepancy arises when a bulk property, such as density, is applied all the way down to molecular dimensions. Our purpose here is not to arrive at a precise estimate of molecular dimensions, so we shall not worry about this difference. However, the discrepancy does point out the fact that the characterization of a material may be sensitive to the size of the sample under consideration. A property such as density depends not only on the mass and volume of the molecules but also on their packing in a bulk sample.

Next, suppose we calculate the average number of water molecules which reside at the surface of the spheres in Tables 1.1 and 1.2. To do this, we estimate the area occupied by each molecule at the surface to be about 10 $Å^2$. If we divide this figure into the total area at various stages of subdivision from Table 1.1, we obtain an estimate of the number of molecules in the surface at each stage of the processs. Our interest is in the order of magnitude of these quantities, so we need not worry about the cross-sectional shape nor the surface packing efficiency of the water molecules in these calculations. Table 1.2 shows the number of water molecules at the surface calculated in this manner; this quantity is also reported as a fraction of the total number of molecules present. Note that this fraction approaches unity as molecular dimensions are approached.

Finally, let us consider the last column in Table 1.2. In this column, the total surface energy of the array of spherical water droplets is reported at each stage of the process. We shall see in Chap. 6 that the surface tension of a substance measures the energy required to make a unit area of new surface. For water, this quantity is about 72 ergs cm^{-2} at room temperature. If we multiply the surface tension of water by the total areas from Table 1.1, we obtain the values listed under "total surface energy" in Table 1.2. It should be recalled that a fixed amount of material (4.19 cm^3 water = 4.19 g water = 0.23 mole water) is involved throughout this entire process. It should also be noted that surface tension, like density, is a macroscopic property; its applicability is highly dubious for very small particles. Nevertheless, as the dimensions of the subdivided units decrease, the total energy associated with the formation of surface takes on values comparable to other chemical energies. At $R = 10^{-4}$ cm,

TABLE 1.2 *Total number of water molecules per sphere and number at surface for spheres of water after n cuts. Also total surface energy of the array of spheres of water*

Cut number	Radius (cm)	Number of water molecules per sphere	Number of water molecules at surface	Fraction of total water molecules at surface	Total surface energy (ergs)
1	5×10^{-1}	1.40×10^{23}	1.26×10^{16}	2.09×10^{-8}	9.07×10^{2}
2	2.5×10^{-1}	1.75×10^{22}	2.51×10^{16}	4.19×10^{-8}	1.81×10^{3}
3	1.25×10^{-1}	2.19×10^{21}	5.03×10^{16}	8.37×10^{-8}	3.63×10^{3}
4	6.25×10^{-2}	2.73×10^{20}	1.01×10^{17}	1.67×10^{-7}	7.26×10^{3}
5	3.13×10^{-2}	3.42×10^{19}	2.01×10^{17}	3.35×10^{-7}	1.45×10^{4}
⋮	⋮	⋮	⋮	⋮	⋮
13.29	10^{-4}	1.40×10^{11}	1.26×10^{20}	2.09×10^{-4}	9.07×10^{6}
16.61	10^{-5}	1.40×10^{8}	1.26×10^{21}	2.09×10^{-3}	9.07×10^{7}
19.93	10^{-6}	1.40×10^{5}	1.26×10^{22}	2.09×10^{-2}	9.07×10^{8}
23·25	10^{-7}	1.40×10^{2}	1.26×10^{23}	2.09×10^{-1}	9.07×10^{9}
26.58	10^{-8}	1.40×10^{-1}	1.26×10^{24}	2.09	9.07×10^{10}

the surface energy is about 9×10^{6} ergs/0.23 mole or about 1 cal mole^{-1}. By $R=10^{-7}$ cm, this quantity equals 1 kcal mole^{-1}.

The increasing importance of the surface area as the linear dimensions of particles decrease is stated concisely in a quantity known as the specific area of a substance A_{sp}. This quantity is determined as the ratio of the area divided by the mass of an array of particles. If the particles are uniform spheres, as we have assumed throughout this section, this ratio equals

$$A_{sp}=\frac{A_{tot}}{m_{tot}}=\frac{n4\pi R^2}{n\frac{4}{3}\pi R^3\rho} \tag{1}$$

where n is the number of spheres having a radius R and made of a material of density ρ. Simplifying Eq. (1)* leads to the result

$$A_{sp}=\frac{3}{\rho R} \tag{2}$$

This formula generalizes the conclusion reached in Tables 1.1 and 1.2. It shows clearly that, for a fixed amount of material, the surface area is inversely proportional to the radius for uniform, spherical particles. At the same time the formula reminds us that some lower limit for R must be imposed since the relationship is undefined for $R=0$. In the event of nonuniform or nonspherical particles, alternate expressions would have to be used. In no case, however, would the conclusion be altered that the surface plays an increasingly important role as the dimensions of the particles decrease.

* This manner of referencing is used for equations occurring in the same chapter.

The concept of specific area defined by Eq. (1) is important because this is a quantity which can be measured experimentally for finely divided solids without any assumptions as to the shape or uniformity of the particles. We shall discuss the use of gas adsorption to measure A_{sp} in Chap. 8. If the particles are known to be uniform spheres, this measured quantity may be interpreted in terms of Eq. (2) to yield a value of R. If the actual system consists of nonuniform spheres, an average value of the radius may be evaluated by Eq. (2). Finally, even if the particles are nonspherical, a quantity known as the radius of an equivalent sphere may be extracted from experimental A_{sp} values. This often proves to be a valuable way of characterizing an array of irregularly shaped particles. We shall have a good deal more to say about average dimensions later in this chapter, and about equivalent spheres in Chap. 3.

In the foregoing discussion, we have emphasized two-phase colloidal systems in which the concept of "surface" plays an important role. Our definition of the colloidal range is based on the linear dimensions of particles, however, and there are numerous natural and synthetic polymer molecules whose dimensions, considered individually, fall within this range. It is clear when we deal with single molecules that the concept of "surface" is greatly different than when we consider a particle made of many molecules. If, in a given situation, we tend to concentrate on the surface characteristics of a material, then we are working in surface chemistry. If, however, we look at the subdivided sample as an array of particles, then we are working in colloid chemistry. As far as we are concerned, these two fields differ primarily in point of view Their mutual concern with finely subdivided material is the common denominator that connects the two disciplines.

1.3 LYOPHILIC AND LYOPHOBIC COLLOIDS

In the preceding section, we saw that either large molecules or finely subdivided bulk matter could be considered colloids inasmuch as both may consist of particles in the range 10^{-7} to 10^{-4} cm in dimension. The difference between these two situations lies in the relationship that exists between the colloidal particle and the medium in which it is embedded. Macromolecular colloids give true solutions in the thermodynamic sense with the medium which surround them. Subdivided bulk matter, on the other hand, forms a two-phase (at least) system with the medium. We have already noted that the word "surface" connotes the existence of a phase boundary and therefore has a specific chemical meaning in the multiphase case which is inapplicable to macromolecular colloids.

The terms lyophilic and lyophobic are used to distinguish between one-phase and two-phase colloidal systems, repectively. These terms mean, literally, "solvent loving" and "solvent fearing." When water is the medium or solvent, the terms hydrophilic or hydrophobic are often used.

The intent of this book is to discuss as wide a variety of colloidal phenomena as possible from a unified point of view and in a single set of terms. We use the words "continuous" and "dispersed" to refer to the medium and to the particles in the colloidal size range, respectively. It should be understood that these are distinctly different phases in lyophobic systems and are solvent and solute in lyophilic systems.

Furthermore, the system as a whole is generally called a dispersion when we wish to emphasize the colloidal nature of the dispersed particles. This terminology is by no means universal. Lyophilic dispersions are true solutions and may be called such, although this term ignores the colloidal size of the solute molecules. Lyophobic colloids are known by a variety of terms, depending on the nature of the phases involved. Some of these are listed in Table 1.3. Some of the terms (e.g., aerosol, gel) are somewhat ambiguous so the reader is warned to make certain that the system is fully understood, particularly when the original literature is consulted. Remember that a common feature of all systems we consider is that some characteristic linear dimension of the dispersed particles will fall in the range defined in Sec. 1.1. When we deal with two-phase colloids in this book, we are primarily concerned with systems in which the dispersed phase is solid and the continuous phase is liquid.

Next, we will consider another difference between lyophobic and lyophilic colloids in addition to the presence or absence of surfaces between the continuous and dispersed species. This difference deals with the "stability" of the dispersion. We shall examine the meaning(s) of this term in more detail in the next section. For now, the following distinction is sufficient.

Lyophilic colloids can form true solutions, and true solutions are produced spontaneously when solute and solvent are brought together. In the absence of chemical changes or changes of temperature, a solution is stable indefinitely. Finely subdivided dispersions of two phases do not form spontaneously when the two phases are brought together. As a matter of fact, if such a dispersion is allowed to stand long enough, the reverse process would spontaneously occur. For example, oil and water can be vigorously mixed to form a nontransparent, heterogeneous mass. However, on standing, it will separate into two clear, homogeneous layers. We know from thermodynamics that spontaneous processes occur in the direction of decreasing Gibbs free energy. Therefore, we may conclude that the separation of a two-phase dispersed system to form two distinct layers is a change in the direction of decreasing Gibbs free energy. In connection with the discussion of Table 1.2, we saw that there is more surface energy in a two-phase system when the dispersed phase is in a highly subdivided state than when it is in a coarser state of subdivision.

TABLE 1.3 *Summary of some of the descriptive names used to designate two-phase colloidal systems*

Continuous phase	Dispersed phase	Descriptive names
Gas	Liquid	Fog, mist, aerosol
Gas	Solid	Smoke, aerosol
Liquid	Gas	Foam
Liquid	Liquid	Emulsion
Liquid	Solid	Sol, colloidal solution, gel, suspension
Solid	Gas	Solid foam
Solid	Liquid	Gel, solid emulsion
Solid	Solid	Alloy

This suggests a correlation between the inherent instability of a highly dispersed lyophobic system and the thermodynamics of the surface. This is discussed in more detail in Chap. 6. For the present it is sufficient to note that lyophobic systems "dislike" their surroundings enough to want to separate out. Lyophilic systems, on the other hand, are perfectly "happy" in a solution. The two categories differ radically in their thermodynamic stability.

It should also be remembered that thermodynamics has nothing to say about the rate at which processes occur. It is a fact that many two-phase dispersions appear unchanged over very long periods of time. The situation is analogous to the thermodynamic instability of diamond with respect to graphite. The kinetics of the diamond–graphite reaction are slow enough that the thermodynamic instability is of very little practical consequence. Likewise, many colloidal dispersions have kinetic stability, even though they are unstable thermodynamically. This type of colloidal "solution" (Table 1.3) resembles a true solution. This is one way in which two-phase dispersions and solutions of macromolecules are very similar and explains, in part, how the two can be grouped together in our study of colloidal phenomena.

1.4 STABLE AND UNSTABLE SYSTEMS

In the preceding section, we saw that two-phase dispersions will always spontaneously change into a smaller number of large particles, given sufficient time. However, many solutions of macromolecules do not undergo spontaneous separation into two phases. Common usage tempts us to describe the first as "unstable" and the second as "stable." Although these terms are very frequently used in colloid chemistry, the reader should realize that the words are meaningless unless the process to which they are applied has been clearly defined. The situation is somewhat analogous to the insistence in thermochemistry that one always keep in mind the balanced chemical equation to which the thermodynamic quantities (ΔH, ΔC_p, etc.) apply. The coarsening described here is only one of a variety of possible processes that a dispersion might undergo.

The following examples will illustrate this caveat and at the same time help define some of the common processes in colloid chemistry. To begin with, we shall rarely be concerned with chemical reactions in the conventional sense, although there are some important areas such as heterogeneous catalysis in which chemical reactions at surfaces are of primary interest. Our approach to colloid chemistry in this book is directed mainly to the physical chemistry rather than the reaction chemistry of these systems. As a matter of fact, chemical reactions are often unintended side effects in colloidal systems. For example, a protein molecule may undergo hydrolysis to alter the size, shape, and/or the solubility of the molecule. In such a case, we may see a molecule that forms a solution which is thermodynamically stable with respect to phase separation but which is not stable with respect to a chemical change. For the most part, however, we shall regard the systems we discuss as stable with respect to chemical change.

In the preceding section, we noted that two-phase colloids are thermodynamically unstable with respect to the coarsening process. It is important that we define this coarsening in more detail. Specifically, we need to differentiate between coalescence and flocculation. By coalescence we mean the process whereby two (or more) small particles fuse together to form a single larger particle. The central feature of coalescence is the fact that total surface area is reduced. Flocculation is the process whereby small particles clump together like a bunch of grapes (a floc), but do not fuse into a new particle. In flocculation, there is no reduction of surface although certain surface sites may be blocked at the points at which the smaller particles touch.

When small particles coalesce, all evidence of the smaller particles is erased: Only the new, larger particle remains. With flocculation, however, the small particles retain their identity; only their kinetic independence is lost: The floc moves as a single unit. The terms aggregation and coagulation are also used to describe the process we have called flocculation. Likewise, the clusters which form as the products of the process may be called flocs or aggregates. The individual particles from which the flocs are assembled are called primary particles. A system may be relatively stable in the kinetic sense with respect to one of these processes, say coalescence, and be unstable with respect to the other, flocculation.

Finally, the word "stability" is also used to describe the extent to which small particles remain uniformly distributed throughout a sample. In Chap. 3, for example, we shall examine the tendency of heavier particles to settle to the bottom of the container, a process known as sedimentation. It is quite possible for a system to be unstable with respect to sedimentation, but relatively stable (kinetically) with respect to flocculation or coalescence. On the other hand, an array of primary particles may be small enough to be relatively stable with respect to sedimentation but unstable with respect to flocculation.

Systems that are "stable" and "unstable" in any of these processes are not equally esteemed in all areas of application. A dispersion which is too fine to settle out may be a source of great frustration in some parts of a manufacturing process. A dispersion that settles too rapidly may be equally troublesome to those who have to pump the stuff around. The chemist doing gravimetric analyses wants coarse precipitates; the one who uses adsorption indicators wants finely subdivided particles to form.

In summary, then, we must keep several things in mind when the word "stability" is used in colloid chemistry. First, whenever we describe a two-phase dispersion in these terms, the words are being used relatively and in a kinetic sense. Second, there is little unanimity among workers about the nomenclature of various processes. Finally, whether a stable or unstable system is desirable depends entirely on the context.

1.5 MICROSCOPY

Because of the particle sizes involved, the microscope is an instrument which is ideally suited for the study of some lyophobic colloids. For a particle to be visible, there must be an acceptable difference between its refractive index and that of its surroundings. This requirement has nothing to do with particle size: A glass rod can be made to "disappear" by immersing it in a liquid of matching refractive index. The fact that a technique is not applicable to all possible systems does not invalidate it; this only means that the worker must have access to more than one technique in order to deal with a variety of problems.

The first thing we tend to think of in connection with microscopes is the magnification they achieve. More important, however, is a quantity known as the resolving power or limit of resolution of the microscope. Magnification determines the size of an image, but the resolving power determines the amount of distinguishable detail. Enlargement without detail is of little value. For example, a row of small spherical particles will appear simply as a line if it is enlarged with an instrument of poor resolving power. Further magnification would increase the thickness of the "line," but would not reveal its particulate nature. If the resolving power is increased, however, the individual spheres would be discernable. One may then choose a magnification which is convenient. Both the depth and area of the in-focus field decrease as the magnification is increased, so one pays a price for enlargement, even though the amount of perceptible detail is not affected much by the magnification.

A beam of light is always diffracted at the edges of an object to produce a set of images of the edge known as a diffraction pattern. The diffracted light is what our eye receives from an object and from which our image of the object is constructed. The diffracted light contains sufficient information to assemble an image; any light that is not incorporated into the image will result in a loss of detail in the image. To estimate the efficiency with which an image reproduces the object, let us consider the situation in which all first-order diffracted light is intercepted by the lens, but light from all higher orders of diffraction is assumed to be lost.

Suppose we have a set of pinholes in an opaque shield, and that this shield is illuminated by a source which is far enough away that the incident light may be regarded as a set of parallel rays, as sketched in Fig. 1.1a. To an observer on the opposite side of the shield, each pinhole will function as a light source from which a hemispherical wave front seems to emerge. If adjoining pinholes are separated by a distance d, the wave from one source, say S_1, must travel a longer distance than light from S_2 to reach a distant screen at point P. If the distance between the shield and the screen is large compared to the wavelength, the extra distance may be equated with $d \sin \theta$. Positive reinforcement through diffraction occurs whenever such an extra distance equals some integral number of wavelengths of light. For first-order diffraction, then, we require

$$d \sin \theta = \lambda' \tag{3}$$

where λ' is the wavelength of the medium between the screen and the shield. If the medium has a refractive index n, then $\lambda' = \lambda/n$ where λ is the wavelength under vacuum. Therefore, Eq. (3) becomes

$$d \sin \theta = \frac{\lambda}{n} \tag{4}$$

Recall that our objective is to consider an image constructed from only first-order diffracted light. To do this, we identify the perforated shield S_1S_2 as the object, the

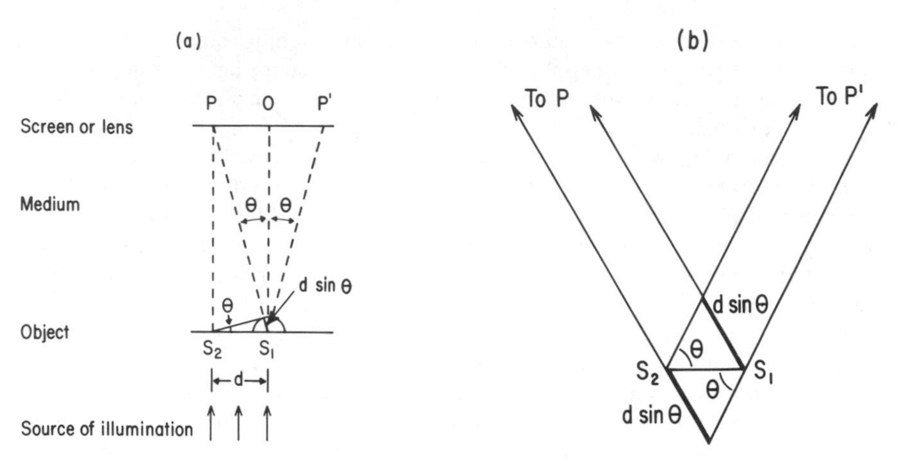

FIGURE 1.1 *(a) The geometry upon which the resolving power d of a microscope is based. (b) Detail showing how light from both sources must be intercepted by lens to become part of the image.*

screen as the objective lens of the microscope, and the distance PP' as the diameter of the objective. It is clear from Fig. 1.1b that light originating at S_1 must travel a distance $d \sin \theta$ longer than the light from S_2 to form an image at P; likewise, light originating at S_2 must travel a distance $d \sin \theta$ longer than light from S_1 to form an image at P'. Thus, to be intercepted by the lens and thereby become part of the image, rays from different parts of the source must travel paths which differ by $2d \sin \theta$. In order to do this and still consist of only first-order diffracted light, that difference in path length must equal λ'. Therefore, the resolving power d is estimated to be

$$d = \frac{\lambda}{2n \sin \theta} \tag{5}$$

which is sometimes written

$$d = \frac{\lambda}{2(\mathrm{NA})} \tag{6}$$

where NA (i.e., $n \sin \theta$) is called the numerical aperture of the lens. The resolving power measures the magnitude of the separation between objects that is required to produce discernably different images, when the angle subtended by the microscope is 2θ.

The significance of Eq. (5) is that any points which are closer together than the distance d will produce a first-order diffraction image with light having a wavelength equal to λ/n at some angle greater than θ. This light would not be intercepted by the lens PP', so a significant amount of detail about distances of this magnitude is lost. It is the wave nature of light that imposes a limit to the amount of detail an image may possess. Equation (5) shows that the resolving power is decreased by increasing θ or by decreasing λ/n. The subtended angle 2θ is increased by increasing the diameter of the lens and by decreasing the distance between the object and the lens; the design of the lens limits the range of these parameters. The ratio λ/n may be decreased by decreasing λ or by increasing n. Although shorter wavelengths improve resolving power, visible light is almost always used in microscopy, primarily because of the absorption of shorter wavelengths by glass. Since the refractive index of some oils is 50% higher than that of air, a significant improvement in the resolving power is achieved by filling the gap between object and lens with so-called immersion oil.

We may use Eq. (5) to estimate the particle size which will be clearly resolved microscopically. As a numerical example, we may take λ to be 500 nm, $n = 1.5$, and $2\theta = 140°$—all of which are attainable but optimal values for these quantities. This leads to a resolving power of 142 nm or 1420 Å, which represents the lower limit of resolved particle size under completely ideal circumstances. Under closer to average circumstances, a figure twice this value may be more typical. These calculations reveal that direct microscopic observation is feasible only for particles at the upper end of the colloidal size range, and only then if the particle contrasts sufficiently with its surroundings in refractive index.

A variation of direct microscopic examination extends the range of microscopy considerably but at the expense of much detail. By this technique, known as dark field microscopy, particles as small as 50 Å may be detected under optimum conditions, with about 200 Å the lower limit under average conditions. In dark field microscopy, the sample is illuminated from the side rather than from below as in an ordinary microscope. If no particles were present, no light would be deviated from the horizontal into the microscope and the field would appear totally dark. The presence of small dispersed particles, however, leads to the scattering of some light from the horizontal into the microscope objective, a phenomenon sometimes called the Tyndall effect. The presence of colloidal particles is indicated by minute specks of light in an otherwise dark field.

In dark field microscopy, the particles are only a blur; no details are distinguishable at all. Some rough indication of the symmetry of the particles is afforded by the twinkling that accompanies the rotation of unsymmetrical particles, but this is a highly subjective observation. However, the technique does permit the rate of particle diffusion to be followed. We shall see in Chap. 3 how to relate this information to particle size and shape. The number of particles per unit volume may also be determined by direct count once the area and depth of the illuminated field have been calibrated. This is an important technique for the study of kinetics of flocculation, a topic we discuss in Chap. 10. In dark field, as in direct observation microscopy, the refractive index difference between the particles and the medium is one of the crucial factors that influences the feasibility of the method. Very good resolving power is obtained with metallic particles which offer maximum contrast in refractive index. The technique of dark field microscopy played an important historic role in colloid chemistry, particularly in the study of metallic colloids. R. Zsigmondy (Nobel Prize, 1925), for example, made extensive use of this technique in his study of colloidal gold. Note that it is the deviation of a beam of light or scattered light that is utilized in dark field microscopy. We shall have a good deal more to say about light scattering in Chap. 5.

It is clear from Eq. (5) that the best prospect for extending the range of microscopy lies in the extension by orders of magnitude of the wavelength used to produce the image. The wave-particle duality principle of modern physics shows us how to make this extension. According to the de Broglie equation, the wavelength of a particle is inversely proportional to its momentum:

$$\lambda = \frac{h}{mv} \tag{7}$$

where h is Planck's constant and m and v are the particle mass and velocity, respectively. In an electron microscope, a beam of electrons replaces light in producing an image. An electron beam is produced by a hot filament, accelerated by an electron gun, and focused by electric or magnetic fields which function as lenses. A schematic comparison of light and electron microscopes is shown in Fig. 1.2. Although wavelengths equaling fractions of angstroms are easily achieved in electron microscopes, the numerical apertures of such microscopes are low. Accordingly, the resolving power of a conventional electron microscope is generally about 10 Å. Because of this remarkable resolution, however, small features may be enormously magnified without loss of detail. An easy conversion factor to remember when examining electron micrographs is that an objective 1 μm in length appears 1 inch long when it is magnified by 25,400 [10^{-4} cm $\times$ (1 in/2.54 cm) $\times$ 25,400 = 1 in].

The intensity of the electron beam that is transmitted through the specimen under observation in an electron microscope depends on the thickness of the sample and concentration of atoms in the sample. Thus, unless there are large differences in these characteristics between the particles of the sample and the support on which they rest, a very low contrast image will be produced. A poor quality image results from these low-contrast situations; the picture is analogous to a landscape, viewed from an airplane, when the sun is directly overhead. The most common way of overcoming this difficulty is by means of a technique known as shadow-casting. The sample to be examined in the electron microscope is placed in a chamber which is

FIGURE 1.2 *Schematic comparison of (a) light and (b) electron microscopes showing components which perform parallel functions in each.*

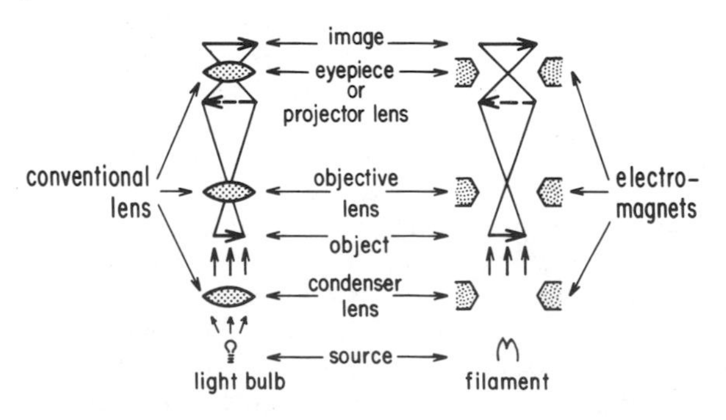

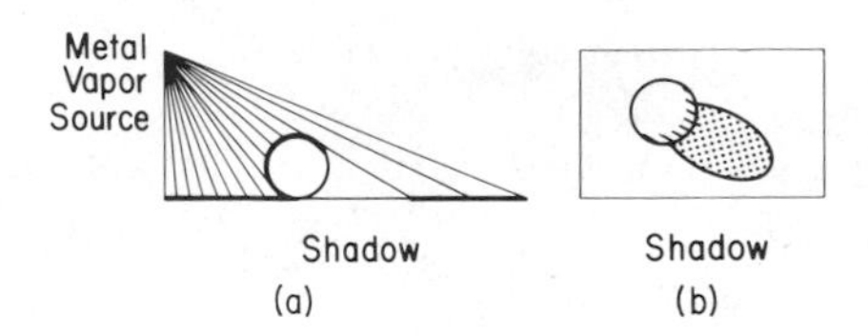

FIGURE 1.3 *(a) Side and (b) top views of a spherical particle shadowed by metal vapor.*

then evacuated. Next some gold or other metal is vaporized in the same chamber. The vapor condenses on all cooler surfaces, including the surface of the sample. If the vapor source is positioned off to the side of the sample, the condensed vapor will deposit unevenly, effectively casting a shadow over the sample. The situation is shown schematically in Fig. 1.3; Figs. 1.9a and 1.9b show actual photomicrographs enhanced by this technique. With shadow-casting, the field shows much more detail, just as the view from an airplane does when the sun is lower in the sky.

When the angular position of the vapor source relative to the sample is known, the thickness of the particle may be evaluated from the length of its shadow by a simple trigonometric calculation. Alternatively, the length of shadow cast by a spherical reference particle, introduced for calibration, may be used as a standard to calculate the thickness of particles.

At the beginning of this section, the microscope was proposed as an ideal tool for the study of colloidal particles. The light microscope is limited in range, but the electron microscope clearly has access to the entire colloidal size range. However, a very real disadvantage still persists even with the electron microscope. The ordinary electron microscope requires that the sample be placed in an evacuated enclosure for examination by the electron beam. Therefore, a dispersion must be evaporated to dryness before examination. The result, although still informative, bears about as much resemblance to the original two-phase dispersion that a pressed flower bears to a blossom on a living plant. In subsequent chapters, we shall discuss some in situ techniques for the characterization of colloidal particles, especially with respect to particle weight.

Despite the limitations just noted, there is no other branch of chemistry that can reasonably expect to see directly the particles of interest in that field, however appealing that prospect may seem to chemists. However, with some colloids this is a possibility. We shall not pass up the opportunity to take a look! Figures 1.4, 1.5, 1.9, and 1.10 are electron micrographs of lyophobic colloids. We shall comment on these individually in the following sections.

1.6 PARTICLE SIZE AND SHAPE

Figure 1.4 shows an electron micrograph of latex particles made from polystyrene cross-linked with divinylbenzene. These particles display a remarkable degree of homogeneity with respect to particle size. Such a sample is said to be monodisperse, in contrast to polydisperse systems which contain a variety of particle sizes. We shall

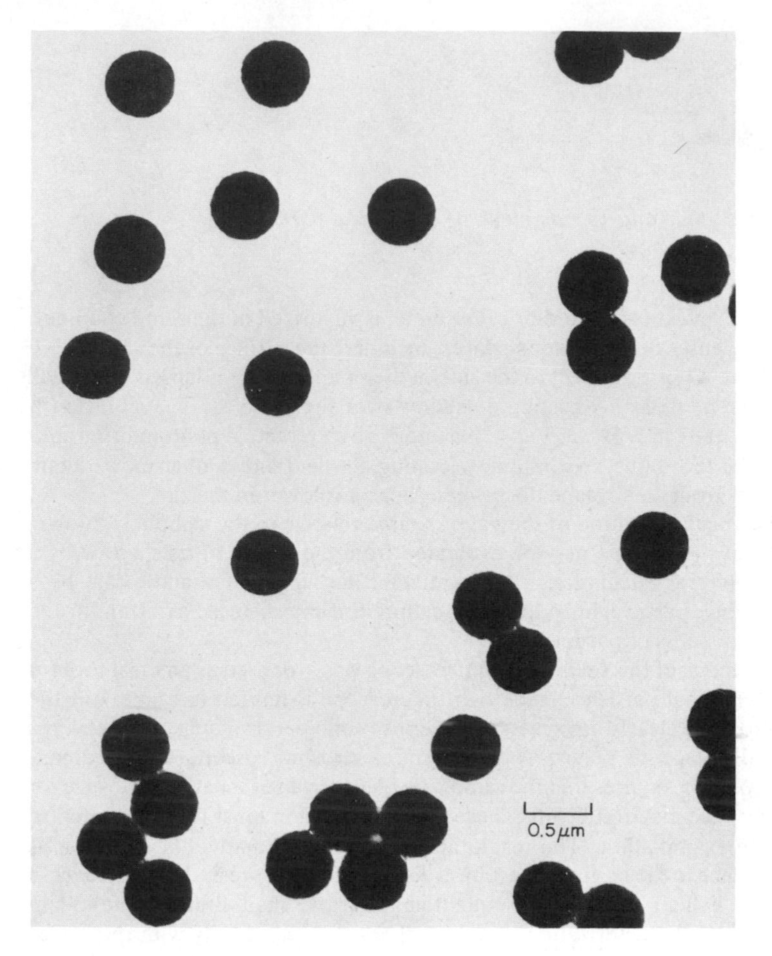

FIGURE 1.4 *Electron micrograph of cross-linked monodisperse polystyrene latex particles. The latex is a commercial product* ($\bar{d} = 0.500\ \mu$m) *sold as a calibration standard. (Photograph courtesy of R. S. Daniel and L. X. Oakford, California State Polytechnic University.)*

have a good deal more to say about polydisperse systems in later sections of this chapter. The particles of Fig. 1.4 are also perfectly spherical. Except for the nature of the material involved, this figure could be a photograph of the hypothetical array of spheres discussed in Sec. 1.2.

FIGURE 1.5 *Electron micrograph* (150,000×) *of carbon black particles* (*a*) *before and* (*b*) *after heating to* 2700°*C in the absence of oxygen.* [*From F. A. Heckman,* Rubber Chem. Technol. **37**: 1243 (1964), *used with permission.*]

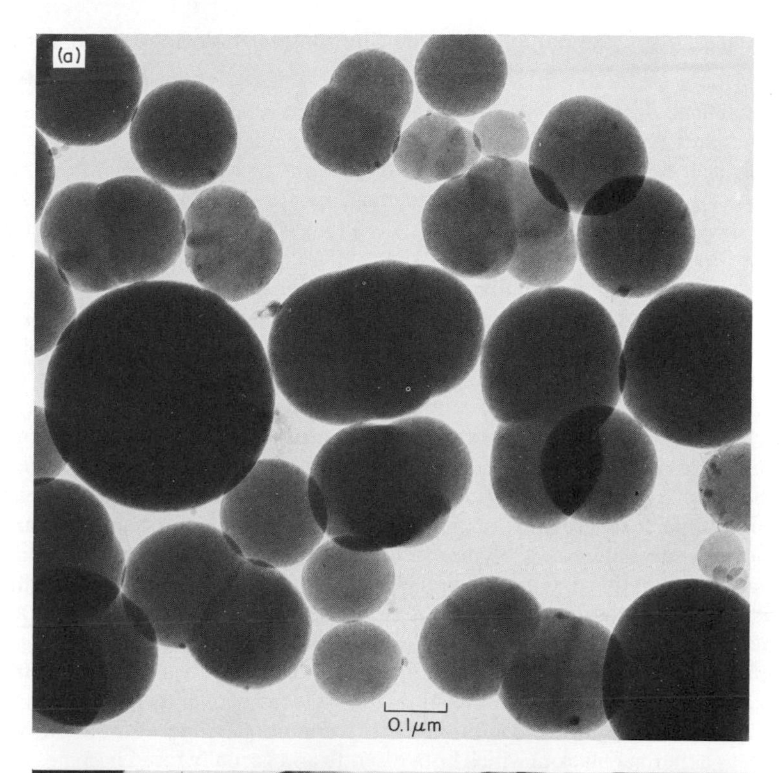
(a)
0.1 μm

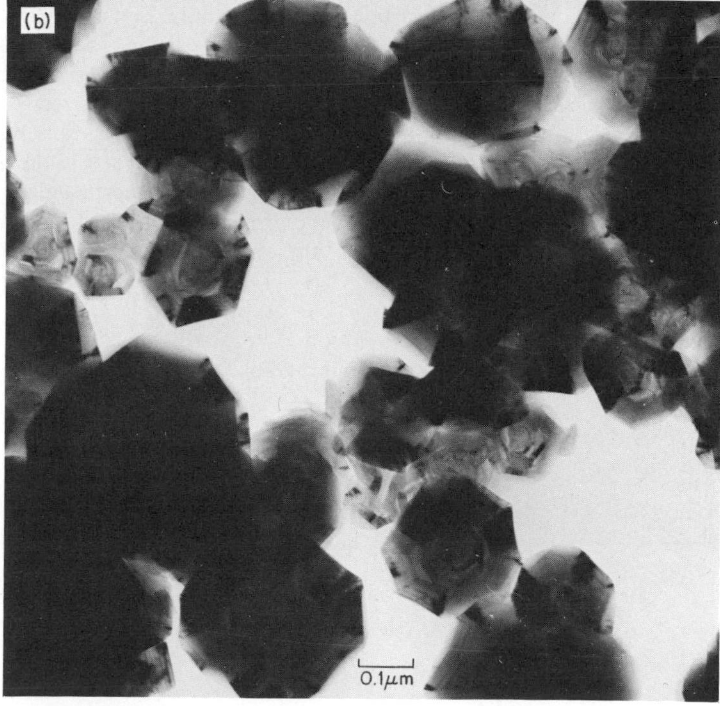
(b)
0.1 μm

It is difficult to imagine an array of particles which would be any easier to describe than the latex particles of Fig. 1.4. A single parameter, such as the radius of the spheres, characterizes the dispersed phase in terms of a linear dimension. If the density of the material is known, the mass of the individual particles is fully determined also. The radius of the particles in Fig. 1.4 is 2.50×10^{-5} cm; therefore, a sphere has a volume of 6.54×10^{-14} cm^3. The density of polystyrene is about 1.05 g cm^{-3}; the mass of an individual particle in Fig. 1.4 is 6.87×10^{-14} g.

When we deal with lyophobic colloids, we shall generally indicate the mass of the individual particle in grams per particle. When we deal with lyophilic colloids, we shall use the molecular weight in grams per mole just as we would for a low-molecular-weight substance. Of course the two differ by Avogadro's number. Since particle weights are so much larger than we are accustomed to in ordinary chemistry, we must be especially careful to consider the units (i.e., per particle or per mole) in which a result is reported since we may not immediately recognize the number itself.

Monodisperse spheres are not only uniquely easy to characterize but they are also very rarely encountered. Polymerization under carefully controlled conditions allows the preparation of the polystyrene latex shown in Fig. 1.4. Latexes of this sort are used as standards for the size calibration of optical and electron micrographs.

In the majority of two-phase colloidal systems, the particles are neither spherical nor monodisperse. Figure 1.5 shows micrographs of carbon black particles. Broadly speaking, this material is soot, but a great deal of control over its properties may be accomplished by varying the conditions of its preparation. Figure 1.5a shows what is known as a thermal black in which both discrete and partially fused particles may be seen. Figure 1.5b shows the same carbon black preparation (not the same field or particles) after heat treatment at 2700°C in the absence of oxygen. The particles take on a distinctly polyhedral shape with this treatment which is known as graphitization.

Many solid particles are not actually spherical but are characterized by a high degree of symmetry like a sphere, and are often approximated as spheres. For example, a polyhedron approximates a sphere more and more closely as the number of its faces increases. The primary particles of the graphitized thermal black shown in Fig. 1.5b are sufficiently symmetrical to be approximated as spheres. Likewise, many substances which display irregular but symmetrical particles are often described by a characteristic dimension which is called a "diameter." This terminology does not necessarily mean that the particles are spherical.

The forms sketched in Fig. 1.6 represent some irregularly shaped particles as they might be observed in a light or electron micrograph. The length of a line which

FIGURE 1.6 *A schematic illustration of Martin's diameters for irregular particles.*

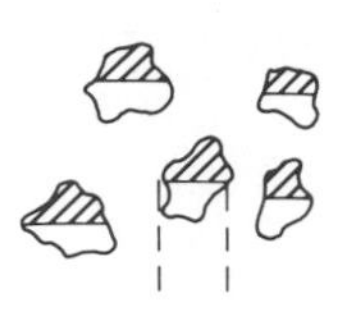

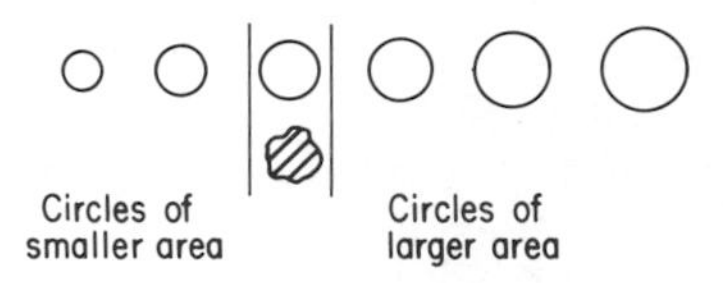

FIGURE 1.7 *The use of a graticule to estimate a characteristic dimension of an irregular particle.*

bisects the projected area of a particle is a parameter known as Martin's diameter. The direction along which the Martin diameter is measured is arbitrary, but it should be used consistently to avoid subjective bias. The lines sketched in the figure are intended to represent this quantity. The Martin diameter is most easily measured on photographs although movable crosshairs in the eyepiece of a microscope also permit such distances to be measured by direct observation.

Another method of characterizing irregular particles consists of reporting the diameter of a circle which projects the same cross section as the particle in question. A technique for doing this is to insert an object known as a graticule into the eyepiece of a microscope. This is merely an assortment of circles of different sizes etched on a transparent slide. The observer decides which circle most closely approximates the projected area of the particle. Figure 1.7 illustrates how a graticule might be used to size an irregular particle. Graticules are available with circles whose diameter increases in a $\sqrt{2}$ progression. Both the graticule dimension and the Martin diameter are extremely tedious to evaluate since a large number of particles must be examined for the values to have any statistical significance. We shall return to the statistical aspects of this presently.

The sphere is favored above all other geometries as a model for actual particles since it is characterized by a single parameter. Sometimes, however, the particles of a dispersion are so unsymmetrical that no *single* parameter, however defined, can begin to describe the particle. In this case, the next best thing is to describe the particle as an ellipsoid of revolution. An ellipsoid of revolution is that three-dimensional body which results from the complete rotation of an ellipse around one of its axes. We shall define a as the "radius" of the ellipsoid measured along the axis

FIGURE 1.8 *(a) Prolate ($a > b$) and (b) oblate ($a < b$) ellipsoids of revolution, showing the relationship between the semiaxes and the axis of revolution.*

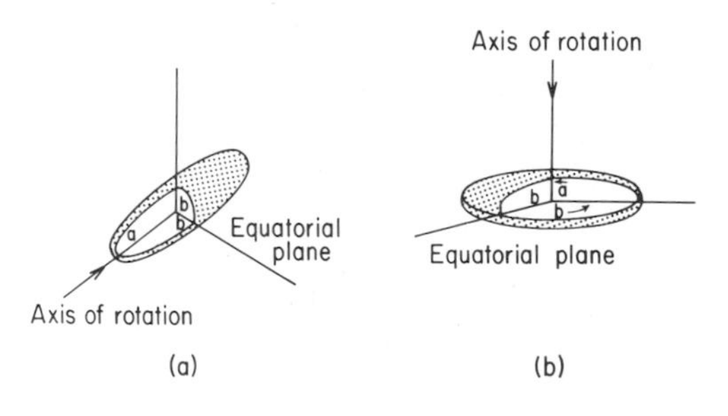

(a)
0.20μm

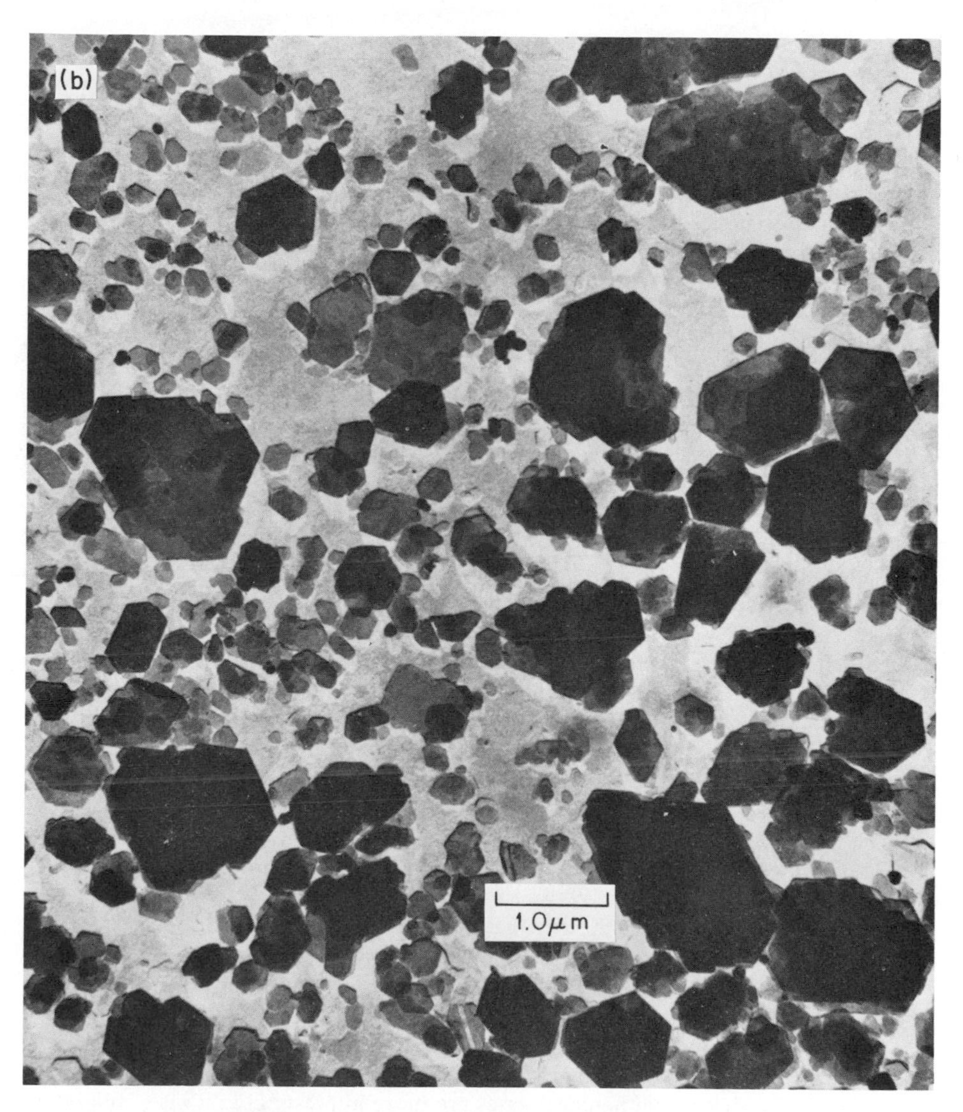

FIGURE 1.9 *Electron micrograph of two different types of particles which represent extreme variations from spherical particles. (a) Tobacco mosaic virus particles. (Photograph courtesy of Carl Zeiss, Inc., New York.) (b) Clay particles (sodium kaolinite) of mean diameter* 0.2 μm *(by matching circular fields). [from M. D. Luh and R. A. Bader,* J. Colloid Interface Sci. **33**: 539 (1970) *used with permission.] In both (a) and (b), contrast has been enhanced by shadow-casting.*

of rotation and b as the "radius" measured in the equatorial plane. Obviously, if these two measurements of radius are equal for a particle, that particle is spherical. If $a > b$, the particle is called a prolate ellipsoid; if $a < b$, it is an oblate ellipsoid. These two geometries are illustrated in Fig. 1.8. The ratio a/b, called the axial ratio of the ellipsoid, is frequently used as a measure of the deviation from sphericity of a particle. It plays an important role, for example, in our discussions of viscosity and sedimentation in Chaps. 2 and 3, respectively. In the event that $a \ggg b$, the prolate ellipsoid approximates a cylinder and, as such, is often used to describe rod-shaped particles such as the tobacco mosaic virus particles shown in Fig. 1.9a. Likewise, if $a \lll b$, the oblate ellipsoid approaches the shape of a disk. Thus, even the irregular clay platelets of Fig. 1.9b may be approximated as oblate ellipsoids.

It should be fairly obvious that the two-parameter ellipsoidal geometry is far more accurate than the single-parameter spherical geometry to describe assymmetric particles. We shall see below that, just as two parameters are generally used to describe unsymmetrical particles, two parameters are also preferable to characterize polydisperse systems. One might argue that in both cases more than two parameters would be better yet. Sometimes this might be true, but often experimental difficulties or uncertainties make this infeasible.

One other particle geometry deserves mention. Suppose we were to take a length of string or other flexible material and allow it to tumble freely for a while in a large container. The string would certainly be expected to emerge from this treatment as a tangled jumble. Many long-chain molecules have sufficient flexibility to take on a random configuration like this under the influence of thermal jostling. This "random coil" is likely to be symmetrical rather than stretched out. We shall, accordingly, refer to the radius of such a coil. The random coil is discussed in detail in Sec. 3.12.

1.7 AGGREGATION

We have already introduced the idea that the primary particles of a dispersed system tend to associate into larger structures known as flocs or aggregates. The nature of the interparticle forces responsible for this aggregation is one of the most interesting areas of colloid chemistry. In the absence of interactions, colloidal dispersions would be analogous in many ways to ideal gases in which the individual particles are also independent of each other. However indebted physical chemistry may be to ideal gases, it is clear that nonideal gases are more interesting! Likewise, colloidal systems which are unstable with respect to flocculation provide a lot more information as to the nature of interparticle forces than stable systems do. We shall defer our discussion of the flocculation process until Chap. 10, but a few remarks about flocs—the kinetic units that result from that process—are in order at this time.

In many situations the dispersed phase is present as flocs, not as primary particles. It is of little importance in such systems that the flocs could be disrupted further without subdividing the primary particles or that they might aggregate more to form still larger flocs. The particles that are actually dispersed in the continuous phase are flocs. It is the size, shape, and concentration of the units actually present which determine the properties of the dispersion itself. As a matter of fact, some substances—for example, those carbon blacks known as channel and furnace blacks—possess rigidly fused, floclike structures as their primary particles. Figure 1.10 shows an example of such a particle.

When an electron micrograph shows evidence of aggregation, we must remember that this may be an artifact arising from the preparation of the sample for microscopy. In other words, the amount of flocculation that a colloid displays in its dispersed state and the amount that appears in an electron micrograph made from the same preparation may be quite different. Optical microscopy is safer in this regard since the actual dispersion may be examined without first evaporating the continuous phase to dryness. Also, in both optical and electron microscopy, it is the projected image of the particle that is observed, and microscopic observation alone is often inadequate to distinguish a particle such as that shown in Fig. 1.10 from a true floc in which the structure is fairly readily disrupted.

FIGURE 1.10 *Transmission electron micrograph* (500,000×) *of an individual carbon black (furnace black) particle (Vulcan 6, ISAF, N-*220) *showing the fused-floc structure of the primary particle of this material. (Photograph courtesy of F. A. Heckman, Cabot Corporation, Billerica, Massachusetts,* 01821.)

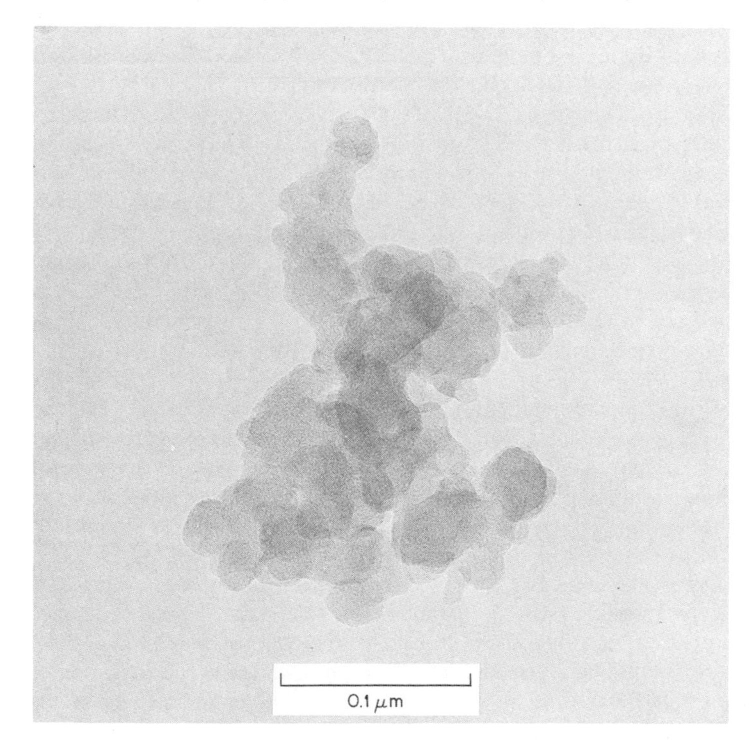

Next, suppose we wish to measure the characteristic dimensions of flocs or floclike particles using micrographs. As a first approximation, we might expect the flocculation process to result in irregular but, on the average, symmetrical particles just as the random coil does. Thus we might characterize a floc in terms of the dimension of an inscribing boundary, such as the Martin diameter or the diameter of an equivalent circle determined from a graticule. Then, suppose we wish to evaluate the mass of the floc enclosed within this boundary. To convert a linear dimension into a particle mass, the particle shape and density must be known. We have already commented on the approximations involved in treating irregular particles as spheres. Here, however, we encounter an additional problem besides the geometrical approximation already discussed. The question is what do we use for the density of a floc to convert the particle volume into its mass. If the "floc" dimensions have been measured, it is clearly the "floc" density which must be used. The latter is intermediate between the density of the dispersed and continuous phases, the exact value depending on the structure of the floc. Of course, one might attempt to estimate the number of primary particles in an aggregate, then use their size and density as an alternate means of evaluating the mass of a floc. The main point of this, however, is the following. Whenever the dispersed particles are flocculated, the properties of the dispersed units are intermediate between those of the two different phases involved. We shall encounter this difficulty again in Chap. 3 where the density of the settling unit, whatever it may be, is involved in sedimentation.

A very interesting approach to research concerning flocs is the technique of computer simulation. By this method, aggregates are "assembled" by a computer which uses random numbers to determine the coordinates from which each primary particle approaches the growing floc. The model that has been most studied consists of spherical primary particles, although linear sets of spheres have been used to simulate asymmetric primary particles. The probability of adhesion on contact has been made a variable quantity in some studies. As might be expected intuitively, more open flocs result when the probability of adhesion at initial contact is high. If, on the other hand, the added particle is permitted to roll along the surface of the growing floc before adhering, a more compact structure results. Figure 1.11 is a sketch of the projection of a random floc assembled in this manner. The resemblance it bears to an actual floclike particle is evident from a comparison of Fig. 1.11 with Fig. 1.10.

A variation of the computer simulation procedure which is even more realistic permits the joining together of small flocs to form larger ones, rather than restricting the addition to primary particles only. This leads to structures that are even more expanded than those resulting from the addition of primary particles alone.

These computer simulations permit the density of primary particles within the floc to be evaluated, important information for relating the properties of the floc to its composition. As might be expected, however, it is difficult to know a priori what model to use for a particular system. However, this technique does allow some interesting a posteriori interpretations of known structures to be made. Another closely related problem that has been studied by computer simulation is the volume occupied by a sediment. As with flocs, it is found that sediments become more voluminous as the probability of adhesion on contact increases.

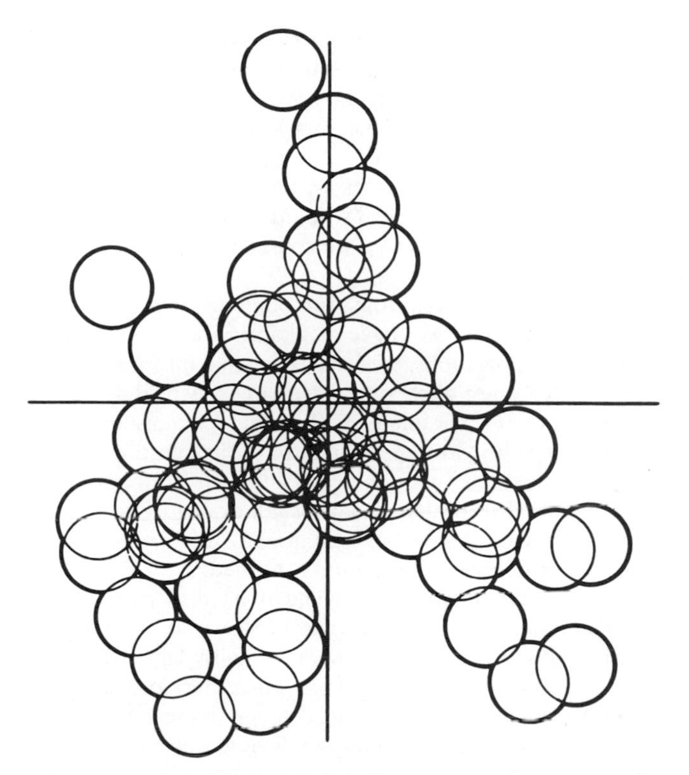

FIGURE 1.11 *Computer simulation of a floc consisting of* 76 *spherical primary particles. Solid point shows center of gravity.* [*From M. J. Vold,* J. Colloid Sci. **18**: 684 (1963), *used with permission.*]

This discussion of flocs leads us to another important characteristic of dispersions which we have not yet considered in sufficient detail: polydispersity. Monodisperse systems are the exception rather than the rule. Even in those rare cases in which a monodisperse system exists, any flocculation that occurs will result in a distribution of particle sizes because of the random nature of the flocculation process.

1.8 POLYDISPERSITY

The only realistic attitude to take toward the dispersed systems we are interested in is to assume that they are polydisperse. Even the particles in Fig. 1.4 which appear remarkably uniform contain a narrow distribution of particle sizes. There are rare cases in which the distribution of dimensions is of negligible width, but generally speaking, a statistical approach is required to describe a colloidal dispersion.

Students of the physical sciences generally encounter statistics in two different places. One of these deals with the treatment of experimental data. From this viewpoint, all measured quantities contain some error which raises questions concerning the best way to report the results of multiple measurements. This is obviously related to our problem of describing, for example, the "characteristic dimension" of the particles in Fig. 1.5 or 1.9. Another place in which the science student encounters statistics is in the theoretical description of large populations or populations that change through a large number of states as, for example, in the kinetic molecular theory of gases. We shall be concerned with this aspect of statistics also. For example, the randomly coiled piece of string we considered changes size and shape continually while it is being shaken.

These two applications of statistics, from our point of view, will differ primarily in the kind of information we have available about our system. Sometimes, as when measuring micrographs, we have individual information on a large number of particles. Our question under these circumstances is how to condense these data into a few key parameters. In other circumstances, the experimental quantity itself will be an "average" quantity. Our question, then, is what kind of distribution is consistent with this average. In both cases, the underlying fact is the existence of a distribution of values for the quantity in question. We shall consider some aspects of these statistical topics here; references in statistics should be consulted if additional information is needed.

Suppose we have just measured the diameters of a field of polydisperse spheres in an electron micrograph. Our objective is to devise reasonable ways of presenting a description of the system in terms of the measured data. A fairly large number of observations is required for any statistical approach to be valid; therefore, to merely tabulate the measurements is inadequate. Some condensation of the data is clearly required. Generally, the first step along these lines is a device known as classification of the data. Classification consists of sorting the observed quantities into 10 to 20

TABLE 1.4 *A hypothetical distribution of* 400 *spherical particles*[a]

Class boundaries $\leq d <$ (μm)	Class mark, d_i (μm)	Number of particles, n_i	Fraction of total number in class, f_i	Total number with $d \leq d_i$, $n_{T,i}$
0–0.1	0.05	7	0.018	7
0.1–0.2	0.15	15	0.038	22
0.2–0.3	0.25	18	0.045	40
0.3–0.4	0.35	28	0.070	68
0.4–0.5	0.45	32	0.080	100
0.5–0.6	0.55	70	0.175	170
0.6–0.7	0.65	65	0.163	235
0.7–0.8	0.75	59	0.148	294
0.8–0.9	0.85	45	0.113	339
0.9–1.0	0.95	38	0.095	377
1.0–1.1	1.05	19	0.048	396
1.1–1.2	1.15	4	0.010	400

[a] The data are classified into 12 classes: The class marks, number, and fraction of particles per class and the total number of particles up to and including each class are listed.

categories called classes. Having fewer than 10 categories results in a loss of detail in the description of the distribution; more categories than 20 does not improve the representation in proportion to the extra effort it requires. If, then, we observed in an electron micrograph that all the particles were less than 1.2 μm in diameter, it would be reasonable to sort them into classes 0.1 μm wide. That is, all particles less than 0.1 μm in diameter would fall into one class, those for which 0.1 $\mu m \leq d <$ 0.2 μm in the second, and so on until class 12, which would include particles for which 1.1 $\mu m \leq d < 1.2$ μm. The frequency distribution of such a sample is a tabulation of the number of particles in each class. Table 1.4 represents the frequency distribution for a hypothetical array of spheres; all the numerical examples of this section are based on this sample of 400 particles. Each class is represented by the midpoint of the interval, a quantity called the class mark, symbolized by d_i for class i. Similarly, we define the number of particles in each class as n_i.

A common graphical representation of a frequency distribution is the histogram, a bar graph in which the class marks are plotted as the abscissa and the height of the bar is proportional to the number of particles in the class. Sometimes the ordinate is defined as the fraction of particles in the class, f_i. Figure 1.12a is a plot of the histogram of the data in Table 1.4. Obviously, as the number of classes approaches infinity, the width of each interval approaches zero and the histogram approaches a smooth curve. Analytical distribution functions give the equation for such smooth curves. However, in practice, a bar graph is a convenient approximation to the smooth function.

Another way in which these kinds of data are sometimes represented is as a cumulative curve in which the total number (or fraction) of particles $n_{T,i}$ having diameters less (sometimes more) than and including a particular d_i are plotted versus d_i. Figure 1.12b shows the cumulative plot for the same data shown in 1.12a as a histogram. The cumulative curve is equivalent to the integral of the frequency

FIGURE 1.12 *Graphical representation of the data in Table 1.4. Data are presented (a) as a histogram, (b) as a cumulative distribution curve.*

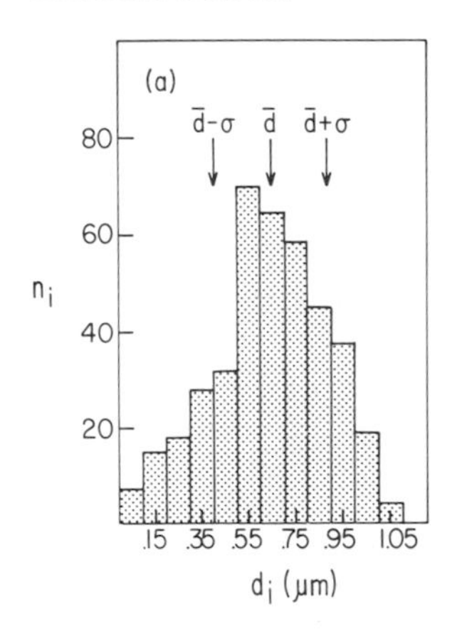

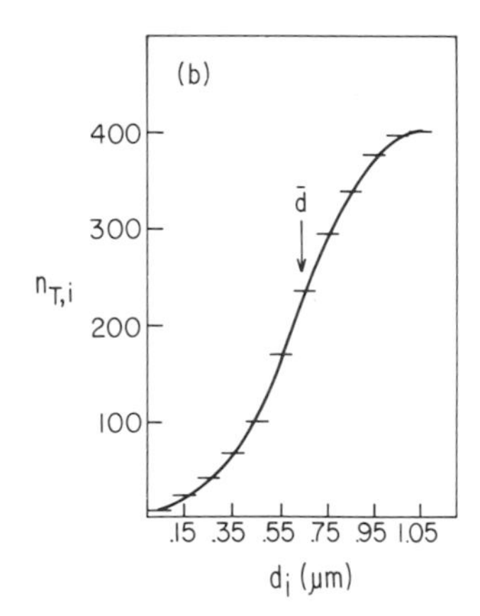

distribution up to the specified class mark. Cumulative distribution curves are discussed in Chap. 3 in connection with sedimentation.

1.9 THE AVERAGE DIAMETERS

Although the histogram is a convenient pictorial way to present data, a more concise representation is often required. The mean and standard deviation are the most familiar numerical parameters used for this purpose. The following equations define the mean diameter $\bar{d}$

$$\bar{d} = \frac{\sum_i n_i d_i}{\sum_i n_i} \tag{8}$$

and the standard deviation σ

$$\sigma = \left[\frac{\sum_i n_i (d_i - \bar{d})^2}{\sum_i n_i - 1}\right]^{1/2} \tag{9}$$

of the distribution. Since $\Sigma\, n_i$ represents the total population size and is constant for a given problem, Eq. (8) may be written

$$\bar{d} = \sum_i f_i d_i \tag{10}$$

since $f_i = n_i / \Sigma_i\, n_i$. A similar result may be written for the standard deviation, provided the total number of particles is sufficiently large to justify the approximation $n_T - 1 = n_T$:

$$\sigma = \left[\sum_i f_i (d_i - \bar{d})^2\right]^{1/2} \tag{11}$$

The significance of the standard deviation should be noted. The quantity $d_i - \bar{d}$ is the deviation of a particular value from the mean. Since such deviations can be either positive or negative, this quantity is squared prior to averaging. The square root of the average, then, is a measure of the spread of the data in a particular sample. From a computational point of view, the following formula provides an easier means for evaluating σ:

$$\sigma = [\overline{d^2} - (\bar{d})^2]^{1/2} \tag{12}$$

At this point, it is convenient to introduce the concept of a moment of a distribution function. The general definition of the kth moment of the distribution about a point d_0 is given by the equation

$$k\text{th moment} = \sum_i f_i (d_i - d_0)^k \tag{13}$$

where the numerical value of the exponent determines precisely which moment we have. This definition is analogous to the definition of "moments" in physics except that here the "weighting factor" f_i rather than mass appears in the formula. The mean is the first moment about the origin: Eq. (10) results when $d_0 = 0$ and $k = 1$ in Eq. (13). Likewise, the standard deviation is the square root of the second moment ($k = 2$) about the mean ($d_0 = \bar{d}$). Therefore, the first moment of a distribution measures the location of a distribution and the second moment measures the spread of the distribution. Higher moments also convey certain information about the shape of a distribution. For example, the third moment is a measure of the "skewness" or lopsidedness of a distribution. It equals zero for symmetrical distributions and is positive or negative depending on whether a distribution contains a higher proportion of particles larger or smaller than the mean. The fourth moment (called "kurtosis") purportedly measures peakedness, but this quantity is of questionable value.

The mean and the standard deviation for the data of Table 1.4 are 0.64 and 0.24 μm, respectively; $\bar{d}$ has been marked in Figs. 1.12a and 1.12b also.

In surface and colloid chemistry, it is particle diameters (or radii) and molecular weights which are most often reported as "averages." Some of the most widely encountered "averages" are listed in Table 1.5 by name, symbol, and definition. The need for more than one definition for "average" arises from the fact that different experimental techniques applied to polydisperse systems perceive the "average" properties of the population differently. We shall acquire a better appreciation of this fact through the discussion of various specific procedures in later chapters. In Chap. 4, for example, we shall discover that osmotic pressure measures the number average molecular weight of a polydisperse system; and in Chap. 5, we shall see that

TABLE 1.5 *Some of the more widely encountered "averages" in surface and colloid chemistry, including their defining equations*

Name	Symbol	Definition	Quantity averaged	Weighting factor
Number average	$\bar{d}$ (also $\bar{M}_n$)	$\frac{\Sigma_i n_i d_i}{\Sigma_i n_i}$	Diameter (also molecular weight)	Number in class
Weight average	$\bar{M}_w$	$\frac{\Sigma_i w_i M_i}{\Sigma_i w_i}$	Molecular weight	Weight in class
Second moment about origin	$\overline{d^2}$	$\frac{\Sigma_i n_i d_i^2}{\Sigma_i n_i}$	Square of diameter	Number in class
Surface average	$\bar{d}_s$	$(\overline{d^2})^{1/2}$	Square of diameter	Number in class
Third moment about origin	$\overline{d^3}$	$\frac{\Sigma_i n_i d_i^3}{\Sigma_i n_i}$	Cube of diameter	Number in class
Volume average	$\bar{d}_v$	$(\overline{d^3})^{1/3}$	Cube of diameter	Number in class

light scattering measures the weight average molecular weight. At this time, we are less concerned with the origin of these differences than with the statistical definition of these quantities. In this section, we are primarily concerned with the various average diameters; average molecular weights are discussed in the next section.

The general formulas for calculating averages are given by Eqs. (8) and (10). These two differ from one another inasmuch as the weighting factor is normalized by definition in the latter and not in the former. A weighting factor is said to be normalized if the sum of weighting factors for the whole population equals unity, a consideration that must be introduced at some point in the calculation. When fractions of the whole are used as weighting factors, normalization is introduced through the definition of the weighting factor. If the content of a class is used as a weighting factor, the summation of products must be divided by the sum of weighting factors so that the condition of normalization is eventually included.

Inspection of Table 1.5 reveals that the two basic variations which generate the various averages reported are

1. averaging quantities other than the simple diameter or molecular weight, and
2. using something other than the number of entries in a class as a weighting factor.

For example, the second moment about the origin is obtained by averaging d_i^2, using n_i as weighting factor. Similarly, $\bar{M}_w$ is obtained by averaging the molecular weights in a distribution, using the weight in each class as the weighting factor. The requirement of normalization guarantees that $\bar{M}_w$ has the same units as $\bar{M}_n$; however, higher moments than the first have units which are the square or cube of the units of the first moment. Therefore, to obtain a "characteristic *linear* dimension" from a higher moment, it is conventional to extract the suitable root. The surface average and volume average diameters in Table 1.5 are thus seen to be the square root and cube root of the second and third moments, respectively. In the next few paragraphs, we shall discuss the specific significance of $\bar{d}_s$ and $\bar{d}_v$.

To understand the physical significance of $\bar{d}_s$, suppose we calculate the mean surface area of a particle from the sample of Table 1.4. To do this, we calculate the area contribution of each size class of spheres, add together the contribution of each class weighted by the number of particles in the class, then finally divide this sum by the total number of particles. This generates the following equation for the average area per particle, $\bar{A}$:

$$\bar{A} = \frac{\pi \sum_i n_i d_i^2}{\sum_i n_i} = \pi \sum_i f_i d_i^2 \tag{14}$$

Except for π, this is the same as the second moment about the origin [$k = 2$, $d_0 = 0$ in Eq. (13)] for the distribution. The diameter of a sphere having this "average" area is the surface average diameter $\bar{d}_s$ and equals

$$\bar{d}_s = \left(\sum_i f_i d_i^2\right)^{1/2} \tag{15}$$

The surface average diameter is always larger than the number average for a polydisperse system since the larger diameters contribute relatively more to the sum

TABLE 1.6 *The number average, surface average, and volume average diameters of the spheres in the distribution of Table 1.4. Also included are comparisons between the average area and average volume calculated by using $\bar{d}$ and either $\bar{d}_s$ or $\bar{d}_v$*

Quantity	Value
Mean diameter, $\bar{d}$	$= 0.64\ \mu m$
Surface average diameter, $\bar{d}_s$	$= 0.68\ \mu m$
Volume average diameter, $\bar{d}_v$	$= 0.72\ \mu m$
Average area based on $\bar{d}$, $\pi \bar{d}^2$	$= 1.29\ \mu m^2$
Average area based on $\bar{d}_s$, $\pi \bar{d}_s^2$	$= 1.46\ \mu m^2$
Average volume based on $\bar{d}$, $\frac{1}{6}\pi \bar{d}^3$	$= 0.14\ \mu m^3$
Average volume based on $\bar{d}_v$, $\frac{1}{6}\pi \bar{d}_v^3$	$= 0.19\ \mu m^3$

of the squares than they would if totaled directly. Table 1.6 shows the number average and the surface average diameters for the dispersion of Table 1.4 along with the mean area per particle associated with each. Note that the area calculated using $\bar{d}$ is too low.

In the foregoing discussion of the surface average diameter, it has been assumed that detailed information concerning the particle size distribution is the data available. Suppose, instead, that what is known about a sample is the number of particles it contains per unit weight (determined by osmotic pressure, Chap. 4) and the total surface area per unit weight (determined by gas adsorption, Chap. 8). The ratio of the area per unit weight to the number per unit weight would give the average area per particle, a quantity comparable to Eq. (14). If we postulate a spherical shape for the particles, we may divide this average area by π and take the square root to give the diameter associated with this average area per particle, $\bar{d}_s$. This is not the same as the mean particle diameter that would be obtained microscopically for the same distribution. Again, the difference arises from the different ways the polydisperse sample is measured by various experimental techniques.

The physical significance of the volume average diameter is analyzed similarly. Using the distribution of Table 1.4 as an example, we calculate the volume of the spheres in each class and take the mean:

$$\bar{V} = \frac{\frac{1}{6}\pi \sum_i n_i d_i^3}{\sum_i n_i} = \frac{1}{6}\pi \sum_i f_i d_i^3 \tag{16}$$

The diameter of a sphere having this average volume is the volume average diameter $\bar{d}_v$:

$$\bar{d}_v = \left(\sum_i f_i d_i^3\right)^{1/3} \tag{17}$$

We see that $\bar{d}_v$ is larger than the surface average diameter because the cubing leads to a greater contribution from the larger classes.

Again, we may consider how this quantity might be approached experimentally. If we measured the number of particles per unit weight (osmotic pressure) and the volume per unit weight (density), the average volume per particle may be measured. Treating the particles as spheres, the diameter associated with this average volume per particle $\bar{d}_v$ may be evaluated.

In summary, note that the relative magnitude of the number, surface, and volume average diameters is given by

$$\bar{d} < \bar{d}_s < \bar{d}_v \tag{18}$$

for a polydisperse system. Only for a monodisperse system would all three parameters have identical values. Therefore, the divergence from unity of the ratio of any two of these, measured independently, is often taken as an indication of the polydispersity of the dispersion. We shall see in the next section that this comparison of averages evaluated by different techniques finds particular application in the characterization of molecular weight distributions.

We conclude this section with the observation that sometimes the frequency of distributed quantities is given by an analytical expression—a theoretical distribution function—rather than by discrete weighting factors. In this case, the summation in Eq. (10) is replaced by an integral and the integration is carried out over the full range of the variable. We shall have more to say about theoretical distribution functions in Sec. 1.11, and shall use this approach to calculate various averages in Sec. 3.11 and 3.12.

1.10 THE MOLECULAR WEIGHT AVERAGES

It is not merely the linear dimensions of particles that may be averaged in more than one way. As a matter of fact, we shall deal more frequently with various averages of particle weight than particle dimension. The latter are more easily visualized, however, and are appropriate to our approach through microscopy to the characterization of polydisperse systems. Let us now turn our attention to two common ways—introduced in Table 1.5—of averaging molecular weights: the number average and weight average.

Suppose a dispersion is classified into a set of categories in which there are n_i particles of molecular weight M_i in the ith class. Then the number average molecular weight $\bar{M}_n$ equals

$$\bar{M}_n = \frac{\sum_i n_i M_i}{\sum_i n_i} = \sum_i f_i M_i \tag{19}$$

Alternatively, we could define the average in such a way that it is the weight of particles in each class w_i rather than their number that is used as the weighting factor. This results in an average known as the weight average molecular weight $\bar{M}_w$:

$$\bar{M}_w = \frac{\sum_i w_i M_i}{\sum_i w_i} \tag{20}$$

The weight of material in a particular size class is given by the product of the number of particles in the class times their molecular weight, however, so Eq. (20) may be written

$$\bar{M}_w = \frac{\Sigma_i (n_i M_i) M_i}{\Sigma_i n_i M_i} = \frac{\Sigma_i n_i M_i^2}{\Sigma_i n_i M_i} = \frac{\Sigma_i f_i M_i^2}{\Sigma_i f_i M_i} \tag{21}$$

The weight average molecular weight is thus seen to equal the ratio of the second moment of the distribution about the origin to the first moment of the distribution about the origin. As already noted, we shall see that measurement of the osmotic pressure of a polydisperse system permits the experimental evaluation of $\bar{M}_n$ (Chap. 4), and light scattering experiments enable us to measure $\bar{M}_w$ (Chap. 5). It follows from the definition of these various averages that

$$\frac{\bar{M}_w}{\bar{M}_n} \geq 1 \tag{22}$$

by analogy with the inequalities (18). Only when the system is monodisperse does the equality apply in (22). Here, too, the deviation of this ratio from unity may be taken as a measure of polydispersity.

The relationship between the ratio $\bar{M}_w/\bar{M}_n$ and the standard deviation of the molecular weight distribution is easily seen as follows. From Eq. (21) it is clear that

$$\sum_i f_i M_i^2 = \bar{M}_w \sum_i f_i M_i = \bar{M}_w \bar{M}_n \tag{23}$$

From the general procedure for defining the mean, the left-hand side may also be written as $\overline{M^2}$. Substituting this result into Eq. (12) permits us to rewrite the latter as

$$\sigma = (\bar{M}_n \bar{M}_w - \overline{M_n}^2)^{1/2} = \bar{M}_n \left(\frac{\bar{M}_w}{\bar{M}_n} - 1 \right)^{1/2} \tag{24}$$

Therefore, the square root of the amount by which the molecular weight ratio exceeds unity measures the standard deviation of the distribution relative to the number average molecular weight.

As a numerical example of these definitions, we may calculate the number and weight average particle masses of the distribution of spheres from Table 1.4 if we stipulate a density for the material. If we assume the material to be water with a density of 1.0 g cm^{-3}, then the statistical description of the dispersion in terms of particle masses is summarized in Table 1.7. The range of masses is so diverse that the results would undoubtedly be classed differently if a sampling by mass rather than by diameter had been done in the first place. However, we shall not bother with a reclassification here. The data of Table 1.7 show that the total mass of the 400 spheres of water is about 7.66×10^{-11} g. The number average particle mass is 1.92×10^{-13} g per particle, and the weight average particle mass is 3.43×10^{-13} g per particle. The ratio $\bar{M}_w/\bar{M}_n$ equals 1.79 for this particular degree of polydispersity, which makes σ about 1.70×10^{-13} g per particle according to Eq. (24). Since the density has been taken as 1.0 g cm^{-3} in this calculation, the number average volume of this array of spheres is 1.92×10^{-13} cm^3 per particle = 0.192 μm^3 per particle, the same as calculated from the volume average diameter in Table 1.6.

TABLE 1.7 *Recalculation of the distribution from Table 1.4 in terms of the masses of particles in each class, m_i, and the total weight of material in each class, W_i^a*

n_i	Class mark d_i (μm)	$m_i = \frac{1}{6}\pi\rho d_i^3$ (g)	$n_i m_i = W_i$ (g)	$w_i m_i = n_i m_i^2$ (g^2)
7	0.05	6.54×10^{-17}	4.58×10^{-16}	2.99×10^{-32}
15	0.15	1.76×10^{-15}	2.64×10^{-14}	4.65×10^{-29}
18	0.25	8.17×10^{-15}	1.47×10^{-13}	1.20×10^{-27}
28	0.35	2.25×10^{-14}	6.30×10^{-13}	1.42×10^{-26}
32	0.45	4.77×10^{-14}	1.53×10^{-12}	7.28×10^{-26}
70	0.55	8.69×10^{-14}	6.08×10^{-12}	5.29×10^{-25}
65	0.65	1.44×10^{-13}	9.36×10^{-12}	1.35×10^{-24}
59	0.75	2.21×10^{-13}	1.30×10^{-11}	2.88×10^{-24}
45	0.85	3.21×10^{-13}	1.44×10^{-11}	4.64×10^{-24}
38	0.95	4.49×10^{-13}	1.71×10^{-11}	7.66×10^{-24}
19	1.05	5.86×10^{-13}	1.11×10^{-11}	6.52×10^{-24}
4	1.15	7.96×10^{-13}	3.18×10^{-12}	2.53×10^{-24}
$\Sigma=400$			$\Sigma=7.66\times10^{-11}$	$\Sigma=2.63\times10^{-23}$

[a] A density of unity has been assumed for the particles.

All these different "averages" are admittedly confusing. Without this information, however, it would be far more confusing to try to rationalize the discrepancy between two different molecular weight determinations on the same sample by methods which yield different averages. The divergence between such values is a direct consequence of polydispersity in the sample. As we have seen it is not only unavoidable but also informative as to the extent of polydispersity.

1.11 THEORETICAL DISTRIBUTION FUNCTIONS

It was noted earlier that histograms approach smooth distribution curves as the number of classes is increased to a very large number. Sometimes it is desirable to represent a distribution function by an analytical expression which applies continuously to the measured variable. The most familiar of such functions is the normal or Gaussian distribution function:

$$f(x)=\frac{1}{\sigma\sqrt{2\pi}}\exp\left[-\frac{1}{2}\left(\frac{x-\bar{x}}{\sigma}\right)^2\right] \tag{25}$$

In this equation $f(x)$ expresses the fraction of particles having x values between x and $x+dx$; it replaces f_i, which plays a corresponding role in discrete distributions. Some characteristics of the normal distribution are the following. The function $f(x)$ has its maximum value at $x=\bar{x}$ and drops off exponentially with the square of the deviation of a value from the mean, where such deviations are measured as fractions or multiples of the standard deviation. The pre-exponential factor accomplishes the normalization of the function. That is, the integral under the curve over all possible values of $x(-\infty$ to $\infty)$ equals unity. In a broad distribution, σ is large and the exponential does not drop off as rapidly as in a narrow distribution (recall that all deviations are measured relative to the standard deviation). Since the area under the curve is always unity, a narrow distribution will show larger values of $f(x)$ at the maximum whereas a broader distribution will have a smaller value for the function at the maximum. This is why the standard deviation appears in the denominator of the pre-exponential, normalization factor. The normal distribution is the "curve" over which students and teachers alike agonize in connection with course grades. We discuss this distribution function in greater detail in Chap. 3. For the present, we are concerned only with its descriptive capabilities. For this purpose, it is sufficient to note that tables are available (e.g., in the *Handbook of Chemistry and Physics*, Chemical Rubber Company) which supply the value of this function in terms of the standard unit, $t=(x-\bar{x})/\sigma$:

$$f(t)=\frac{1}{\sqrt{2\pi}}\exp\left(-\frac{t^2}{2}\right)dt \tag{26}$$

The normal distribution is commonly encountered in the cumulative form, that is, as the fraction of particles larger (oversize) or smaller (undersize) than a particular t_i value. Since the total area under the normal curve equals unity, the area under one "tail" of the curve from t_i to ∞ gives the fraction of the population having t values greater than the integration limit t_i:

$$f_{t>t_i}=\frac{1}{\sqrt{2\pi}}\int_{t_i}^{\infty}\exp\left(-\frac{t^2}{2}\right)dt \tag{27}$$

Likewise, the cumulative fraction of particles smaller than t_i equals

$$f_{t<t_i}=1-\frac{1}{\sqrt{2\pi}}\int_{t_i}^{\infty}\exp\left(-\frac{t^2}{2}\right)dt \tag{28}$$

Tables are also available (e.g., *C.R.C. Handbook*) for the area under the normal curve between $\bar{x}\,(t=0)$ and one value of t_i (i.e., they apply to one tail only). This area is known as the error function and is often symbolized as $\text{Erf}(t_i)$:

$$\text{Erf}(t_i)=\frac{1}{\sqrt{2\pi}}\int_{0}^{t_i}\exp\left(-\frac{t^2}{2}\right)dt \tag{29}$$

In terms of the error function, Eq. (26) becomes

$$f_{t>t_i}=\tfrac{1}{2}\pm\text{Erf}(t_i) \tag{30}$$

where the positive value is used if t is negative and the negative value is used if t is positive. These signs are reversed if the cumulative fraction of undersize particles is to be determined. Consulting the tables, we see that, in a normally distributed sample, 15.87% of the particles will have t values greater than $+1.0$. This is the percentage of particles for which the deviation from the mean is greater than one standard deviation unit.

Suppose a polydisperse system is investigated experimentally by measuring the number of particles in a set of different classes of diameter or molecular weight. Suppose further that these data are believed to follow a normal distribution function. To test this hypothesis rigorously, the chi-squared test from statistics should be applied. A simple graphical examination of the hypothesis can be conducted by plotting the cumulative distribution data on probability paper as a rapid, preliminary way to evaluate whether the data conform to the requirements of the normal distribution.

Probability paper is a commercially available graph paper which has one coordinate subdivided in ordinary arithmetic units and the other coordinate subdivided into cumulative probability units. The latter are spaced in such a way that normally distributed data will produce a straight line graph when the cumulative percentage of undersize (or oversize) particles is plotted on the probability coordinate and the size variable (diameters, weights, etc.) is plotted on the arithmetic scale. Figure 1.13 shows schematically how normally distributed data (a) are transformed when

FIGURE 1.13 *Three representations of a normal or Gaussian distribution: (a) as a frequency function, (b) as a cumulative function, and (c) as a cumulative function linearized by plotting on probability paper.*

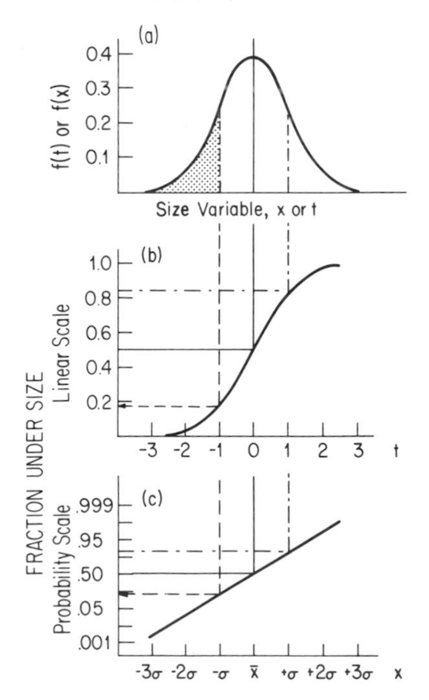

replotted as a cumulative distribution (b) and, finally, when graphed on probability paper (c). The x value corresponding to $y=50\%$ on probability paper gives the mean value. The x value at $y=15.87\%$ gives $\bar{x}-\sigma$ and the x value at $y=84.13\%$ $(100-15.87)$ gives $\bar{x}+\sigma$ when the percentage of undersize particles is plotted (the signs are reversed when percent oversize is plotted). From these values, σ may be determined. Thus a linear plot on probability paper suggests conformity to the normal distribution and also permits the graphical evaluation of $\bar{x}$ and σ.

As we shall see in Chap. 3, the normal distribution comes about when a large number of purely random factors are responsible for the distribution. It is mainly applicable to particles which are produced by condensation, precipitation, or polymerization processes which are purely random.

Dispersions which are produced by comminution—mechanical subdivision of larger chunks—are more likely to produce a linear graph on probability paper if the logarithm of the variable rather than the variable itself is plotted against the probability. Graph paper graduated this way is called log-probability paper. The logarithmic scale implies a much wider range of values for the variable, and also an unsymmetrical distribution function may be written by analogy with Eq. (25):

$$f(\ln x)=\frac{1}{\ln \sigma_g\sqrt{2\pi}}\exp\left(-\frac{\ln x_i-\ln \bar{x}_g}{2\ln \sigma_g}\right)d(\ln x) \tag{31}$$

However, an important difference also emerges from this analogy. The quantities which are normally distributed are logarithms of variables, not the variables themselves. This means that the mean and standard deviation which are obtained from log-probability plots are geometrical averages rather than arithmetical averages. This is the significance of the subscript g in Eq. (31). This is most easily understood by writing the expression for the number average value for the quantity $\ln x_i$:

$$\overline{\ln x}=\sum_i f_i \ln x_i=\sum_i \ln x_i^{f_i} \tag{32}$$

Taking the antilog of this quantity converts the summation into a product over all terms, indicated by Π_i:

$$\text{antiln}(\overline{\ln x})=\prod_i x_i^{f_i}=\bar{x}_g \tag{33}$$

When the averaging is carried out in this way, the result is known as the geometric mean $\bar{x}_g$. The coordinate corresponding to the 15.87% undersize y value equals $\ln \bar{x}_g-\ln \sigma_g$. Because of the properties of logarithms, this is the same as the logarithm of the ratio $\bar{x}_g/\sigma_g$. From this, σ_g may be evaluated. We shall not concern ourselves further with these geometrical averages except to note that

$$\bar{x}_g<\bar{x}_n \tag{34}$$

for any polydisperse system. The validity of this relationship is easily demonstrated by calculating the two averages for a hypothetical distribution.

REFERENCES

1. R. D. Cadle, *Particle Size*, Reinhold, New York, 1965.
2. G. Herdan, *Small Particle Statistics* (2nd ed.), Academic Press, New York, 1960.
3. B. Jirgensons and M. E. Straumanis, *Colloid Chemistry*, Macmillan, New York, 1962.
4. H. R. Kruyt (ed.), *Colloid Science*, Vols. 1 and 2, Elsevier, Amsterdam, 1949, 1952.
5. A. I. Medalia, in *Surface and Colloid Science*, Vol. 4 (E. Matijević, ed.), Wiley, New York, 1971.
6. K. J. Mysels, *Introduction to Colloid Chemistry*, Wiley-Interscience, New York, 1959.
7. H. F. Schaeffer, *Microscopy for Chemists*, Dover, New York, 1953.
8. D. J. Shaw, *Introduction to Colloid and Surface Chemistry*, Butterworth, London, 1966.

PROBLEMS*

1. The specific area of dust particles from the air over Pittsburgh, Pennsylvania, has been determined by gas adsorption.†

Treatment	A_{sp} ($m^2\ g^{-1}$)
4 h under vacuum at 200°C	5.61
8 h under vacuum at 25°C	2.81

   (a) Calculate the radius of these particles if they are assumed to be uniform spheres of density 2.2 g cm^{-3}.
   (b) Propose an explanation for the effect on particle size of degassing.
   (c) What kind of average is obtained for the radius by this procedure?

2. Colloidal palladium particles in an alumina matrix catalyze the hydrogenation of ethene to ethane. The following data describe various catalyst preparations.‡

Diameter of Pd particles (Å)	55	75	75	115	145
ppm Pd in catalyst	170	250	200	250	250
% conversion per 25 mg catalyst	50	45	40.5	38.5	29

* The data for many of the problems in this book are taken from graphs appearing in the original literature. As a result, the values given do not necessarily reflect the accuracy of the original experiments. Likewise, the number of significant figures cited may not be justified in terms of the approximations involved in graph reading.

† M. Corn, T. L. Montgomery, and R. J. Reitz, *Science* **159**: 1350 (1967).

‡ J. Turkevich and G. Kim, *Science* **169**: 873 (1970).

(a) Calculate the activity of these catalysts on the basis of the weight of Pd and the area of Pd in the preparations.
(b) Does the catalytic role of Pd seem to be a bulk or surface phenomenon?

3. The accompanying table shows how the trace metal content of coal-ash aerosols depends on particle size.*

Range of diameters (μm)	g trace element per g ash					
	Pb	Tl	Sb	Cd	Se	As
30–40	300	5	9	<10	<15	160
5–10	820	20	25	<10	<50	800
1.1–2.1	1600	76	53	35	59	1700

Discuss the implications of these results on human health in view of the following considerations: (a) intrinsic toxicity of these trace elements; (b) small particles travel to lung, larger particles are stopped in nose, pharynx, and so on; (c) trace elements are absorbed from the alveoli 7 to 10 times more efficiently than from the upper respiratory spaces; (d) possible mechanism for particle size effect on trace element content.

4. Select a field containing about 30 particles from Fig. 1.9b and measure the Martin diameters of the population parallel to the bottom of the photograph (a photocopy can be made and the particles checked off to avoid duplication and omissions). Classify the data and calculate the mean and standard deviation. Repeat, measuring the Martin diameter parallel to the side of the photograph. Discuss the agreement or discrepancy between the two means in terms of (a) bias in choice of the field, (b) systematic orientation effects, and (c) size of population.

5. Suppose that the particles in Fig. 1.4 were actually oblate ellipsoids (all lying in their preferred orientation) rather than spheres. Would their volume be over- or underestimated by assuming the particles to be spheres? In terms of their axial ratio, calculate the factor by which the mass is under- or overestimated when the particles are assumed to be spheres. (Consult a handbook for the volume of an ellipsoid.)

6. Fifty grams of ZnO–TiO_2 mixtures are shaken with 250 ml water and allowed to settle. After equilibrating for 14 days, the sediment volume is 1.65 times larger when the weight ratio ZnO/TiO_2 is 1.0 than when the ratio is 100.† Some particle characteristics are the following:

	Diameter (μm)	Density (g ml^{-1})	Charge in water
ZnO	1.0	5.6	Positive
TiO_2	2.2	4.2	Negative

* D. F. S. Natusch, J. R. Wallace, and C. A. Evans, Jr., *Science* **183**: 202 (1974).
† L. H. Princen and M. J. DeVena, *J. Am. Oil Chem. Soc.* **39**: 269 (1962).

(a) Assuming the particles are uniform spheres, calculate the ratio of the number of particles of ZnO to TiO_2 for each of the weight ratios given.
(b) Propose an explanation for the more voluminous sediment that results when the weight ratio is 1.0 than when it is 100.

7. The following data describe the particle size distribution in a dispersion of copper hydrous oxide particles in water*:

d_i (μm)	0.426	0.401	0.376	0.351	0.326	0.301	0.276	0.251	0.226	0.201
n_i	1	0	6	6	17	14	11	12	6	6

Calculate $\bar{d}_n$, σ, $\bar{d}_s$, and $\bar{d}_v$ for this dispersion.

8. A graticule was used to size sand particles and glass spheres.† The percentage by weight of particles less than the stated size was found to be as follows:

d (μm)	0.4	0.8	1.6	2.4	3.0	4.0	8.0	12.0
(a) Sand:	0.01	0.07	0.23	0.56	1.23	2.35	11.77	18.06
(b) Glass:	0.01	0.11	0.26	0.43	0.72	1.43	17.84	28.07
d (μm)	16.0	20.0	30.0	40.0	60.0	75.0	90.0	120.0
(a) Sand:	24.62	32.11	52.33	64.14	83.60	—	98.17	100.0
(b) Glass:	36.89	47.68	59.47	61.78	91.49	100.0		

Plot the results on probability and log-probability coordinates. Use the best representation to evaluate (the appropriate) average and standard diviation for the samples.

9. Particle size distributions were measured on aerosols collected in a New York highway tunnel and the following results were obtained‡:

Cumulative % mass $< d_i$	30	40	50	60	70	80
d_i, weekend	0.5	1.0	2.5	5.0	—	—
d_i, weekday	—	—	0.07	0.2	0.9	4.0

Plot these results on log-probability coordinates and estimate the mean and standard deviation for each distribution. What kind of "averages" are these quantities? How do the weekend and weekday particle size distributions compare with respect to location and width of the maximum? The weekend results are attributed to automobile exhaust, whereas the weekday results are assumed to be "diluted" by aerosols from outside the tunnel.

* P. McFadyen and E. Matijević, *J. Colloid Interface Sci.* **44**: 97 (1973).
† G. L. Fairs, *Chem. Ind.* **62**: 374 (1943).
‡ R. E. Lee, Jr., *Science* **178**: 567 (1972).

10. The mass of bull sperm heads in a sample was determined by interference microscopy and the following results were obtained*:

n_i	4	2	27	37	32	26	20	8	3	3	1
$w_i \times 10^{12}$ (g)	5	6	7	8	9	10	11	12	13	14	15

Calculate the number average and weight average weights of these particles.

* G. F. Bahr and E. Zeitler, *J. Cell Biol.* **21**: 175 (1964).

The Viscosity of Dilute Dispersions

2

... imagine that your Tradesman drags behind his regular and respectable vertex, a parallelogram of twelve or thirteen inches in diagonal:—What are you to do with such a monster sticking fast in your house door? [From Abbott's *Flatland*]

2.1 INTRODUCTION

The way liquids flow is one of their most obvious properties. We use a variety of terms in everyday and technical language to describe this property. Thus, we speak of the "thickness" of cream, the "weight" of oil, and the "leveling" of paint. The science student will probably recognize that all these terms describe in one way or another the property known as the viscosity of the liquid. But substituting a technical term does little to make this somewhat elusive property more intelligible.

Our primary interest, of course, is to examine the effect on viscosity of the presence of dispersed particles in a system. Preliminary to doing this, however, we must arrive at a clear understanding of what viscosity means in the more simple case of pure liquids. Doing this will take us into fluid mechanics, an area in which most chemistry students have had very little exposure. It is also an area filled with complicated mathematics which we shall avoid as much as possible. However, to avoid mathematics completely is impossible. We shall discuss the equations of motion and continuity which hold a place in fluid mechanics analogous to the place held by the Schrödinger equation in quantum mechanics. That is, the equation of motion contains the answer to all problems of fluid flow, except that we can rarely solve it because of mathematical difficulties.

After establishing this background, we shall turn to a consideration of the flow properties of dispersed systems. An equation derived by Einstein will be our point of departure here. The Einstein equation relates the viscosity of an assembly of spherical particles to the concentration of the particles. Finally we shall see how certain systems behave which do not conform to the assumptions of the Einstein derivation.

2.2 THE MATHEMATICS OF FLUID FLOW: GENERAL

Suppose we imagine a liquid flowing smoothly past a stationary planar wall and moving parallel to the surface of the wall. As a result of the viscous properties of that liquid, the velocity of the liquid will vary with distance from the wall. This variation is quantitatively described by the velocity gradient dv/dx where v is the velocity of the liquid and x is the distance from the wall. It is convenient to picture the liquid as a set of layers which extend parallel to the wall. Each moves with a uniform velocity which is determined, in part, by its distance from the wall. With this simple picture in mind, we may write the following expressions, either of which may be regarded as a definition of the viscosity of the liquid:

Force acting per unit area in a layer of flowing liquid

$$= \eta \frac{dv}{dx} \tag{1}$$

and

Rate at which energy is deposited in a unit volume of flowing liquid

$$= \eta \left(\frac{dv}{dx} \right)^2 \tag{2}$$

We shall have a good deal more to say about Eqs. (1) and (2) later in the chapter, including a statement of the limits on their applicability. For now, however, they introduce the following ideas about viscosity:

1. Either a force or energy context may be used to discuss flow phenomena.
2. The viscosity is a factor of proportionality in Eqs. (1) and (2); therefore, the more viscous a liquid is, the larger will be the forces and the rate of energy deposition in a flowing liquid for any velocity gradient.
3. The presence of a colloidal size particle in the liquid increases the viscosity because of the effect it has on the flow pattern. Two effects which readily come to mind are illustrated in Fig. 2.1. In Fig. 2.1a, the velocity profile near a wall is shown for a pure liquid. The variation of velocity among layers is indicated by the arrows of different length. In Fig. 2.1b, a nonrotating particle is pictured, cutting across several layers in the flowing liquid. Since the particle does not rotate (by hypothesis), it must slow down the fluid so that layers on opposite sides of the particle have the same velocity, that of the particle itself. The overall velocity gradient is thus reduced. Since the applied force is presumably the same in both Figs. 2.1a and 2.1b, the reduced velocity gradient must be offset by an increase in η. Alternatively, we might consider a particle which is induced to rotate by its position in the velocity gradient. Such a situation is shown in Fig. 2.1c. In this case, some of the energy which would otherwise keep the liquid flowing is deposited in the particle, causing it to rotate. In both cases, the presence of the particle increases the viscosity of the fluid.
4. The increase in viscosity due to dispersed particles is expected to increase with the concentration of the particles, a dependence we may tentatively describe in terms

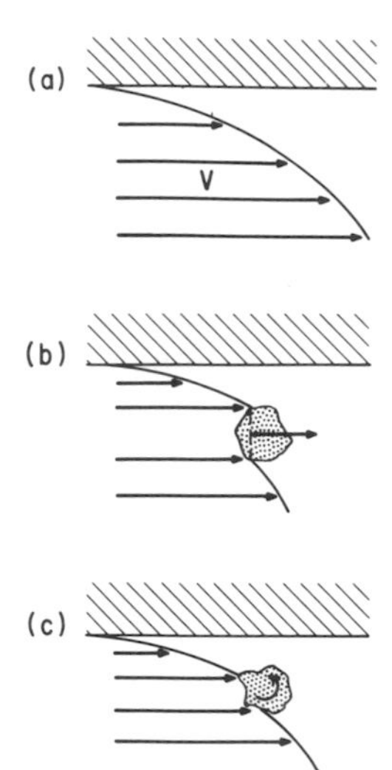

FIGURE 2.1 *Schematic illustration of the flow pattern for (a) a pure liquid near a stationary wall and for a dispersion that contains (b) nonrotating and (c) rotating particles.*

of a power series in concentration c:

$$\eta = A + Bc + Cc^2 + \cdots \tag{3}$$

In this equation A, B, C, . . . are constants whose values are to be determined. This much is evident: As the concentration of a dispersion goes to zero, its viscosity must go to that of the continuous phase. Therefore, $A = \eta_0$, the viscosity of the medium. Furthermore, the constants B, C, . . . might resonably be expected to depend on the size, shape, orientation, and so on, of the dispersed units.

These points give a qualitative indication of the subject matter of this chapter. Our approach to these considerations is in terms of general principles, but we shall not attempt derivations with the maximum generality possible. Instead, we assume some conditions which simplify the mathematics and at the same time describe a large number of systems of practical interest. The more general approach is the subject matter of rheology, the science of flow and deformation of materials. The following is the particular set of restrictions which apply to the majority of our discussions:

1. The density and viscosity of the fluid are uniform in all parts of the sample. Among other things, this implies constant temperature throughout.
2. The velocity is sufficiently low for the fluid to flow without turbulence. This condition is called laminar flow.

3. The velocity is independent of time at any location in the sample, although it may vary widely from location to location. This is called the stationary state condition.
4. Forces due to viscosity are often assumed to be very much larger than inertial forces in the same system—so much so that the latter are often neglected in comparison with the former. This is called the Stokes approximation.
5. In dealing with dispersions, we shall generally assume that the colloidal units are totally free from interactions with one another. This means that the relationships we generate are expected to apply to real systems only in the limit of low concentrations.
6. For the present, we assume the size and shape of the dispersed units to be unaffected by the flow. Thus, we specifically exclude the possibility that flocs are formed or disrupted by the velocity gradients in a flowing dispersion. In fact, there are many important colloidal phenomena which are sensitive to the velocity gradient. We discuss some of these in Chap. 10 in connection with flocculation.

Assumptions 5 and 6 are characteristics of what is known as Newtonian flow. Non-Newtonian behavior generally implies some modification of interaction or structure arising from the conditions of flow.

These assumptions and/or approximations are cited throughout the chapter at the various places where they affect a derivation. It seems worthwhile to collect the most important of them together in one place, however, since the major results of this chapter are subject to these constraints.

With the foregoing ideas as background, we may pose the following as the central question of rheology: What is the velocity and what is the pressure at various locations and various times in a flowing fluid? As we shall see in the next few sections, an analysis of this question provides us with a set of differential equations—the equations of continuity and motion—the solutions of which supply the desired information. The solutions of these and all differential equations depend on the boundary conditions of the problem. That is, any general solution for the pressure p and the velocity v must have the correct values at the surface coordinates of the system and at the initial and final times of the experiment.

The differential equations we will consider are sufficiently complex that, for the most part, we will leave them for the experts. It is important, however, that we understand some of the principal boundary conditions. Therefore, let us consider a few general remarks about p and v at the boundaries of a hypothetical system.

Pressure is a force per unit area which is always directed normal or perpendicular to a surface. If the fluid were static, the pressure would be the same in all directions at any particular location in the liquid, although it would vary from place to place within the sample. If the pressure on the top surface of a liquid is p_0, then the pressure some distance h beneath the surface equals

$$p = p_0 + \rho g h \tag{4}$$

where ρ is the density of the liquid and g is the acceleration due to gravity.

The velocity of a portion of liquid next to a stationary solid is zero; if the solid is moving with a velocity v, then the liquid adjacent to the solid will also have the velocity v. There is nothing particularly self-evident about these statements, but they are experimental facts. Some commonplace evidence that suggests this is the thin layer of dust which accumulates on the blades of a fan. However stiff the breeze may be some distance in front of the blades, the air is still and travels with the blades at the surface. We shall see the kind of physical interactions between solid and liquid which are fundamentally responsible for this when we discuss adsorption in Chaps. 7 and 8 and van der Waals forces in Chap. 10. The nonslip condition between solid and fluid is regarded as an experimental fact in this chapter.

Now, let us describe a condition known as stationary state flow. When a fluid is first induced to flow there is a period of time during which the velocity changes rapidly with time as well as location. It is the nature of viscosity to oppose flow; hence, opposing viscous forces quickly balance the initial accelerating force. Once this condition is met, the velocity will continue to vary with position, but will be independent of time. This is called the stationary state velocity, and represents a boundary condition that applies at $t = \infty$. Since the stationary state velocity is, in fact, achieved very rapidly, it is often assumed to hold throughout an experiment.

In the preceding paragraphs we have enumerated some of the boundary conditions that are useful in solving the differential equations of fluid mechanics. This may seem a bit premature since we do not even have the differential equations as yet. An advantage of thinking first about boundary conditions is that we must define the boundaries of our system in order to apply them. Problems with simple, symmetrical boundaries are always easier to solve. It is no coincidence that Einstein's theory predicts the viscosity of a dispersion of *spheres*. Likewise, we shall see that most pieces of apparatus used to measure viscosity are designed with highly symmetrical geometries.

2.3 THE EQUATION OF CONTINUITY

Consider a small rectangular volume element within a flowing liquid, as shown in Fig. 2.2. A coordinate system has been introduced so that the axes are parallel to the

FIGURE 2.2 *A volume element in a liquid flowing with velocity v.*

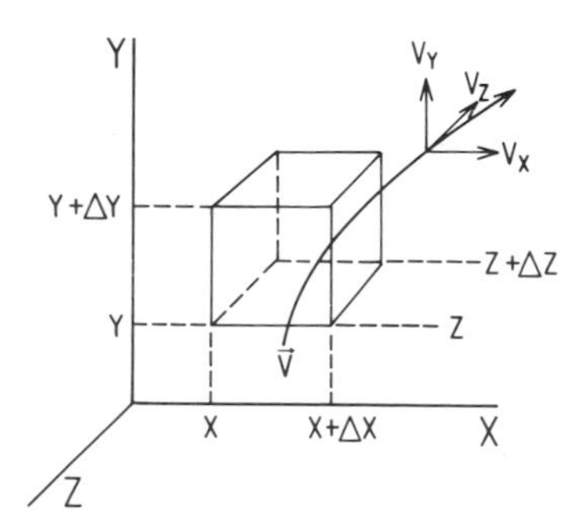

sides of the volume element. In terms of these coordinates, the volume of the element is $\Delta x\ \Delta y\ \Delta z$.

Now let us consider the net rate of flow of matter into this volume element. The rate of mass accumulation equals the difference between the rate at which mass enters the element and the rate at which it leaves. The rate at which matter enters the volume element at x is given by $\rho v_x\ \Delta y\ \Delta z$. Evaluated at x, it equals $\rho(v_x)_x\ \Delta y\ \Delta z$. Of course, v_x is the x component of velocity at the location of interest and $\Delta y\ \Delta z$ is the area of the face perpendicular to x. Likewise, the rate of mass exit at $x+\Delta x$ is $\rho(v_x)_{x+\Delta x}\ \Delta y\ \Delta z$. The difference between these two quantities gives the net matter deposited in the volume element as a result of the x component of flow:

$$[\rho(v_x)_{x+\Delta x} - \rho(v_x)_x]\ \Delta y\ \Delta x \tag{5}$$

Similar expressions arise from the y and z components of velocity. Therefore, the net rate at which matter is deposited in the volume element, $(\partial\rho/\partial t)\ \Delta x\ \Delta y\ \Delta z$, is

$$\left(\frac{\partial\rho}{\partial t}\right)\Delta x\ \Delta y\ \Delta z = [\rho(v_x)_{x+\Delta x} - \rho(v_x)_x]\ \Delta y\ \Delta z + [\rho(v_y)_{y+\Delta y} - \rho(v_y)_y]\ \Delta x\ \Delta z + [\rho(v_z)_{z+\Delta z} - \rho(v_z)_z]\ \Delta x\ \Delta y \tag{6}$$

If the fluid is incompressible, then $\rho = \text{const}$ and $\partial\rho/\partial t = 0$. Therefore,

$$[(v_x)_{x+\Delta x} - (v_x)_x]\ \Delta y\ \Delta z + [(v_y)_{y+\Delta y} - (v_y)_y]\ \Delta x\ \Delta z + [(v_z)_{z+\Delta z} - (v_z)_z]\ \Delta x\ \Delta y = 0 \tag{7}$$

If we now divide through by $\Delta x\ \Delta y\ \Delta z$ and consider the limit in which Δx, Δy, and Δz approach zero, we obtain

$$\frac{\partial v_x}{\partial x} + \frac{\partial v_y}{\partial y} + \frac{\partial v_z}{\partial z} = 0 \tag{8}$$

a result which is often encountered in vector notation as

$$\boldsymbol{\nabla} \cdot \mathbf{V} = 0 \tag{9}$$

where $\boldsymbol{\nabla} = \partial/\partial x + \partial/\partial y + \partial/\partial z$ and the dot product is taken. Equation (8) is known as the equation of continuity for an incompressible fluid, since the constancy of ρ was explicitly assumed in its derivation.

2.4 THE EQUATION OF MOTION

Now let us examine the rate at which momentum accumulates in the same volume element. To begin with, we recognize that the rate of change of momentum equals a force $\mathbf{F}$:

$$\frac{\partial(m\mathbf{v})}{\partial t} = m\left(\frac{\partial\mathbf{v}}{\partial t}\right) = m\mathbf{a} = \mathbf{F} \tag{10}$$

Further, the net rate of change of momentum *per unit volume* is given by $\rho\ \partial\mathbf{v}/\partial t$. Therefore, the net rate of change in momentum per unit volume equals the sum of all forces acting per unit volume. This last quantity is made up of three contributions: external forces, pressure forces, and viscous forces. Let us now consider each of these contributions in turn. Since momentum and force are both vector quantities, it will be convenient to resolve each into x, y, and z components.

As far as we are concerned, the only external force to be considered is the force of gravity. Since our coordinate system has been set up to parallel the walls of the volume element, we must recognize that the force of gravity will not necessarily consist of just one contribution, but may have x, y, and z components as well. For the x component of the gravitational force per unit volume we write

$$\frac{(F_x)_{\text{ext}}}{\Delta x\ \Delta y\ \Delta z} = \frac{mg_x}{\Delta x\ \Delta y\ \Delta z} = \rho g_x \tag{11}$$

where similar relations hold for $(F_y)_{\text{ext}}$ and $(F_z)_{\text{ext}}$.

Next, let us consider the force per unit volume arising from pressure. The net force in the x direction equals

$$(F_x)_{\text{press}} = (p_{x+\Delta x} - p_x)\ \Delta y\ \Delta z \tag{12}$$

with analogous contributions in the y and z directions. If we divide Eq. (12) by the volume of the element, and take the limit, we obtain for the x component of the pressure force per unit volume:

$$\frac{(F_x)_{\text{press}}}{\Delta x\ \Delta y\ \Delta z} = \frac{\partial p}{\partial x} \tag{13}$$

Again, similar expressions hold for $(F_y)_{\text{press}}$ and $(F_z)_{\text{press}}$.

Finally, we turn our attention to the contribution of the viscous forces per unit volume in the system. Imagine two parallel plates of area A between which is sandwiched a liquid of viscosity η. If a force F_x parallel to the x direction is applied to one of these plates, it will move in the x direction as shown in Fig. 2.3. Our concern is the description of the velocity of the fluid enclosed between the two plates. In order to do this, it is convenient to visualize the fluid as consisting of a set of layers stacked parallel to the boundary plates. At the boundaries, those layers which are in contact with the plates are assumed to possess the same velocities as the plates themselves: That is, $v_x = 0$ for the lower plate and equals v_x for the upper plate.

FIGURE 2.3 *Illustration of the relationship between applied force per unit area and fluid velocity.*

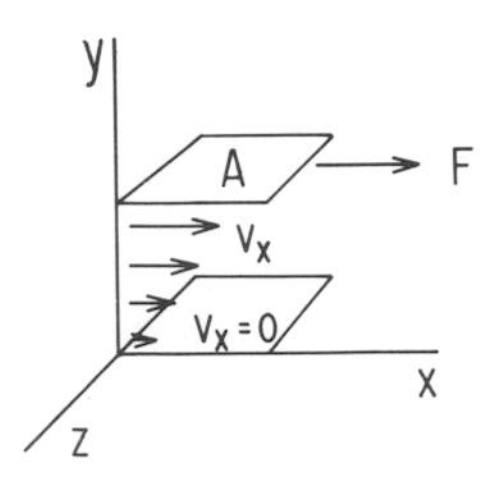

Intervening layers have intermediate velocities. This condition is known as laminar flow and is limited to low velocities.

Newton's law of viscosity [Eq. (1)] describes the behavior of a large number of fluids, including dilute colloidal systems, under stationary state conditions. Writing this result in the notation of Fig. 2.3 gives

$$\frac{F}{A} = \eta \frac{dv_x}{dy} \tag{14}$$

Those fluids which obey the form predicted by Eq. (14) are said to be Newtonian. Figure 2.4 is a sketch of F/A versus the velocity gradient (also called the rate of shear since dv/dy has units of reciprocal time) for several different modes of behavior. For a Newtonian fluid, this representation gives a straight line of zero intercept and slope equaling η. Non-Newtonian fluids generally show nonlinear plots; their "viscosity," the slope of the tangent to the curve at various points, is a function of the rate of shear. As already mentioned, we shall concentrate on Newtonian behavior in this chapter, and return to non-Newtonian colloids in Chap. 10. It might be noted, in passing, that most actual representations of experimental results display the data with the coordinates interchanged from the way they are shown in Fig. 2.4. In that case, it is the cotangent of the angle which describes the slope of the line at any point that determines the true (if Newtonian) or apparent (if non-Newtonian) viscosity of the system. Figure 10.15 is an example of this alternative representation.

Inspection of Eq. (14) reveals that η has units of mass time^{-1} length^{-1}. In cgs units, 1 g s^{-1} cm^{-1} is defined to equal one poise. At room temperature, pure water has a viscosity of about 0.01 P. It is important to realize that viscous forces act parallel to fixed surfaces and not perpendicular to them as does pressure. As such, they are called shearing forces; we shall define F/A by the symbol τ for shearing forces.

With these definitions in mind, let us return to the question of the force density at a volume element submerged in the flowing liquid. If we apply Eq. (14) to the x component of velocity, for example, and take the difference between the shear on opposite parallel faces, we obtain

$$(F_x)_{vis} = (\tau_{x+\Delta x} - \tau_x)\,\Delta y\,\Delta z + (\tau_{y+\Delta y} - \tau_y)\,\Delta x\,\Delta z + (\tau_{z+\Delta z} - \tau_z)\,\Delta x\,\Delta y \tag{15}$$

FIGURE 2.4 *Comparison of Newtonian liquids with several forms of non-Newtonian behavior.*

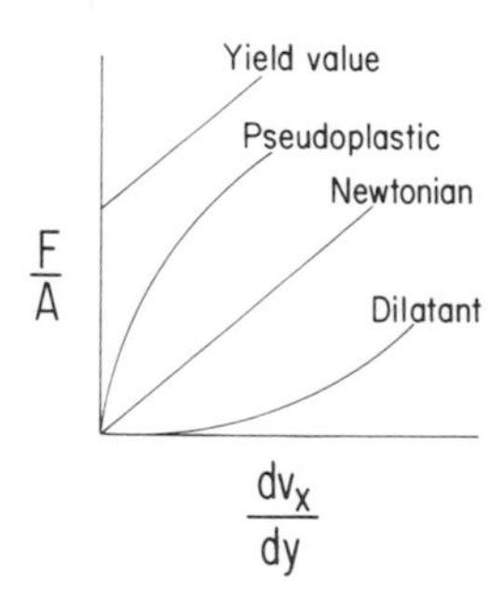

In this equation, for the x component of the viscous force, only v_x is involved in the various expressions for τ, and the subscripts on τ indicate the direction of the gradient. If we divide Eq. (15) through by the volume of the element and consider the limit, we obtain

$$\frac{(F_x)_{\text{vis}}}{\Delta x\ \Delta y\ \Delta z} = \eta\left[\frac{\partial^2 v_y}{\partial x^2} + \frac{\partial^2 v_y}{\partial y^2} + \frac{\partial^2 v_z}{\partial z^2}\right] \tag{16}$$

if the value of η is constant. The plausibility of this result is suggested by differentiating Eq. (14) with constant η and substituting into (15). The vectorial nature of these quantities complicates matters considerably, however, and the rigorous proof of Eq. (16) is outside the scope of this book. In vector notation, Eq. (16) becomes

$$\frac{(F_x)_{\text{vis}}}{\Delta x\ \Delta y\ \Delta z} = \eta \nabla^2 v_x \tag{17}$$

where $\nabla^2 = \partial^2/\partial x^2 + \partial^2/\partial y^2 + \partial^2/\partial z^2$. Analogous expressions may be written for F_y and F_z. Equations (11), (13), and (17) provide us with expressions for the x components of the three kinds of force operating on an element of volume in the flowing liquid. Analogous expressions exist also for the y and z components of each. The sum of all these contributions equals the rate of momentum accumulation in the volume element, $d(\rho v)/dt$:

$$\frac{d(\rho \mathbf{v})}{dt} = \mathbf{F}_{\text{ext}} + \mathbf{F}_{\text{press}} + \mathbf{F}_{\text{vis}} \tag{18}$$

Next, let us consider further development of the left-hand side of Eq. (18).

In the most general case, both ρ and $\mathbf{v}$ may be regarded as functions of x, y, z, and t. Although we shall impose limitations on this generality in the next few paragraphs, we shall begin our evaluation of $d(\rho\mathbf{v})/dt$ by considering the general case. Using the chain rule for differentiation, we write

$$\frac{d(\rho \mathbf{v})}{dt} = \rho\left(\frac{d\mathbf{v}}{dx}\frac{dx}{dt} + \frac{d\mathbf{v}}{dy}\frac{dy}{dt} + \frac{d\mathbf{v}}{dz}\frac{dz}{dt} + \frac{d\mathbf{v}}{dt}\right)$$

$$+\mathbf{v}\left(\frac{d\rho}{dx}\frac{dx}{dt} + \frac{d\rho}{dy}\frac{dy}{dt} + \frac{d\rho}{dz}\frac{dz}{dt} + \frac{d\rho}{dt}\right) \tag{19}$$

Now, several special cases of this relationship can be written. If ρ depends on time but not on location, Eq. (19) becomes

$$\frac{d(\rho \mathbf{v})}{dt} = \rho\left(\frac{d\mathbf{v}}{dx}v_x + \frac{d\mathbf{v}}{dy}v_y + \frac{d\mathbf{v}}{dz}v_z + \frac{d\mathbf{v}}{dt}\right) + \mathbf{v}\frac{d\rho}{dt} \tag{20}$$

since $dx/dt = v_x$, and so on. Equation (20) may be expressed more concisely in vector notation as

$$\frac{d(\rho \mathbf{v})}{dt} = \rho(\mathbf{v}\cdot\nabla)\mathbf{v} + \mathbf{v}\frac{d\rho}{dt} \tag{21}$$

If ρ is independent of both time and location, Eq. (19) simplifies to

$$\frac{d(\rho\mathbf{v})}{dt}=\rho\left(v_x\frac{d\mathbf{v}}{dx}+v_y\frac{d\mathbf{v}}{dy}+v_z\frac{d\mathbf{v}}{dz}+\frac{d\mathbf{v}}{dt}\right) \tag{22}$$

We may substitute Eq. (22) into (18) and (11), (13), and (17) into (18) as well to obtain

$$\rho\left(v_x\frac{d\mathbf{v}}{dx}+v_y\frac{d\mathbf{v}}{dy}+v_z\frac{d\mathbf{v}}{dz}+\frac{d\mathbf{v}}{dt}\right)=\rho\mathbf{g}+\frac{\partial p}{\partial x}+\frac{\partial p}{\partial y}+\frac{\partial p}{\partial z}+\eta\left(\frac{\partial^2\mathbf{v}}{\partial x^2}+\frac{\partial^2\mathbf{v}}{\partial y^2}+\frac{\partial^2\mathbf{v}}{\partial z^2}\right) \tag{23}$$

or in vector notation

$$\rho(\mathbf{v}\cdot\boldsymbol{\nabla})\mathbf{v}+\rho\frac{\partial\mathbf{v}}{\partial t}=\rho\mathbf{g}+\boldsymbol{\nabla}p+\eta\nabla^2\mathbf{v} \tag{24}$$

This result is called the equation of motion for an incompressible fluid.

The first term on the left-hand side of Eq. (24) is called the inertial term and the second is called the viscous term. For low velocities, the latter term is more important than the former. For conditions under which the first term can be neglected entirely, a situation called the Stokes approximation, Eq. (24) simplifies to

$$\rho\frac{\partial\mathbf{v}}{\partial t}=\rho\mathbf{g}+\boldsymbol{\nabla}p+\eta\nabla^2\mathbf{v} \tag{25}$$

Under stationary state conditions, the velocity is independent of time and Eq. (25) becomes

$$\rho\mathbf{g}+\boldsymbol{\nabla}p+\eta\nabla^2\mathbf{v}=0 \tag{26}$$

This form of the equation of motion is known as the Stokes–Navier equation.

The sequence of assumptions introduced between Eqs. (19) and (26) clearly has a twofold effect. The resulting equations become more limited in their applicability as they become simpler to solve. This trade-off is justified for our purposes since the Stokes–Navier equation applies with sufficient accuracy to many colloidal phenomena. Recall that the Stokes–Navier result is a differential equation, the solutions to which are expressions which give p and $\mathbf{v}$ as functions of x, y, and z and also satisfy the boundary conditions of the problem.

We conclude this section by defining a quantity called the substantial derivative for which we use the symbol D/Dt. Ordinarily, the coordinate system that is used to describe a flow problem is defined with axes which are stationary with respect to the flow. When we are talking about dispersions, however, another frame of reference is advantageous. In this case, a coordinate system which moves with the particle along a line of flow is very useful. When we use the convention of moving with a volume element along a streamline, all the terms on the right-hand side of Eq. (20) may be replaced by the substantial derivative:

$$\frac{D(\rho\mathbf{v})}{Dt}=\rho\left(v_x\frac{d\mathbf{v}}{dx}+v_y\frac{d\mathbf{v}}{dy}+v_z\frac{d\mathbf{v}}{dz}\right)+\mathbf{v}\frac{\partial\rho}{\partial t} \tag{27}$$

where the differential notation D/Dt reminds us that we are *moving with* the particle. Otherwise, the substantial derivative is treated like any other derivative.

Many of the equations we have written in this section are quite complicated. Nevertheless, it is important to note that they merely state some very elementary physics about complex phenomena. The equation of continuity [Eq. (8)] is an abstract statement of the principle of conservation of matter. The equation of motion [Eq. (23), and (26), which is a special case of (23)] states that the acceleration of a volume element is proportional to the sum of the forces acting upon it: Newton's second law! The equations of motion and continuity are the basic relationships of fluid mechanics; our primary reason for taking them up is the fact that they are encountered so often in the literature which describes the flow properties of colloidal systems. It should be remembered that the basic facts described by Eqs. (8) and (23) remain the same even though these equations are often written in forms that look very different. We have already written each of these in the equivalent vector notation: Eqs. (9) and (24). From this point they are often taken through a multitude of transformations from vector field theory which alter the appearance of the equations completely. In addition, the equations of continuity and motion are often transformed into either cylindrical or spherical coordinates in anticipation of certain geometrical features of the boundaries. For this reason it is important to remember the fundamentally simple physical meaning of these two relationships and not be overly concerned with the various forms in which they may be written.

2.5 THE POISEUILLE EQUATION AND CAPILLARY VISCOMETERS

As an application of the foregoing principles to a practical problem, let us consider the stationary state flow of a liquid through a long cylindrical tube of radius R and length l as shown in Fig. 2.5a. It is clear that this problem has cylindrical symmetry and will be more readily analyzed in terms of v_r, v_θ, and v_z than in terms of v_x, v_y, and v_z. Transforming the equations of continuity and motion, Eqs. (8) and (23), respectively, into cylindrical coordinates involves quite a bit of tedious mathematics which serves more to obscure our objectives than to advance them. The following simplified analyses will lead us to the same result as would be obtained by solving the equation of motion for this flow problem.

FIGURE 2.5 *Definition of the coordinates of (a) a cylindrical capillary and (b) a cylindrical shell of flowing liquid.*

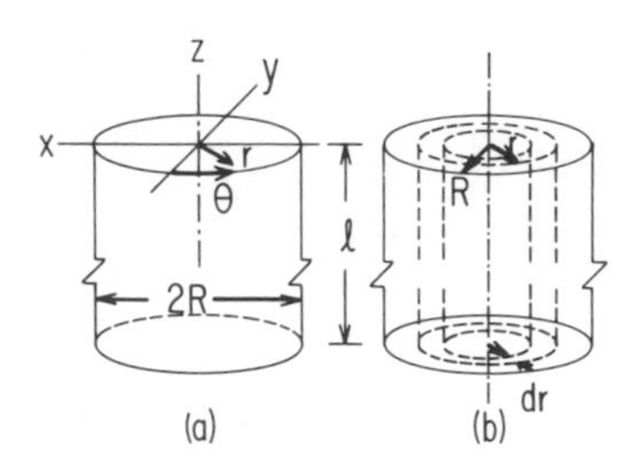

If the tube is long enough for end effects to be neglected and for stationary state conditions to prevail, then the flow is totally in the z direction. The radial and angular components of the flow equal zero. The streamlines, therefore, consist of a bundle of lines running parallel to the z axis. Because of the cylindrical symmetry of the arrangement, the value of $\mathbf{v}$ is independent of θ, although it will vary with r, the radial distance from the center of the tube. Thus, all the streamlines within a volume element shaped like a cylindrical shell will have the same value of v_z. This situation is shown in Fig. 2.5b. Let us attempt to write an expression for v_z as a function of r, the radius of the cylindrical shell. Since stationary state conditions hold within the volume element, the external and pressure forces must exactly balance the viscous force, according to Eq. (26).

The increment in viscous force acting on this element is the difference between the viscous forces on the outer and inner surfaces of the element, where each of these is given by Eq. (14):

$$\Delta F_{\text{vis}} = (F_{\text{vis}})_{\text{out}} - (F_{\text{vis}})_{\text{in}} = 2\pi(r+dr)\eta l\left(\frac{dv_z}{dr}\right)_{r+dr} - 2\pi r\eta l\left(\frac{dv_z}{dr}\right)_r \tag{28}$$

Next, we must relate $(dv_z/dr)_{r+dr}$ to $(dv_z/dr)_r$. The following expression accomplishes this, provided dr is small:

$$\left(\frac{dv_z}{dr}\right)_{r+dr} = \left(\frac{dv_z}{dr}\right)_r + \left(\frac{d^2v_z}{dr^2}\right) dr \tag{29}$$

If this result is substituted into Eq. (28), expanded, and only terms linear in dr retained, the expression becomes

$$\Delta F_{\text{vis}} = 2\pi\eta l\left(r\frac{d^2v_z}{dr^2}\,dr + \frac{dv_z}{dr}\,dr\right) = 2\pi\eta l\frac{d}{dr}\left(r\frac{dv_z}{dr}\right) \tag{30}$$

This force is counterbalanced by the increments in external and pressure forces:

$$\Delta(F_{\text{ext}} + F_{\text{press}}) = 2\pi l\rho g r\,dr + 2\pi\,\Delta p r\,dr \tag{31}$$

where the first term equals the weight of the shell and the second is the force on the shell if a pressure difference of Δp exists across the ends of the tube. Setting Eqs. (30) and (31) equal to each other yields

$$\eta\frac{d}{dr}\left(r\frac{dv_z}{dr}\right) = \left(\rho g + \frac{\Delta p}{l}\right) r\,dr \tag{32}$$

Integration of Eq. (32) yields

$$\eta r\frac{dv_z}{dr} = \tfrac{1}{2}\left(\rho g + \frac{\Delta p}{l}\right) r^2 \tag{33}$$

where the condition that $r\,dv_z/dr$ equals zero at $r = 0$ is used to evaluate the integration constant.

Equation (33) may be integrated again:

$$\eta \int dv_z = \tfrac{1}{2}\left(\rho g + \frac{\Delta p}{l}\right) \int r\,dr \tag{34}$$

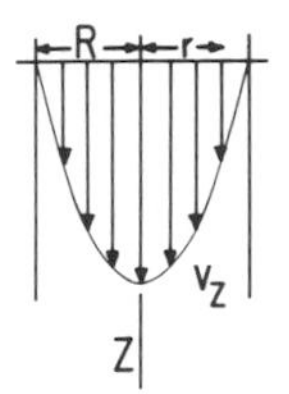

FIGURE 2.6 *The parabolic profile of velocity in a cylindrical tube.*

Using the boundary condition that $v_z = 0$ at $r = R$ to evaluate the constant yields

$$v_z = \frac{(\rho g + \Delta p/l)(r^2 - R^2)}{4\eta} \tag{35}$$

This equation describes the velocity of a fluid element to be a parabolic function of its radial distance from the center of the tube as shown in Fig. 2.6. Equation (35) is also a solution to Eq. (23) when the latter is written in cylindrical coordinates.

The rate of volume flow through the tube, V/t, equals the summation of the cross-sectional area of each shell multiplied by the velocity of that shell where the latter is given by Eq. (35):

$$\frac{V}{t} = \int_0^R \frac{(\rho g l + \Delta p)}{4\eta l}(r^2 - R^2) 2\pi r \, dr \tag{36}$$

or

$$\frac{V}{t} = \frac{(\rho g l + \Delta p)\pi R^4}{8\eta l} \tag{37}$$

Equation (37), known as Poiseuille's law, provides the basis for the most common technique for measuring the viscosity of a liquid or a dilute colloidal system, namely, the capillary viscometer.

Most capillary viscometers are designed with a relatively large bulb at both ends of the capillary as shown in Fig. 2.7. A constant volume in the upper bulb is designated by two lines etched at either end of the bulb. The viscometer is used by measuring the time required for the liquid level to drop from one line to the other as the fluid drains through the capillary. In such an apparatus, the difference in height of the two liquid columns is relatively constant during the time required for flow. Generally, the only pressure difference across the liquid is due to the weight of the liquid. Under these conditions, Eq. (37) can be written

$$t = \text{const}\,\frac{\eta}{\rho} \tag{38}$$

in which the constant incorporates all the parameters which characterize the apparatus. Comparison of the flow times of two substances, one known and one unknown, through the same apparatus provides an easy way to evaluate η for the

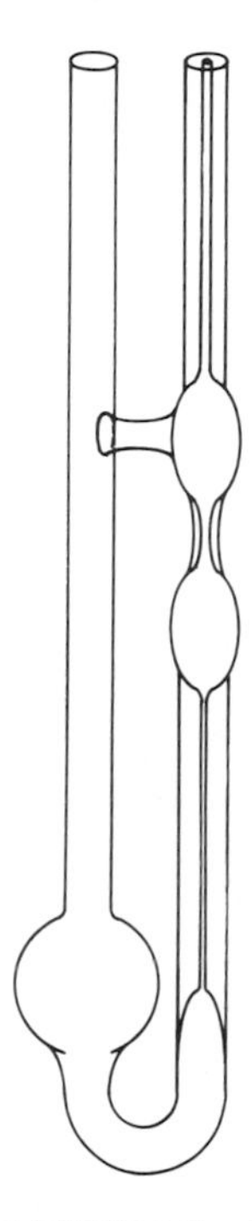

FIGURE 2.7 *Schematic of a capillary viscometer.*

unknown. For greater accuracy, correction factors which allow for end effects may also be included in this type of calculation.

In conclusion, let us examine the velocity gradient or rate of shear at various positions within the capillary. Equation (35) shows that dv_z/dr is directly proportional to r, so in a given viscosity measurement the magnitude of gradient varies between zero and its maximum value at the wall. We could use capillaries of different radii to investigate the type of effect shown in Fig. 2.3. This is not the best way to conduct such an investigation, however, because the gradient varies so much with distance from the center. The concentric cylinder viscometer and the cone-and-plate viscometer described in the following section are preferable when a variable rate of shear is desired.

2.6 CONCENTRIC CYLINDER AND CONE-AND-PLATE VISCOMETERS

Figure 2.8 shows schematic diagrams of two additional types of viscometers: (a) the concentric cylinder and (b) the cone-and-plate viscometers. In both, the liquid under investigation is placed in contact with a rotating component of the apparatus. Because of the viscosity of the fluid, there is resistance to the rotational motion which induces a measurable torque on the suspension mechanism. Through suitable calibration, the force needed to produce a given torque can be determined. In either

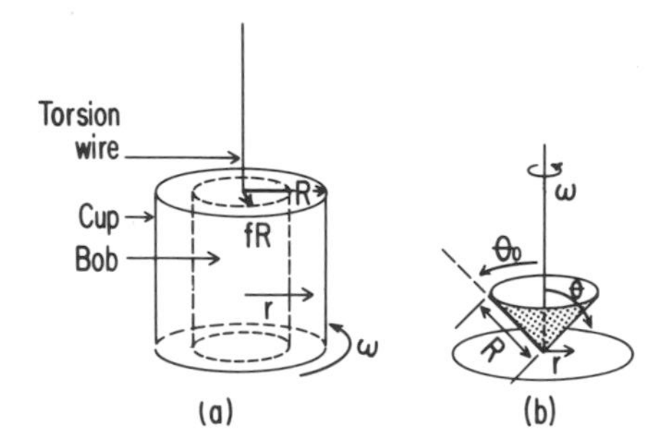

FIGURE 2.8 *Schematic of (a) concentric cylinder and (b) cone-and-plate viscometers.*

apparatus, the amount of torque increases with the viscosity of the fluid. If the relationship between the fluid flow and the instrument parameters is known, torque measurements permit the evaluation of η.

A mathematical analysis of both the concentric cylinder and cone-and-plate viscometers begins with the equations of continuity and motion. We shall not discuss these viscometers in as much detail as the capillary viscometer; only a few general remarks on each analysis are provided.

The concentric cylinder viscometer, also known as the Couette viscometer, clearly has radial symmetry. When it is operating under stationary state conditions, v_r and v_z are zero; therefore, we only need to consider the θ component of the equation of motion to characterize the fluid flow.

Let us suppose it is the outer cylinder of the apparatus that rotates and that it has a radius R, a height l, and an angular velocity ω (in rads per second). Furthermore, we express the radius of the inner stationary cylinder as a fraction f of the radius of the outer cylinder ($R_{in} = fR_{out}$). The boundary conditions for v_θ (in centimeters per second) are

$$\text{at} \quad r = R: \qquad v_\theta = R\omega = (v_\theta)_{max} \tag{39}$$

$$\text{at} \quad r = fR: \qquad v_\theta = 0 \tag{40}$$

Under these conditions, a solution to the θ component of the equation of motion is

$$v_\theta = \frac{\omega R(fR/r - r/fR)}{(f - 1/f)} \tag{41}$$

Equation (41) describes the variation of v_θ as a function of r. The force transmitted to the outer cylinder is given by

$$F = 4\pi\eta l\omega R\left(\frac{f^2}{1-f^2}\right) \tag{42}$$

A satisfying physical picture of this result is obtained as follows. We define the distance of separation between the two cylinders δ by the expression

$$\delta = R(1-f) \tag{43}$$

and consider the case where it is very small, that is, f approaching unity. The area of contact between the liquid and wall is $2\pi Rl$; if we divide both sides of Eq. (42) by this area A, we obtain

$$\frac{F}{A} = 2\eta\omega\left(\frac{f^2}{1-f^2}\right) \tag{44}$$

Since $(1-f^2) = (1+f)(1-f) \simeq 2(1-f)$, Eq. (44) becomes

$$\frac{F}{A} = \frac{\eta\omega f^2}{1-f} \tag{45}$$

Multiplying numerator and denominator of this expression by R and substituting $(v_\theta)_{max}$ from Eq. (39) and δ from (43) gives

$$\frac{F}{A} = \frac{\eta (v_\theta)_{max} f^2}{\delta} \tag{46}$$

Since the velocity at the stationary inner wall is zero, this expression becomes

$$\lim_{f \to 1} \frac{F}{A} = \eta \frac{dv_\theta}{dr} \tag{47}$$

which is equivalent to Eq. (14), the defining equation for η. Thus in the limiting case when the separation between the cylinders is negligible compared to their radius of curvature, the concentric cylindrical surfaces approximate the infinite parallel plates of the model used to define the coefficient of viscosity.

Figure 2.9 shows a commercially available device which measures viscosity by essentially the concentric cylinder method. In the apparatus shown, the inner cylinder rotates and the torque is transmitted to the outer cylinder, and is recorded on the drum which rotates as ω varies. In this way, F/A is measured at different rates of shear. In most commercial instruments, it is the inner cylinder that rotates, even though there is a theoretical advantage to having the outer cylinder rotate rather than the inner one. The reason is that laminar flow is stabilized by centrifugal forces in the former case, whereas centrifugal forces tend to induce turbulent flow in the latter.

Finally, inspection of Eq. (41) reveals that dv_θ/dr is no more constant here than in the capillary viscometer. In both the capillary and concentric cylinder viscometers, either the mean or the maximum rate of shear is reported in describing the conditions under which the viscosity is measured. In neither case is this a constant quantity throughout the fluid. Since both δ and ω can be varied over a wide range of values, however, a broad spectrum of shear rates can be conveniently studied by this type of apparatus.

We conclude this section with a few remarks about the cone-and-plate type of viscometer, sketched schematically in Fig. 2.8b. In this viscometer, the fluid is placed between a stationary plate and a cone which touches the plate at the apex of the cone. This apparatus also possesses cylindrical symmetry, but this time in order to indicate a location within the fluid we must specify not only r, the distance from the axis of rotation, but also the location within the gap between the cone and the plate,

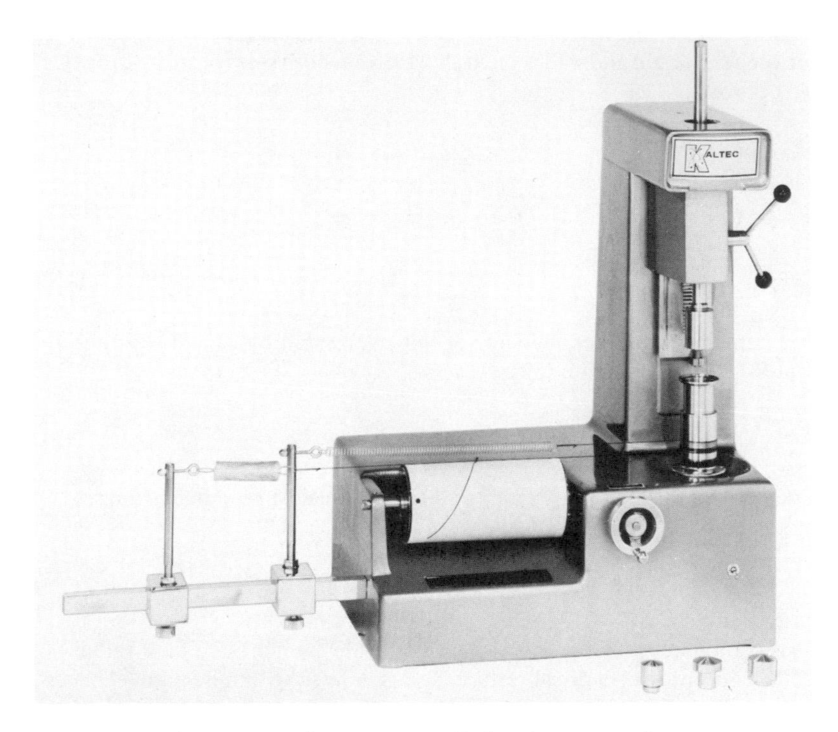

FIGURE 2.9 *A commercial concentric cylinder viscometer, the Hercules Hi-Shear viscometer, Kaltec Scientific Instruments, Inc., Kalamazoo, Michigan (used with permission).*

conveniently measured by θ, the angle from the vertical (see Fig. 2.8b). Application of the equation of continuity and the equation of motion to this situation, followed by considerable mathematical manipulation, leads to the result that

$$\frac{dv_\theta}{dr} \simeq \omega \frac{\cos \theta}{\cos \theta_0} \tag{48}$$

where θ_0 is the angle between the vertical and the wetted surface of the cone. This approximation holds for the case in which both θ and θ_0 are close to 90°. This condition is always met in actual practice where the angle $90 - \theta_0$ is generally less than 5°. Some commercial concentric cylinder viscometers have interchangeable parts; this means that a conical rotor may be substituted for a cylindrical one. Equation (48) shows that the primary velocity gradient within the fluid is independent of radial position to a very good approximation. This feature distinguishes cone-and-plate viscometers from both of the viscometers previously discussed.

To determine the viscosity of the fluid, the torque T necessary to turn the cone at an angular velocity ω is measured. This torque depends on the viscosity of the fluid according to the equation

$$T = \frac{\frac{4}{3}\pi R^3 \eta \omega \sin \theta_0}{\cot \theta_0 + \frac{1}{2}\{\ln[(1+\cos \theta_0)/(1-\cos \theta_0)]\} \sin \theta_0} \tag{49}$$

where R is the radius of the cone measured along the wetted surface. Like the Couette viscometer, this apparatus can be operated, in principle, at a variety of different angular velocities, and thus η can be studied at different rates of shear. Furthermore, this last quantity is reasonably constant throughout the fluid in the apparatus.

As noted previously, dilute colloidal systems display Newtonian behavior; that is, their apparent viscosity is independent of the rate of shear. Accordingly, the capability to measure η under conditions of variable shear is relatively superfluous in these systems. However, non-Newtonian behavior is commonplace in flocculated colloids, as we shall see in Chap. 10.

2.7 DISSIPATION OF ENERGY AND VISCOSITY

In the preceding sections, we examined viscous phenomena from a largely mechanical viewpoint. This is convenient when our interest is in the measurement of viscosity. Having looked at this in some detail, however, we are now ready to turn our attention to other aspects of viscous behavior. Specifically, we are interested in the relationship between the viscosity of a dispersion and the nature of the dispersed phase. We shall examine this relationship in two stages. First, Einstein's theory for the viscosity of a dispersion of spheres is discussed. For particles which do not conform with the Einstein model, a more empirical approach is followed. To proceed, we need to translate our relationships involving viscosity from force to energy expressions.

Energy may be described in terms of a force operating through a distance. If we know the rate at which that distance is covered, that is, the velocity, we know the rate of energy production. In the case of viscous flow, it is more accurate to relate viscous forces to the rate of energy dissipation, but the mathematical difference between energy production and energy dissipation is merely a matter of sign. A flowing liquid possesses a certain amount of translational kinetic energy; viscous forces operating over a length of the flow path would clearly damp out an unsustained flow entirely. Viscosity is responsible for the transformation of translational kinetic energy into other forms of energy. To see this quantitatively, let us return to the equation of motion.

Following the motion of a volume element in a flowing liquid, we may combine Eqs. (23) and (29) to write

$$\frac{D\mathbf{v}}{Dt} = \mathbf{g} + \frac{1}{\rho}\boldsymbol{\nabla} p + \frac{\eta}{\rho}\nabla^2 \mathbf{v} \tag{50}$$

where $D\mathbf{v}/Dt$ is the substantial derivative. Suppose we now multiply both sides of this equation by $\mathbf{v}$, using the dot product when other vector quantities are involved:

$$\mathbf{v}\cdot\frac{D\mathbf{v}}{Dt}=\mathbf{v}\cdot\mathbf{g}+\frac{1}{\rho}\mathbf{v}\cdot\nabla p+\frac{\eta}{\rho}\mathbf{v}\cdot\nabla^2\mathbf{v} \tag{51}$$

Since $\mathbf{v}\,D\mathbf{v}/Dt=\frac{1}{2}Dv^2/Dt$, this may be written

$$\frac{D(\frac{1}{2}\rho v^2)}{Dt}=\rho\mathbf{g}\cdot\mathbf{v}+\mathbf{v}\cdot\nabla p+\eta\mathbf{v}\cdot\nabla^2\mathbf{v} \tag{52}$$

recalling that density is assumed to be a constant. The left-hand side of Eq. (52) gives the rate of change in translational kinetic energy per unit volume as viewed from the perspective of the moving volume element.

One of the advantages of mathematical statements such as Eq. (52) is that they are very general. A disadvantage is that they often achieve this generality by being very abstract, and this abstraction frequently obscures the meaning of the statement. Some additional discussion will help us understand the significance of this result.

Equation (52) describes the rate of energy dissipation *per unit volume.* To get rid of this last stipulation, multiply both sides of the equation by the volume of the element $dx\,dy\,dz$:

$$\frac{D(\mathrm{KE})}{Dt}dx\,dy\,dz=\rho\mathbf{g}\cdot\mathbf{v}\,dx\,dy\,dz+\mathbf{v}\cdot\nabla p\,dx\,dy\,dz+\eta\mathbf{v}\cdot\nabla^2\mathbf{v}\,dx\,dy\,dz \tag{53}$$

The three terms on the right-hand side of this equation are clearly traceable to external, pressure, and viscous forces. Now let us examine each of these in turn to get a better feeling for the physical significance of the result.

The first term on the right-hand side of Eq. (53) describes the rate of energy dissipation arising from movement in a gravitational field. The product of the density times the volume equals a mass. This times the acceleration of gravity equals a force, the weight of a fluid particle. If we postulate that this force is directed in the z direction, the weight times the z component of velocity equals the rate of change of the potential energy of the fluid particle. It is obvious that the potential energy of a particle will change as its z coordinate changes, and that the rate of change of that potential energy equals the rate of change in z, dv_z/dz. Since the total energy of the system is constant, the kinetic energy must also change at an equal (and opposite) rate. The first term of Eq. (53) states this result in very general terms.

Now let us examine the second term in Eq. (53). Since the notation ∇p represents three terms, we only need to look at one part of this quantity, say dp/dz, to discover the physical significance of this term. The product of dp/dz and dz equals dp, the pressure increment in the z direction. Since $\mathbf{v}$ multiplies this as a dot product, it is also the z component of velocity which is involved in this term. The product of v_z and $dy\,dx$ equals the rate at which a volume element is swept by the pressure increment. From thermodynamics, we know that the product of a pressure times volume defines work. Therefore, we see that the second term in Eq. (53) describes the work done on a fluid particle as it follows the flow pattern.

Finally, we consider the third term which explicitly contains the viscous contribution. Again, we will only look at one term of the quantity $\nabla^2\mathbf{v}$, say $\partial^2\mathbf{v}/\partial z^2$. We encountered a second derivative like this previously in Eq. (29). It should be recognized that $(\partial^2 \mathbf{v}/\partial z^2)\, dz$ describes the increment in the velocity gradient in the z direction across the fluid particle. Since shear forces are proportional to velocity gradients according to Eq. (14), η times this quantity equals the increment in shear across the fluid particle. Like pressure, shear forces are expressed per unit area, a stipulation that is removed by multiplying by the area of the face $dx\, dy$. When the resulting force is multiplied by a velocity, the rate at which shear forces do work on the fluid particle results.

Not only is the viscous contribution the most complex of the three quantities we have considered, it is also the one in which we are most interested in this chapter. It seems worthwhile, therefore, to look at this viscous contribution a bit further. However, this time we shall specify a simple but definite flow pattern. Suppose we consider a volume element situated in a streamline which has only an x component of velocity. Further, we shall examine only a portion of that fluid adjacent to a stationary wall in the xy plane. The situation is sketched in Fig. 2.10. The volume element, moving with the fluid, will be subject to shearing forces. The bottom face, resting against a stationary wall, has zero velocity whereas the top surface has a velocity $(dv/dz)\,\Delta z$. According to Figs. 2.3 and 2.10, the shear force on the upper face equals $\eta(dv/dz)\,\Delta x\,\Delta y$. This force times the velocity of the top face $(dv/dz)\,\Delta z$ equals the rate at which shearing forces deposit energy in the volume element. Multiplying these factors yields

$$\frac{D\,(\text{shear energy})}{Dt} = \eta\left(\frac{dv}{dz}\right)^2 \Delta x\, \Delta y\, \Delta z \tag{54}$$

If we divide through by the volume of the element and take the limit, we obtain

$$\frac{D\,(\text{shear energy/volume})}{Dt} = \eta\left(\frac{dv}{dz}\right)^2 \tag{55}$$

This is essentially the same result as Eq. (2), a defining equation for viscosity in energy terms. Equation (55) is equivalent to the third term of Eq. (53) for the flow pattern under consideration. All the mathematical implications of $\eta\mathbf{v}\cdot\nabla^2\mathbf{v}$ are needed to express the equivalent result for more complex flow.

In summary, then, we see that the three terms of Eq. (53) describe the rate at which the kinetic energy of a volume element of flowing fluid changes. One term merely describes how this is affected by gravitational potential energy changes. The

FIGURE 2.10 *Sketch of viscous forces in a rectangular volume element adjacent to a stationary surface.*

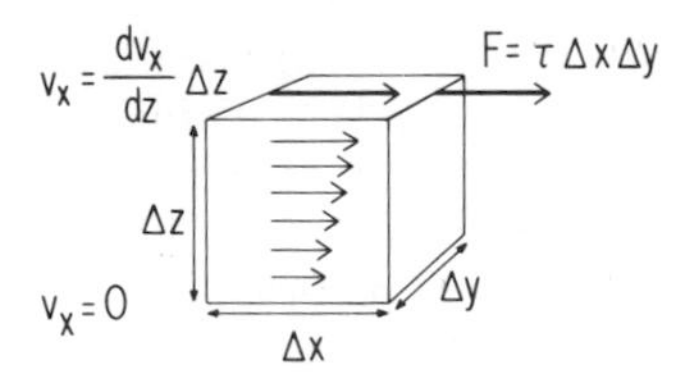

more complex terms describe the rate of change in energy due to forces perpendicular and parallel to the faces of the volume element. Just like the equation of motion from which it follows, Eq. (53) may be transformed into a variety of unrecognizable forms through coordinate transformations and vector identities. If a problem requires that Eq. (52) be integrated over macroscopic boundaries, there are mathematical procedures for transforming volume integrals into surface integrals. These are almost always invoked since they reduce a triple integral to a double integral. At the same time, however, the appearance of the contributing terms is altered substantially. As with the equation of motion itself, it is the physical significance of this result rather than any one particular mathematical form that should be appreciated.

2.8 EINSTEIN'S LAW OF VISCOSITY: THEORY

In 1906, A. Einstein (Nobel Prize, 1921) published his first derivation of an expression for the viscosity of a dilute dispersion of solid spheres. The initial theory contained errors that were corrected in a subsequent paper which appeared in 1911. It would be no mistake to infer from the historical existence of this error that the theory is complex. Therefore, we will restrict our discussion to an abbreviated description of the assumptions of the theory and some highlights of the derivation.

The Einstein derivation is based on the hydrodynamic equations of the preceding sections. As such, it is limited to those cases in which ρ and η for the fluid are constant and in which the flow velocity is low. Furthermore, the theory postulates an extremely dilute dispersion of rigid spheres with no slippage of the liquid at the surface of the spheres. Finally, the spheres are assumed to be large enough compared to the solvent molecules to permit us to regard the solvent as a continuum, but small enough compared to the dimensions of the viscometer to permit us to ignore wall effects. These size restrictions make this result applicable to particles in the colloidal size range.

Einstein considered a fluid in laminar flow through a dilute, random array of spherical particles. The equations of continuity and motion, Eqs. (8) and (23), describe the flow, and Eq. (52) describes the rate at which energy enters a volume element of this fluid. The obvious difficulty in applying these equations to the case in question is the enormous number of surfaces at which boundary conditions must be specified. This leads to the first reason why an infinitely dilute dispersion must be considered. If the particles are sufficiently far apart, each will modify the flow pattern in its environment as if it alone were present. This introduces two boundary conditions: no slippage at the surface and unperturbed flow at larger distances from the surface.

Einstein was able to solve the equation of motion, first for the case with a single sphere present, to give values of v_x, v_y, v_z, and p that satisfied the boundary conditions just described. These quantities were then substituted into Eq. (52). The integration of Eq. (52) is carried out over the surface of a hypothetical sphere of radius r. The distance r is so much larger than the radius R of the spherical particle that the disturbance of flow due to the particle has vanished at this surface. Einstein has shown the result of this integration to be

$$\left(\frac{DE}{Dt}\right)_{\text{sphere}} = K\eta_0\left(\tfrac{8}{3}\pi r^3 + \tfrac{4}{3}\pi R^3\right) \tag{56}$$

where η_0 is the viscosity of the solvent and K is a constant whose significance does not concern us since it cancels out presently. Since $\frac{4}{3}\pi r^3$ represents the volume V of the system and $\frac{4}{3}\pi R^3$ represents the volume occupied by the spherical particle, we can rewrite this result:

$$\left(\frac{DE}{Dt}\right)_{\text{sphere}} = K\eta_0 V(2+\phi) \tag{57}$$

where ϕ is the volume fraction occupied by the sphere. Examining the limit of Eq. (56) as $R \to 0$ permits us to evaluate the rate of deposition of energy in the same volume element in the absence of the sphere. That is, if solvent alone is present

$$\left(\frac{DE}{Dt}\right)_{\text{solvent}} = 2K\eta_0 V \tag{58}$$

Next, we can use Eqs. (57) and (58) to consider the increment per sphere in the rate of energy deposition. Subtracting Eq. (58) from (57) and rearranging gives

$$\left(\frac{DE}{Dt}\right)_{\text{sphere}} - \left(\frac{DE}{Dt}\right)_{\text{solvent}} = \frac{\phi_i}{2}\left(\frac{DE}{Dt}\right)_{\text{solvent}} \tag{59}$$

in which the subscript i has been added to ϕ to remind us that this is the volume fraction occupied by a single sphere. In keeping with this notation, the left-hand side of Eq. (59) may be written $\Delta(DE/Dt)_i$, the change due to one sphere in the rate of energy deposition. If the dispersion contains N spheres sufficiently far part for the effects of each to be independent of the others, we may write

$$\left(\frac{DE}{Dt}\right)_{\text{dispersion}} = \left(\frac{DE}{Dt}\right)_{\text{solvent}} + \sum_{i=1}^{N} \Delta\left(\frac{DE}{Dt}\right)_i = \left(1 + \frac{N\phi_i}{2}\right)\left(\frac{DE}{Dt}\right)_{\text{solvent}} \tag{60}$$

Of course, the number of spheres times the volume fraction occupied per sphere equals the total volume fraction of the dispersed phase, ϕ. This fact combined with Eq. (58) transforms (60) into

$$\left(\frac{DE}{Dt}\right)_{\text{dispersion}} = 2K\eta_0 V\left(1 + \frac{\phi}{2}\right) \tag{61}$$

Einstein was able to derive another equation describing the rate of energy deposition per unit volume for a dispersion of spheres:

$$\left(\frac{DE}{Dt}\right)_{\text{dispersion}} = K\eta(1-\phi)^2 V \tag{62}$$

in which η (without a subscript) is the measured viscosity *of the dispersion* and K has the same significance as in Eq. (61). Again in the derivation of this result, the particles are assumed to be far apart. Therefore, Eqs. (61) and (62), two expressions for the same quantity derived with comparable assumptions, may be equated:

$$K\eta(1-\phi)^2 V = 2K\eta_0 V\left(1+\frac{\phi}{2}\right) \tag{63}$$

Rearranging this equation yields

$$\frac{\eta}{\eta_0} = \frac{1+\phi/2}{(1-\phi)^2} = \left(1+\frac{\phi}{2}\right)(1+\phi+\phi^2+\cdots)^2 \tag{64*}$$

where only the leading terms in ϕ have been retained since ϕ must be small to satisfy the requirement of large distances between spheres. If no term higher than first order in ϕ is retained, Eq. (64) becomes

$$\frac{\eta}{\eta_0} = 1 + 2.5\phi + \cdots \tag{65}$$

a result known as Einstein's law of viscosity. This derivation was not only an important accomplishment in its own right but also served as a model for many subsequent derivations. Before considering these, however, let us examine Einstein's law from an experimental point of view.

2.9 EINSTEIN'S LAW OF VISCOSITY: EXPERIMENTAL

The Einstein equation is one of those pleasant surprises that occasionally emerges from complex theories: a remarkably simple relationship between variables, in this case the viscosity of a dispersion and the volume fraction of the dispersed spheres. A great many restrictive assumptions are made in the course of the derivation of this result, but the major ones are (a) that the particles are solid spheres and (b) that their concentration is small.

These conditions are relatively easy to meet experimentally so Eq. (65) has been tested in numerous studies. Figure 2.11 is an example of the sort of verification that has been obtained. In the work summarized in this figure, a variety of model systems were investigated, using both capillary and Couette viscometers. The solid line in the figure is the Einstein prediction; the agreement between theory and experiment is seen to be very good.

Note that the applicability of the Einstein equation seems equally good regardless of the size sphere used. Although the range of particle radii in Fig. 2.11 is relatively narrow, this conclusion has been verified for particles as small as individual molecules and as large as grains of sand. It might also be noted that experiments of this sort are carried out in mixed solvents or electrolyte solutions which equal the dispersed particles in density. Thus, there is no tendency for the spheres to settle under the influence of gravity (cf. Chap. 3).

The simplicity of the Einstein equation makes it relatively easy to test, but also limits its usefulness rather sharply. With so few variables involved, the quantities we

* There are many places in this book where functions are expanded as infinite series. These series expansions are listed in Appendix A.

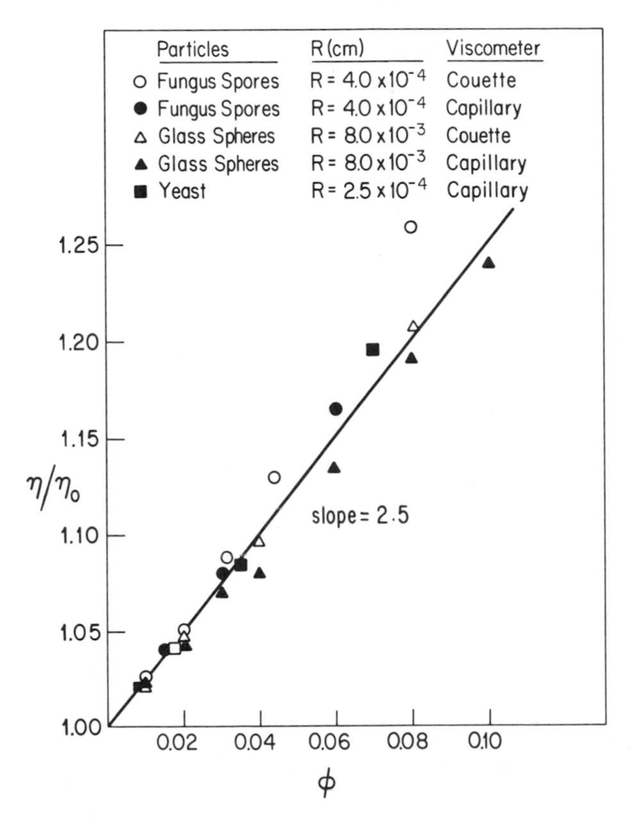

FIGURE 2.11 *Experimental verification of Einstein's law of viscosity for spherical particles of several different sizes and materials. [Data from F. Eirich, M. Bunzl, and H. Margaretha, Kolloid Z.* **74**:276 (1936).]

may evaluate by Eq. (65) are few in number. If we know that a dispersion contains spherical particles, we may calculate its viscosity if we know the volume fraction occupied by the particles, at least if ϕ is small. Although there are situations in which such a calculation is required, most of the time it is the inverse calculation that is needed. That is, viscosity is a measurable quantity from which we try to extract information about the dispersion. All that Eq. (65) offers directly in this area is the evaluation of ϕ from viscosity measurements, again provided ϕ is small and the particles are spheres.

The data shown in Fig. 2.11 are limited to concentrations below 10 vol %. We might ask: What are the consequences of increasing the concentration to higher volume fractions of spheres? The data of Fig. 2.12a show that the viscosity of a system of glass spheres shows significant positive deviations from Einstein's law above $\phi \simeq 0.1$.

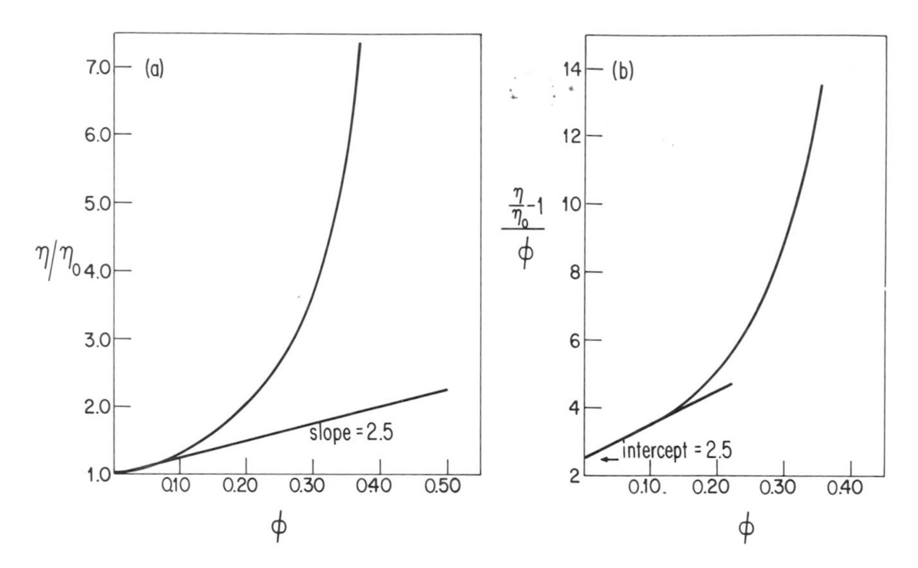

FIGURE 2.12 (*a*) *Relative viscosity versus volume fraction for a dispersion of glass spheres* ($R = 6.5 \times 10^{-3}$ *cm*) *up to* $\phi \simeq 0.40$. (*b*) *Same data plotted as reduced viscosity versus* ϕ. [*Data from V. Vand*, J. Phys. Colloid Chem. **52**:300 (1948).]

Equation (64) suggests that the range of applicability of Einstein's equation might be extended by retaining terms higher than first order in the power series. That is,

$$\frac{\eta}{\eta_0} = 1 + 2.5\phi + k_1\phi^2 + \cdots \tag{66}$$

might give a better fit to the data from dispersions whose concentrations are more than infinitely dilute. There are several points about this result which should be noted:

1. Equations (3) and (66) are identical in form. Equation (66) shows that volume fraction is the theoretically preferred unit of concentration as far as viscosity is concerned.
2. The theoretical evaluation of k_1 and higher coefficients in Eq. (66) requires more than merely retaining additional terms in (64). Einstein's derivation of Eqs. (56) and (62) is the origin of the restriction to very dilute systems.
3. A number of theoretical attempts have been made to evaluate k_1 by going back through the Einstein derivation and superimposing the effects of neighboring particles, rather than treating them as independent. A few of the theoretical values that have been obtained for k_1 from different models are 14.1, 12.6, and 7.35.

Rather than attempting to choose among these theoretical approaches, let us examine an empirical approach to the problem of deviations from the Einstein equation at high concentrations. Toward this end, it is convenient to rearrange Eq. (66) as follows:

$$\frac{(\eta/\eta_0 - 1)}{\phi} = 2.5 + k_1\phi + \cdots \tag{67}$$

This has the effect of reducing the order of each term on the right-hand side of the equation. Now if we plot the left-hand side of Eq. (67) versus ϕ, the result should be a straight line of slope k_1 and intercept 2.5, at least at low concentrations before still higher order terms become important. Figure 2.12b shows the data of Fig. 2.12a replotted in this manner.

Several conclusions are evident from the data replotted in this way. First, the Einstein coefficient, 2.5, is clearly the intercept toward which the quantity $(\eta/\eta_0 - 1)/\phi$ extrapolates as $\phi \to 0$. The initial slope in Fig. 2.12b suggests that k_1 is almost 10·0 for these data, a reasonable value in the light of theoretical predictions.

The strategy used in going from Eq. (66) to (67) is one that is used frequently in colloid and surface chemistry to reduce the order of an equation. We shall use it again in connection with osmotic pressure and also with light scattering.

Before we proceed any further, it should be noted that the results of experiments in viscometry are routinely reported in a variety of forms; the ones we have used in Figs. 2.12a and 2.12b are only two of the possibilities. In Table 2.1 are listed some of the functional forms in which data are often presented. Also listed are the symbols, common and IUPAC names, and the limiting values for these quantities as the concentration of the colloid goes to zero. In Table 2.1, the symbol c has been used to

TABLE 2.1 *Symbol, common and IUPAC names, and limiting values for a variety of forms commonly used to present viscosity data*

Functional form	Symbol	Common name	IUPAC name	$\lim_{c\to 0}$
—	η	Viscosity	—	η_0
η/η_0	η_r	Relative viscosity	Viscosity ratio	1
$\eta/\eta_0 - 1$	η_{sp}	Specific viscosity	—	0
$(\eta/\eta_0 - 1)/c$	η_{red}	Reduced viscosity	Viscosity number	$[\eta]$
$c^{-1} \ln \eta/\eta_0$	η_{inh}	Inherent viscosity	Logarithmic viscosity number	$[\eta]$
$\lim_{c\to 0} \eta_{red}$ or $\lim_{c\to 0} \eta_{inh}$	$[\eta]$	Intrinsic viscosity	Limiting viscosity number	—

represent the concentration of the dispersed phase. The reader is advised to be particularly attentive to the units of c in an actual problem. In theoretical work, concentrations are expressed as volume fractions ($c = \phi$). Experimental results, on the other hand, may be expressed in a variety of practical units. The most common units for c are grams per deciliter (g/dl = g/100 ml); IUPAC recommends grams per milliliter as the units for c. We shall examine the implications of these units for $[\eta]$ in Sect. 2.11. For the present, however, we will continue to use ϕ as the unit of concentration.

The quantity $[\eta]$—the intrinsic viscosity or limiting viscosity number—is seen from either Eq. (3) or (67) to be the coefficient of the first-order concentration term in a series expansion for the viscosity of a dispersion. It may always be evaluated experimentally for any system by measuring η as a function of c. As an empirical parameter, $[\eta]$ requires no knowledge of the nature of the dispersed units. The significance of the Einstein equation is that it predicts from first principles what the value of $[\eta]$ should be for a dilute dispersion of spherical particles.

If the experimental value of $[\eta]$ turns out to be 2.5, as in Fig. 2.12b, the particles are shown to be unsolvated spheres. If the value of $[\eta]$ is something other than 2.5, the dispersed units do not meet the requirements of the Einstein derivation.

Our next task is to examine experimental values of $[\eta]$ to see what correlation we can find between this parameter and such dispersion characteristics as polydispersity, solvation, and particle ellipticity. Each of these sources of deviation from the Einstein result is taken up in the following sections.

FIGURE 2.13 *(a) Relative viscosity and (b) specific viscosity versus volume fraction for polymethyl methacrylate particles having the average diameters shown. [Data from S. G. Ward and R. L. Whitmore,* Brit. J. Appl. Phys. **1**:325 (1950).]

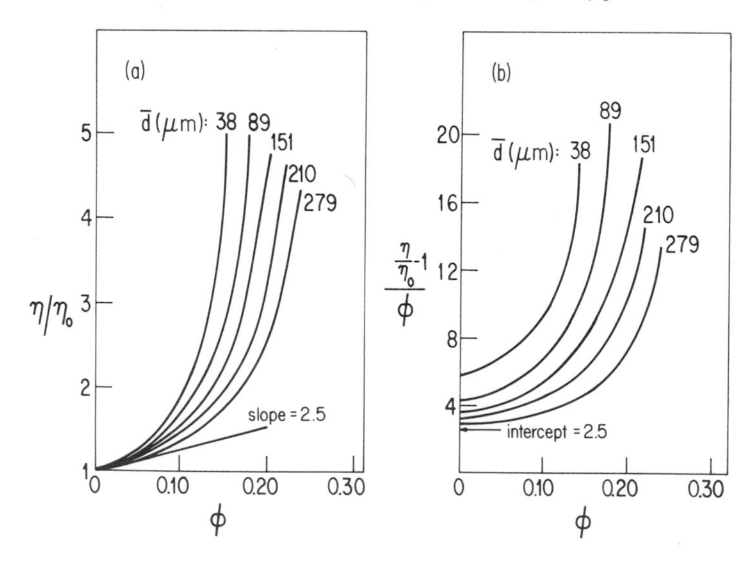

To prepare for these next considerations, examine the data contained in Figs. 2.13a and 2.13b which show the relative and reduced viscosity, respectively, plotted versus volume fraction for dispersions of polymethyl methacrylate particles. These particles are irregularly shaped and have been fractionated into several different size classes. The average equivalent diameter is labeled as a parameter for each of the curves.

Several different observations are immediately evident from these data. First, there are obviously positive deviations from Einstein's law, even at very low concentrations. In fact, from Fig. 2.13a it is not clear whether Eq. (65) applies at all, even in the limit. When the data are replotted (in Fig. 2.13b) according to Eq. (67) to give a more critical test of the results, it is clear that the intrinsic viscosity approaches the Einstein value only for the largest particles. The intrinsic viscosity values increase as the particle size decreases. In addition, the deviations from the linear form predicted by Eq. (67) set in at lower concentrations the smaller the particle size. Neither the Einstein law nor any of the theories developed to extend its concentration range predict this size effect. The samples from which the data of Fig. 2.13 were collected were (a) polydisperse, (b) irregular in shape, and (c) fairly symmetrical in axial ratio. In the following sections, we shall consider the implications of these conditions for viscosity to see whether the observations of Fig. 2.13 may be explained by these facts.

2.10 DEVIATIONS FROM EINSTEIN'S LAW: THE EFFECT OF POLYDISPERSITY

Let us consider the effect of polydispersity on the viscosity of an array of spheres. The Einstein theory contains no explicit dependence on the size of the spheres. Since its derivation treats each sphere as influencing the flow of solvent independently, Eq. (65) should apply equally to monodisperse and polydisperse systems. Figure 2.14 shows the increase in viscosity with volume fraction for an array of

FIGURE 2.14 *The effect of polydispersity on viscosity. Solid line drawn according to Eq.* (70); *points are experimental results.* [*Data from R. Roscoe*, Brit. J. Appl. Phys. **3**:267 (1952).]

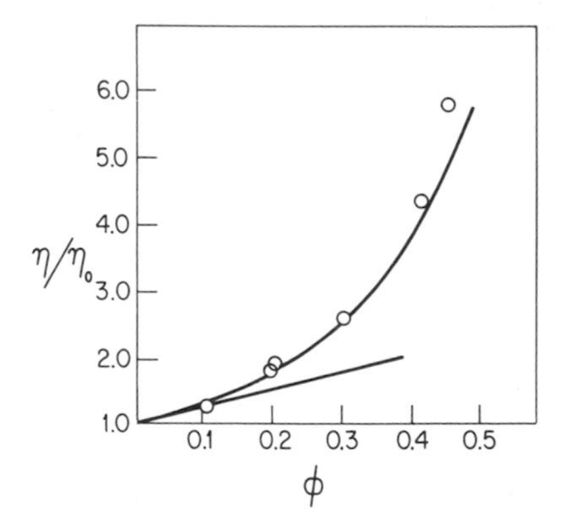

spheres of very diverse sizes. It is seen that the Einstein limit is approached at extreme dilution, with the expected deviations occurring at higher concentrations. The present question is whether polydispersity might have any effect on the magnitude of these deviations.

To do this, we imagine a mixture made up of spheres of several different sizes, each at a volume fraction ϕ_i that, for each of the components, is sufficiently small so that the Einstein equation holds. Further, we assume that the increment in viscosity that accompanies the addition of each successive component is given by Eq. (65), that is,

$$d\eta = 2.5\eta\, d\phi_i \tag{68}$$

In this case, η is the viscosity of the dispersion prior to the addition of the last component. It can be shown that $d\phi_i$ is also equal to $d\Phi/(1-\Phi)$ where Φ is the *total* volume fraction of all spheres in the system. Therefore,

$$d\eta = 2.5\eta \frac{d\Phi}{1-\Phi} \tag{69}$$

This result can be integrated, recalling that $\eta = \eta_0$ at $\Phi = 0$, to give

$$\frac{\eta}{\eta_0} = (1-\Phi)^{-2.5} \tag{70}$$

This result reduces to the Einstein equation as $\Phi \to 0$. The solid line in Fig. 2.14 was drawn according to Eq. (70).

Therefore, we may conclude that polydispersity contributes to positive deviations from the Einstein equation, although its effects vanish at infinite dilution. This means that the size dependence of $[\eta]$ in Fig. 2.13 cannot be attributed to polydispersity. A frustrating aspect of Eq. (70) is that it introduces no new parameter to account quantitatively for the polydispersity, although it does account for positive deviations from the Einstein equation for some systems.

2.11 DEVIATIONS FROM EINSTEIN'S LAW: THE EFFECT OF SOLVATION

Until now, we have assumed that the volume fraction of the colloid in a dispersion is a known quantity. Although the weight or possibly even the number of particles per unit volume may be known, the volume occupied by the particles is generally not known. In this section, we shall examine the relationship between ϕ and other units of concentration—especially for dilute dispersions—and shall give particular attention to the possibility that the dispersion particles are solvated.

A perfectly general expression for volume fraction is given by

$$\phi_2 = \frac{n_2 \bar{V}_2}{n_1 \bar{V}_1 + n_2 \bar{V}_2} \tag{71}$$

in which n_i and $\bar{V}_i$ are the number of moles and the partial molal volume, respectively, of the ith component. We shall designate the continuous phase as component 1 and the dispersed phase as component 2. If there is no interaction between the components such that ΔV for the mixing process equals zero, then $\bar{V}_1$ and $\bar{V}_2$ may be equated with V_1 and V_2, the molar volume of the pure components. Such a substitution cannot be made in general, however, and the proof that it is justified must be supplied in each case. The reader is referred to physical chemistry texts for a discussion of the evaluation of partial molal volumes from experimental data for nonideal solutions.

Since we are interested in the viscosity of dispersions in the limit of extreme dilution where $n_2 \ll n_1$, Eq. (71) may be simplified to

$$\phi_2 = \frac{n_2 \bar{V}_2}{n_1 \bar{V}_1} \tag{72}$$

With colloidal solutes $\bar{V}_2 \gg \bar{V}_1$; therefore, the approximation involved in writing Eq. (72) is limited to even more dilute solutions than would be required for the same approximation in solutions of comparably sized components. The rest of our discussion of concentrations is developed from Eq. (72) so it is important to remember that the results apply to dilute dispersions only.

Next, let us consider the relationship between ϕ and the mole fraction unit of concentration and between ϕ and the weight per unit volume concentration unit. The mole fraction x_2 is defined to be

$$x_2 = \frac{n_2}{n_1 + n_2} \simeq \frac{n_2}{n_1} \tag{73}$$

where the approximation holds for dilute solutions. Comparison of Eqs. (72) and (73) reveals

$$\phi_2 \simeq \frac{\bar{V}_2}{\bar{V}_1} x_2 \tag{74}$$

Note that $x \simeq \phi$ if the two components are comparable in size, but that $x \ll \phi$ when $\bar{V}_2 \gg \bar{V}_1$ as will be the case with colloids.

A practical unit of concentration, c, reports the number of grams of component 2 per milliliter of solution. Therefore,

$$c_2 = \frac{n_2 M_2}{n_1 \bar{V}_1 + n_2 \bar{V}_2} \tag{75}$$

where M_2 is the molecular weight of component 2. This reduces to the following approximations in dilute solution:

$$c_2 \simeq \frac{n_2 M_2}{n_1 \bar{V}_1} = \frac{M_2}{\bar{V}_1} x_2 = \frac{M_2}{\bar{V}_2} \phi_2 \tag{76}$$

We have seen from the Einstein theory that volume fraction is the concentration variable of choice for a discussion of the viscosity of a dispersion. The approximations given by Eqs. (72), (74), and (76) show that, for dilute systems, ϕ is directly proportional to other concentration units:

$$\phi = K \text{ (concentration)} \tag{77}$$

Table 2.2 summarizes these results by listing the values of K for several concentration units, as well as the limitations of the approximation.

When particles are solvated, a certain volume of the solvent must be counted as part of the dispersed phase rather than as part of the continuous phase. For dilute systems, the depletion of the available solvent may be ignored when solvation occurs because the amount of solvent bound is negligible compared to the total amount of solvent present. However, the effect of the bound solvent on the volume of the dispersed units cannot be ignored. If certain models for the mode of solvation are assumed, the effect of solvation on the volume fraction of the particles may be calculated. That is, the value of K in Eq. (77) may be evaluated to relate the actual volume fraction of the solvated particles to their volume fraction if unsolvated, ϕ_{dry}. This is easily done for two models for solvation: (a) solvation limited to the surface of spherical particles and (b) solvation occurring uniformly throughout the dispersed particle. The former probably applies best to lyophobic colloids; the latter, to lyophilic. We shall consider each of these models for solvation in the next few paragraphs. It should be remembered that these are only models for solvation; other possibilities may exist in actual practice.

Suppose we begin by assuming that solvation is restricted to the surface of a particle so that the amount of bound solvent is proportional to the specific surface area A_{sp} of the particles. Then the volume of the solvated particles equals the volume of the unsolvated particles V_2, plus the volume of the bound solvent, $V_{1,b}$:

$$\text{Volume solvated particle} = V_2 + V_{1,b} \tag{78}$$

TABLE 2.2 *Values of the proportionality constant K in Eq.* (77): $\phi = K$ *(concentration). Definitions of symbols and limitations of Eq.* (77) *are also included*

Concentration	K	Definitions	Limitations
Weight fraction, m_2/m_1	ρ_1/ρ_2	m_i = grams component i; ρ_i = density of pure component	Unsolvated particles, dilute
Mole fraction, n_2/n_1	$\bar{V}_2/\bar{V}_1$	$\bar{V}_i$ = partial molal volume	Dilute
Weight per unit volume, c	$\bar{V}_2/M_2$	M_i = molecular weight	Dilute
Volume fraction unsolvated, ϕ_{dry}	$1+3\,\Delta R/R$	R = radius of sphere; ΔR = thickness of bound surface layer	Spherical particles, solvation occurs at surface only
Volume fraction unsolvated, ϕ_{dry}	$1+m_{1,b}/m_2(\rho_2/\rho_1)$	$m_{1,b}$ = mass of bound solvent	Solvation proportional to mass of particle

This may be written

$$\text{Volume solvated particle} = V_2 + A_{sp} m_2 \,\Delta R \tag{79}$$

where ΔR is the thickness of the bound surface layer and m_2 is the mass of the dispersed material. If the dispersed particles are spheres, Eq. (1.2)* may be substituted for A_{sp} to give

$$\text{Volume solvated particle} = V_2 + \frac{3m_2 \,\Delta R}{\rho_2 R} = V_2\left(1 + \frac{3\,\Delta R}{R}\right) \tag{80}$$

The effect of this mode of solvation on the volume fraction of spherical particles is given by

$$\phi_{\text{solvated}} = \left(1 + \frac{3\,\Delta R}{R}\right)\phi_{\text{dry}} \tag{81}$$

which shows that surface solvation increases ϕ by a factor $1 + 3\,\Delta R/R$.

Next, consider the case of a lyophilic colloid which may solvate at numerous positions along the molecule. In this case, it is not the surface area of the molecule but the total volume or mass that determines the extent of solvation. Equation (78) still applies; this time, however, we write

$$V_2 + V_{1,b} = V_2\left(1 + \frac{V_{1,b}}{V_2}\right) = V_2\left(1 + \frac{m_{1,b}}{m_2}\left[\frac{\rho_2}{\rho_1}\right]\right) \tag{82}$$

where $m_{1,b}$ is the mass of bound solvent and the last equality requires the density of the bound solvent to be identical to the free solvent. The effect of this mode of solvation is given by

$$\phi_{\text{solvated}} = \left(1 + \frac{m_{1,b}}{m_2}\left[\frac{\rho_2}{\rho_1}\right]\right)\phi_{\text{dry}} \tag{83}$$

which shows that solvation throughout the dispersed particles increases the volume fraction by the factor $1 + m_{1,b}/m_2[\rho_2/\rho_1]$.

In actual practice, the exact mode of solvation is not usually known. Therefore, deviations of intrinsic viscosity from the value expected for nonsolvated particles may be used to evaluate the extent of solvation of the particles. Equations (81) and (83) not only postulate two different models for solvation but they also predict a different dependence of solvation on particle size. In Eq. (81) the importance of solvation is predicted to decrease with increasing particle size, at least if the thickness of the surface layer is independent of particle size. Equation (83), on the other hand, suggests that $m_{1,b}/m_2$ may be independent of particle size. We shall see presently and again in Chap. 3 that the second model is quite reasonable for aqueous solutions of proteins.

In any experimental check of Einstein's equation, the volume fraction of the unsolvated material would probably be used. However, Einstein's equation

* This manner of referencing is used to describe equations occurring in other chapters. The number to the left of the point is the chapter number.

requires that the volume fraction of the *actual* hydrodynamic species be used. Now suppose we measure the viscosity of a dispersion of spherical particles which are solvated at their surface. The combination (65) and (81) predicts

$$\frac{\eta}{\eta_0} = 1 + 2.5\left(1 + \frac{3\,\Delta R}{R}\right)\phi_{\text{dry}} \tag{84}$$

That is, the intrinsic viscosity based on the volume fraction of the dry material is expected to be larger than the Einstein result by precisely that factor which increases the volume fraction. Equation (84) shows that this factor increases with decreasing particle size, all other things being equal. This is the probable explanation of the size dependence shown in Fig. 2.13. Those particles for which d equals 38 μm in Fig. 2.13 show a value of $[\eta]$ equaling about 5.8. Interpreting this result according to Eq. (84) leads to

$$5.8 = 2.5\left(1 + 3\frac{\Delta R}{R}\right) \tag{85}$$

from which it follows that $\Delta R/R = 0.44$. According to this model, the bound solvent layer is about 44% the radius of the particle in thickness. Obviously, this calculation attributes all the deviation from the simple Einstein result to solvation. In this particular example, both the ellipticity and the polydispersity of the particles should also be considered.

It might be recalled that the particles were rough in the experiment summarized by Fig. 2.13. Such particles might reasonably entrain a layer of solvent near their surface. This points out the fact that no specific mechanism for the binding of solvent to surface is involved in Eqs. (81) and (84). The binding may be chemical, physical, or mechanical; the effect on viscosity is independent of the origin of the solvation.

By the same procedure, it is easily shown that the effect of solvation of the entire particle also increases the intrinsic viscosity by precisely the same factor by which the volume fraction is increased, $1 + m_{1,b}/m_2[\rho_2/\rho_1]$. Again we see positive deviations from the Einstein equation; this time, however, the magnitude of the deviation may be interpreted to yield quantitative data about solvation of the *entire* particle.

Protein particles have an unsolvated density of about 1.34 g cm^{-3}. Therefore, the intrinsic viscosity of hydrated (solvent = water) spherical protein particles is $2.5[1 + 1.34(m_{1,b}/m_2)]$. A spherical protein particle, therefore, which binds about 0.60 g water (g protein)$^{-1}$ will have an intrinsic viscosity of 4.50, 1.8 times greater than predicted by the Einstein equation. Conversely, when $[\eta]$ exceeds 2.5 for spherical particles and the solvation is proportional to the mass of the particle, then viscosity measurements provide quantitative information about the extent of that solvation. In fact, in his original work, Einstein observed that sugar molecules affect the viscosity of water as if their volume were 1.5 times larger than is actually the case. He attributed this to the binding of a certain amount of water to the sugar molecules, thereby increasing their effective volume fraction in solution.

Figure 2.15 shows how this sort of effect has been applied to protein solutions. Our interest for the present is in spherical particles which correspond to the case of $a/b = 1$. In Fig. 2.15, the contour lines are drawn for different values of $[\eta]$. Those

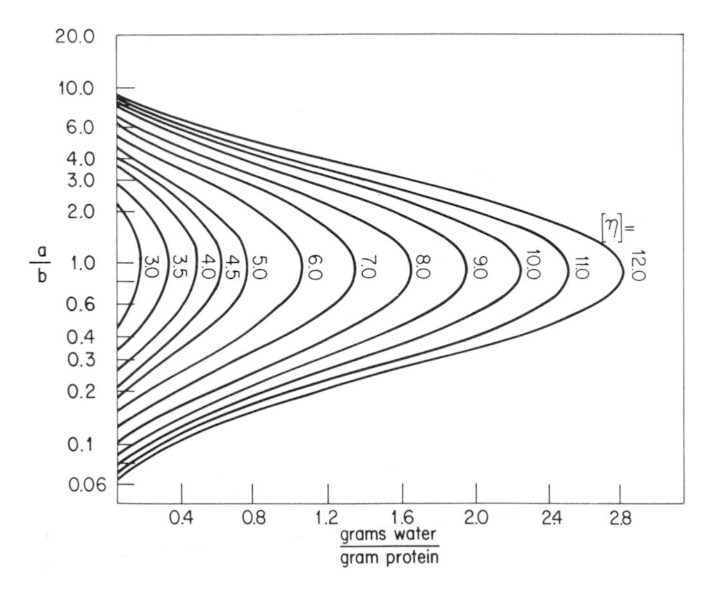

FIGURE 2.15 *Variation of the intrinsic viscosity of aqueous protein solutions with axial ratio a/b and extent of hydration $m_{1,b}/m_2$. [Redrawn from L. Oncley,* Ann. N.Y. Acad. Sci. **41**:121 (1941), *used with permission.*]

points at which the contours cross the line $a/b = 1.0$ indicate the degree of hydration, reported here as grams of water per gram of protein, required to account for the observed viscosity. Note that the value calculated earlier, namely $[\eta] = 4.5$ when $m_{1,b}/m_2 = 0.6$, is shown in Fig. 2.15. That is, the 4.5 contour line crosses the $a/b = 1$ axis at 0.6 g water (g protein)$^{-1}$.

2.12 NONSPHERICAL PARTICLES: THE SIMHA EQUATION

Spherical geometry, favored by theoreticians for its symmetry, applies perfectly to relatively few systems. The sphere represents a sort of "ideal" shape, and we often discuss other shapes in terms of deviations from this ideal. As discussed in Chap. 1, the next simplest particle geometry is that of the ellipsoid of revolution. The ellipticity of these particles is characterized by their axial ratio, a/b. In the case in which this ratio equals unity, the particle is a sphere.

Changing the shape of the dispersed particles from spheres to ellipsoids introduces a complication into the theory of viscosity. Since these particles are unsymmetrical, the possibility of their taking on preferred orientations in the streamlines must be taken into account. We will consider only the case of particles so small that randomization due to Brownian motion outweighs any tendency to orient with the streamlines. It should be clear that the viscosity will be greater in the presence of this tumbling motion (greater dissipation of energy) than in its absence, where the particles align themselves with the streamlines. R. Simha derived the following equation for the viscosity of a dispersion of prolate ellipsoids:

$$[\eta]=\frac{14}{15}+\frac{(a/b)^2}{15(\ln 2a/b-\lambda)}+\frac{(a/b)^2}{5(\ln 2a/b-\lambda+1)} \tag{86}$$

where λ is 1.5 for an ellipsoid and 1.8 for a cylindrical rod.

With the Simha equation (or other theoretical results which have been derived for nonspherical particles) one may extract a numerical value for the axial ratio from an experimental value of $[\eta]$, assuming that all the deviation from the Einstein value is due to ellipticity. Thus the sample with $d=38\ \mu m$ in Fig. 2.13 is also consistent with unsolvated particles having an axial ratio of about 5.0 according to Eq. (86).

The intrinsic viscosities of oblate and prolate ellipsoids of various axial ratios may be estimated from Fig. 2.15 from the intercepts of the contours at the line corresponding to zero hydration.

The principal value of Fig. 2.15, however, is that it allows us to evaluate the combined effects of solvation and ellipticity on the intrinsic viscosity of protein solutions. The contour labeled $[\eta]=8.0$, for example, is consistent with each of the following possibilities:

1. A spherical particle ($a/b=1$) hydrated to the extent of about 1.6 g H_2O (g protein)$^{-1}$.
2. An unhydrated prolate ellipsoid of axial ratio about 7.
3. An unhydrated oblate ellipsoid with $a/b=0.1$.
4. Any other pair of a/b-hydration values which are coordinates of the $[\eta]=8.0$ contour.

Viscosity data are plotted in Fig. 2.16 for several different preparations of tobacco mosaic virus (TMV) particles. TMV particles are cylindrical in shape, and the three lines in the figure correspond to fractions of differing ratios of length to diameter. In this case, there is clearly no approach to Eq. (65) even in the limit of infinite dilution: The Einstein equation is a limiting law for spheres; it does not apply to ellipsoids. As might be expected, however, the deviation from the Einstein theory becomes more pronounced as the axial ratio increases. The straight lines in Fig. 2.16 are drawn on the basis of the Simha equation (86). The curves are labeled by the a/b ratios used. These values agree quite well with values obtained directly from electron micrographs of the TMV particles.

In summary, we have seen that the effects of concentration and polydispersity on viscosity may be eliminated to a certain extent by using intrinsic viscosity as the working variable, since these effects vanish at infinite dilution. The effects of solvation and asymmetry cannot be eliminated nor, for that matter, separated by viscometry alone. Figure 2.15 shows that for any measured value of intrinsic viscosity there is a whole range of ellipticity–solvation combinations that fit the data. Furthermore, the range of possibilities increases as the intrinsic viscosity increases above 2.5. Only additional data from an independent source permit these factors to be evaluated separately. This situation will prevail in many experiments in colloid chemistry; what must be sought, therefore, is some additional characterization of the

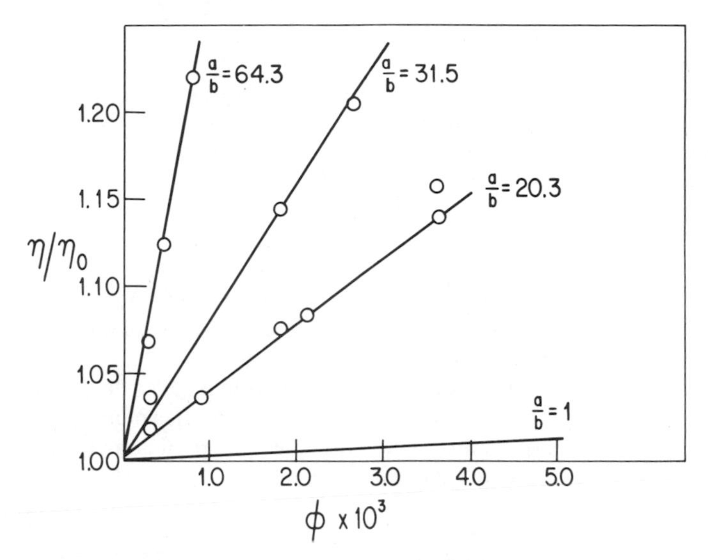

FIGURE 2.16 *Experimental values of relative viscosity versus volume fraction for TMV particles of different a/b ratios. Lines drawn according to Eq.* (86). [*M. A. Lauffer*, J. Am. Chem. Soc. **66**:1188 (1944).]

system. Generally, only one model will successfully account for two entirely different types of measurements on the same system.

2.13 THE STAUDINGER EQUATION

In the preceding sections, we have focused our attention on polydispersity, solvation, and ellipticity as the primary causes for departures of $[\eta]$ from the Einstein value of 2.5. These are but a few of the modifications of the original theory that have actually been investigated. Other particle geometries have also been investigated, an extremely important one being the random coil. In Chap. 3, we describe this geometry in greater detail. For the present, however, it is sufficient to make only the following comments.

Macromolecules often possess sufficient flexibility to take on a variety of configurations as described in Sec. 1.6. The protein molecules we have discussed until now are exceptional in this regard, since disulfide bonds and hydrogen bonding impart to these molecules a rather definite shape. Synthetic polymers, on the other hand, are far better described in statistical terms as a random coil.

A model that is frequently invoked to describe the long chains of high-molecular-weight polymer molecules in solution is a string of beads. The flexibility of such a string permits it to assume many configurations, a situation that makes it a difficult model for a theory of viscosity. Instead of attempting any sort of survey of the efforts that have been directed toward this problem, we shall content ourselves with a cursory view of two extreme configurations followed by an empirical approach to the viscosity of these systems.

Picturing a polymer molecule as a string of beads enables us to visualize two extreme configurations easily. One of these is the case in which the chain is more or less fully extended, so that flowing solvent sweeps past each bead on the string. In this case, the chain is said to be free draining. At the other extreme, we have the nondraining chain. This is the situation in which the chain is wrapped into a tight, jumbled coil. Although there is definitely some solvent enclosed within it, the entire coil domain is regarded as the hydrodynamic unit. It is not too difficult to make predictions about the effect on solution viscosity of free draining or nondraining particles. Most real molecules will be partly draining, so these two extremes should delineate the range of possibilities.

Suppose that we express c as grams of material per unit volume of dispersion. According to Table 2.2, $\phi = (\bar{V}_2/M_2)c$ or c times the partial specific volume of the disperse phase $\bar{V}_{sp}$. The latter equals the volume occupied per gram of material at the concentration in question. According to Eq. (67), then,

$$[\eta] = 2.5\,\bar{V}_{sp} \tag{87}$$

Now let us consider two alternate values for $\bar{V}_{sp}$ as this quantity might be visualized for (a) free draining and (b) nondraining chains.

We shall consider the free draining case first. The partial specific volume of the molecule as a whole would be given by the number of beads times the specific volume of each bead. The weight of the entire chain divided by the weight per bead equals the number of beads in the chain and the reciprocal of the density equals the specific volume of the bead. If the weight and density of the individual beads are constant, we see that $\bar{V}_{sp}$ and therefore $[\eta]$ will be proportional to the weight of the chain, or the molecular weight M of such a chainlike molecule:

$$[\eta] \propto M \tag{88}$$

For the nondraining molecule, on the other hand, we evaluate $\bar{V}_{sp}$ for the molecule as a whole, dividing the volume of the coil by its weight. If the coil is approximately spherical, its volume will be proportional to R^3. As we shall see in the next chapter [Eq. (3.76)], R is proportional to $M^{1/2}$ for such a coil. This makes the volume of the coil proportional to $M^{3/2}$, $\bar{V}_{sp}$ proportional to $M^{1/2}$, and the intrinsic viscosity proportional to $M^{1/2}$:

$$[\eta] \propto M^{1/2} \tag{89}$$

Equations (88) and (89) are two special cases of the following two-parameter equation, known as the Staudinger (Nobel Prize, 1953) equation:

$$[\eta] = kM^a \tag{90}$$

in which we would expect $a = 1$ for free draining chains and $a = \frac{1}{2}$ for randomly coiled molecules.

In practice, both k and a are treated as quantities to be evaluated experimentally. Their values, of course, tell something about the amount of interaction between the polymer molecules and the solvent molecules. Most importantly, however, this relationship may be used to determine the molecular weight of a polymer. The method requires calibration: Intrinsic viscosities are measured on samples of known molecular weight to evaluate the parameters a and k in Eq. (90). These quantities are found to be independent of molecular weight to quite a good approximation, although they vary with the nature of both the continuous and the dispersed phase and the temperature. Once the a and k parameters are known for a given system, however, the Staudinger equation provides a relatively easy and widely used method to evaluate the molecular weight of long-chain molecules.

Table 2.3 compares two polymers which correspond to opposite extremes for a values. For purposes of illustration, the intrinsic viscosities are calculated for samples of each polymer having a molecular weight of 10^5. The intrinsic viscosity of the cellulose triacetate is about eight times larger than that of polyisobutene, even though both have identical molecular weights. Although we are now very far from the model for which the Einstein equation was derived, we continue to see that viscosity is especially sensitive to the volume occupied by the particles.

The results shown in Table 2.3 are fully consistent with the nature of the polymers involved. In cellulose triacetate, for example, the individual "beads" are six-membered rings with three bulky acetate side groups. Such a unit would be rigid internally and because of the side groups the individual units would probably be severely hindered in their rotation with respect to one another. As a consequence, such a molecule is expected to exist in solution in quite an extended form, and its viscous behavior to be approximated by the free draining model ($a = 1$). For the polyisobutene–benzene system, 24°C corresponds to the so-called Θ temperature, that temperature at which phase separation would occur in the system if the polymer were of infinite molecular weight. At the threshold of phase separation the molecule is tightly twisted around itself, so the nondraining model applies. As a matter of fact, examination of many polymer–solvent systems at their Θ temperatures (which vary widely from system to system) reveals that the ratio $[\eta]/M^{1/2}$ averages about 1.2×10^{-3} dl mole$^{1/2}$ g$^{-3/2}$ (±75%). Table 2.3 shows that the polyisobutene–benzene system is in agreement with this value.

The Staudinger equation as a means to determine molecular weight is essentially a comparative method, depending on prior knowledge of the molecular weights of the calibrating species. In the following chapters, we shall consider three different

TABLE 2.3 *Comparison of the intrinsic viscosity of two polymer-solvent systems which differ widely in a and k values*

Polymer:	Cellulose triacetate	Polyisobutene
Solvent:	Acetone	Benzene
Temperature (°C):	25	24 (= Θ)
a:	0.90	0.50
k (dl g^{-1}):	8.97×10^{-5}	107×10^{-5}
Molecular weight (g mole^{-1}):	10^5	10^5
$[\eta]$ (dl g^{-1}):	2.84	0.34
$[\eta]/M^{1/2}$ (dl mole$^{1/2}$ g$^{-3/2}$):	—	1.08×10^{-3}

experimental techniques whereby molecular weights may be determined for a colloid: sedimentation, osmotic pressure, and light scattering. Each of these techniques reveals additional information about the dispersed particles besides their molecular weight, but a substantial part of the discussion in the next three chapters is oriented toward molecular weight determination.

REFERENCES

1. R. B. Bird, W. E. Stewart, and E. N. Lightfoot, *Transport Phenomena*, Wiley, New York, 1960.
2. E. J. Cohn and J. T. Edsall, *Proteins, Amino Acids and Peptides*, ACS Monograph, Hafner, New York, 1965.
3. A. Einstein, *Investigations on the Theory of the Brownian Movement*, Dover, New York, 1956.
4. H. L. Frisch and R. Simha, "The Viscosity of Colloidal Suspensions and Macromolecular Solutions," in *Rheology*, Vol. 1 (F. R. Eirich, ed.), Academic Press, New York, 1956.
5. J. J. Hermans, *Flow Properties of Disperse Systems*, North-Holland Publ., Amsterdam, 1953.
6. H. Lamb, *Hydrodynamics*, Dover, New York, 1945.

PROBLEMS

1. Gillespie and Wiley* used a cone-and-plate viscometer to measure F/A versus dv/dx for dispersions of silica and cross-linked polystyrene in dioctyl phthalate. At a volume fraction of 0.35 for both solids, the following results were obtained:

Silica	$F/A \times 10^3$ (dynes cm^{-2})	2.2	1.4	1.0	0.50	0.25
	dv/dx (s^{-1})	500	325	235	125	60
Polystyrene	$F/A \times 10^3$ (dynes cm^{-2})	1.6	0.80	0.55	0.25	
	dv/dx (s^{-1})	500	235	160	100	

   Use these data to determine either η or the yield value for these dispersions, depending on whether or not the system is Newtonian. Are these results consistent with the fact that the axial ratio was nearer unity and the particle size distribution narrower for the polystyrene than the silica? Explain.
2. An aqueous polybutyl methacrylate latex ($\bar{r} = 200$ Å) has a viscosity of 50,500 cP at $\phi = 0.25$ and a viscosity of 36.7 cP at the same rate of shear when 1.71×10^{-5} g NaCl is added per gram of polymer.† Assuming that these

* T. Gillespie and R. Wiley, *J. Phys. Chem.* **66**:1077 (1962).
† J. G. Brodnyan and E. L. Kelley, *J. Colloid Sci.* **19**:488 (1964).

charged particles must be surrounded by a layer of dissolved ions in solution, what conclusions can you draw about the dependence of the thickness of this layer of ions on the electrolyte content of the continuous phase?

3. A copolymer of vinylpyridine and methacrylic acid (62 and 38 mole %, respectively, in polymer) was studied in 90% methanol–10% water solution. The specific viscosity was measured* as a function of added NaOH or HCl and the following results were obtained:

η_{sp}	0.1	0.25	0.28	0.24	0.21	0.21	0.20
meq added per gram	0	2	4	6	2	4	6
			HCl			NaOH	

Discuss these results in terms of the apparent effect of acid and base on the configuration of the polymer chain. Be sure your explanation is consistent with the chemical nature of the copolymer.

4. A dispersion of polydisperse spheres shows a relative viscosity of about 2.6 at a volume fraction of about 0.31.† Compare this result with the predictions of Eq. (70).

5. The viscosity of cross-linked polymethyl methacrylate spheres in benzene was measured‡ and found to be

ϕ	0.050	0.035	0.028	0.019	0.014	0.010
η/η_0	2.15	1.61	1.41	1.19	1.18	1.12

Calculate $[\eta]$ for these spheres. Is there evidence in these results that the polymer particles may be swollen by the solvent? Explain.

6. The viscosity of uniform, cross-linked polystyrene spheres of two different diameters was measured in benzyl alcohol at 30°C§:

$d = 0.382\ \mu m$	ϕ	0.013	0.030	0.059	0.075
	η_{sp}	0.036	0.086	0.178	0.233
$d = 0.433\ \mu m$	ϕ	0.02	0.04	0.08	
	η_{sp}	0.056	0.116	0.251	

Evaluate the intrinsic viscosity for each size of spherical particle and comment on the results in terms of the Einstein prediction that $[\eta]$ should be independent of particle size. Is the fact that benzyl alcohol is only a moderately good solvent for linear polystyrene consistent with the observed deviation between the experimental and theoretical values for $[\eta]$? Explain.

*T. Alfrey, Jr., and H. Morawitz, *J. Am. Chem. Soc.* **74**:436 (1952).
† H. Eilers, *Kolloid Z.* **97**:313 (1941).
‡ A. Kose and S. Hachisu, *J. Colloid Interface Sci.* **46**:460 (1974).
§ Y. S. Papir and I. M. Krieger, *J. Colloid Interface Sci.* **34**:126 (1970).

7. Criticize or defend the following proposition using the data of Problem 6 and the fact that the intrinsic viscosities of cross-linked polystyrene spheres in solvents such as benzene and CCl_4 (good solvents for linear polystyrene) lie in the range 5.8 to 7.5: "Cross-linked polystyrene spheres swell by imbibing solvent, the effect being more extensive the better the solvent properties of the continuous phase for the non-cross-linked polymer."
8. Use the following data to evaluate the Staudinger a and k constants for cellulose acetate in acetone at 25°C*:

M_n	c (g dl^{-1})	η_{sp} (dl g^{-1})
130,000	0.094	0.289
	0.273	0.99
	0.546	2.77
86,000	0.114	0.286
	0.351	1.10
	0.703	3.12
76,000	0.118	0.247
	0.353	0.89
	0.775	2.70
61,000	0.138	0.239
	0.275	0.52
	0.428	0.88
48,000	0.152	0.209
	0.303	0.45
	0.684	1.23

9. Various molecular weight fractions of cellulose nitrate were dissolved in acetone and the intrinsic viscosity was measured at 25°C†:

$M \times 10^{-3}$ (g mole^{-1})	77	89	273	360	400	640	846	1550	2510	2640
c (dl g^{-1})	1.23	1.45	3.54	5.50	6.50	10.6	14.9	30.3	31.0	36.3

Use these data to evaluate the constants k and a in the Staudinger equation for this system.

10. The relative viscosity of solutions of cellulose nitrate in acetone was measured and extrapolated to zero rate of shear‡:

η/η_0	1.45	1.53	1.67	1.89	2.31	3.41
c (g dl^{-1})	0.0151	0.0176	0.0212	0.0264	0.0352	0.0528

Use these data to evaluate the intrinsic viscosity for the polymer and use the a and k values from Problem 9 to calculate the molecular weight from the value of $[\eta]$.

* A. M. Sookne and M. Harris, *Ind. Eng. Chem.* **37**:475 (1945).
† A. M. Holtzer, H. Benoit, and P. Doty, *J. Phys. Chem.* **58**:624 (1954).
‡ A. M. Holtzer, H. Benoit, and P. Doty, *J. Phys. Chem.* **58**:624 (1954).

11. The following intrinsic viscosity values of some high-molecular-weight polystyrene fractions have been reported*:

$\bar{M} \times 10^{-6}$ (g mole^{-1})	$[\eta]$ (dl g^{-1})	
43.8	5.5	Cyclohexane at 35.4°C
27.4	4.4	
43.5	67.7	Benzene at 40°C
26.8	36.5	

Use these data to evaluate the constants in the Staudinger equation. Are the values obtained consistent with the known facts that 35.4° is the Flory (Θ) temperature for polystyrene in cyclohexane while benzene is a good solvent for polystyrene at 40°C?

12. Solutions of nylon-66 were studied in 90% formic acid solutions at 25°C and the following data were obtained for two different molecular weight fractions†:

c (g dl^{-1})	0.744	0.527	0.368	0.164
η_{sp}/c (dl g^{-1})	0.485	0.477	0.478	0.450

and

c (g dl^{-1})	0.742	0.640	0.537	0.436	0.332	0.225	0.132	0.058
η_{sp}/c (dl g^{-1})	0.897	0.892	0.886	0.876	0.864	0.847	0.805	0.778

Calculate the intrinsic viscosity for these two polymers. The Staudinger constants for this system are known to be $a = 0.72$ and $k = 11 \times 10^{-4}$ dl g^{-1}; calculate the molecular weights of the two nylon fractions.

* D. McIntyre, L. J. Fetlers, and E. Slagowski, *Science* **176**:1041 (1972).
† G. B. Taylor, *J. Am. Chem. Soc.* **69**:635 (1947).

Sedimentation and Diffusion, and Their Equilibrium

3

Even if you had completed your third year . . . in the University, and were perfect in the theory of the subject, you would still find that there was need of many years of experience, before you could move in a fashionable crowd without jostling against your betters. [From Abott's *Flatland*]

3.1 INTRODUCTION

In this chapter we shall examine the effects on particles in the colloidal size range of sedimentation and diffusion, first taken separately, then combined. Sedimentation occurs under the influence of gravity and, considerably faster, in a centrifuge. Both gravitational and centrifugal sedimentation are discussed herein. A key relationship in understanding the rate of sedimentation is Stokes' law, a hydrodynamic relationship that follows from the considerations of the preceding chapter. Stokes' law will appear again when we discuss the kinetics of flocculation in Chap. 10 and electrokinetic phenomena in Chap. 11.

Our analysis of diffusion entails statistical arguments, some of which may be familiar from kinetic molecular theory. In discussing diffusion, our attention is focused on the diffusion coefficient. The latter is defined experimentally by Fick's laws and theoretically by two equations derived by Einstein. Like Einstein's equation of viscosity, the equations for diffusion are remarkably simple results. In contrast to the viscosity equation, however, both can be derived without too much difficulty.

As our educated intuition might lead us to expect, larger particles sediment more rapidly and diffuse more slowly than smaller particles do. The effects of sedimentation and diffusion, therefore, are not of comparable magnitude for all sizes of particles. There is a range of particle sizes, however, for which the two are comparable and equilibrium between them is established. This equilibrium between sedimentation and diffusion has been extensively studied by means of the ultracentrifuge. Since many of the particles thus investigated are of biological importance, we shall frequently use protein molecules as our examples in this chapter.

Much of our discussion of sedimentation alone and sedimentation combined with diffusion is focused on the determination of the mass or molecular weight of the dispersed particles.

3.2 SEDIMENTATION AND THE MASS-TO-FRICTION FACTOR RATIO

To see how sedimentation comes about, consider the gravitational forces which operate on a particle of volume V and density ρ_2 which is submerged in a fluid of density ρ_1. The situation is shown in Fig. 3.1a for a spherical particle, but the discussion is independent of the actual shape of the particle. The particle experiences a force due to gravity F_g, taken to be positive in the downward direction. At the same time, a buoyant force F_b acts in the opposite direction. A net force equal to the difference between these forces results in the acceleration of the particle:

$$F_{\text{net}} = F_g - F_b = V(\rho_2 - \rho_1)g \tag{1}$$

This force will pull the particle downward if $\rho_2 > \rho_1$, and the particle is said to sediment. If, on the other hand, $\rho_1 > \rho_2$, then the particle will move upward; the latter is called "creaming." As the net velocity of the particle is increased, the viscous force opposing its motion increases also. Soon this force, shown in Fig. 3.1b, equals the net driving force responsible for the motion. Once the forces acting on the particle balance, the particle experiences no further acceleration and a stationary state velocity is reached. It may be shown that, under stationary state conditions and for small velocities, the force of resistance is proportional to the stationary state velocity v:

$$F_v = fv \tag{2}$$

where the proportionality constant is called the friction factor. We shall consider some aspects of the proof of Eq. (2) for spherical particles in the following section. However, Eq. (2) is independent of any particular geometry. The stationary state velocity is positive for sedimentation and negative for creaming.

FIGURE 3.1 *The forces acting on a spherical particle due to (a) gravity alone and (b) gravity and the viscosity of the medium* ($\rho_2 > \rho_1$).

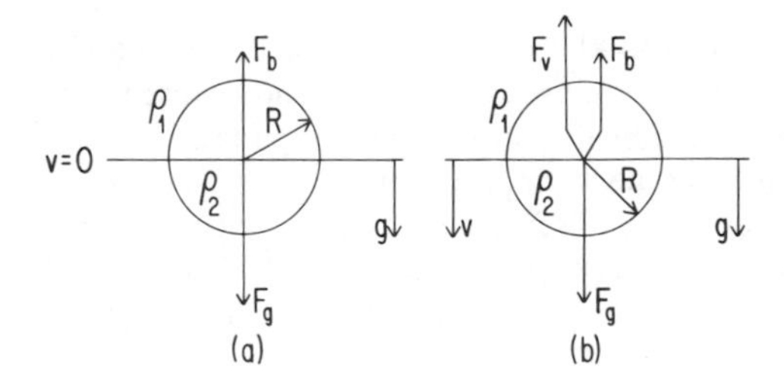

Since the net force of gravity and the viscous force are equal under stationary state conditions, Eqs. (1) and (2) may be equated to give

$$V(\rho_2 - \rho_1)g = fv \tag{3}$$

The stationary state is quite rapidly achieved, so Eq. (3) describes the velocity of a settling particle over much of its fall. Equation (3) may also be written

$$m\left(1 - \frac{\rho_1}{\rho_2}\right)g = fv \tag{4}$$

This equation has the following features:

1. It is independent of particle shape.
2. It assumes that the bulk density of the pure components applies to the settling units (i.e., no solvation).
3. It permits the evaluation of v for a situation in which m/f is known.
4. It permits the evaluation of m/f in a situation where v is known.

The stationary state sedimentation velocity of a particle is an experimentally accessible quantity for some systems, so item (4) summarizes much of our interest in sedimentation.

Unfortunately, it is the ratio m/f rather than m alone that is obtained from sedimentation velocity in the general case of particles of unspecified geometry. The situation is comparable to the result of the classical experiment of J. J. Thomson in which the charge-to-mass ratio of the electron was determined. What is needed is a method for arriving at one of the quantities in the ratio independently. This information, in addition to the ratio, permits the evaluation of both the mass and the friction factor of the particle. In general, it takes two experiments to evaluate numerically these two parameters which characterize the dispersed particles. There are two ways of proceeding to overcome this impasse:

1. The particle may be assumed to be a sphere, in which case its friction factor may be calculated theoretically and thus eliminated from Eq. (4). We shall discuss the friction factor of spherical particles in Sec. 3.3.
2. An experiment may be conducted which permits the evaluation of f from measured quantities. Diffusion studies are ideally suited for this purpose as we shall see later in this chapter.

Before turning to these two procedures to eliminate f from the sedimentation equation, one other consideration inherent in the use of Eq. (4) should be discussed. As previously noted, Eq. (4) uses the bulk density of the pure components. If the continuous and bulk phases are totally noninteracting, this may be justified. However, we have already seen several situations in which such an assumption would not work. In Chap. 1, for example, we discussed flocs and in Chap. 2 we discussed solvation. In both cases, the density of the settling unit is intermediate between the density of the two pure components.

Equation (3) shows that the sedimentation velocity is larger the larger the density difference between the particle and the medium. Any situation which brings the density of the settling unit closer to that of the solvent will decrease the sedimentation velocity. To an observer who is unaware of its cause, however, the smaller velocity would be interpreted by Eq. (4) as indicating a smaller value of m/f. Since the actual mass of colloidal material is unaffected by the solvation, it is more correct to attribute the reduced sedimentation velocity to an increase in the value of the fraction factor. We shall designate the friction factor for an unsolvated particle by f_0 and that of a solvated particle by f. Because of the lower sedimentation velocity of the solvated particle, it is clear that the ratio f/f_0 is greater than unity. Thus, for two otherwise identical particles, the one with the greater solvation will display the smaller sedimentation velocity and the larger f/f_0 ratio. Further, if the correct particle mass is to be obtained from the combination of sedimentation and diffusion experiments, then diffusion experiments must perceive the kinetic unit the same way that sedimentation does. This is indeed the case.

Until now, the friction factor has been merely a proportionality factor of rather ill-defined origin. We shall not undertake a derivation of Eq. (2) in any general sense. In the next section, however, we shall outline the derivation of an important result due to G. G. Stokes—the friction factor for an unsolvated sphere.

3.3 THE STOKES EQUATION

We begin our discussion of the Stokes equation by assuming that the particle is stationary and that the fluid is flowing in the $+z$ direction with a velocity v_z. What is actually relevant is the relative velocity between the particle and the fluid, so this situation is the same as a particle moving in the $-z$ direction through a stationary fluid. The latter situation describes the sedimentation of the sphere, our primary concern, but the former picture is more consistent with the mathematical models of Chap. 2.

In many ways, the assumptions of this derivation resemble those of the Einstein problem in the preceding chapter. In deriving both the Stokes and the Einstein equations, we restrict ourselves to a fluid of constant density and constant viscosity, in slow, stationary state, and laminar flow past a spherical particle. We continue to examine systems which are so dilute that each particle may be treated independently. However, in other ways the two derivations are very different. In the preceding chapter our interest was in calculating the viscosity of the entire system, which included *both* the continuous and the dispersed phases. In the present derivation, the system consists of the continuous phase only. The sphere enters the picture only inasmuch as it defines a boundary of the system. In the Einstein derivation, we sought a result that was independent of the rate of flow; in the present problem, we are explicitly looking for the dependence of the viscous force on the rate of flow. An important difference between the two derivations appears only in the final results. The viscosity of a dispersion of spheres, according to Einstein, is independent of the radius of the particles. According to Stokes' law, the force on a sphere is directly proportional to its radius.

The equations of continuity and motion, Eqs. (2.8) and (2.23), are the starting points once again. Since the spherical surface represents an important boundary, a spherical coordinate system is set up with its origin at the center of the sphere as shown in Fig. 3.2. Since the fluid is flowing in the $+z$ direction, it is clear that the flow streamlines around the particle will be symmetrical with respect to the z axis. Because of this, the present problem bears a certain resemblance to the analyses of the cylindrically symmetrical viscometers of Sec. 2.6. That is, v_z is not a function of ϕ.

Now let us consider the boundary conditions that must be satisfied in this problem. First, there is no slipping at the surface of the sphere, which means that all components of velocity equal zero at $r = R$. Next, the disturbance caused by the presence of the sphere must diminish with distance, specifically, p and v_z must take on their unperturbed values as $r \to \infty$. Stokes solved this problem in 1850 by deriving an equation, called the stream function, which describes the trajectory of a particle of fluid past the stationary sphere. The following expressions for pressure and velocity are obtained from this:

$$v_r = v_z\left[1 - \frac{3}{2}\frac{R}{r} + \frac{1}{2}\left(\frac{R}{r}\right)^2\right]\cos\theta \tag{5}$$

$$v_\theta = -v_2\left[1 - \frac{3}{4}\frac{R}{r} - \frac{1}{4}\left(\frac{R}{r}\right)^3\right]\sin\theta \tag{6}$$

$$p = -\frac{3}{2}\eta\frac{v_z}{R}\left(\frac{R}{r}\right)^2\cos\theta - \rho g z + \text{const} \tag{7}$$

It is important to remember that these expressions are nothing more than solutions to the equations of continuity and motion in spherical coordinates. These equations satisfy the boundary conditions of the problem. Consider, for example, a position in the plane perpendicular to the z axis, where $\theta = 90°$, $\cos\theta = 0$, $\sin\theta = 1$. Since $R/r \to 0$ as $r \to \infty$, we see that $v_\theta \to -v_z$, $v_r \to 0$, and $p \to$ const in the horizontal plane at large distances from the particle.

The next step in the derivation of Stokes' law consists of evaluating the rate of energy deposition in the system using Eq. (2.52). Since the integration is over the entire fluid, the limits are from $r = R$ to $r = \infty$ and over all values of the angular

FIGURE 3.2 *Definition of coordinates for a liquid flowing with a velocity v_z past a sphere of radius R.*

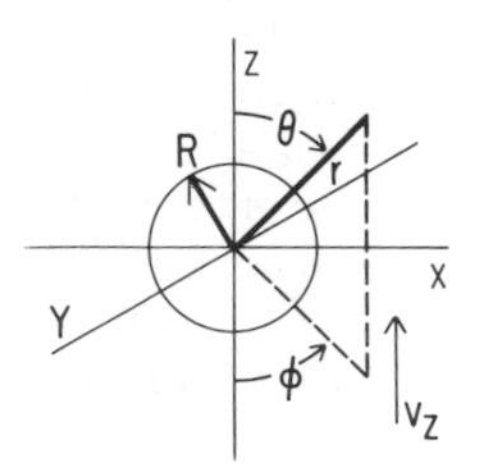

coordinates. Omitting the details of the integration, this leads to the following expression for the rate of energy deposition:

$$\frac{DE}{Dt} = 6\pi\eta v_z^2 R \tag{8}$$

The product of the force F exerted by the particle and the velocity of the fluid also equals the *rate* at which work is done on the fluid by the particle. Equating this product to (8) gives

$$Fv_z = 6\pi\eta v_z^2 R \tag{9}$$

It is clear that the force acting on the fluid acts in opposition to the flow. Therefore, if the fluid is flowing in the plus direction, the sphere resists the flow by a force in the negative direction. Conversely, if the particle is moving in one direction, the fluid resists the movement by a force in the opposite direction. This force is given by Stokes' law

$$F = 6\pi\eta R v \tag{10}$$

where the subscript on v is no longer needed, provided we remember that F and v are in opposing directions.

Equation (10) expresses, for a spherical particle, the general result presented as Eq. (2). Two aspects of the result are important:

1. For small stationary state velocities, the viscous force on a particle is directly proportional to the velocity.
2. For a spherical particle, the friction factor is given by

$$f = 6\pi\eta R \tag{11}$$

It was just noted that the case of spherical particles is one in which the friction factor can be eliminated from Eq. (4), yielding a result which permits the mass of a spherical particle to be evaluated from sedimentation data alone. To see how this works, we return to Eq. (3), using (11) to evaluate f and $\frac{4}{3}\pi R^3$ as a substitution for V. Thus for a spherical particle, Eq. (3) becomes

$$\tfrac{4}{3}\pi R^3(\rho_2 - \rho_1)g = 6\pi\eta R v \tag{12}$$

Equation (12) may be solved for the stationary state sedimentation velocity

$$v = \frac{2}{9}\frac{R^2(\rho_2 - \rho_1)g}{\eta} \tag{13}$$

and for the radius of the spherical particle

$$R = \left[\frac{9\eta v}{2(\rho_2 - \rho_1)g}\right]^{1/2} \tag{14}$$

Since the particle to which it applies is a sphere, Eq. (14) may also be used to calculate the mass and friction factor of the particle:

$$m = \rho_2 \tfrac{4}{3}\pi R^3 = \tfrac{4}{3}\pi\rho_2\left[\frac{9\eta v}{2(\rho_2-\rho_1)g}\right]^{3/4} \tag{15}$$

and

$$f = 6\pi\eta\left[\frac{9\eta v}{2(\rho_2-\rho_1)g}\right]^{1/2} \tag{16}$$

Equation (12) is an important result since it describes the relationship between R, v, η, and $\Delta\rho$, the density difference. Any one of these quantities may be evaluated by the equation when the other three are known. Thus, Eq. (12) can be used to determine the density difference between two phases or to determine the viscosity of a liquid (the basis of so-called falling ball viscometers). In this chapter, however, our interest is in the characterization of colloidal particles by means of observations of their sedimentation behavior. Therefore, we shall be primarily concerned with Eqs. (14) and (15), which are specifically directed toward this objective.

Stokes' law and the equations developed from it apply to spherical particles only. In the preceding chapters we have seen that the dispersed units in systems of actual interest often fail to meet this shape requirement. Equation (15) is sometimes used in these cases anyway. The lack of compliance of the system to the model is acknowledged by labelling the mass, calculated by Eq. (15), as the mass of an "equivalent sphere." As the name implies, this is a fictitious particle with the same density as the unsolvated particle which settles with the same velocity as the experimental system. If the actual settling particle is an unsolvated polyhedron, the equivalent sphere may be a fairly good model for it, and the mass of the equivalent sphere may be a reasonable approximation to the actual mass of the particle. The approximation clearly becomes poorer if the particle is unsymmetrical or solvated, or both. Characterization of dispersed particles by their mass as equivalent spheres at least has the advantage of requiring only one experimental observation, the sedimentation rate, on the system. We shall see in later sections that the equivalent sphere calculations still play a useful role even in systems for which supplementary diffusion studies have also been conducted.

Suppose the actual settling particles in an experiment are unsolvated, but rather than being spheres they are ellipsoids of axial ratio a/b. Does the equivalent sphere overestimate or underestimate the mass of these particles? To answer this qualitatively, consider the following argument. Small particles of the size we encounter in colloid chemistry undergo continuous jostling due to Brownian movement. This means that small ellipsoidal particles fluctuate through every possible orientation during their settling. Whereas larger particles may adopt a hydrodynamically favorable orientation in settling, the small particles tumble through all orientations. It is possible to imagine a spherical surface surrounding one of these tumbling particles. The radius $\bar{R}$ of this encompassing sphere will naturally be larger than the radius R of a spherical particle of the same mass. The relationship between $\bar{R}$ and the dimensions of the actual particle will depend on the shape of the particle and on the procedure used to average over all orientations. For our purposes, it is sufficient to recognize that the friction factor for such a particle may be written

$$f = 6\pi\eta\bar{R} \tag{17}$$

by analogy with Eq. (11). To distinguish between the friction factor for an actual sphere and the spherical envelope surrounding a tumbling ellipsoid, we shall attach the subscript 0 to the former. Since $\bar{R} > R$, it follows that $f > f_0$. Therefore, we have used a friction factor which is too small in evaluating the mass of an equivalent sphere. Because we have underestimated the viscous resistance it experiences, we also underestimate the mass of the settling ellipsoid by treating it as a sphere.

We have concluded that for an unsolvated ellipsoid f/f_0 is greater than unity. In the case of solvated spheres in Sec. 3.2 we reached a similar conclusion. Thus either solvation or ellipticity causes the friction factor of a particle to be larger than it would be in the absence of these effects. We have used the same ratio f/f_0 to describe the effects of solvation and ellipticity because it is impossible to differentiate between these two effects by sedimentation and diffusion alone. As a matter of fact, the mass calculated from experimental results will be low by a factor f/f_0 in each of the following situations:

1. For a solvated particle of any shape which is analyzed in terms of sedimentation and diffusion experiments, using the density of the unsolvated particles.
2. For an unsolvated ellipsoid, using the formulas for an equivalent sphere.

We shall see how to use these facts quantitatively in Sec. 3.8.

Since the mass of an equivalent sphere is an important concept in its own right, let us next consider how this quantity may be extracted from experimental data.

3.4 SEDIMENTATION UNDER GRAVITY: EXPERIMENTAL

We begin our discussion of the experimental aspects of sedimentation by considering settling caused by gravity; in the next section we shall examine centrifugation. Table 3.1 lists some sedimentation velocities, calculated by Eq. (13), for several values of the ratio $\Delta\rho/\eta$, and a range of particle radii. It is clear from this table that the range of particle sizes that can be conveniently studied by sedimentation under gravity is quite limited, the precise limits being determined by the factor $\Delta\rho/\eta$. Although sedimentation under gravity applies only to particles at the upper size limit of the usual colloidal range, there are still a number of important applications of the technique. We will consider only two of the procedures that have been developed for such studies.

If a particle is large enough to be visible to the unaided eye or in a traveling microscope, its velocity can be measured directly. However, smaller particles present more of a problem. If a dispersion consists of particles of uniform size, all will settle at the same rate so their positions relative to each other do not change until they reach the bottom of the container. This means that a sharp boundary will exist between the domain occupied by the settling particles and that part of the system which has already been swept free of particles. Even though the individual particles

TABLE 3.1 *Sedimentation velocities* (cm s^{-1}) *for several different values of* $\Delta\rho/\eta$ *and various R values spanning the colloidal size range*

R (cm)	$\Delta\rho/\eta$ (g cm^{-3} P^{-1})		
	34	100	476
10^{-4}	7.40×10^{-5}	2.18×10^{-4}	1.02×10^{-3}
10^{-5}	7.40×10^{-7}	2.18×10^{-6}	1.02×10^{-5}
10^{-6}	7.40×10^{-9}	2.18×10^{-8}	1.02×10^{-7}
10^{-7}	7.40×10^{-11}	2.18×10^{-10}	1.02×10^{-9}
Particles in water at 20°C:	Protein	Sulfur ($v>0$) Gas bubbles ($v<0$)	AgI

may not be visible, this boundary may be visible if the particles differ sufficiently in color or refractive index from the continuous phase. The velocity of the boundary and the velocity of the particles are the same. If the particles differ in size, no such boundary will develop. Each size fraction in a polydisperse system will settle at its own characteristic velocity so that, at any given time, one end of the column may be free of some size fractions but not of others. The "boundary zone" between the domain which clearly contains settling particles and the domain which is totally free of particles will appear to be a diffuse region across which the concentration of the disperse phase gradually diminishes. Since no sharp boundary develops, some other method must be devised to follow sedimentation velocity. Clearly, we would like as much information as possible about the particle size distribution since the system is polydisperse. The following example shows a procedure whereby this information may be obtained.

Suppose the pan of a balance is positioned at some convenient location below the surface of a dispersion. As sedimentation occurs, the settled material collects on the balance pan. The total weight W of the material on the pan is measured at various times t, either by adding counterweights to a second pan or by noting the displacement of a calibrated fiber which supports the pan.

Figure 3.3a shows the type of data obtained when this method is applied to a clay sample. Clearly, at any given time, the weight of material that has accumulated equals the weight of all particles large enough to have settled onto the pan in the time of the experiment. Such particles come from two categories. They include all those particles large enough to have fallen through the full length of the container *and* a fraction of smaller particles which have only fallen from a fraction of the column but still land on the pan. Both of these contribute to the total weight observed at any time. Since the cumulative weight equals the sum of two contributions, we may write

$$W = w + t\frac{dW}{dt} \tag{18}$$

where w is the weight of the particles which have fallen through the full height of the column at that particular time, and $t\,dW/dt$ represents the cumulative contribution

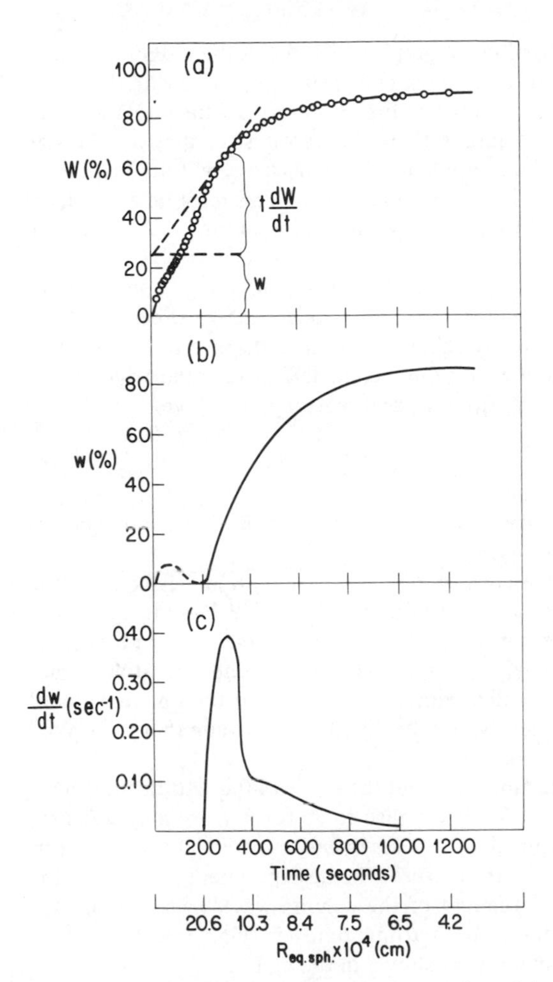

FIGURE 3.3 *Sedimentation of clay particles*: (*a*) *cumulative weight (as percent) versus time*, (*b*) *cumulative weight of oversize material (as percent) versus time*, (*c*) *frequency by weight versus time and radius, assuming equivalent spheres.* [*Data from S. Oden*, Proc. Roy. Soc. Edinburgh **36**:219 (1915).]

of smaller size particles at that time. The quantity dW/dt is the slope of the cumulative curve at a particular point. As such, it measures the rate of incidence of particles smaller than the cutoff size which is associated with that time: The larger particles, after all, have already settled out. Furthermore, these smaller particles have been collecting on the pan throughout the entire time of the run t. Therefore, their total weight equals $t\,dW/dt$. The graphical significance of Eq. (18) is shown in Fig. 3.3a. This shows that the intercepts defined by tangent lines drawn at different times give w directly, without actually requiring the calculation suggested by (18).

Next, we plot w versus time to obtain a graph which shows the weight contribution at any time due to particles larger than the cutoff size associated with that time. Such a plot for the data of Fig. 3.3a is shown in Fig. 3.3b. This is the integrated size distribution curve for all particles larger than the cutoff size, that is, oversize particles. Since Fig. 3.3b is an integrated distribution curve, it follows that its derivative gives a distribution function for particle sizes. This must be a graphical differentiation since we do not know the analytical expression for the cumulative curve.

Figure 3.3c shows the results of a graphical differentiation of Fig. 3.3b. That is, the ordinate values in Fig. 3.3c are slopes from Fig. 3.3b, dw/dt. In effect, then, this is the second time the original data have been differentiated: once to prepare Fig. 3.3b and again for Fig. 3.3c. That dw/dt is the second derivative of the original data may also be shown by differentiating Eq. (18) and rearranging to give

$$\frac{dw}{dt} = -t\frac{d^2W}{dt^2} \tag{19}$$

Since the second derivative is negative for a curve which opens downward, such as Fig. 3.3a, dw/dt is a positive quantity.

There is a loss of accuracy with each graphical differentiation. Therefore, the original data must be very good for a particle size distribution such as Fig. 3.3c to be prepared from cumulative sedimentation results. At both long and short times, the total weight increases almost linearly, which means that the second derivative cannot be determined with any accuracy at all at either end of the distribution. In spite of these limitations, Fig. 3.3c shows clearly that the distribution is quite sharply peaked at a particle size which settles in about 300 s.

Until now, no assumptions whatsoever about the shape of the settling units have been made in the analysis of Fig. 3.3. If we wish to go further, we may translate settling times into the radius of equivalent spheres by means of Eq. (14). However, at this point an assumed model enters the picture. In the experiment from which the data of Fig. 3.3 were obtained, the height of the column was 20 cm and $\Delta\rho$ was 2.17 g cm^{-3}. With this information the settling times in Fig. 3.3c have been converted to radii of equivalent spheres, as shown in the figure.

This type of analysis gives not only the radius of the most probable equivalent sphere but also an indication of the distribution of the particle sizes. As already noted, the equivalent sphere becomes a progressively poorer model for particles as the system deviates from the model in terms of either asymmetry or solvation, or both.

The equivalent sphere is a very poor model for the system in Fig. 3.3, since clay particles are generally dispersed as platelets as illustrated in the electron micrograph of Fig. 1.9b.

A slight variation of this method for analyzing particle size distributions is particularly convenient if the system contains particles of several discretely different particle sizes rather than a continuous distribution of sizes. A thin band of such a dispersion is layered on top of a column of pure solvent. Particles in the different size categories will settle through the solvent at different rates so that, if the column is long enough, the first fraction collected at the bottom of the column will contain only the largest particles, with smaller sizes showing up in progressively later fractions. The clear solvent is literally the column within which the various particles sizes are resolved. The "resolving power" of such an arrangement increases with the length of the column at least up to some optimum length. It might be noted that there is a certain formal similarity between this situation and column chromatography.

Provided the column is long enough to separate the different sizes of particles, this method gives w versus time directly. Any additional interpretations of the results are made in a manner entirely analogous to the one just described. Since only a narrow band of the dispersion is used in this method, the weight of the dispersed phase in each fraction will be relatively low. Fairly sensitive analytical techniques are required for the fractions collected.

Sedimentation runs should be conducted at constant temperature, not only so that $\Delta\rho$ and η are known but also to minimize disturbances due to convection. Any sort of disturbance will obviously disrupt the segregation of the particles by size which has occurred as a result of sedimentation. An intrinsic difficulty with the balance method lies in the fact that the liquid below the balance pan is less dense than the liquid with dispersed particles above the pan. Thus, there is a tendency for a counterflow of pure solvent which would introduce an error in the particle size analysis.

The methods just discussed are only two of a wide variety of techniques which provide essentially the same kinds of information. In general, any measurement can be interpreted in terms of particle size distribution which gives (a) the amount of suspended material a fixed distance below the surface at various times or (b) the amount of material at various depths at any one time. Pressure, density, and absorbance are additional measurements that have been analyzed this way. A basic limitation of all these methods is the narrow range of particle sizes that can be investigated feasibly by sedimentation under gravity. Therefore, we turn next to a consideration of centrifugation, particularly the ultracentrifuge, as a means of extending the applicability of sedimentation measurements.

3.5 SEDIMENTATION IN A CENTRIFUGAL FIELD

A well-known fact from elementary physics is that a particle traveling in a circular path of radius R at an angular velocity ω (in radians per second) is subject to an acceleration in the radial direction equal to $\omega^2 R$. Since there are 2π radians per revolution, ω equals the number of revolutions per second (rps) times 2π or the number of revolutions per minute (rpm) times the ratio $2\pi/60$. It is not particularly difficult to produce accelerations by centrifugation which are many times larger than the acceleration due to gravity. It is conventional, in fact, to describe the radial acceleration achieved by a centrifuge as some multiple of the standard gravitational acceleration or as being so many g's, where 1 g is about 980 cm s^{-2}.

The ultracentrifuge is an instrument in which a cell is rotated at very high speeds in a horizontal position. As we shall see presently, the gravitational acceleration is easily increased by a factor of 10^5 in such an apparatus. Accordingly, the size of particle which may be studied by sedimentation is decreased by the same factor. The ultracentrifuge has been used extensively for the characterization of colloidal materials, particularly those of biological origin such as proteins, nucleic acids, and viruses.

Because of the extreme importance of the ultracentrifuge, it seems appropriate to describe it in some detail. The particular apparatus described is the Beckman Analytical Ultracentrifuge, Model E (Beckman Instruments, Inc., Palo Alto, California). Although other models differ in details, the essential features are present in all ultracentrifuges. The actual sedimentation takes place in a cell mounted within an aluminum or titanium rotor. The cell is sector-shaped; its side walls converge toward the center along radial lines. The depth of the cell in the direction of sedimentation is 1.4 cm. Since the radial acceleration is proportional to the distance from the axis of rotation, we see that this quantity varies from top to bottom in the cell. Although this variation is considered explicitly in a later section, it is sufficient for the present for us to consider the average acceleration at the midpoint of the cell which is located 6.5 cm from the center of the rotor. The accelerations reported in Table 3.2 describe the acceleration at this location for various rpm values.

An important part of the ultracentrifuge is the optical system which makes observation during operation a possibility; it is shown schematically in Fig. 3.4. The sample compartment fits into a hole in the rotor and is positioned so as to intersect the light path of the optics. Those faces of the cell which are perpendicular to the light path are transparent to visible and ultraviolet light, so that various optical methods of chemical analysis may be employed to measure the distribution of material along the sedimentation path. Schlieren refractometry, interferometry, and spectrophotometry are all employed for this purpose, the last being particularly useful when very low concentrations are involved. The location of the settling particles may be followed by one of these methods and the results may be recorded either photographically or on a chart recorder.

The Schlieren system of optics is an analytical method that is particularly well suited to following the location of a chemical boundary with time. It is routinely employed in ultracentrifuges and also in electrophoresis experiments, as we shall see in Chap. 11. Schlieren optics produce an effect which depends on the way the refractive index varies with position, the refractive index gradient, rather than on the

TABLE 3.2 *The angular acceleration corresponding to rotation at various rpm in the Spinco Model E Ultracentrifuge. Evaluated for* $R = 6.5$ cm

RPM	Acceleration (cm s^{-2})	Acceleration (g)
1×10^3	7.128×10^4	7.274×10^1
5×10^3	1.782×10^6	1.818×10^3
10×10^3	7.128×10^6	7.274×10^3
20×10^3	2.851×10^7	2.909×10^4
40×10^3	1.140×10^8	1.164×10^5
60×10^3	2.566×10^8	2.618×10^5

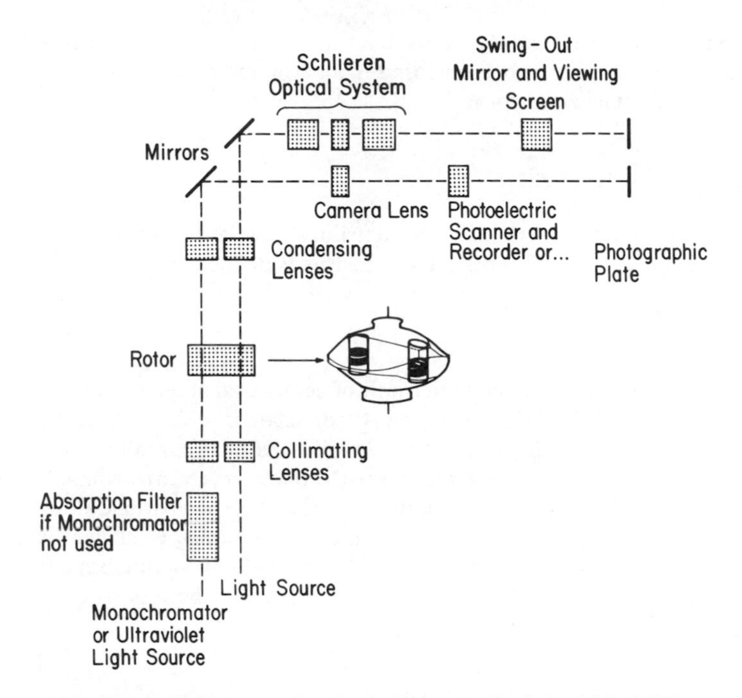

FIGURE 3.4 *Schematic of optical systems in the Spinco Model E Ultracentrifuge. (Beckman Instruments, Inc., redrawn with permission.)*

refractive index itself. Therefore, the Schlieren effect is the same at all locations along the axis of sedimentation except at any place where the refractive index is changing. In such a region, it will produce an optical effect which is proportional to the refractive index gradient. The boundary between two layers is thus perceived as a sharp peak on a flat baseline by Schlieren optics. The displacement of such a peak with time measures the velocity of the boundary. Likewise, a band of settling particles is seen as a broad Schlieren peak whose width and velocity measure the width and velocity of the band. The Schlieren optical system works by using a diaphragm to cut off from a photographic plate that light which is deviated from the optical path by the refractive index gradient. The same effect may be produced on a ground-glass screen for instantanous viewing by an ingenious system which combines a diagonal slit and a cylindrical lens to produce an image of the refractive index gradient. We shall see in Chap. 5 that the change of refractive index with concentration is also involved when the intensity of scattered light is measured as a method for characterizing dispersed particles.

The usual precautions regarding temperature and vibration control also apply to the ultracentrifuge. To overcome air resistance and the attendant frictional heating, the rotor spins in an evacuated compartment which may be thermostated over a wide range of temperatures. The rotor is mounted on a flexible drive shaft which minimizes the need for precise balancing as a requirement for vibration-free operation. Finally, the rotor assembly is enclosed in an armored steel chamber for safety. At this speeds, a runaway rotor is deadly!

3.6 THE SEDIMENTATION COEFFICIENT

The results of a sedimentation experiment in a centrifugal field are conventionally reported as the sedimentation coefficients. This quantity is defined as

$$s = \frac{dR/dt}{\omega^2 R} \tag{20}$$

that is, it equals the sedimentation velocity per unit of centrifugal acceleration. In the cgs system, this ratio has the units cm s^{-1}/cm s^{-2} or seconds. In practice, the quantity 10^{-13} s is defined to be 1.0 svedberg (symbol S) after T. Svedberg, the originator of the ultracentrifuge and poineer in its use (Nobel Prize, 1926). Sedimentation coefficients are usually reported in this unit. If the location of a particle along its settling path is measured as a function of time, the sedimentation coefficient is readily evaluated by integrating Eq. (20), using the following limits of integration: The component is at radial position R_1 at time t_1 and at R_2 at t_2. Therefore,

$$\omega^2 s \int dt = \int \frac{dR}{R} \tag{21}$$

or

$$s = \frac{\ln R_2/R_1}{\omega^2(t_2 - t_1)} \tag{22}$$

The sedimentation coefficient is concentration dependent, consequently it is usually measured at several different concentrations and the results are extrapolated to zero concentration.

Under stationary state conditions, the force due to the centrifugal field and the viscous force of resistance will be equal. Therefore, $\omega^2 R$ replaces g in Eq. (2) to give

$$m\left(1 - \frac{\rho_1}{\rho_2}\right)\omega^2 R = f\frac{dR}{dt} \tag{23}$$

where f is the friction factor for the settling particle. Equations (23) and (20) may be combined to yield

$$\frac{m}{f}\left(1 - \frac{\rho_1}{\rho_2}\right) = s \tag{24}$$

which shows that the sedimentation coefficient is directly proportional to the ratio of the mass-to-friction factor. As with sedimentation under the acceleration of gravity, any further interpretation of m/f depends either on independent determination of the friction factor from diffusion or on the assumption of spherical particles with Eq. (11) used to evaluate f. Thus, experimental sedimentation coefficients may be analyzed to yield the mass, radius, and friction factor of an equivalent sphere. In the absence of supplementary data, this is as far as sedimentation alone can be interpreted.

In the preceding sections of this chapter we have considered sedimentation as if it were the only process which influenced the spatial distribution of particles. If this were the case, all systems of dispersed particles, even gases, would eventually settle out. In practice, convection currents arising from temperature differences keep many systems well stirred. Even in carefully thermostated laboratory samples, however, there is another factor operating which prevents the complete sedimentation of small particles, namely diffusion. Diffusion and sedimentation are opposing processes inasmuch as the former tends to keep things dispersed whereas the latter tends to collect them in one place. Diffusion is more important for smaller particles: Remember that sedimentation is negligible for gases. For larger particules, diffusion is negligible. Of course, there is a range of particle sizes in which both effects are comparable; we will examine the combined effects of sedimentation and diffusion in later sections. For the present, however, let us examine thc phenomenon of diffusion by itself.

3.7 DIFFUSION AND FICK'S LAWS

If external forces such as gravity can be neglected, the composition of a single equilibrium phase will be macroscopically uniform throughout. This means that the concentration or density is constant throughout the phase. It should be noted that we are talking about the macroscopic description of the phase. At the molecular level, there will be local fluctuations from the mean value which are important in light scattering, as we shall see in Chap. 5. However, for the present, we restrict ourselves to the macroscopic description. Fundamentally, it is the second law of thermodynamics that is responsible for the uniform distribution of matter at equilibrium, since entropy is maximum when the molecules are distributed randomly through the space available to them. If for some reason a nonuniform distribution of matter should exist, the particles of the system will experience a force which tends to distribute them uniformly. Consider, for example, the situation shown in Fig. 3.5a in which two solutions of different concentration are separated by a hypothetical porous plug of zero thickness. In this case there will be a migration of solute from the high concentration side to the low concentration side until a uniform distribution is obtained, that is, until equilibrium is reached.

Now let us define the following quantities. Suppose Q represents the amount of material that flows through a cross section of area A in Fig. 3.5. The rate of change in the quantity Q/A is called the flux of solute across the boundary. That is, the flux J is defined as

$$J = \frac{d(Q/A)}{dt} \tag{25}$$

Therefore, the amount of material which crosses A in a time interval Δt is

$$Q = AJ\,\Delta t \tag{26}$$

(a)

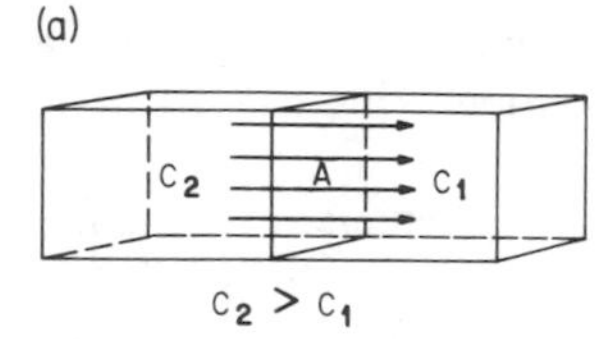

(b)

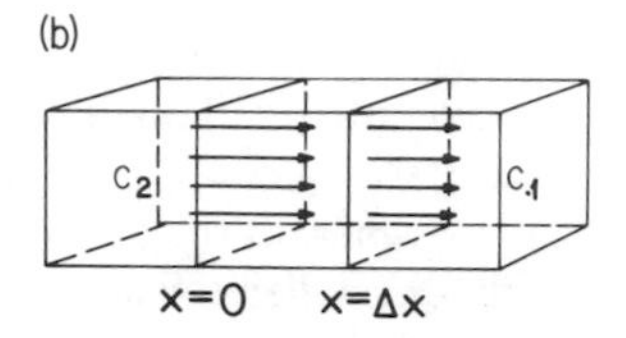

FIGURE 3.5 *Solutions of two different concentrations separated by a porous plug of thickness (a) zero and (b) Δx. Arrows indicate direction of flux.*

The phenomenological equation which relates the flux of material across the boundary to the concentration gradient at that location is given by Fick's first law:

$$J = D\frac{\partial c}{\partial x} \tag{27}$$

where D is defined to be the diffusion coefficient of the solute. Assuming that the amount of material is measured in the same units both in J and in c, Eqs. (25) and (27) show that the cgs units of D are square centimeters per second.

The resemblance between Eq. (27) and Eq. (2.1) should not be overlooked. The diffusion coefficient is the proportionality constant between the flux of matter and the gradient of concentration whereas the coefficient of viscosity establishes the proportionality between the flux of momentum and the velocity gradient. Both of these phenomena along with heat and charge transport are known collectively as transport processes.

Now, instead of a boundary of zero thickness, let us consider the concentration changes which occur within a zone of cross section A which has a thickness Δx as shown in Fig. 3.5b. Any change in the amount of material in this zone will equal the difference between the amount of material which enters the zone and the amount of material which leaves it:

$$\Delta Q = Q_{\text{in}} - Q_{\text{out}} = (J_{\text{in}} - J_{\text{out}})A\ \Delta t \tag{28}$$

The quantity ΔQ also equals the product of the volume of the zone and the concentration change that occurs in it:

$$\Delta Q = A\ \Delta x\ \Delta c \tag{29}$$

Equating (28) and (29) and substituting Fick's first law [Eq. (27)], we obtain

$$D\frac{[(\partial c/\partial x)_{x=0} - (\partial c/\partial x)_{x=\Delta x}]}{\Delta x} = \frac{\Delta c}{\Delta t} \tag{30}$$

assuming that the cross section is uniform throughout the compartment and that D is independent of small changes of concentration. Finally, if we take the limit of Eq. (30) as Δx and Δt approach zero, we obtain

$$\frac{\partial c}{\partial t} = D\frac{\partial^2 c}{\partial x^2} \tag{31}$$

a result known as Fick's second law. The formal similarity between this result and Eq. (2.2) should also be noted. In the present relationship the rate of mass accumulation in a volume element is considered; in Eq. (2.2), it is the rate of energy accumulation that is involved. Note the similar roles of the coefficients of diffusion and viscosity in the two relationships and the part played by the gradients in both cases.

Equation (31) is a differential equation whose solution describes the concentration of a system as a function of time and position. The solution depends on the boundary conditions of the problem as well as on the parameter D. This is the basis for the experimental determination of the diffusion coefficient. Equation (31) is solved for the boundary conditions which apply to a particular experimental arrangement. Then the concentration of the diffusing substance is measured as a function of time and location in the apparatus. Fitting the experimental data to the theoretical concentration function permits the evaluation of the diffusion coefficient for the system under consideration. Rather than getting deeply involved in the mathematics of differential equations, we shall use a statistical model to find a solution to Eq. (31) for a system with simple boundary conditions. This will be sufficient to illustrate the experimental technique whereby diffusion coefficients are determined and will also lead to a better understanding of the random processes underlying diffusion. This statistical discussion and the experimental procedure it suggests are taken up in Sec. 3.10.

Next, it will be helpful to anticipate a description of experimental procedures and consider the magnitude of measured diffusion coefficients. The self-diffusion coefficients for ordinary liquids with small molecules are of the order of magnitude of 10^{-5} cm^2 s^{-1}; for colloidal substances, they are typically of the order of 10^{-7} cm^2 s^{-1}. In the next section, we shall see that for spherical particles the diffusion coefficient is inversely proportional to the radius of the sphere. Therefore, every increase by a factor of 10 in size decreases the diffusion coefficient by the same factor. Qualitatively, this same inverse relationship applies to nonspherical particles as well. Once again, we see that diffusion decreases in importance with increasing particle size, precisely those conditions for which sedimentation increases in importance. For larger particles, where D is very small, the diffusion coefficient also becomes harder to measure. For spherical particles, the time required for the particle to diffuse a distance of 1.0 cm is directly proportional to its radius. For small molecules, this time requirement is an experimentally convenient one; it becomes experimentally inconvenient for particles at the upper end of the colloidal range.

3.8 THE DIFFUSION COEFFICIENT AND THE FRICTION FACTOR

As already noted, the driving force underlying diffusion is primarily thermodynamic in origin. A very general way to describe forces is to write them as gradients of a potential. In the context of diffusion, the potential to be used is the chemical potential μ_i, the partial molal Gibbs free energy of the component of interest. Thus, the driving force per particle behind diffusion may be written

$$F_{\text{diff}} = \frac{1}{N_A}\frac{\partial \mu_i}{\partial x} \tag{32}$$

It is necessary to divide by Avogardro's number since μ_i is a molar quantity. Thermodynamics shows that

$$\mu_i = \mu_i^0 + RT \ln a_i = \mu_i^0 + RT \ln \gamma_i c_i \tag{33}$$

where a_i, c_i, and γ_i are the activity, concentration, and activity coefficient, respectively, of the ith component. Since we are interested in infinitely dilute systems, the activity coefficient may be assumed to equal unity. Substituting Eq. (33) into (32) gives

$$F_{\text{diff}} = kT\frac{\partial \ln c_i}{\partial x} = \frac{kT}{c_i}\frac{\partial c_i}{\partial x} \tag{34}$$

where k is Boltzmann's constant R/N_A. Under stationary state conditions this force will be equal to the force of viscous resistance, given by fv according to Eq. (2). Therefore, the velocity of diffusion equals

$$v = \frac{kT}{fc}\frac{\partial c}{\partial x} \tag{35}$$

where the subscript has been omitted from the concentration of the solute c since this is now the only quantity involved in the relationship. Finally, we make the following observation. The flux of material through a cross section equals the product of its concentration and its diffusion velocity:

$$J = cv_{\text{diff}} \tag{36}$$

Combining Eqs. (35) and (36) and comparing with (27) leads to the important result

$$D = \frac{kT}{f} \tag{37}$$

It should be noted that this derivation contains no assumptions about the shape of the particles.

Many of the relationships of this chapter have involved the friction factor f which, until now, has been an unknown quantity except for spherical particles. Equation (37) breaks this impasse and points out the complementarity between sedimentation and diffusion measurements. For example, substitution of Eq. (37) into (4) gives the following results:

$$m = \frac{kTv}{D(1-\rho_1/\rho_2)g} \tag{38}$$

and substituting (37) into (24) yields

$$m = \frac{kTs}{D(1-\rho_1/\rho_2)} \tag{39}$$

Equations (38) and (39) show that diffusion studies combined with sedimentation studies, either under gravity or in a centrifuge, yield information about particle masses with no assumptions about the shape of the particle.

Human hemoglobin, for example, has a sedimentation coefficient of 4.48S and a diffusion coefficient of 6.9×10^{-7} cm^2 s^{-1} in aqueous solution at 20°C. The density of this material is 1.34 g cm^{-3}. Substituting these values into Eq. (39) shows the particle mass to be

$$m = \frac{(1.38 \times 10^{-16})(293)(4.48 \times 10^{-13})}{(6.9 \times 10^{-7})(1-1.0/1.34)} = 1.03 \times 10^{-19} \text{ g particle}^{-1} \tag{40}$$

or, in terms of molecular weight,

$$M = (1.03 \times 10^{-19})(6.02 \times 10^{23}) = 62{,}300 \text{ g mol}^{-1} \tag{41}$$

Since the friction factor was measured experimentally, this value is correct regardless of the state of solvation or ellipticity of the hemoglobin molecules in solution. We shall see presently how the combination of these two experimental approaches may be interpreted further to yield some information about solvation and ellipticity.

We have already seen that the ratio f/f_0 describes the effect on the friction factor of either solvation or ellipticity, or both. This ratio equals unity for an unhydrated sphere and increases with both the amount of bound solvent and the axial ratio of the particles. We are now in a position to see how this ratio may be evaluated experimentally. The steps of the procedure are itemized below and are illustrated numerically for human hemoglobin particles in Table 3.3:

1. The diffusion coefficient and sedimentation velocity (or sedimentation coefficient) are used to evaluate m by Eq. (38) or (39).
2. The friction factor is evaluated from the diffusion coefficient by Eq. (37).
3. The mass of the particle is divided by the density of the dry material to determine its volume, assuming the particle to be unsolvated.
4. The radius of the particle is calculated from its volume, assuming it to be a sphere.
5. The friction factor f_0 of the equivalent, unsolvated sphere is evaluated by Eq. (11).
6. The ratio of the experimental friction factor to f_0 is determined.

As noted, the larger this ratio is than unity, the more the particle deviates from the unsolvated spherical shape assumed to calculate f_0. Although this statement is qualitatively accurate, we realize that the f/f_0 value reflects some actual quantitative condition of the particles, and we search for additional ways to interpret this ratio.

The effect of solvation on the ratio f/f_0 for spheres is easily evaluated if the mode of solvation is that assumed in deriving Eq. (2.82). In that case—where the solvent

TABLE 3.3 *A summary of the characterization of human hemoglobin at 20°C based on sedimentation and diffusion measurements*[a]

Quantity	Value	Equations
Sedimentation coefficient, s	4.48×10^{-13} s	Experimental
Diffusion coefficient, D	6.9×10^{-7} cm^2 s^{-1}	Experimental
Density, ρ	1.34 g cm^{-3}	Experimental
Mass of particle, m	1.03×10^{-19} g molecule^{-1}	(40)
Molecular weight	6.23×10^{4} g mol^{-1}	(41)
f	5.86×10^{-8} g s^{-1}	(37)
Volume of particle	7.69×10^{-20} cm^3	$V_{\text{unsolv}}=m/\rho$
Radius of equivalent sphere	2.64×10^{-7} cm	$R_{\text{sph}}=(3V/4\pi)^{1/3}$
f_0	4.98×10^{-8} g s^{-1}	(11)
f/f_0	1.18	—
$m_{1,b}/m_2$, if spherical	0.48 g H_2O (g protein)$^{-1}$	(42)
a/b if unsolvated and prolate	4.0	(43)
a/b if unsolvated and oblate	0.24	(44)

[a] Data from Cohn and Edsall [2].

was bound throughout the entire particle in proportion to the mass of the particle—the ratio of friction factors is given by

$$\frac{f}{f_0}=\frac{6\pi\eta R}{6\pi\eta R_0}=\left(\frac{V_{\text{solv}}}{V_{\text{unsolv}}}\right)^{1/3}=\left(1+\frac{m_{1,b}\rho_2}{m_2\rho_1}\right)^{1/3} \tag{42}$$

Using the bulk densities for protein and water, we obtain a value for f/f_0 of 1.18—the ratio for human hemoglobin—for spheres hydrated in this mode to the extent of 0.48 g water (g protein)$^{-1}$.

F. Perrin derived expressions for the ratio f/f_0 for ellipsoids of revolution in terms of the ratio of the equatorial semiaxis to the semiaxis of revolution, b/a. The following expressions were obtained:

1. For prolate ellipsoids ($b/a<1$):

$$\frac{f}{f_0}=\frac{\left[1-\left(\frac{b}{a}\right)^2\right]^{1/2}}{\left\{\left(\frac{b}{a}\right)^{2/3}\ln\frac{1+(1-[b/a]^2)^{1/2}}{b/a}\right\}} \tag{43}$$

2. For oblate ellipsoids ($b/a>1$):

$$\frac{f}{f_0}=\frac{\left[\left(\frac{b}{a}\right)^2-1\right]^{1/2}}{\left(\frac{b}{a}\right)^{2/3}\tan^{-1}\left[\left(\frac{b}{a}\right)^2-1\right]^{1/2}} \tag{44}$$

These expressions may also be used to account for an observed f/f_0 ratio. The f/f_0 value listed in Table 3.3 for human hemoglobin corresponds either to an unhydrated prolate ellipsoid of axial ratio $a/b = 4.0$ or to an unhydrated oblate ellipsoid with $a/b = 0.24$.

The effects of solvation and asymmetry on intrinsic viscosity are similar to the effects they have on f/f_0. As with viscosity, intermediate cases between solvated spheres and unsolvated ellipsoids can also be calculated which are consistent with a given axial ratio. Figure 3.6 is a plot of possible combinations of hydration and asymmetry for protein particles in water. Similar curves could be drawn for other materials as well. For the human hemoglobin molecule discussed in Table 3.3, the combination of sedimentation and diffusion measurements gives an f/f_0 value which lies within the domain defined by the 1.15 and 1.20 contours of Fig. 3.6. The current picture of the structure of hemoglobin, deduced from x-ray diffraction studies, suggests that the molecule may be regarded as an ellipsoid with height, width, and depth equal to 64, 55, and 50 Å, respectively. Applying these dimensions to the dispersed unit leads us to describe the particle as being hydrated to the extent of about 0.4 to 0.5 g water (g protein)$^{-1}$.

Sedimentation and diffusion data allow for the unambiguous determination of particle mass and also allow the suspended particles to be placed on a contour in a plot such as that of Fig. 3.6. This is as far as these experiments can take us toward the characterization of the particles. Of course, additional data from other sources, such as the x-ray diffraction results just cited, may lead to still further specification of the system. One such source of information is intrinsic viscosity data for the same dispersion.

FIGURE 3.6 *Variation of the ratio f/f_0 with asymmetry and hydration for aqueous protein dispersions. [Redrawn from L. Oncley,* Ann. N.Y. Acad. Sci. **41**:121 (1941), *used with permission.]*

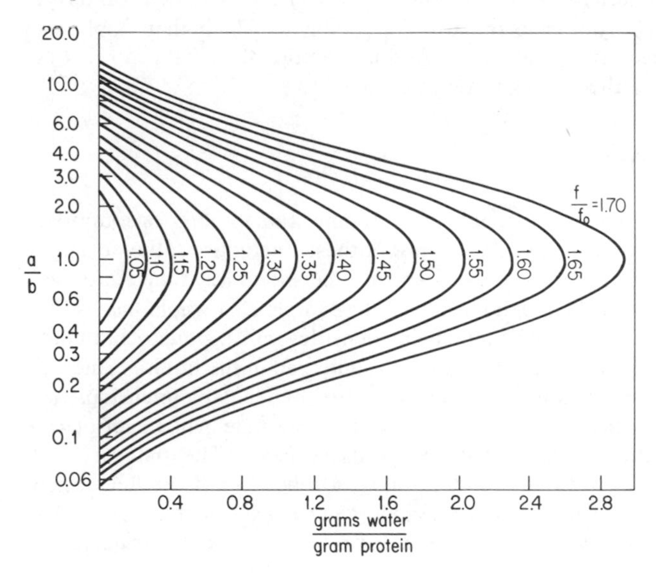

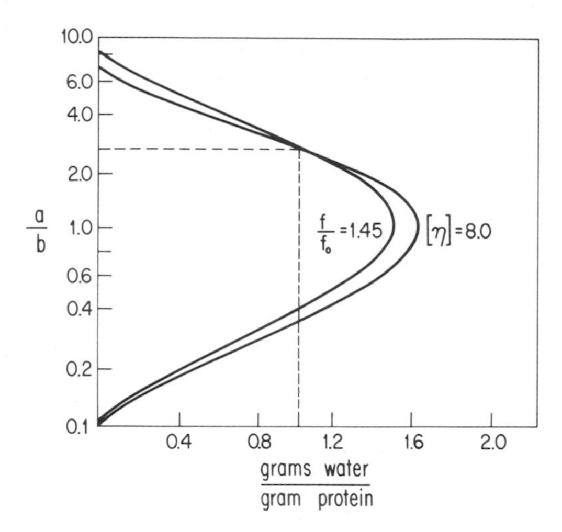

FIGURE 3.7 *Superposition of the* $[\eta]=8.0$ *contour from Fig.* 2.15 *and the* $f/f_0=1.45$ *contour from Fig.* 3.6. *Crossover characterizes particles unambiguously with respect to hydration and axial ratio.*

The complementarity between viscosity and sedimentation-diffusion is fairly evident from a comparison of Figs. 2.15 and 3.6. Let us suppose for the moment that we have evaluated f/f_0 by the six-step calculation just outlined. This places us on one of the contours of Fig. 3.6. If we also have intrinsic viscosity data on the same system, we can also identify the relevant contour in Fig. 2.15. Figure 3.7 shows how two lines from these different sets of contours might overlap. The two contours in Fig. 3.7 were selected arbitrarily, but for the system to which they apply they characterize the particles unambiguously. In this example, the axial ratio is about 2.5 and the hydration is about 1.0 g water (g protein)$^{-1}$.

3.9 THE RANDOM WALK

A liquid that is totally homogeneous on a macroscopic scale undergoes continuous fluctuations at the molecular level. As a result of these fluctuations, the density of molecules at any location in the liquid varies with time and at any time varies with location in such a way that the mean density of the sample as a whole has its bulk value. This pattern of "flickering" molecular densities will produce continually varying pressures on the surface of any particle submerged in the liquid. Since the fluctuations are confined to domains of the order of molecular dimensions, this randomly variable pressure is quite small. A small particle will be displaced, however, by the resulting force unbalance at its surface. The pattern of its displacements will also be a totally random thing, a reflection of the fluctuations which cause the motion. As the size of the submerged particles increases, the effect of fluctuations on them decreases. In this section, we shall consider the trajectory of a particle in the colloidal size range which is engaged in this pattern of motion. Such movements have been studied microscopically and are called Brownian motion after Robert Brown who described them in 1828.

In general, a dispersed particle is free to move in all three dimensions. For the present, however, we are going to restrict our consideration to the motion of a particle undergoing random displacements in one dimension only. The model used to describe this motion is called a one-dimensional random walk. Its generalization to three dimensions is taken up in a later section.

To begin with, the statistical nature of this phenomenon should be apparent. We might watch the pattern of displacements of thousands of otherwise identical particles, and find no uniformity in the zigzag steps they follow. Only statistical quantities such as the "average" displacement after a certain number of steps or after a certain elapsed time make any sense. Let us consider how to calculate such a quantity.

Suppose we consider a game in which a marker is moved back and forth along a line, say the x axis, in a direction determined by the toss of a coin. The rules of the game provide that we move the marker one distance l in the plus direction every time "heads" is tossed, and one distance l in the negative direction every time "tails" is tossed. If n_+ and n_- represent the number of heads and tails, respectively, in a game consisting of a total of n tosses, then we may write

$$n_+ + n_- = n \tag{45}$$

We may also write

$$x = (n_+ - n_-)l \tag{46}$$

where x is the net displacement of the marker after n tosses. These two equations may be solved simultaneously for n_+ and n_- to give

$$n_+ = \tfrac{1}{2}\left(n + \frac{x}{l}\right) \tag{47}$$

and

$$n_- = \tfrac{1}{2}\left(n - \frac{x}{l}\right) \tag{48}$$

The problem of interest may now be expressed as follows: What is the probability $P(n, x)$ that the marker will be at position x after n moves? The answer to this problem is supplied by the well-known binomial distribution formula:

$$P(n, x) = \frac{n!}{n_+!\,n_-!} p_+^{n_+} p_-^{n_-} \tag{49}$$

in which p_+ and p_- represent the probability of throwing a head or a tail, respectively, in a single toss. Although this formula is often used, its validity is not always fully appreciated. Accordingly, let us briefly examine the origin of Eq. (49).

Each time we toss an unbiased coin, the outcome is independent of all other tosses. The probability of tossing n_+ heads is therefore $p_+^{n_+}$, since the probabilities compound by multiplication when we require a specified set of outcomes. Therefore, n_+ events, each of which occur with a probability p_+, will occur with a probability $p_+^{n_+}$. If we specify the outcome further by requiring n_- tails in addition to the heads already specified, then the probability is given by $p_+^{n_+}p_-^{n_-}$. What we have calculated is the probability of a particular, fully specified sequence of outcomes such as the following for $n = 6$: HHHTTT. However, the same net number of heads and tails could come about in other ways. For example, HTTTHH, HHTHTT, and HTHTHT are also consistent with $n = 6$ and $n_+ = n_- = 3$, along with the distribution previously given and other possibilities. Therefore, we must multiply the probability of one specified sequence by the number of other sequences which have the same net composition of heads and tails.

At first glance, this factor may appear to be $n!$, the number of different ways the n items can be rearranged, giving a probability of $n!p_+^{n_+}p_-^{n_-}$. This quantity overcounts, however, since not all rearrangements are recognizably different. For example, we can interchange the first two members of the sequence TTHHHT to produce the sequence TTHHHT; the interchange obviously leaves the sequence unaltered. Therefore, we must divide $n!p_+^{n_+}p_-^{n_-}$ by the number of ways that identical members of the series can be interchanged without altering the sequence. Any of the n_+ heads (or the n_- tails) may be interchanged among themselves with this result. There are $n_+!$ permutations of the heads and $n_-!$ permutations of the tails. Therefore, these are the factors by which the previous result overcounted. Dividing by these factors leaves us with the binomial equation. The student should be so familiar with this argument as to be able to justify each factor in Eq. (49).

Since the probability of tossing either a head or a tail is equal to $\frac{1}{2}$ for a fair coin, Eq. (49) may be rewritten

$$P(n, x) = \frac{n!}{[\frac{1}{2}(n + x/l)]![\frac{1}{2}(n - x/l)]!}\left(\frac{1}{2}\right)^n \tag{50}$$

by incorporating Eqs. (47) and (48). The length l by which the marker is moved in our hypothetical game is equivalent to the displacement of a particle in a single fluctuation. Because of the nature of the fluctuations underlying the whole process, these individual steps are very small. Observable diffusion is always the result of a very large number of such steps. For the case in which n is a large number, the factorials of Eq. (50) may be expanded according to Sterling's approximation:

$$\ln y! = y \ln y - y \tag{51}$$

which is valid for large values of y. Taking the logarithm of Eq. (50) and applying approximation (51) leads to the result

$$-\ln P = \frac{nl + x}{2l} \ln\left(1 + \frac{x}{nl}\right) + \frac{nl - x}{2l} \ln\left(1 - \frac{x}{nl}\right) \tag{52}$$

The *net* displacement x is always small compared to the total distance traveled nl, since a good deal of the back-and-forth motion cancels out. Therefore, the logarithmic terms in Eq. (52) may be expanded as power series (see Appendix A) in x/nl with all terms higher than second order in x/nl disregarded as negligible. This leads to the result

$$\ln P \simeq -\frac{x^2}{2nl^2} + \cdots \tag{53}$$

or

$$P(n, x) = k \exp\left(-\frac{x^2}{2nl^2}\right) \tag{54}$$

where the factor k represents a normalization constant.

When we discussed normalization in Sec. 1.11 we saw that a well-behaved probability function adds up to unity when the probabilities for all possible outcomes are totaled. Therefore, in order to evaluate the constant k in Eq. (54) we should integrate (54) over all possible values of x—which is equivalent to summing the probabilities—and set the result equal to unity. This leads to the expression

$$k = \left[\int_{-\infty}^{\infty} \exp\left(-\frac{x^2}{2nl^2}\right) dx\right]^{-1} \tag{55}$$

which may be evaluated from a table of integrals* to give

$$k = (2\pi nl^2)^{-1/2} \tag{56}$$

Substitution of this result into Eq. (54) gives a continuous analytical expression for $P(n, x)$ which equals the binomial result for large values of n:

$$P(n, x) = (2\pi nl^2)^{-1/2} \exp\left(-\frac{x^2}{2nl^2}\right) \tag{57}$$

Strictly speaking, we should multiply both sides of Eq. (57) by dx. The equation now expresses the probability of a displacement between x and $x + dx$ after n random steps of length l.

3.10 RANDOM WALK STATISTICS AND THE DIFFUSION COEFFICIENT

The application of Eq. (57) to at least one set of boundary conditions for a diffusion problem is easy. We shall let a modification of Fig. 3.5a describe the system. This time, instead of having a solution on one side of the barrier and pure solvent on the other side, suppose we imagine that both sides of the barrier are filled with solvent. Furthermore, suppose that the solute under investigation is introduced into the system in the pores of the plug assuming the latter to be infinitesimally thin. Thus, at the beginning of the experiment, all the material is present at $x = 0$, in the notation of the derivation above, at a concentration c_0. With the passage of time, the material will gradually diffuse outward; the number of diffusion steps taken will be directly proportional to the elapsed time:

$$n = Kt \tag{58}$$

* This integral is a gamma function. A table of gamma functions is included in Appendix B.

Equation (57) may be transformed into an expression that gives the probability as a function of x and t by substituting (58) into (57):

$$P(x, t)\,dx = (2\pi Ktl^2)^{-1/2} \exp\left(-\frac{x^2}{2Ktl^2}\right) dx \tag{59}$$

This suggests that the concentration as a function of x and t in the diffusion cell just described is given by multiplying c_0 by $P(x, t)$:

$$c(x, t)\,dt = c_0(2\pi Ktl^2)^{-1/2} \exp\left(-\frac{x^2}{2Ktl^2}\right) dx \tag{60}$$

Since $P(x, t)$ is normalized, its integral over all values of x equals unity. Likewise, integrating Eq. (60) over all values of x gives c_0: The same quantity of solute is present at all times whether it is concentrated at the origin or is spread out from $-x$ to $+x$. The function given by Eq. (60) is a solution to Fick's second law [Eq. (31)] for a one-dimensional problem in which all the material is present initially at $x = 0$ and at a concentration c_0.

The only ambiguity remaining at this point is the value of K in Eqs. (58) to (60). This constant is easily evaluated as follows. If Eq. (60) is, indeed, a solution to (31), then the right- and left-hand sides of the latter must be equal when the indicated differentiations are carried out. Differentiating Eq. (60), substituting into (31), and simplifying leads to the result

$$K = \frac{2D}{l^2} \tag{61}$$

Substituting this expression into Eq. (60) yields

$$\frac{c}{c_0}\,dx = (4\pi Dt)^{-1/2} \exp\left(-\frac{x^2}{4Dt}\right) dx \tag{62}$$

This relationship describes the diffusion of solute in the x direction when it is concentrated initially in an infinitesimally thin layer at the origin. Figure 3.8 shows how the relative concentration c/c_0 varies with time and position according to Eq. (62). The approach to equilibrium is evidenced by the fact that the concentration gradually becomes more uniform as time increases.

Equation (62) may be devloped somewhat further as follows. We define a parameter z to be

$$z = \frac{x}{(2Dt)^{1/2}} \tag{63}$$

then rewrite Eq. (62) in terms of this quantity:

$$\frac{c}{c_0}\,dx = \frac{1}{\sqrt{2\pi}} \exp\left(\frac{-z^2}{2}\right) dz = P(z)\,dz \tag{64}$$

Thus, the concentration ratio c/c_0 is seen to be described at all times as a function of the single parameter z. The function $P(z)$ defined by Eq. (64) is the normal

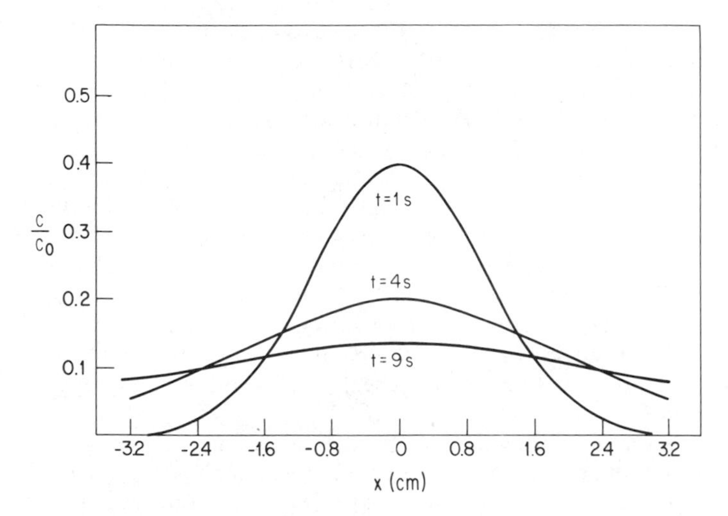

FIGURE 3.8 *Variation of* c/c_0 *with distance from origin at* 1, 4, *and* 9 *s if* $D = \frac{1}{2}\ \text{cm}^2\ \text{s}^{-1}$. *Drawn according to Eq.* (62).

distribution function, Eq. (1.27). In Chap. 1, the function z is defined to be δ/σ where δ is the deviation of a particular value from the mean of a distribution and σ is the standard deviation of the distribution. Since the net displacement of a diffusing particle is analogous to δ, we may infer that the quantity $\sqrt{2Dt}$ is also analogous to the standard deviation. The point of this is the following. It is well known that the width of the normal error curve at its inflection point (where the second derivative changes sign or the curve changes from concave to convex) is equal to σ. Therefore, we can conclude that the width of the curves shown in Fig. 3.8, measured at their inflection points, increases in proportion to $\sqrt{t}$. This is one method whereby D could be measured in an experiment which corresponds to these boundary conditions.

With the mathematics of the one-dimensional random walk as background, we may visualize the following experimental arrangement whereby D could be measured. Suppose that a narrow band of dispersion is layered between two portions of solvent in a long tube. We shall not worry about the practical difficulties in doing this since, in practice, other initial conditions which are easily obtained are actually used. The narrow band we have pictured, however, approximates the initial state of the system described by Eq. (62). We might further imagine observing this band by means of a Schlieren optical system (Sec. 3.5). The two edges of the band, where the refractive index gradient is large, would define the positively and negatively sloped branches of the Schlieren peak. With the passage of time, the material in the band diffuses outward; the Schlieren peak would also broaden. In other words, the Schlieren pattern observed at successive times would generate a family of curves which very much resemble the theoretical curves shown in Fig. 3.8.

The curves in Fig. 3.8 are drawn for an arbitrary value of the diffusion coefficient. The experimental profiles produced by the Schlieren optics are characterized by the diffusion coefficient of the experimental system. The remaining question is how to extract the appropriate D value from the experimental observations.

Remember that the normal distribution function has an inflection point (where the second derivative changes sign) at $z = 1.0$. Therefore, the x value at which the inflection point occurs at any time equals $\sqrt{2Dt}$ according to Eq. (63). By locating the inflection point at different times during a diffusion experiment, the appropriate D value may be evaluated for the diffusing species. For example, on the $t = 1$ s curve in Fig. 3.8, the inflection point lies at $x = 1.0$ cm. Substituting $x = 1.0$ cm and $t = 1.0$ s when $z = 1$ into Eq. (63) enables us to calculate the value of the diffusion coefficient that was used ($D = \frac{1}{2}$ cm^2 s^{-1}) to draw the curves in Fig. 3.8. A similar analysis may be conducted on experimental curves obtained by the Schlieren method.

As an experimental procedure, this method is less precise than others which have been developed for the evaluation of D. It does point out, however, the intimate connection between diffusion and the random events at the molecular level which cause it.

A more practical experimental method for the determination of D is based on Fig. 3.5b. The theoretical arrangement represented by Fig. 3.5b is implemented experimentally in an apparatus like that sketched in Fig. 3.9. One side of the sintered glass barrier contains solution, the other side contains pure solvent. The entire apparatus is thermostated and both compartments are magnetically stirred. Samples are withdrawn at various times and the quantity of material that has diffused into the solvent compartment is measured. What is obtained experimentally, therefore, is a record of the approach toward a uniform distribution of material on both sides of the barrier.

The situation represented by the apparatus in Fig. 3.9 has also been analyzed theoretically. The function $c(x, t)$ which satisfies Fick's second law when there is a solution of concentration c_0 on one side of a boundary at $t = 0$ and pure solvent on the other side is given by

$$c = \frac{c_0}{2}\left[1 - 2\int_0^z P(z)\,dz\right] = \frac{c_0}{2}[1 - 2\,\mathrm{Erf}\,(z)] \tag{65}$$

We can verify the plausibility of this expressions as follows. Recall that z and t are

FIGURE 3.9 *Laboratory apparatus equivalent to Fig. 3.5b. Entire apparatus is rotated between the poles of a magnet for stirring.*

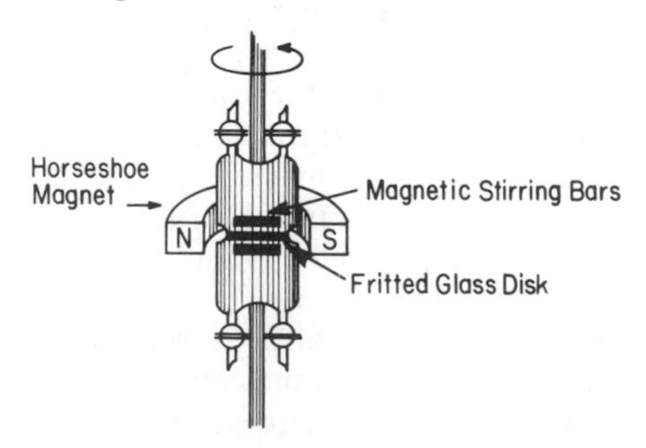

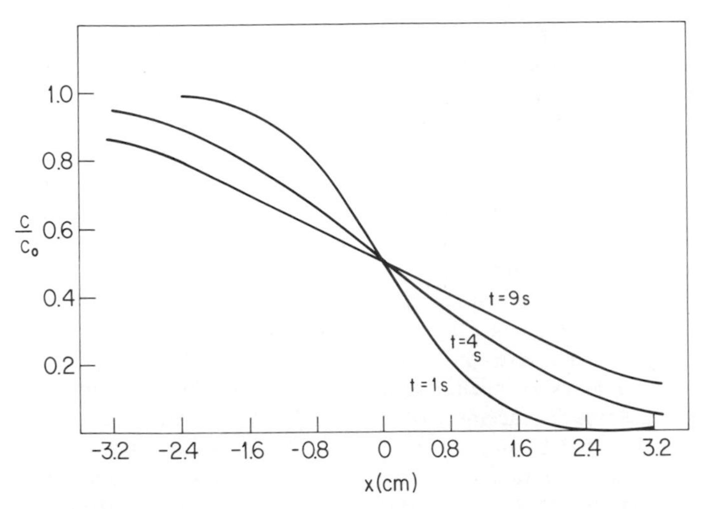

FIGURE 3.10 *The diffusion of solute from solution into pure solvent.* c/c_0 *plotted versus distance at* 1, 4, *and* 9 *s for* $D = \frac{1}{2}\ \text{cm}^2\ \text{s}^{-1}$. *Drawn according to Eq.* (65).

inversely related and that the integral gives the area under one-half of the error curve, between its midpoint ($z = 0$) and some specified value of z. Therefore, at $t = 0$, z is infinite, the integral equals $\frac{1}{2}$, and $c = 0$. At $t = \infty$, z is zero, the integral equals zero, and $c = \frac{1}{2}c_0$. Thus, Eq. (65) makes sense at either extreme. The detailed profile of c/c_0 versus z is obtained by reading values of the integral in Eq. (65) from tables which give the area under the normal error curve. The results of this procedure are plotted in Fig. 3.10 where c/c_0 is shown versus x at several different times. The approach toward a uniform distribution of material is evident.

As in the case of diffusion from an initially thin layer, experimental concentration data and theoretical concentration profiles may be compared. From this comparison, the value of D which is consistent with the observed diffusion behavior may be evaluated. The diffusion coefficient is a function of concentration; therefore, it is measured at a series of different concentrations and extrapolated to zero concentration.

The primary difficulty with this procedure is the tendency of the pores to plug due to the entrapment of air bubbles, clogging by solid particles, or adsorption of the diffusing molecules themselves.

3.11 "AVERAGE" DISPLACEMENTS FROM RANDOM WALK STATISTICS

Having examined the connection between the phenomenological equations of diffusion and the statistics of the random walk, let us now return to the random walk as a model for Brownian motion. The problem we wish to consider is the "average" displacement of a marker after an n-step, one-dimensional random walk. The foregoing discussion already supplies the answer to this problem: We have seen that the probability function for the one-dimensional walk is symmetrical about the origin, implying a mean displacement of zero. This simply reflects the fact that on the average the number of heads and tails will be equal. While we accept this result, we feel that there is something unsatisfactory about it. The average displacement is zero because positive and negative displacements are equally probable and effectively cancel one another. It is not because the marker scarcely moves from its initial location, a conclusion suggested by the value of the mean displacement.

This shows that the mean displacement is simply not a useful parameter to describe the trajectory of the particle, and suggests we should seek an alternate quantity. Instead of averaging the displacements directly, suppose we first square them to eliminate the differences in sign, then average and take the square root. This quantity, called the root mean square (rms) displacement, will give a better measure of the meanderings of the marker since the sign differences have been eliminated.

The calculation of the rms displacement is quite simple. We recall from Sec. 1.9 that any average is calculated by multiplying the quantity to be averaged by an appropriate probability function, then integrating the result over all possible values of the variable. Applying this procedure to the problem at hand gives the result

$$\overline{x^2} = \int_{-\infty}^{\infty} x^2 P(n, x)\, dx \tag{66}$$

in which $P(n, x)$ is given by eq. (57). Making this substitution gives

$$\overline{x^2} = (4\pi Dt)^{-1/2} \int_{-\infty}^{\infty} x^2 \exp\left(\frac{-x^2}{4Dt}\right) dx \tag{67}$$

the value of which is found in the tables of Appendix B. Evaluating the integral leads to the conclusion that

$$\overline{x^2} = 2Dt \tag{68}$$

This important equation, also derived by Einstein, provides us with a means for measuring the diffusion coefficient for particles which are visible in a miscroscope. This is a particle size range for which measurement of D by following concentration changes with time is very difficult. Instead what is done is to measure microscopically the actual displacement of a particle in a time t. The rms displacement, evaluated from a statistically meaningful number of observations, permits D to be calculated. J. Perrin (Nobel Prize, 1926) interpreted observations of Brownian motion in terms of Eqs. (11), (37), and (68) as a means of determining the first precise value of Avogadro's number.

Equation (68) also permits us to assign a physical interpretation to the diffusion coefficient in addition to the macroscopic meaning it has from Fick's laws. Rearranging and factoring in a way that admittedly ignores the averaging procedure, we write Eq. (68) as

$$D = \frac{1}{2}\left(\frac{x}{t}\right)x \tag{69}$$

Since D is a constant for a particular substance, the ratio x/t must vary inversely with x. The distance traveled by a diffusing particle divided by the time required for the displacement, x/t, gives the apparent diffusion velocity of the particle. Equation (69) shows that the diffusion velocity is inversely proportional to the length of path over which it is measured. This apparently paradoxical conclusion becomes less mysterious when we concentrate on the distinction between the diffusion velocity and the actual velocity of the particle as it travels along its zigzag path. The latter quantity reflects the average translational kinetic energy of the particle; its average value depends on the absolute temperature but is independent of the distance traveled. The diffusion velocity, on the other hand, decreases as the distance traveled increases. This simply means that large displacements are so much less probable than small displacements that they require disproportionately longer times to occur. Note that if $x = 1$ cm, D has the significance of being equal to half the diffusion velocity measured over that distance. This is a result that was anticipated at the end of Sec. 3.7.

3.12 THE RANDOM COIL AND RANDOM WALK STATISTICS: A DIGRESSION

The dimensions of a randomly coiled polymer molecule are a topic which appears to bear no relationship to diffusion. However, both the coil dimensions and diffusion can be analyzed in terms of random walk statistics. Therefore, we may take advantage of the statistical arguments we have developed to consider this problem.

Suppose we visualize a polymer molecule as consisting of n segments of length l, connected by a completely flexible linkage at each joint. Imagine, furthermore, that the placement of each successive segment is determined by some sort of purely random criterion. We could anchor one end of the chain to the origin of a hypothetical coordinate system, for example, and position successive segments as follows. Suppose we toss a single die and agree to orient the next segment in the $+x$ direction if the die shows a 1, and in the $-x$ direction if a 6 shows. Likewise, 2, 5, 3, and 4 correspond to $+y$, $-y$, $+z$, and $-z$, respectively. Our problem is this: *On the average*, what will be the end-to-end distance $\bar{R}$ for a chain of n segments? To calculate this, we must assume that n is very large and must ignore the sites excluded by previously positioned segments.

We recognize first that the end-to-end distance can be resolved into x, y, and z components which obey the relationship

$$R^2 = x^2 + y^2 + z^2 \tag{70}$$

The probability that the loose end of the chain is in a volume element located at x, y, and z is given by

$$P(x, y, z)\,dx\,dy\,dz = P(x)P(y)P(z)\,dx\,dy\,dz \tag{71}$$

where $P(x)$ is the probability that the x coordinate has a value between x and $x + dx$, with similar definitions for $P(y)$ and $P(z)$. Equation (57) gives the expression for $P(x)$, except we must remember that now only $n/3$ segments will be aligned with the x axis since each of the directions is independent and equally probable. Since the x, y, and z directions are equivalent, the same expression holds for $P(y)$ and $P(z)$ with the appropriate change of variables. Therefore, incorporating Eq. (57), Eq. (71) becomes

$$P(x, y, z)\,dx\,dy\,dz = \left(2\pi\frac{n}{3}l^2\right)^{-3/2} \exp\left(\frac{-3R^2}{2nl^2}\right) dx\,dy\,dz \tag{72}$$

This expression gives the probability that the loose end of the molecule is in a volume element located at some particular values of x, y, and z, as shown in Fig. 3.11a. Our interest is not in any *specific* x, y, z coordinates, but in all combinations of x, y, z coordinates which result in the end of the chain being a distance R from the origin. This can be evaluated by changing the volume element in Eq. (72) to spherical coordinates, then integrating over all angles. This amounts to replacing the volume element in Eq. (72) with the volume of a spherical shell of radius R and thickness dR:

$$dx\,dy\,dz \rightarrow 4\pi R^2\,dR \tag{73}$$

A geometrical representation of this situation is shown in Fig. 3.11b. Incorporating this expression into Eq. (72) gives the probability that the loose end of the molecule lies between R and $R + dR$, irrespective of the direction:

$$P(R)\,dR = 4\pi\left(2\pi\frac{n}{3}l^2\right)^{-3/2} R^2 \exp\left(\frac{-3R^2}{2nl^2}\right) dR \tag{74}$$

FIGURE 3.11 *Coordinate systems in which one end of a flexible random coil lies at the origin and the other end is (a) in a volume element dx dy dz and (b) in a spherical volume element $4\pi R^2\,dR$.*

(a)

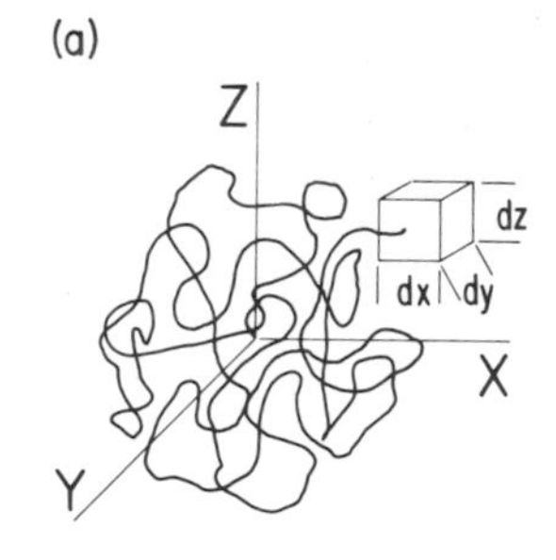

(b)

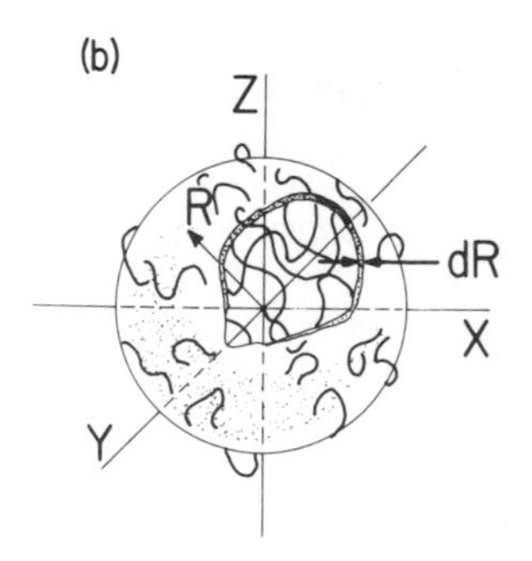

With this distribution function, it is an easy matter to calculate average quantities. Again, since positive and negative displacements are equally probable, we shall evaluate the average of R^2. Following the same procedure as used in Eq. (66), we write

$$\overline{R^2} = \left(\frac{2\pi n l^2}{3}\right)^{-3/2} 4\pi \int_0^\infty R^4 \exp\left(\frac{-3R^2}{2nl^2}\right) dR \tag{75}$$

Evaluation of this integral (Appendix B) leads to the result

$$\overline{R^2} = nl^2 \tag{76}$$

which is the desired quantity.

Equation (76) shows that the rms end-to-end distance in a polymer coil equals the square root of the number of steps times the length of each step. We might ask, therefore, what is the physical significance of n and l for a polymer molecule? A general formula for a typical vinyl polymer is

$$\left(\begin{array}{c} \text{H}\quad\text{H} \\ |\quad\;\;| \\ -\text{C}-\text{C}- \\ |\quad\;\;| \\ \text{H}\quad\text{X} \end{array}\right)_n$$

where n is called the degree of polymerization. Since this quantity measures the number of repeated segments in the chain, it seems reasonable to equate this with the number of steps in the three-dimensional random walk. If M represents the molecular weight of the polymer and M_0 is the molecular weight of the vinyl monomer, then M/M_0 equals n. Since M_0 and l are constants, we see that the radius of the coil is predicted to be proportional to $M^{1/2}$. This result was anticipated in connection with Eq. (2.89).

The physical significance of l in Eq. (76) is somewhat harder to define. At first glance it appears to be the length of the repeating unit, about 2.5 Å for a vinyl polymer. We must remember, however, that the derivation of Eq. (76) assumed that the coil was connected by completely flexible joints. Molecular segments are attached at definite bond angles, however, so an actual molecule has less flexibility than the model assumes. Any restriction on the flexibility of a joint will lead to an increase in the dimensions of the coil. The effect of fixed bond angles on the dimensions of the chain may be incorporated into the model as follows.

A 360° rotation around any carbon–carbon bond in a vinyl chain will cause the *next* bond in the chain to trace out a cone with one carbon atom at the apex and the other carbon atom along the rim of the cone. Ignoring hindered rotation for the moment, we see that each position on the rim of such a cone is an equally probable site for the apex of cone generated by the next bond in the chain. This situation is illustrated schematically in Fig. 3.12. This effect has been shown to increase the actual length of the repeating unit, l_0, by the factor

$$l^2 = l_0^2\left(\frac{1-\cos\theta}{1+\cos\theta}\right) \tag{77}$$

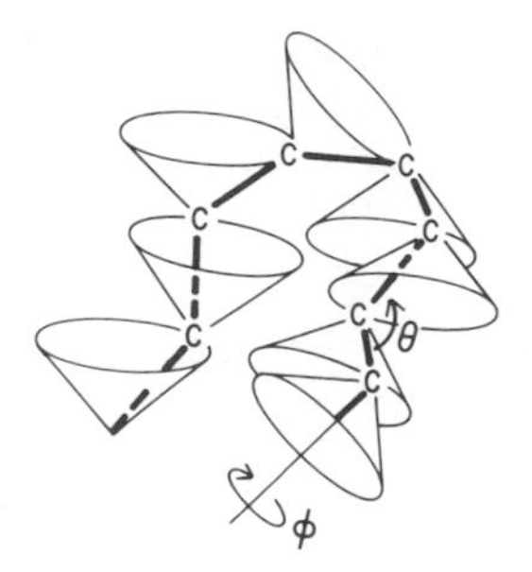

FIGURE 3.12 *The effect of fixed bond angle in restricting the flexibility of a polymer coil. For tetrahedral bonds,* $\theta = 109°$.

where θ is the bond angle. For a tetrahedral bond angle, $\cos\theta = -\frac{1}{3}$, so the additional factor in Eq. (77) equals 2.

In Fig. 3.12 it is implied that the terminal carbon atom can occupy any position on the rim of the cone. That is, there is assumed to be perfectly free rotation around the penultimate carbon–carbon bond. This is equivalent to saying that all values of ϕ, the angle that describes the rotation (see Fig. 3.12), are equally probable. Any hindrance to free rotation will block certain configurations, expanding the coil dimensions still further.

Finally, interactions between the polymer segments and the solvent molecules may introduce a bias which tends to position segments as close or as far as possible from other segments depending on the nature of the interaction. As a matter of fact, those properties of polymer solutions which are sensitive to the dimensions of the molecules, such as viscosity, vary widely with solvent "goodness." A good solvent may be defined as one in which polymer–solvent contacts are favored; a poor solvent is one in which polymer–polymer contacts are favored. Therefore, the value of l in Eq. (77) will be increased in a good solvent and decreased in a poor solvent.

In concluding this section we note that all the statistical equations of this chapter could have been borrowed directly from kinetic molecular theory by simply changing the variables. We illustrate this now by going in the opposite direction. For example, if we replace the quantity $3/nl^2$ by m/kT and replace R by v in Eq. (72), we obtain the Boltzmann distribution of molecular velocities in three dimensions. If we make the same substitutions in Eq. (76), we obtain an important result from kinetic molecular theory:

$$\overline{v^2} = \frac{3kT}{m} \tag{78}$$

This shows that Eq. (76) occupies a position for the random coil which is analogous to the position of average kinetic energy in the kinetic molecular theory. This is not just a fortuitous similarity, but a reflection of the statistical basis of both. A little self-test: Did you recognize the similarity between the formalisms of the last few sections and kinetic theory as we went along?

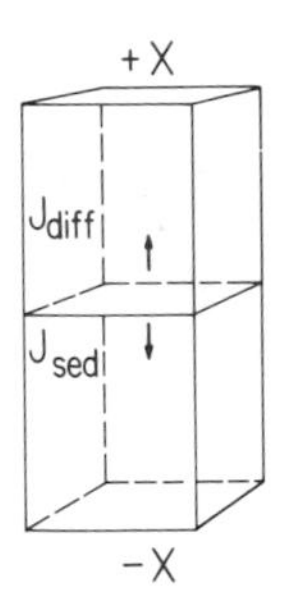

FIGURE 3.13 *The relationship between the flux due to sedimentation and that of diffusion. At equilibrium, the two are equal.*

3.13 EQUILIBRIUM BETWEEN SEDIMENTATION AND DIFFUSION

We have already noted that sedimentation and diffusion are opposing processes, the one tending to collect, the other to scatter. Let us now consider the circumstances under which these two tendencies equal each other. Once this condition is reached, of course, there will be no further macroscopic changes; the system is at equilibrium. In order to formulate this problem, consider the unit cross section shown in Fig. 3.13, in which the x direction is in the direction of either a gravitational or centrifugal field. Suppose this field tends to pull the particles in the $-x$ direction. Gradually the concentration of the particles will increase in the region below the cross section of interest.

Back-diffusion occurs at a rate which increases with the buildup of a concentration gradient. When equilibrium is finally reached, we may write

$$J_{\text{sed}} = J_{\text{diff}} \tag{79}$$

where J_{sed} is the flux across the area due to sedimentation and J_{diff} is the flux due to diffusion. The latter is given by Eq. (27) and the former by

$$J_{\text{sed}} = vc \tag{80}$$

in which v is the sedimentation velocity and c is the concentration at the plane. Substituting Eqs. (27) and (80) into (79) gives

$$vc = D\frac{dc}{dx} \tag{81}$$

If we substitute the value for the rate of sedimentation under gravity, Eq. (4), we obtain

$$\frac{m}{f}\left(1-\frac{\rho_1}{\rho_2}\right)g = D\frac{dc}{dx} \tag{82}$$

If the sedimentation occurs in a centrifugal field, on the other hand, g must be replaced by $\omega^2 x$ in Eq. (4):

$$\frac{m}{f}\left(1-\frac{\rho_1}{\rho_2}\right)\omega^2 x = D\frac{dc}{dx} \tag{83}$$

Equations (82) and (83) are easily integrated to produce expressions which give c as a function of x at equilibrium. Defining c_1 and c_2 to be the equilibrium concentrations at x_1 and x_2 and then integrating, we obtain for Eq. (82):

$$\frac{m}{fD}\left(1-\frac{\rho_1}{\rho_2}\right)g(x_2-x_1) = \ln\frac{c_2}{c_1} \tag{84}$$

and for (83):

$$\frac{m}{2fD}\left(1-\frac{\rho_1}{\rho_2}\right)\omega^2(x_2^2-x_1^2) = \ln\frac{c_2}{c_1} \tag{85}$$

Finally, we recall Eq. (37), which permits us to substitute kT for Df, giving (a) for sedimentation equilibrium under gravity:

$$\ln\frac{c_2}{c_1} = \frac{m}{kT}\left(1-\frac{\rho_1}{\rho_2}\right)g(x_2-x_1) \tag{86}$$

and (b) for sedimentation equilibrium in a centrifuge:

$$\ln\frac{c_2}{c_1} = \frac{m}{2kT}\left(1-\frac{\rho_1}{\rho_2}\right)\omega^2(x_2^2-x_1^2) \tag{87}$$

Note that sedimentation equilibrium studies permit the evaluation of particle mass with no assumptions about particle shape.

It should now be apparent why the ultracentrifuge is such an important tool in molecular biology. Although the method is in no way restricted to particles of biological significance, these particles are of a size and density which are ideally suited to the ultracentrifuge. An ultracentrifuge permits the evaluation of particle mass through equilibrium studies [Eq. (87)] and the evaluation of the ratio m/f through studies of the rate of sedimentation [Eq. (23)]. Combining these data permits the separate evaluation of f. From the mass and density of the material, the volume and radius of an equivalent sphere and, hence, f_0 can be calculated. Then Fig. 3.6 may be consulted to determine particle characterizations which are consistent with the ratio f/f_0. Although both utilize the same instrument, sedimentation rate and sedimentation equilibrium are two different experiments which complement one another very nicely.

It might be noted that sedimentation equilibrium is approached very slowly; however, techniques which permit equilibrium conditions to be estimated from pre-equilibrium measurements have been developed by W. J. Archibald. Equations (86) and (87) predict a linear semilogarithmic plot of c versus x or x^2, for gravitational and centrifugal studies, respectively. The slope of such a plot is proportional to the mass of the particles involved. Remember that monodispersity was assumed in the derivation of these equations. If this condition is not met for an experimental system, the plot just described will not be linear. If each particle size present is at equilibrium, however, each component will follow the equations and the experimental plot will be the summation of several straight lines. Under certain conditions these may be resolved to give information about the polydispersity of the system. In any event, nonlinearity implies polydispersity once true equilibrium is reached.

We conclude this chapter with a final observation about Eq. (86). If the particles in question are gas molecules instead of suspended particles, then the concentration ratio equals the ratio of pressures measured at two locations and the particle mass requires no correction for buoyancy. Under these conditions, Eq. (86) becomes

$$\ln \frac{p_2}{p_1} = \frac{mg(x_2 - x_1)}{kT} = \frac{Mg(x_2 - x_1)}{RT} \tag{88}$$

This familiar equation gives the variation of barometric pressure with elevation. Once again, we are reminded of the connection between the material of this chapter and kinetic molecular theory.

REFERENCES

1. R. B. Bird, W. E. Stewart, and E. N. Lightfoot, *Transport Phenomena*, Wiley, New York, 1960.
2. E. J. Cohn and J. T. Edsall, *Proteins, Amino Acids and Peptides*, American Chemical Society Monograph, reprinted by Hafner Publ., New York, 1965.
3. J. M. Dallavalle, *Micromeritics*, Pitman, New York, 1948.
4. A. Einstein, *Investigations in the Theory of the Brownian Movement*, Dover, New York, 1956.
5. K. J. Mysels, *Introduction to Colloid Chemistry*, Wiley-Interscience, New York, 1959.
6. T. Svedberg and K. O. Pederson, *The Ultracentrifuge*, Oxford University Press, London and New York, 1940.

PROBLEMS

1. The following results describe the rate of accumulation of rutile (TiO_2; $\rho = 4.2$ g cm^{-3}) particles on a submerged balance pan.*

Time (min)	W(g)	t (cont.)	W (cont.)
0	0	35	0.310
1.5	0.045	60	0.403
2.0	0.060	84	0.480
2.5	0.075	127	0.610
3.0	0.090	159	0.702
4.0	0.110	226	0.880
5.5	0.125	274	0.998
8.0	0.160	327	1.110
14	0.210	384	1.240
24	0.260	420	1.320

* W. F. Sullivan and A. E. Jacobson, *Symposium on Particle Size Measurement*, ASTM Publication No. 234, 1959.

(a) Plot W versus t and use this to evaluate w as a function of time. Prepare a plot of w versus t.

(b) At what time (t_{max}) does the greatest increment in w appear to occur?

(c) Calculate the radius of an equivalent sphere corresponding to t_{max} ($\rho_{soln} = 0.997\ g\ cm^{-3}$, $\eta = 0.00894$ P, $h = 12$ cm).

2. A preparation of reduced and carboxymethylated Mouse–Elberfeld virus protein particles reached sedimentation equilibrium after 40 h at 12,590 rpm.* The following data show the recorder displacement (proportional to concentration) versus r for this protein:

c (arbitrary units)	2.29	2.51	2.79	3.09	3.51	3.89	4.47	5.01	5.89	6.61	7.41	8.51
r (cm)	6.55	6.58	6.60	6.65	6.67	6.69	6.71	6.74	6.76	6.79	6.81	6.84

(a) Use these results to evaluate the mass of the particles present ($\rho_{protein} = 1.370\ g\ cm^{-3}$) and estimate R_0 and f_0 for the molecules. Does the sample appear to be monodisperse?

(b) The sedimentation coefficient is known as be 2.7S for this preparation. Evaluate f and f/f_0.

(c) What can be said about the possible axial ratio–hydration combinations of this protein in terms of Fig. 3.6?

3. Southern bean mosaic virus (SBMV) particles are centrifuged at 12,590 rpm and the absorbance at 260 nm is measured along the settling direction as a function of time. The center of the absorption band varies with distance from the center of the rotor as follows†:

t (min)	R (cm)
16	6.22
32	6.32
48	6.42
64	6.52
80	6.62
96	6.72
112	6.82
128	6.92
144	7.02

Calculate the sedimentation coefficient of SBMV particles from these results.

4. Phosphatidylcholine micelles are spherical particles having a molecular weight of 97,000 g mole^{-1}. Assuming that the density of the dry lipid ($\rho = 1.018\ g\ cm^{-3}$) applies to the micelles, calculate R and D for these particles in water at 20°C. The experimental value of the diffusion coefficient is 6.547 cm^2 s^{-1} under these conditions.‡

*R. R. Rueckert, *Virology* **26**: 345 (1965).

†J. Vinograd, R. Bruner, R. Kent, and J. Weigle, *Proc. Natl. Acad. Sci. U.S.* **49**:902 (1963).

‡H. Hasser, in *Water, a Comprehensive Treatise*, Vol. 4 (F. Franks, ed.), Plenum Press, New York, 1975, Chapter 4.

5. Calculate the diameter of a spherical particle ($\rho = 4$) for which the rms displacement due to diffusion at 25°C is 1% the distance of sedimentation in a 24-h period through a medium for which $\rho = 1$ g cm^{-3} and $\eta = 9 \times 10^{-3}$ P. For what diameter is the diffusion distance 10% of the settling distance?
6. The diffusion of alkyl ammonium ions into clay pellets has been studied by bringing the pellet into superficial contact with an isotopically labeled salt, and then, after a suitable time, using a microtome to slice the pellet. The radioactivity is then measured in successive thin slices of the pellet. Assuming that Eq. (65) describes the diffusion process, estimate how long it would take for 1% of the initial activity to appear in the 15th slice inward from the exposed surface of the dry clay if each slice is 40 μm thick. The diffusion coefficients for the methyl and trimethyl ammonium cations under these conditions are 7.03×10^{-12} and 2.65×10^{-11} cm^2 s^{-1}, respectively.*
7. Suppose two reservoirs 4 cm apart are cut into an agar gel in a Petri dish. Solutions of $Pb(NO_3)_2$ and Na_2CrO_4 are introduced simultaneously into the two reservoirs. At what distance into the gel does $PbCrO_4$ precipitate if the diffusion coefficients of Pb^{2+} and CrO_4^{2-} in agar are 0.657×10^{-5} and 0.752×10^{-5} cm^2 s^{-1}, respectively?† Where would Prussian blue precipitate if the reservoirs contained Fe^{3+} ($D = 0.434 \times 10^{-5}$ cm^2 s^{-1}) and $Fe(CN)_6^{4-}$ ($D = 0.557 \times 10^{-5}$ cm^2 s^{-1})?
8. Use a compass to inscribe a circle around the floc in Fig. 1.11 such as a graticule might be used to define an average diameter for the floc. Compare the diameter of the inscribed circle to the diameter of the primary particles in the floc. How does the ratio of these radii compare with the value that would be obtained for the case $n = 76$ if the floc were built up according to a random walk model? Compare the features of the two models with the objective of accounting for the relative magnitude of the floc radius relative to the primary particle radius in the two cases.
9. The molecular weights and sedimentation coefficients of human plasminogen and plasmin ($\rho = 1.40$ g cm^{-3}) are shown below‡:

	Plasminogen	Plasmin
M (g mole^{-1})	81,000	75,400
s at 20°C (S)	4.2	3.9

(a) Calculate the diffusion coefficient for each.
(b) Prepare a graph such as that of Fig. 3.8 showing quantitatively how an initially thin band of these proteins widens with time. Show at least three different times.

10. Colloids (casein micelles) of two different particle sizes are isolated from skim milk by centrifugation under different conditions. The sedimentation and diffusion coefficients of the two preparations are as follows§:

* R. G. Gast and M. M. Morfland, *J. Colloid Interface Sci.* **37**:80 (1971).
† *R. E. Lee and F. R. Meeks, J. Colloid Interface Sci.* **35**:584 (1971).
‡ G. H. Barlow, L. Summaria, and K. C. Robbins, *J. Biol. Chem.* **244**:1138 (1969).
§ C. V. Morr, S. H. C. Lin, R. K. Dewan, and V. A. Bloomfield, *J. Dairy Sci.* **56**:415 (1973).

Preparation conditions	$D \times 10^8$ (cm^2 s^{-1})	$s \times 10^{13}$ (s^{-1})
5 min at 5,000 rpm	0.97	2200
40 min at 20,000 rpm	2.82	800

Calculate the mass per particle and the gram molecular weight of the two micelle fractions, assuming $\rho = 1.43$ g cm^{-3} for the dispersed phase.

11. The following data give the number of gold particles (as $\log_{10}$) versus depth beneath the surface for an aqueous dispersion allowed to reach sedimentation equilibrium under the influence of gravity*.

depth (mm)	4.44	5.06	5.67	6.30	6.90	7.53	8.15	8.65
log n	10.36	10.51	10.63	10.75	10.89	11.05	11.22	11.39

Calculate the radius of the gold particles ($\rho_{Au} = 19.3$ g cm^{-3}), treating the dispersed units as equivalent spheres.

12. At equilibrium at 20°C, the concentration of tobacco mosaic virus (TMV) shows a linear semilogarithmic graph when plotted against the square of the distance to the axis of rotation in a centrifuge. Evaluate the molecular weight of the TMV particles if $\ln c = -1.7$ at $r^2 = 44.0$ cm^2 and -2.8 at 41.2 cm^2 when the dispersion is centrifuged at 6.185 rps. The density of the TMV is 1.36 g cm^{-3}.†

13. Verify that the expansion (see Appendix A) of Eq. (52) leads to (53). Verify that the integration (see Appendix B) of Eqs. (67) and (75) leads to (68) and (76), respectively.

* C. M. McDowell and F. L. Usher, *Proc. Roy. Soc. London* **138A**:133 (1932).
† F. N. Weber, Jr., R. M. Elton, H. G. Kim, R. D. Rose, R. L. Steere, and D. W. Kupke, *Science* **140**:1090 (1963).

Osmotic and Donnan Equilibrium

All faults or defects, . . . Pantocyclus attributed to some deviation from perfect Regularity in the bodily figure, caused perhaps . . . by some collision in a crowd; by neglect to take exercise, or by taking too much of it; or even by a sudden change of temperature

[From Abbott's *Flatland*]

4.1 INTRODUCTION

This chapter is concerned primarily with the osmotic pressure of colloidal systems. There are two major reasons for examining osmotic phenomena in considerable detail. First, in the systems to which it is applicable, osmotic pressure provides one of the least ambiguous ways available for characterizing dispersed particles, especially with respect to molecular weight. Second, much of the material developed in this chapter will also be applied to light scattering (in Chap. 5) and to monolayers (in Chap. 7).

Osmotic pressure is already a familiar topic to most students in a course in colloid and surface chemistry. It is easily included in general chemistry courses because of the resemblance between the osmotic pressure of a dilute solution and the pressure of an ideal gas. Likewise, it is discussed in physical chemistry as a straightforward application of solution thermodynamics. In this chapter, these aspects are reviewed again, but presumably in greater depth than in an ordinary physical chemistry course.

We shall begin our discussion of osmotic pressure by considering this property in an ideal solution, particularly as a means of determining the molecular weight of a colloid. After this, we shall examine sequentially osmotic phenomena arising from dilute nonideal colloids, nonideal colloids that are not too dilute, dilute charged colloids, and charged colloids that are not too dilute.

This is the first place in this book that we have had any occasion to mention charged colloids. The introduction of the notion that colloidal particles may carry an electric charge is an important aspect of the chapter. The last three chapters of the book are concerned almost exclusively with various aspects of charged systems. Several effects of low-molecular-weight electrolytes on the osmotic pressure of a charged colloid are completely consistent with the salt effects we discuss in detail in Chaps. 9, 10, and 11.

4.2 THE OTHER COLLIGATIVE PROPERTIES

The boiling temperature, the freezing temperature, the vapor pressure, and the osmotic pressure are those characteristics of solutions which are collectively known as the colligative properties of the solution. We shall defer for the moment any explicit comments on osmotic pressure, and examine the other three colligative properties for purposes of orientation. All the colligative properties originate from the same thermodynamic considerations, a discussion of which is the topic of the next section. All of them share as common features a dependence on certain characteristics of the solvent, but an independence of the nature of the solute. It is only through the number of particles present that the solute influences the value of the colligative property under consideration.

In reviewing the colligative properties other than osmotic pressure, we shall assume that the solute is nonvolatile and does not form solid solutions with the solvent. Further, we shall consider only those situations in which the concentration of the solution is sufficiently dilute that activity corrections may be disregarded. Throughout this chapter, the subscripts 1 and 2 are used to identify the solvent and solute, respectively. For now, we shall consider only one solute; later, when a second solute species is present, the subscript 3 is used to identify it.

For ideal solutions the vapor pressure of component i, p_i, is related to the mole fraction of that component, x_i, and the vapor pressure of the pure liquid, $p_i°$, by Raoult's law:

$$p_i = x_i p_i° \tag{1}$$

Since x_i is always less than unity for actual solutions, the vapor pressure of each component is always less above a solution than it would be above the pure liquid. The amount by which the vapor pressure is lowered, Δp_i, is

$$\Delta p_i = p_i° - p_i = p_i°(1 - x_i) \tag{2}$$

When just two components are present, $1 - x_1 = x_2$. Therefore,

$$\Delta p_1 = p_1° x_2 \tag{3}$$

This form of Raoult's law emphasizes the fact that the vapor pressure lowering depends on the nature of the solvent as well as the concentration of the solute. In two-component solutions

$$x_2 = \frac{n_2}{n_1 + n_2} \simeq \frac{n_2}{n_1} \tag{4}$$

in which the n terms equal the number of moles of the indicated component. The approximate form of Eq. (4) applies in the case of dilute solutions for which $n_2 \ll n_1$.

Combining Eqs. (3) and (4) gives the following result for dilute solutions:

$$n_2 = n_1 \frac{\Delta p_1}{p_1^\circ} \tag{5}$$

This form of Raoult's law emphasizes the point that the number of moles of solute in a dilute solution may be determined by measuring the amount by which the solvent vapor pressure is lowered.

Solutions of nonvolatile solutes boil at higher temperatures than the pure solvent because of this vapor pressure lowering. Application of thermodynamic relationships to this situation gives the equation

$$\ln x_1 = \frac{\Delta H_v^\circ}{R}\left(\frac{1}{T_b} - \frac{1}{T_b^\circ}\right) \tag{6}$$

in which T_b° is the boiling point of the pure solvent, T_b is the boiling point of the solution, ΔH_v° is the heat of vaporization of the pure solvent, and R is the gas constant. In this as in all thermodynamic equations, T is always expressed in degrees Kelvin. Since only two components are present

$$\ln x_1 = \ln(1 - x_2) \simeq -x_2 \tag{7}$$

where the approximation arises from expanding the logarithm (Appendix A) and retaining only the first term because of our interest in dilute solutions. Under these conditions, the difference between the boiling point of the solution and the boiling point of the solvent, ΔT_b, will be small so the product $T_b T_b^\circ \simeq T_b^{\circ 2}$. Combining this result with Eqs. (6) and (7) yields

$$\Delta T_b = \frac{R T_b^{\circ 2}}{\Delta H_v^\circ} x_2 \tag{8}$$

We see that the boiling point elevation depends on the properties of the solvent and the concentration of the solute. In dilute solutions, the approximation given by Eq. (4) holds, so a convenient working form of Eq. (8) is given by consolidating all the solvent-determined parameters as follows:

$$\Delta T_b = \frac{R T_b^{\circ 2} m_2}{\Delta H_v^\circ 1000/M_1} = k_{b,1} m_2 \tag{9}$$

In this equation, m_2 is the number of moles of solute per kilogram of solvent or the molality of the solution and M_1 is the molecular weight of the solvent. The aggregate of constants $k_{b,1}$ is called the boiling point constant and is tabulated for many solvents.

Again using the dilute solution approximation of Eq. (4), the expression for the boiling point elevation may be written

$$n_2 = \frac{n_1 \Delta H_v^\circ}{R T_b^{\circ 2}} \Delta T_b \tag{10}$$

This form stresses the fact that measurement of the boiling point elevation of a solution is a means of measuring the number of solute particles present in the solution.

The thermodynamic relationships which permit the increase in boiling point to be calculated also allow us to evaluate the freezing point depression of a solution. As a matter of fact, the theoretical basis for these relationships applies to any equilibrium in which one component (solvent in these examples) is partitioned between two phases, one of which is pure solvent and the other is solution. The nature of the physical states involved is immaterial as far as the mathematical formalism is concerned. This fact permits us to write the freezing point analogs to Eqs. (6), (9), and (10) directly, after allowing for the fact that ΔT_f is negative:

$$\ln x_1 = -\frac{\Delta H_f^\circ}{R}\left(\frac{1}{T_f} - \frac{1}{T_f^\circ}\right) \tag{11}$$

$$\Delta T_f = -\frac{RT_f^{\circ 2} m_2}{\Delta H_f^\circ 1000/M_1} = k_{f,1} m_2 \tag{12}$$

and

$$n_2 = -\frac{n_1\,\Delta H_f^\circ}{RT_f^{\circ 2}}\,\Delta T_f \tag{13}$$

In these expressions ΔH_f° is the heat of fusion ($S \rightarrow L$) of the solvent, T_f° is the freezing point of the pure solvent, T_f is the freezing point of the solution, and the aggregate constant $k_{f,1}$ is the freezing point depression constant for the solvent. The latter are also tabulated for easy use.

This section is not intended to review in any depth the colligative properties other than osmotic pressure. Rather, the objective is to point out that measurements of the various colligative properties permit the number of solute particles in a solution to be evaluated. Occasionally, learning this information may be an end in itself. More frequently, however, it is a step toward the evaluation of the molecular weight of the solute species. This is a very common application of ΔT_f and ΔT_b experiments. The weight of solute in a given portion of solution is almost always a known quantity, either from the preparation of the solution or from the weight of residue after evaporation. If the weight of solute and the number of moles of solute are known for the same solution, then the apparent molecular weight of the solute equals the ratio of the former to the latter. The word "apparent" in the preceding sentence is a safeguard to allow for such contingencies as dissociation and solvation, to mention only two. A second objective of this section is to point out that the colligative properties—other than osmotic pressure—are generally ineffective as a means of measuring the molecular weights of particles in the colloidal size range. To illustrate this quantitatively, let us consider the following concrete example. Suppose we have an aqueous solution of a monodisperse colloid in which the particles are of molecular weight M_2. To be specific, let us consider the colligative properties of a solution which is 1.0% by weight in the colloidal solute. Table 4.1 lists several relevant constants for water, as well as the values of the vapor pressure lowering, the boiling point elevation, and the freezing point depression for various values of M_2 in the colloidal range. At one time or another, most students have done a molecular

TABLE 4.1 *Physical constants of water which are required for colligative property calculations. Calculated values of* Δp, ΔT_b, *and* ΔT_f *for a* 1% *aqueous (ideal) solution for a range of solute molecular weights*

$p_1°$ at 25°C: 23.8 torrs
$k_{b,1}$: 0.514 deg kg mole^{-1}
$T_b°$: 373 K
$k_{f,1}$: 1.855 deg kg mole^{-1}
$T_f°$: 273 K

	M_2 (g mole^{-1})			
	10^3	10^4	10^5	10^6
Δp_1 (torrs)	4.33×10^{-3}	4.33×10^{-4}	4.33×10^{-5}	4.33×10^{-6}
ΔT_b (K)	5.19×10^{-3}	5.19×10^{-4}	5.19×10^{-5}	5.19×10^{-6}
ΔT_f (K)	1.87×10^{-2}	1.87×10^{-3}	1.87×10^{-4}	1.87×10^{-5}

weight determination by means of one of these measurements. Recollection of those experiences should underline the practical difficulties associated with even the most favorable of the situations described in Table 4.1. Of course, the choice of both the solvent and the concentration was arbitrary in developing the table; it is conceivable that different choices might increase one or more of the Δ quantities by an order of magnitude or so. Even then, the upper limit of molecular weight which may be evaluated by means of these colligative properties appears to be about 1000 under optimum conditions.

The need for a more sensitive means of counting particles is evident if this information is to be useful for determining the molecular weights of colloidal particles. It is the fourth of the colligative properties, osmotic pressure, that will extend into the colloidal size range the utility of these counting methods for molecular weight determination.

4.3 THE THEORY OF OSMOTIC PRESSURE

An extremely useful quantity in the thermodynamic treatment of multicomponent phase equilibria is the chemical potential. The chemical potential for component i, μ_i, is the partial molal Gibbs free energy with respect to component i at constant pressure and temperature:

$$\mu_i = \left(\frac{\partial G}{\partial n_i}\right)_{p,T,n_{j \neq i}} \tag{14}$$

Although the notation of this mathematical definition of chemical potential is somewhat cumbersome, the physical significance is fairly clear. The chemical potential is the coefficient which describes the way the Gibbs free energy of a system changes per mole of component i, if the temperature, pressure, and number of moles of all components other than the ith are held constant. Although it is expressed on a molar basis, it is important to note that μ_i is a differential quantity. That is, it represents the local slope of the line which shows the variation of G with n_i. The line itself arises from slicing across the complex surface which describes G at the specified values of p, T, and n_j. For a pure substance, μ_i is identical to the Gibbs free energy per mole of that substance. For the present, we shall regard the pure substance as the standard state for μ_i and shall represent it by the symbol $\mu_i°$.

The great utility of the chemical potential in phase equilibrium problems arises in the following way. In open systems where the number of moles of any component may increase or decrease, any change in the Gibbs free energy of the system as a whole may be expressed as the sum of the following contributions:

$$dG = \left(\frac{\partial G}{\partial T}\right)_{p,n} dT + \left(\frac{\partial G}{\partial p}\right)_{T,n} dp + \sum_i \left(\frac{\partial G}{\partial n_i}\right)_{p,T,n_j} dn_i \tag{15}$$

This equation may be written

$$dG = -S\,dT + V\,dp + \sum_i \mu_i\,dn_i \tag{16}$$

by substituting into (15) the definition of chemical potential and the familiar relationships

$$\left(\frac{\partial G}{\partial T}\right)_{p,n} = -S \tag{17}$$

and

$$\left(\frac{\partial G}{\partial p}\right)_{T,n} = V \tag{18}$$

For any equilibrium which occurs at constant temperature and pressure the first two terms from the right-hand side of Eq. (16) equal zero, reducing the equation to the form

$$dG = \sum_i \mu_i\,dn_i \tag{19}$$

If the total system consists of several different phases, which we shall designate by Greek subscripts α, β, . . . , then we may also write

$$dG = dG_\alpha + dG_\beta + \cdots \tag{20}$$

Finally, the equilibrium condition requires that

$$dG = 0 \tag{21}$$

Therefore, the equilibrium between two phases means that

$$dG = 0 = dG_\alpha + dG_\beta = \sum_i \mu_{i\alpha}\,dn_{i\alpha} + \sum_i \mu_{i\beta}\,dn_{i\beta} \tag{22}$$

That is, for each of the components the following holds:

$$\mu_{i\alpha}\,dn_{i\alpha} + \mu_{i\beta}\,dn_{i\beta} = 0 \tag{23}$$

If the entire system consists of only two phases, the hypothesis here, the conservation of matter requires that any substance lost from one phase must appear in the other:

$$dn_{i\alpha} = -dn_{i\beta} \tag{24}$$

Combining these last two results leads to the conclusion that the condition for phase equilibrium at constant temperature and pressure is

$$\mu_{i\alpha} = \mu_{i\beta} \tag{25}$$

It is important to realize that the chemical potential of each component must be the same in all equilibrium phases, although the value for this quantity will, in general, be different for each component.

The foregoing argument made no stipulations whatsoever as to the nature of the phases in equilibrium. Basically, it is this generality that is responsible for the similarity between the four colligative properties. In addition to liquid–vapor and liquid–solid equilibrium, Eq. (25) applies also when equilibrium exists in the system shown in Fig. 4.1. This figure shows schematically two liquid phases—one solution, the other pure solvent—separated by a partition known as a semipermeable membrane. This membrane is, in many ways, the central feature of osmometry; we shall have more to say about it presently. For now, it is sufficient to define a semipermeable membrane as one which is permeable to the solvent and impermeable to the solute. Thus, the semipermeable membrane in Fig. 4.1 prevents the two liquids from mixing and at the same time allows both sides of the membrane to come to equilibrium.

In order for solvent and solution to be in equilibrium in an apparatus such as that shown in Fig. 4.1, the solution side must be at a higher pressure than the solvent side. This excess pressure is what is known as the osmotic pressure of the solution. If no external pressure difference is imposed, solvent will diffuse across the membrane until an equilibrium hydrostatic pressure head has developed on the solution side. In order to prevent too much dilution of the solution as a result of the solvent flow into it, the column in which the pressure head develops is generally of very narrow diameter. We shall return to the details of osmotic pressure experiments in the next section. First, however, the theoretical connection between this pressure and the concentration of the solution must be established.

Since the two sides of the membrane are in true isothermal equilibrium in an osmotic pressure experiment, the chemical potential of the solvent must be the same on both sides of the membrane. On the side containing pure solvent, μ_1 equals $\mu_1°$. On the solution side of the membrane, the chemical potential of the solvent must also

FIGURE 4.1 *Schematic of an osmotic pressure experiment.*

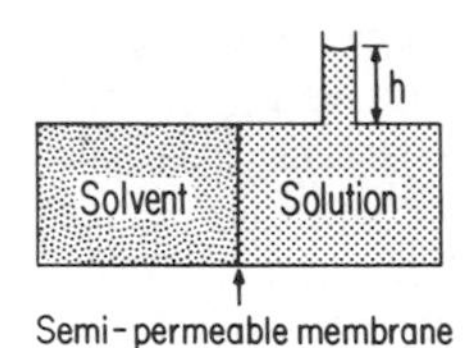

equal the same value, according to the equilibrium criterion of Eq. (25). To see how this comes about, we must examine some other relationships that apply to the chemical potential.

Perhaps the most basic equation involving the chemical potential is the one which relates this quantity to the activity of component i in the solution, a_i:

$$\mu_i = \mu_i{}^\circ + RT \ln a_i \tag{26}$$

For the present, a_i is expressed in mole fraction units. We see, therefore, that μ_i approaches $\mu_i{}^\circ$ as a_i approaches unity. Further, since μ_i is the partial molal Gibbs free energy, Eq. (18) also applies to μ_i, provided we replace V with $\bar{V}_i$, the partial molal volume of component i:

$$\left(\frac{\partial \mu_i}{\partial p}\right)_{T,n} = \bar{V}_i \tag{27}$$

Suppose we apply these relationships to the equilibrium of a liquid mixture and its vapor. At equilibrium, μ_i must have the same value for each component in both the liquid and vapor phase. Therefore,

$$\mu_{1,\mathrm{L}} = \mu_{1,\mathrm{V}} \tag{28}$$

or

$$\left(\frac{\partial \mu_{i,\mathrm{L}}}{\partial p}\right)_{T,n} = \left(\frac{\partial \mu_{i,\mathrm{V}}}{\partial p}\right)_{T,n} = \bar{V}_i = \frac{RT}{p_i} \tag{29}$$

where the last relationship assumes the vapor to behave ideally. If Eq. (26) is used to evaluate the left-hand side of this equation, we obtain

$$RT\, \partial \ln a_i = RT\frac{\partial p_i}{p_i} \tag{30}$$

Again recalling that $p_i = p_i{}^\circ$, the normal vapor pressure of the pure liquid, when $a_i = 1$, Eq. (30) may be integrated to give

$$a_i = \frac{p_i}{p_i{}^\circ} \tag{31}$$

This relationship constitutes the basic definition of the activity. Furthermore, if the solution behaves ideally, $a_i = x_i$ and Eq. (31) defines Raoult's law.

This brief review of solution thermodynamics alerts us to two facts which explain osmotic equilibrium. First, Eq. (26) reminds us that the chemical potential has its greatest value for a pure substance, $\mu_i{}^\circ$. Any value of a_i less than unity will cause μ_i to be altered from $\mu_i{}^\circ$ by an amount $RT \ln a_i$, which will be negative for $a_i < 1$. Second, any pressure on a liquid which exceeds $p_i{}^\circ$ increases μ_i above $\mu_i{}^\circ$. This is seen from the combination of Eqs. (26) and (31). Thus, consideration of the chemical potential of the solvent makes it clear how osmotic equilibrium comes about. The presence of a solute lowers the chemical potential of the solvent. This is offset by a positive pressure on the solution, the osmotic pressure, π, so that the net chemical potential on the solution side of the membrane equals that of the pure solvent on the other side of the membrane. This is summarized by the expression

$$\mu_1{}^{\circ} = \mu_1{}^{\circ} + RT \ln a_1 + \int_{p_1{}^{\circ}}^{p_1{}^{\circ}+\pi} \bar{V}_1 \, dp \tag{32}$$

In order to relate π to the concentration of the solution, then, we must find a way to integrate Eq. (32). The easiest way of doing this is to assume that $\bar{V}_1$ is constant. This approximation is justified since the solution is a condensed phase and shows negligible compressibility. Making this assumption and integrating Eq. (32) gives

$$\ln a_1 = -\frac{\pi \bar{V}_1}{RT} \tag{33}$$

Combining Eqs. (26) and (33) permits us to express the chemical potential in terms of osmotic pressure instead of activity:

$$\mu_1 = \mu_1{}^{\circ} - \pi \bar{V}_1 \tag{34}$$

This relationship will be useful in Chap. 5.

Equation (33) provides the relationship we have sought between osmotic pressure and concentration. If the solution is ideal, we may replace activity by mole fraction. Then Eq. (33) becomes

$$\ln x_1 = -\frac{\pi \bar{V}_1}{RT} = \ln(1 - x_2) \simeq -x_2 - \frac{x_2^2}{2} - \cdots \tag{35}$$

This form reminds us of the expansions that were performed on the analogous expressions for freezing and boiling equilibria [Eqs. (6) and (11)]; therefore, for dilute solutions we write

$$x_2 = \frac{\pi \bar{V}_1}{RT} \tag{36}$$

or

$$n_2 = \frac{n_1 \pi \bar{V}_1}{RT} = \frac{\pi V_1}{RT} \tag{37}$$

where V_1 is simply the volume of the solvent in the solution, $n_1 \bar{V}_1$. This relationship, known as the van't Hoff equation (after J. H. van't Hoff, first Nobel Prize in chemistry, 1901), is analogous in form to the ideal gas law and, like the latter, is a limiting law which applies perfectly only in the limit of $\pi \to 0$ or $n_2/V_1 \to 0$. Equation (37) shows that osmotic pressure experiments provide a means of measuring the number of solute particles in a solution, as was the case with the other colligative properties as well. If the weight of solute in the solution is also known, this information may be used to evaluate molecular weights. As a numerical example, we may calculate the osmotic pressure of a 1% aqueous solution of the colloidal solutes considered in Table 4.1. At 25°C and for a 1% solution, Eq. (37) may be written

$$\pi(\text{atm}) = \frac{(0.082)(298)}{(M_2)(0.099)} = \frac{247}{M_2} \tag{38}$$

where M_2 is the molecular weight of the solute. In millimeters of mercury, this becomes

$$\pi\ (\text{mm Hg}) = \frac{247}{M_2}\ \text{atm} \times \frac{760\ \text{mm Hg}}{1\ \text{atm}} = \frac{1.88 \times 10^5}{M_2} \tag{39}$$

and in millimeters of solution

$$\pi\ (\text{mm soln}) = \frac{1.88 \times 10^5}{M_2} \times \frac{\rho_{\text{Hg}}}{\rho_{\text{soln}}} = \frac{1.88 \times 10^5}{\text{M}_2} \times \frac{13.5}{1.0} = \frac{2.53 \times 10^6}{M_2} \tag{40}$$

The final numerical value in Eq. (40) is based on the assumption that the density of the solution equals 1.0 g cm^{-3}, the same value as water. This last result shows that 1% aqueous solutions of solutes having molecular weights of 10^3, 10^4, 10^5, and 10^6 would produce osmotic pressures corresponding to liquid columns of 2530, 253, 25.3, and 2.53 mm, respectively. This calculation, like those of Table 4.1, assumes that the 1% solution is ideal. The purpose of this numerical example is to show that the values of osmotic pressures are very much larger than the other colligative properties of the same solution.

All the colligative properties decrease per unit weight of solute as the molecular weight of the solute increases. The very much larger magnitude of the osmotic pressure makes it the only one of the colligative properties which is useful for the determination of the molecular weight of colloidal solutes.

4.4 EXPERIMENTAL OSMOMETRY

To carry out an osmotic pressure experiment, we need to prepare a solution, to find a suitable semipermeable membrane, to achieve isothermal equilibrium, and to measure the equilibrium pressure. Aside from noting that the pressures produced by colloidal solutes are large enough to be measurable, we have not yet considered any of the experimental aspects of osmometry. This is our present task.

First, a suitable solvent and membrane must be found. The solvent must dissolve enough solute to produce an adequate pressure. The results of measurements made at relatively high concentrations may be extrapolated to zero concentration, so we need not worry about the effects of nonideality. We shall discuss the extrapolation procedure in Sec. 4.5. As low a solvent viscosity as possible is desirable to minimize the time required for equilibration.

It is important that the materials be free from contaminants in osmotic pressure experiments. Suppose, for example, that the solvent contains a small amount of impurity which, like the colloidal solute, is retained by the membrane. Then, as far as the osmotic pressure is concerned, that impurity will contribute to the osmotic pressure in the same way that the colloid does. Since the osmotic pressure responds to the number of solute particles present, a low-molecular-weight impurity in extremely small amounts may contain as many or more *particles* as a dilute solution of a colloidal solute of very high molecular weight. Quite large errors in molecular weight may arise in this way. The confusion may be compounded if the same system is investigated using a different membrane material. It is conceivable that another membrane would be permeable to the impurity. This would result in a different apparent molecular weight for the colloid. We have approached the issue of impurities from the viewpoint of the solvent. Actually, the colloid is more likely to be the source of the impurity since these materials are often difficult to purify. We shall discuss some additional aspects of this problem in the sections on average molecular weight (Sec. 4.7), on charged colloids (Sec. 4.9), and on dialysis (Sec. 4.11).

The membrane is the source of most of the difficulties in osmometry. There is no general way to select a membrane material that will be permeable to one component and impermeable to another for any conceivable combination of chemicals. Very high molecular weight components are generally more easily retained, however, so we have a slight advantage in this regard. The membrane must be sufficiently thin to permit equilibration at a reasonable rate. At the same time, it must be strong enough to withstand the considerable pressure differences which may exist across it. This problem may generally be overcome, at least in part, by suitable mechanical support of the membrane. A more serious problem is the preparation of thin membranes that are free from minute imperfections which would constitute a "leak" between the two compartments. Such a leak would totally invalidate the experimental results. One peculiarity of membranes is their tendency to display what is known as an asymmetry pressure. That is, an equilibrium pressure difference may exist across a membrane even when there is solvent on both sides. This must be measured and subtracted as a "blank correction."

Equilibrium osmometry is a thermodynamic phenomenon. As such, it makes no difference what mechanism the membrane uses to retain the solute nor do we learn anything about the mechanism from equilibrium studies. It is easy to visualize that some solutes, especially those in the colloidal size range, are retained by a sieve effect. That is, the molecules are simply too large to pass through the pores in the membrane material. Another possibility is a mechanism whereby the membrane displays selective solubility. This means that the membrane dissolves the solvent but not the solute. In this way, the solvent can pass through the membrane, while the solute is retained. An analogous mechanism for charged particles may arise by the membrane repelling (and thus retaining) particles of one particular charge.

A great many different materials have been used in osmotic pressure experiments. Various forms of cellophane and animal membranes are probably the most common membrane materials. Various other polymers, including polyvinyl alcohol, polyurethane, and polytrifluorochloroethylene, have also been used along with such inorganic substances as $CuFe(CN)_6$ precipitated in a porous support.

Once a suitable membrane and solvent are selected, an experimental arrangement must be devised which measures the equilibrium pressure under isothermal conditions. Many variations in apparatus design have been studied. Two particularly instructive pieces of apparatus are shown in Fig. 4.2. The assembly shown in

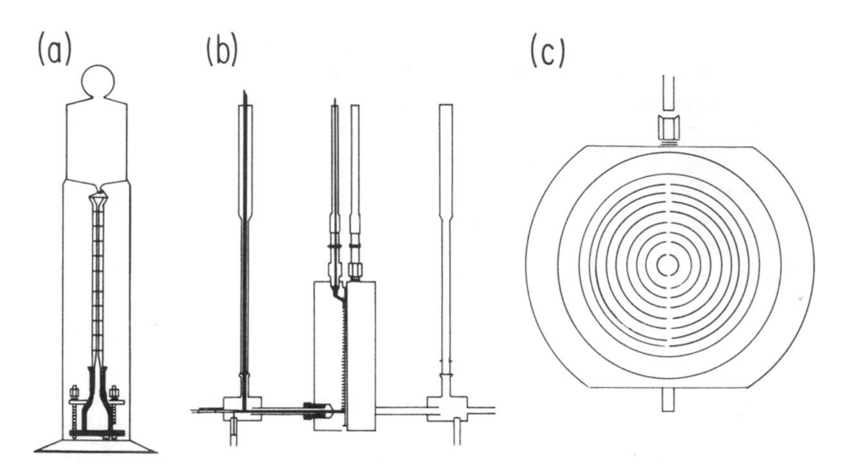

FIGURE 4.2 *Two osmometers: in (a), solution compartment is submerged in solvent. [Reprinted with permission from D. M. French and R. H. Ewert,* Anal. Chem. **19**:165 (1947), *copyright by the American Chemical Society.] In (b), solution and solvent occupy grooves on opposite faces of central unit as shown in detail in (c). [Reprinted with permission from R. M. Fuoss and D. J. Mead,* J. Phys. Chem. **47**:59 (1943), *Copyright by the American Chemical Society.]*

Fig. 4.2a consists of an inner solution compartment with a relatively large opening at the membrane end and a capillary at the small end. The entire solution chamber is then immersed in a tube containing the solvent. Once assembled, the entire apparatus is placed in a constant temperature bath for equilibration.

Another osmometer design is shown assembled in Fig. 4.2b and in detail in Fig. 4.2c. The membrane is pressed between two grooved faces which contain solvent on one side and solution on the other. The grooves are attached to filling tubes and to capillaries where the pressure head develops. This apparatus permits a large contact area between liquids with a minimum volume of liquids involved.

Several times in this discussion, we have noted the importance of experimental conditions which permit as rapid an equilibration as possible. The implication of these remarks is that osmotic equilibrium is reached slowly. In some cases, as much as one week may be required for equilibrium to be achieved. To shorten this time, procedures based on measuring the rate of approach to equilibrium have been developed. The osmometer of Fig. 4.2b is especially suited for this procedure. In successive runs, the capillary on the solution side of the membrane is filled with solution to some initial setting which will be above or below the equilibrium location of the meniscus. At various times after the initial settings, the height of the liquid column is measured. It is found that the ascending and descending branches of the curve converge to the same point—a value which equals the equilibrium osmotic pressure. This is shown in Fig. 4.3. The rate at which equilibrium is achieved

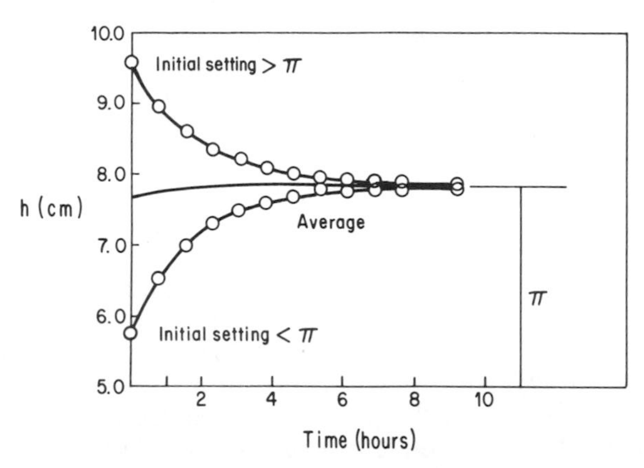

FIGURE 4.3 *Data showing the approach to osmotic equilibrium from initial settings above and below the equilibrium column height.* (*From Bonnar et al.* [2], *redrawn with permission.*)

decreases as equilibrium is approached. Therefore, it is desirable to bracket the true value as narrowly as possible to take full advantage of this approach-to-equilibrium extrapolation procedure. With some judicious planning, the time for an osmotic experiment may be shortened considerably by this dynamic method.

Once equilibrium has been reached, the height difference between the two liquid surfaces is all that remains to be measured. The primary factor to note here is that capillaries are used to minimize the dilution effects. This means that corrections for capillary rise must be taken into account unless the apparatus allows the difference between two carefully matched capillaries to be measured. We shall discuss capillary rise in detail in Sec. 6.6. Finally, there is an extremely important practical reason for good thermostating in osmometry experiments in addition to the theoretical requirement of isothermal conditions. The apparatus consists of a large liquid volume attached to a capillary and, therefore, has the characteristics of a liquid thermometer: The location of the meniscus is quite sensitive to temperature fluctuations.

4.5 OSMOTIC PRESSURE OF NONIDEAL SOLUTIONS

In Sec. 4.3, we abandoned a completely general discussion of osmotic pressure after reaching Eq. (33) in favor of the simpler assumption of ideality. The ideal result applies to real solutions in the limit of infinite dilution. The objective of this section is to examine the extension of Eq. (33) to nonideal solutions or, more practically, to solutions whose concentrations are greater than infinitely dilute.

Since both the osmotic pressure of a solution and the pressure–volume–temperature behavior of a gas are described by the same formal relationship [Eq. (37)], it seems plausible to approach nonideal solutions along the same lines that are used in dealing with nonideal gases. The behavior of real gases may be written as a power series in one of the following forms for n moles of gas:

$$\frac{pV}{nRT} = 1 + Bp + Cp^2 + \cdots \tag{41}$$

or

$$\frac{pV}{nRT} = 1 + B\left(\frac{n}{V}\right) + C\left(\frac{n}{V}\right)^2 + \cdots \tag{42}$$

Equations of this type are known as virial equations and the constants they contain are called the virial coefficients. It is the second virial coefficient B which describes the earliest deviations from ideality. It should be noted that B would have different but related values in Eqs. (41) and (42) even though the same symbol is used in both cases. One must be especially attentive to the form of the equation involved, particularly with respect to units, when using literature values of quantities such as B. The virial coefficients are temperature dependent and vary from gas to gas. Clearly, both Eqs. (41) and (42) reduce to the ideal gas law as $p \to 0$ or as $n/V \to 0$. Finally, it might be recalled that the second virial coefficient in Eq. (42) is related to the van der Waals a and b constants as follows:

$$B = b - \frac{a}{RT} \tag{43}$$

This last relationship points out that for gases the second virial coefficient arises both from the finite size and from the interactions of the molecules of the gas since these are the origins of b and a, respectively.

As we extend these ideas to nonideal solutions, a similar set of statements will apply to the resulting power series:

1. The second virial coefficient is our primary concern since we shall focus attention on the first deviations from ideality.
2. The value of B will depend in part on the units chosen for concentration as well as on the temperature and the nature of the system.
3. The virial equation for osmotic pressure must reduce to the van't Hoff equation in the limit of infinite dilution.
4. The second virial coefficient may be expected to reflect both the finite size and the interactions of the molecules.

Each of these points is taken up in the following discussion.

The easiest way to extend these considerations to the osmotic pressure of nonideal solutions is to return to Eq. (35) which relates π to a power series in mole fraction. This equation applies to ideal solutions, however, since ideality is assumed in replacing activity by mole fraction in the first place. To retain the form yet extend its applicability to nonideal solutions, we formally include in each of the concentration terms a correction factor defined to permit the series to be applied to nonideal solutions as well:

$$\frac{\pi \bar{V}_1}{RT} = A' x_2 + \tfrac{1}{2} B' x_2{}^2 + \cdots \tag{44}$$

The coefficients A', B', . . . , must all equal unity in ideal solutions in order to recover Eq. (35). Since the van't Hoff equation is a limiting law, the coefficient A' must equal unity in *all* solutions. Therefore, Eq. (44) becomes

$$\frac{\pi \bar{V}_1}{RT} = x_2 + \tfrac{1}{2}B' x_2^2 + \cdots \tag{45}$$

Next, let us consider the transformation of mole fraction units to weight per volume concentration units c in Eq. (45). Examination of Table 2.2 reminds us that for dilute solutions the following expression relates these two systems of units:

$$x_2 = \frac{\bar{V}_1}{M_2} c \tag{46}$$

Substituting this result into Eq. (45) gives

$$\frac{\pi \bar{V}_1}{RT} = \frac{\bar{V}_1}{M_2} c + \tfrac{1}{2}B' \left(\frac{\bar{V}_1}{M_2}\right)^2 c^2 + \cdots \tag{47}$$

Equation (47) may be rearranged to yield

$$\frac{\pi}{RTc} = \frac{1}{M_2} + \frac{1}{2}\frac{B' \bar{V}_1}{M_2^2} c = \frac{1}{M_2} + Bc \tag{48}$$

Note that this rearrangement reduces the order of the equation by the same procedure that was used on Eq. (2.66). This form suggests that a plot of π/RTc versus c should be a straight line, the intercept and slope of which have the following significance:

$$\text{Intercept} = \left(\frac{\pi}{RTc}\right)_0 = \frac{1}{M_2} \tag{49}$$

$$\text{Slope} = B = \frac{1}{2}\frac{B' \bar{V}_1}{M_2^2} \tag{50}$$

A plot of π/RTc versus c is the most commonly encountered form in which osmotic pressure results are presented.

Note that the value of the intercept, the value of π/RTc at infinite dilution, obeys the van't Hoff equation [Eq. (36)]. At infinite dilution, even nonideal solutions reduce to this limit. The value of the slope is called the second virial coefficient by analogy with Eq. (42). Note that the second virial coefficient is the composite of two factors: B' and $\frac{1}{2}\bar{V}_1/M_2^2$. The factor B' describes the first deviation from ideality in a solution; it equals unity in an ideal solution. The second cluster of constants in B arises from the conversion of practical concentration units to mole fractions. Although it is the nonideality correction in which we are primarily interested, we shall discuss it in terms of B rather than B' since the former is the quantity which is measured directly.

Figures 4.4a and 4.4b are examples of two plots of π/RTc versus concentration. In Fig. 4.4a the data all describe different molecular weight fractions of the same solute, cellulose acetate, in acetone solutions. The parameters listed on the curves indicate the molecular weights of the different fractions as calculated by Eq. (49);

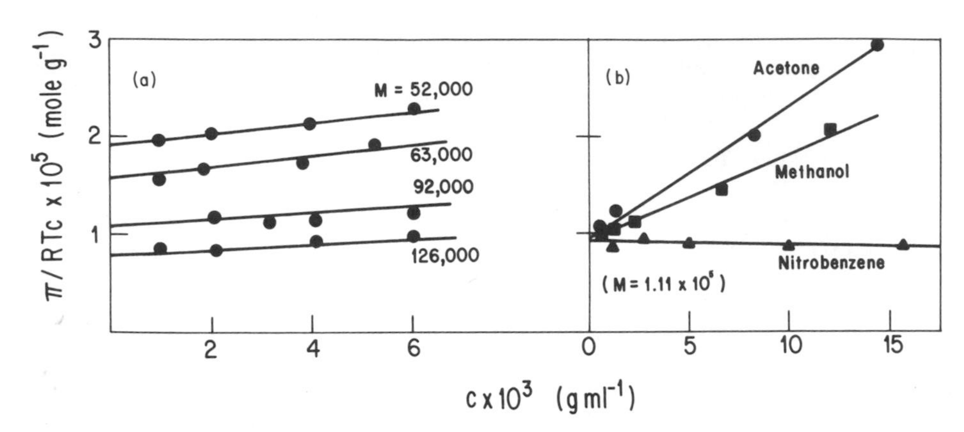

FIGURE 4.4 π/RTc *versus concentration* (*a*) *for various cellulose acetate fractions in acetone* [*data from A. Bartovics and H. Monk*, J. Am. Chem. Soc. **65**:1901 (1943)] *and* (*b*) *for nitrocellulose in three different solvents* [*data from A. Dobry*, J. Chem. Phys. **32**:50 (1935)].

they range from 52,000 to 126,000 g mole^{-1}. Since the lines in this plot all have essentially the same slope, B must be the same for each.

Figure 4.4b shows data for a sample of nitrocellulose in three different solvents. All show the same intercept corresponding to a molecular weight of 1.11×10^5 g mole^{-1} as required by Eq. (49). Note, however, that the slopes are different, including even a negative slope, indicating that wide variations in B are possible. We shall examine the factors that determine B in Sec. 4.7.

4.6 THE NUMBER AVERAGE MOLECULAR WEIGHT

As we saw in Chap. 1, the condition of polydispersity is quite normal with colloidal solutes. In that chapter, we took the position that the entire distribution of particle sizes was known and sought to characterize such a distribution in terms of suitable statistical parameters. Our discussion of osmometry has shown that it is possible to evaluate the molecular weight of a colloidal solute by osmotic pressure measurements. Next, we consider the fact that the sample on which such a measurement is made is, more than likely, polydisperse and that the molecular weight obtained from such an experiment is some average quantity. The objective of this section is to show that it is the number average molecular weight which is determined in an osmotic pressure experiment on a polydisperse system. For the purposes of this demonstration, the solution is assumed to be ideal.

Experimental results from a polydisperse system may be related as follows:

$$\pi_{\text{exp}} = \frac{c_{\text{exp}} RT}{\bar{M}} \tag{51}$$

where $\pi_{\exp}$ and $c_{\exp}$ represent the experimental osmotic pressure and concentration, respectively, and $\bar{M}$ is the average molecular weight. It is the method of averaging in Eq. (51) that we seek to determine. This relationship applies to the observable quantities. Precisely the same equation may be written, however, for each of the molecular weight fractions of the solute, designated here by the subscript i:

$$\pi_i = \frac{c_i RT}{M_i} \tag{52}$$

Two additional relationships are fairly evident. The experimental osmotic pressure is the sum of the pressure contributions of the individual components:

$$\pi_{\exp} = \sum_i \pi_i \tag{53}$$

and the experimental concentration is the sum of the concentrations of the components:

$$c_{\exp} = \sum_i c_i \tag{54}$$

Next, Eqs. (51) through (54) can be combined as follows:

$$\pi_{\exp} = \sum_i \pi_i = RT \sum_i \frac{c_i}{M_i} = RT \frac{c_{\exp}}{\bar{M}} = RT \frac{\Sigma_i c_i}{\bar{M}} \tag{55}$$

Equation (55) may be simplified to yield

$$\bar{M} = \frac{\Sigma_i c_i}{\Sigma_i (c_i/M_i)} \tag{56}$$

This result still fails to resemble any of the standard averages listed in Table 1.5. However, if we introduce the following expression for c_i:

$$c_i = \frac{n_i M_i}{V} \tag{57}$$

we obtain

$$\bar{M} = \frac{(1/V)\, \Sigma_i n_i M_i}{(1/V)\, \Sigma_i (n_i M_i / M_i)} = \frac{\Sigma_i n_i M_i}{\Sigma_i n_i} \tag{58}$$

Equation (58) shows the average molecular weight determined from osmometry to be the number average molecular weight as defined by Eq. (1.20).

This same conclusion may also be reached by the following argument. The product $n_i M_i$ in Eq. (57) equals the weight of component i in the solution; the total weight of solute in the solution equals $\Sigma_i n_i M_i$. The experimental osmotic pressure depends on and, therefore, measures the total number of moles of solute $\Sigma_i n_i$. The ratio of the total weight to the total number of moles of solute defines the number average molecular weight. This argument also makes it clear that all the colligative properties—not just osmotic pressure—yield number average molecular weights since they all share the feature of being proportional to the number of solute molecules present.

This discussion of the nature of the average obtained from colligative property measurements probably seems to the reader like "much ado about nothing." The number average is, after all, the mean of a distribution and exactly what is ordinarily meant by the word "average" in common usage. We shall see in Chap. 5, however, that light scattering also yields an average molecular weight for polydisperse systems, but that the averaging is done quite differently. The fact that the colligative properties "count" solute particles also has some interesting consequences for charged colloids, as we shall see in Sec. 4.9.

4.7 THE SECOND VIRIAL COEFFICIENT FOR UNCHARGED PARTICLES

By analogy with Eq. (43), we might expect the second virial coefficient to depend on the size and/or the interactions of the molecules in solution. Although this expectation is basically correct, we must not take the form of Eq. (43), a gas equation, too literally in discussing solutions. Furthermore, it must be recalled that in gases only interactions between the molecules of the gas are possible. In solution, we may consider solvent–solvent, solute–solute, and solvent–solute interactions. The first of these is the same on both sides of the membrane in an osmotic pressure experiment and the second is unimportant in relatively dilute solutions, so only the third will concern us here. This is a good reminder that the analogy with gases cannot be pushed too far.

By itself, thermodynamics provides us with no information on the molecular origin of the second virial coefficient. The latter is merely a phenomenological coefficient from an exclusively thermodynamic viewpoint. Statistical thermodynamics undertakes the task of providing molecular interpretations to such quantities. Without presenting the proofs in detail, we shall consider the statistical mechanical interpretation for the second virial coefficient according to several different models. Table 4.2 lists the factors B' and B as calculated by several theoretical models. We shall discuss each of these briefly in the following paragraphs. First, however, the following general observations about the contents of Table 4.2 will be helpful. Three different models for nonideality are considered in the table: noninteracting spheres, noninteracting rods, and interacting random coils. Statistical mechanical values of B' are presented for each. These values of B' are combined with Eq. (50) to give the values of B according to each model. It should be noted that B' is dimensionless as required by Eq. (44) and that cgs units are indicated for B. Next, the expressions for B are rearranged to relate explicitly the experimental B values in cgs units to some parameter which characterizes the dispersion. The nature of these parameters depends on the model involved, so each of them is considered specifically below.

Now let us examine the results presented in Table 4.2 in terms of the specific models involved. The principal difference lies between the interacting and noninteracting models. Accordingly, we shall comment on the noninteracting particles of two different geometries first, then turn our attention to the case in which the particles have a random coil configuration with extensive solvent interaction.

TABLE 4.2 *Values of the nonideality correction factor B' and the second virial coefficient B for three different models of nonideal colloidal solutions*

General description of model	Noninteracting spheres	Noninteracting rods	Interacting random coils
B' (dimensionless)	$4\frac{\bar{V}_2}{\bar{V}_1}$	$2\frac{L}{d}\frac{\bar{V}_2}{\bar{V}_1}$ L: Length; d: Diameter	$2\frac{\Delta G}{RT}\left(\frac{\bar{V}_2}{\bar{V}_1}\right)^2$ ΔG: Free energy of solute–solvent interaction
B (moles cm^3 g^{-2})	$2\frac{\bar{V}_2}{M_2^2}=\frac{2}{\rho_2 M_2}$	$\frac{L\bar{V}_2}{dM_2^2}=\frac{L}{\rho_2 M_2 d}$	$\frac{\Delta G}{V_1 RT}\left(\frac{\bar{V}_2}{M_2}\right)^2$
B Expressed in terms of characteristic parameters (cgs units)	$\frac{6}{\rho_2{}^2 4\pi R^3 N_A}$ R: Radius of sphere	$\frac{4}{\rho_2{}^2 \pi d^3 N_A}$ d: Diameter of rod	$\frac{(\frac{1}{2}-\chi)}{\bar{V}_1}\left(\frac{\bar{V}_2}{M_2}\right)^2$ $\frac{\psi(1-\Theta/T)}{\bar{V}_1}\left(\frac{\bar{V}_2}{M_2}\right)^2$ ψ, χ, Θ: Parameters of interest for the solution

Table 4.2 shows that the parameter B' is proportional to the ratio $\bar{V}_2/\bar{V}_1$ for both the noninteracting sphere and the noninteracting rod. This ratio describes how much larger the solute particles are than those of the medium; for colloidal solutes, it can be appreciable. The factor that multiplies this ratio converts the actual volume to an *excluded* volume. The student may recall that the excluded volume of spherical gas molecules is larger than the actual volume of the molecules by a factor of 4. This same factor describes the excluded volume of spherical molecules in solution. For rod-shaped particles, the factor $2L/d$ performs a similar function. Note that the nonideality increases with the ratio L/d in the latter case. Therefore, even when the shape of the particles is unknown, their asymmetry may be estimated from the second virial coefficient if the particles are known to possess some fixed geometry.

In the case of the noninteracting particles, $M_2/\bar{V}_2$ may be replaced by the density of the dry material. Alternatively, the mass of the particles in these cases can be replaced by the product of the density times the volume of the individual particles. When these latter substitutions are made for the volume of a sphere and a cylindrical rod, the B values are seen to be related to the radius and the diameter for these two geometries. Thus experimental values of the second virial coefficient may be interpreted in terms of these particle dimensions if the shape of the particles is known.

Note that all the preceding interpretations require particles which are uncharged and unsolvated, and possess a definite geometry. Particles that are held in a definite three-dimensional structure, for example by disulfide or hydrogen bonds, are most likely to meet this last requirement. Molecules that are flexible tend to assume the random coil configuration which we discussed in Sect. 3.12. The dimensions of such a coil are highly sensitive to the relative magnitude of solute–solute, solute–solvent, and solvent–solvent interactions. Table 4.2 shows that the second virial coefficient in these systems also depends on the ratio $\bar{V}_2/\bar{V}_1$ in addition to the value of ΔG for these interactions. The partial molal volume of the colloidal solute must be determined experimentally rather than evaluated from the density of the dry material because of these interactions. Clearly, the value of ΔG also depends on the interactions. For interacting random coils, therefore, we conclude that the second virial coefficient provides more information about interactions than about particle geometry.

In polymer chemistry, the interaction free energy ΔG is usually expanded in one of two ways. In both cases, it is written as the difference between two quantities—essentially enthalpy and entropy terms. For our purposes, it is sufficient to observe that the free energy of interaction may be expressed in terms of either of the following formalisms:

$$\frac{\Delta G}{RT} = \frac{1}{2} - \chi \tag{59}$$

or

$$\frac{\Delta G}{RT} = \psi\left(1 - \frac{\Theta}{T}\right) \tag{60}$$

The parameters χ and Θ are most commonly used to characterize the interacting system. Note that Θ has the dimensions of temperature and is sometimes called the Flory temperature after P. J. Flory (Nobel Prize, 1974) who has made great contributions to polymer chemistry. The parameter ψ is related to the partial molal entropy of the solvent and does not concern us in this discussion. Both Eqs. (59) and (60) have been related to the second virial coefficient in Table 4.2.

The usefulness of the parameters Θ and χ may be seen as follows. When several different solvents are being compared at a single temperature, as in Fig. 4.4b, the differences in the second virial coefficient may be interpreted in terms of the interaction parameter χ. On the other hand, when a single solvent is being examined at several different temperatures, the second virial coefficient can be used to identify the temperature Θ.

Substitution of either Eqs. (59) or (60) into the second virial coefficient shows that the latter may be either positive or negative for interacting random coils. The condition $B = 0$ may be interpreted to mean either $\chi = \frac{1}{2}$ or $T = \Theta$. Positive values for the second virial coefficient, on the other hand, correspond either to $\Theta < T$ or to $\chi < \frac{1}{2}$. The inequalities are reversed for negative B values. Thus, depending on the nature of the experiment conducted and the relationship used for its interpretation, a polymer–solvent system may be quantitatively characterized in its interactions by means of either the Θ temperature or the interactions parameter χ.

On the theoretical side, there is a considerable literature devoted to the analysis and interpretation of both Θ and χ. We shall make only a few remarks about these quantities here. The interested reader is referred to other sources, for example Flory [3], for additional information. We have already noted in Sec. 2.15 that the Θ temperature corresponds to the critical temperature for precipitation of a polymer sample of infinite molecular weight in the given solvent. Now we discover a second and, for present purposes, even more meaningful interpretation of Θ. Since it is the temperature at which the slope of the π/RTc versus c curve is zero, Θ is analogous to the Boyle temperature of a gas. It represents a condition in which excluded volume and interaction effects cancel and in which the system behaves as if it were ideal at all concentrations. From a practical point of view, there are obvious advantages to working at the Θ temperature for polymers since a single osmotic pressure may be confidently used in Eq. (49) for a molecular weight determination.

The interaction parameter χ is related to the enthalpy of mixing for the polymer solution according to the Flory–Huggins theory of polymer solubility. When this theory is applied to the phenomenon of phase separation, it is found that $\chi = \frac{1}{2}$ at the critical point for phase separation for a polymer of infinite weight. Thus the conditions $\chi = \frac{1}{2}$ and $T = \Theta$ both describe, in alternate theoretical forms, the same physical situation. In terms of osmotic pressure, $\chi = \frac{1}{2}$ also corresponds to a situation in which the slope of a π/RTc versus c curve equals zero.

Both the interaction parameter χ and the Θ temperature play an important role in the theory of polymer solutions. As a matter of fact, the terms "good" and "poor" used in Sec. 3.12 to describe the solvent in various polymer solutions may be defined quantitatively in terms of these parameters. A solvent is said to be "good" for a particular polymer if $T > \Theta$ or $\chi < \frac{1}{2}$ for the polymer–solvent system. If the opposite inequalities hold, the solvent is said to be "poor." Osmometry provides a convenient way to measure these parameters.

In the next chapter in connection with our discussion of light scattering, we shall discover another way to evaluate B and the parameters derivable therefrom.

4.8 DONNAN EQUILIBRIUM AND ELECTRONEUTRALITY

We have had no occasion as yet in this book to note that colloidal solutes may possess an electrical charge just like their low-molecular-weight counterparts. Chapters 9, 10, and 11 are concerned with those properties of colloids which are direct consequences of the charge of the particles. For the present, we shall introduce the idea of charged particles by examining the effect of the charge on the osmotic pressure of the system.

The charge on a colloidal particle may originate either from the dissociation of functional groups which are covalently bonded to the colloid or from the preferential adsorption of ions to the surface. We shall defer our discussion of adsorption until Chap. 7. The charge of a colloid cannot be regarded as a fixed quantity like molecular weight, but must be treated as a variable whose value depends on the nature and concentration of other components of the system. For example, proteins are positively charged at very low pH levels and negatively charged at very high pH levels; the point of electroneutrality varies from one protein to another. In the case of proteins, it is clearly the ionization of acidic and basic functional groups attached to the polypeptide chain that is primarily responsible for the charge characteristics of the molecules.

At this point, we shall not concern ourselves any further with the origin of the charge of a colloidal system. Rather, our attitude is that charge is one more characteristic that must be measured and understood in order to characterize certain systems fully.

Although it is not particularly difficult to formulate the thermodynamics of charged systems in perfectly general terms, the resulting notation is cumbersome. Instead of the completely general form, therefore, we shall consider a very specific case. The principal features will emerge clearly from this example; other situations may be readily derived by parallel arguments. The system we shall be concerned with consists of three components: Component 1 is the solvent, usually water; component 2 is the colloidal electrolyte; and component 3 is a low-molecular-weight uni-univalent electrolyte, MX. We shall (arbitrarily) designate the colloidal electrolyte PX_z, consisting of a positively charged macroion, having a valence number $+z$, paired with z X^- ions. It could be the negative ion of the colloidal electrolyte that is the macroion and the low-molecular-weight solute could have a different stoichiometry, but the essential features would remain the same.

In physical chemistry, it is convenient to express concentrations as molalities and to use molality units to express the activity of the components. This is the convention we follow in this section. Accordingly, the standard state for a component consists of a solution in which that component has an activity of 1.0 mole (kg solvent)$^{-1}$.

That specific situation we wish to consider is the osmotic equilibrium that develops in an apparatus which has a semipermeable membrane that is impermeable to the macroion only. That is, the membrane is assumed to be permeable not only to the solvent but also to both of the ions of the low-molecular-weight electrolyte, but not to the colloidal ion, P^{+z}. At equilibrium, the low-molecular-weight ions will be found on both sides of the membrane, but not in equal concentrations because of the presence of the macroions on one side of the membrane. We shall designate that side of the membrane which contains the macroions as the α phase and the solution from which the macroions are withheld as the β phase.

Equation (25) continues to describe the equilibrium condition; applying it to component 3 leads to the following:

$$\mu_{3,\alpha} = \mu_{3,\beta} \tag{61}$$

Substituting Eq. (26) for the β phase and (32) for the α phase which is under an osmotic pressure π yields

$$\mu_3^\circ + RT \ln a_{3,\beta} = \mu_3^\circ + RT \ln a_{3,\alpha} + \int_0^\pi \bar{V}_3 \, dp \tag{62}$$

At sufficiently low concentrations of the macroion the osmotic pressure term will be negligible compared with $RT \ln a_{3,\alpha}$, so Eq. (62) becomes

$$a_{3,\alpha} = a_{3,\beta} \tag{63}$$

It may be recalled from physical chemistry that the activity of a 1 : 1 electrolyte is given by the product of the activities of the positive and negative ions of the compound; therefore,

$$(a_{M,\alpha})(a_{X,\alpha}) = (a_{M,\beta})(a_{X,\beta}) \tag{64}$$

This expression describes what is known as the Donnan equilibrium. It does *not* say that the activity of M^+ and/or X^- is the same on both sides of the membrane, but that the ion activity product is constant on both sides of the membrane. In the sense that an ion product is involved, the Donnan equilibrium clearly resembles all other ionic equilibria.

Remembering that $a_+ = \gamma_+ m_+$ and $\gamma_\pm{}^2 = \gamma_+\gamma_-$ where $\gamma_\pm$ is the mean ionic activity coefficient (appropriate to molality units) enables us to rewrite Eq. (64) as

$$(m_{M,\alpha})(m_{X,\alpha})\gamma_{\pm,\alpha}^2 = (m_{M,\beta})(m_{X,\beta})\gamma_{\pm,\beta}^2 \tag{65}$$

Of course, in the limit of infinite dilution $\gamma_\pm \to 1$. For the present, we shall restrict our attention to sufficiently dilute solutions so that activity coefficients may be neglected and molalities may be used instead of activities. It might also be noted that in dilute solutions where this simplification is apt to be valid, molality and molarity are almost equal.

Another factor which we have not yet taken into account is the requirement that both sides of the membrane be electrically neutral. For the α phase which contains the macroion, this condition is expressed by

$$zm_{P,\alpha} + m_{M,\alpha} = m_{X,\alpha} \tag{66}$$

In the β phase which contains only low-molecular-weight ions, electroneutrality requires

$$m_{M,\beta} = m_{X,\beta} \tag{67}$$

The significance of the Donnan equilibrium is probably best seen as follows. Combining Eqs. (65) and (67) yields

$$m_{M,\beta}^2 = m_{X,\beta}^2 = (m_{M,\alpha})(m_{X,\alpha}) \tag{68}$$

Next, we use Eq. (66) to substitute for either $m_{M,\alpha}$ or $m_{X,\alpha}$ in Eq. (68), obtaining the following quadratic equations for $m_{M,\alpha}$ and $m_{X,\alpha}$:

$$m_{M,\alpha}^2 + zm_P m_{M,\alpha} - m_{M,\beta}^2 = 0 \tag{69}$$

and

$$m_{X,\alpha}^2 + zm_P m_{X,\alpha} - m_{X,\beta}^2 = 0 \tag{70}$$

These expressions permit us to evaluate the concentration of the low-molecular-weight ions in the compartment with the macroions, the α phase, in terms of z and the concentration of ions in the other compartment.

The situation is most easily understood by considering a numerical example. Table 4.3 lists values of $m_{M,\alpha}$ and $m_{X,\alpha}$ calculated using Eqs. (69) and (70). These values have been determined for two different values of $m_{M,\beta} = m_{X,\beta}$: 10^{-3} and 10^{-2}. The parameter zm_P has been selected at six evenly spaced intervals between 10^{-3} and

TABLE 4.3 *Values of $m_{M,\alpha}$ and $m_{X,\alpha}$ for two values of m_β and a range of values of zm_P. Also listed are the ratios $(m_\alpha/m_\beta)_M$ and $(m_\alpha/m_\beta)_X$*[a]

	$m_{M,\beta} = m_{X,\beta} = 10^{-3}$				$m_{M,\beta} = m_{X,\beta} = 10^{-2}$			
zm_P	$m_{M,\alpha}$	$m_{X,\alpha}$	$(m_\alpha/m_\beta)_M$	$(m_\alpha/m_\beta)_X$	$m_{M,\alpha}$	$m_{X,\alpha}$	$(m_\alpha/m_\beta)_M$	$(m_\alpha/m_\beta)_X$
10^{-3}	6.18×10^{-4}	1.62×10^{-3}	0.62	1.62	9.51×10^{-3}	1.05×10^{-2}	0.95	1.05
2×10^{-3}	4.14×10^{-4}	2.41×10^{-3}	0.41	2.41	9.05×10^{-3}	1.11×10^{-2}	0.91	1.11
4×10^{-3}	2.36×10^{-4}	4.24×10^{-3}	0.24	4.24	8.20×10^{-3}	1.22×10^{-2}	0.82	1.22
6×10^{-3}	1.62×10^{-4}	6.16×10^{-3}	0.16	6.16	7.44×10^{-3}	1.34×10^{-2}	0.74	1.34
8×10^{-3}	1.23×10^{-4}	8.12×10^{-3}	0.12	8.12	6.77×10^{-3}	1.48×10^{-2}	0.67	1.48
10^{-2}	9.9×10^{-5}	1.01×10^{-2}	0.10	10.09	6.18×10^{-3}	1.62×10^{-2}	0.62	1.62

[a] All concentrations are in moles per kilogram of solvent. The α phase contains the positive macroions.

10^{-2}. A solution containing 1 g of colloidal electrolyte of molecular weight 10^5 per 100 g of water, for example, would have a value of $m_P = 10^{-4}$; if the macroion carries a charge of +10, the parameter zm_P equals 10^{-3}. It is evident from the table that the concentration of low-molecular-weight positive ions is larger in the β phase than in the α phase (which contains the macroions), and that the situation is reversed for the negative ions. The requirement of electroneutrality brings this about. To better show the uneven distribution of low-molecular-weight ions on the two sides of the membrane, Table 4.3 also lists the ratio of the concentrations on both sides of the membrane for both the positive and the negative ions.

Two conclusions may be drawn readily from an inspection of the results of Table 4.3. First, the uneven distribution of the simple ions becomes more pronounced as the quantity zm_P increases. The low-molecular-weight ions are free, after all, to pass through the membrane; it is only electroneutrality that holds them back. The more macroions present or the higher charge they carry, the more unsymmetrically the simple electrolyte will be distributed. Table 4.3 also shows that the uneven distribution of low-molecular-weight electrolyte becomes less pronounced as the concentration of this electrolyte is increased. We return to this point in the next section when we discuss the osmotic pressure of charged systems.

The combined effects of electroneutrality and the Donnan equilibrium permit us to evaluate the distribution of simple ions across a semipermeable membrane. If electrodes reversible to either the M^+ or the X^- ions were introduced to both sides of the membrane, there would be no potential difference between them; the system is at equilibrium and the ion activity is the same in both compartments. However, if calomel reference electrodes are also introduced into each compartment in addition to the reversible electrodes, then a potential difference will be observed between the two reference electrodes. This potential, called the membrane potential, reflects the fact that the membrane must be polarized because of the macroions on one side. It might be noted that polarized membranes abound in living systems, but the polarization there is thought to be primarily due to differences in ionic mobilities for different solutes rather than the sort of mechanism that we have been discussing. We shall return to a more detailed discussion of the electrochemistry of colloidal systems in Chap. 9.

4.9 THE OSMOTIC PRESSURE OF CHARGED COLLOIDS

Now let us turn our attention to the osmotic pressure generated by the macroion in this system. Since we have already restricted ourselves to dilute solutions, it is adequate for our purposes to substitute into Eq. (49), making allowance for the fact that we have been expressing concentrations as molality in this section. The volume of 1 kg of solvent equals $1000/\rho_1$ or $1000V_1°/M_1$, so Eq. (49) becomes

$$\pi \frac{1000V_1^0}{M_1} = mRT \tag{71}$$

where m is the molality of the solute responsible for the osmotic pressure and V_1^0 is the molar volume of the solvent. Since there are solute molecules on both sides of the membrane, the osmotic pressure will be due to the excess solute on the side of the membrane that carries the macroion (the α phase). Therefore, we may replace m in Eq. (71) by

$$m = m_{P,\alpha} + m_{M,\alpha} + m_{X,\alpha} - m_{M,\beta} - m_{X,\beta} \tag{72}$$

Substituting from the electroneutrality equations (66) and (67) transforms (72) into

$$m = m_{P,\alpha} + m_{M,\alpha} + zm_{P,\alpha} + m_{M,\alpha} - 2m_{M,\beta} \tag{73}$$

This is further modified by the Donnan relationship, Eq. (65), also rewritten to include electroneutrality, to give

$$m = m_{P,\alpha}(1+z) + 2m_{M,\alpha} - 2[m_{M,\alpha}(zm_{P,\alpha} + m_{M,\alpha})]^{1/2} \tag{74}$$

Combining Eqs. (71) and (74) yields for the osmotic pressure of the system:

$$\pi = \frac{M_1RT}{1000V_1^\circ}\left[(1+z)m_P + 2m_M - 2m_M\left(1+\frac{zm_P}{m_M}\right)^{1/2}\right] \tag{75}$$

It should be noted that all concentrations have been expressed in terms of the molality of the solute in the compartment which contains the macroion, so the α subscript is no longer necessary.

Although Eq. (75) is rather awkward as written, several highly informative variations of it are obtained by considering different limiting situations. These are summarized in Table 4.4 for the cases in which $m_M = 0$, $m_M \gg m_P$, and $m_M > m_P$. In the table, the expressions for π which follow from Eq. (75) in each of these cases are written both in terms of the molality of the macroion m_P and with the concentration of the latter expressed in weight per unit volume c_2, specifically grams of colloid per liter of solution.

We shall not present the algebraic manipulations which lead to the various forms presented in Table 4.4. However, two relationships which are involved in generating these forms might be noted. First, the quantity $(1 + zm_P/m_M)^{1/2}$ in Eq. (75) may

TABLE 4.4 *Special cases of Eq. (75) for the osmotic pressure of a charged colloid in the presence of a low-molecular-weight salt*

	m_P (moles kg^{-1})	c_2 (g liter^{-1})
$m_M = 0$	$\pi = \frac{M_1RT}{1000V_1^\circ}(1+z)m_P$	$\pi = \frac{RT(1+z)c_2}{M_2}$
$m_M \gg m_P$	$\pi = \frac{M_1RT}{1000V_1}m_P$	$\pi = \frac{RTc}{M_2}$
$m_M > m_P$	$\pi = \frac{M_1RT}{1000V_1}\left[m_P + \frac{z^2}{4}\frac{m_P^2}{m_M}\right]$	$\pi = \frac{RTc}{M_2} + \frac{z^2}{4}\frac{1000V_1^\circ RTc^2}{M_1M_2^2m_M}$
		$\frac{\pi}{RTc} = \frac{1}{M_2} + \frac{z^2 1000c}{4\rho_1 M_2^2 m_{MX}}$

be approximated by the binomial series expansion (Appendix A) to give

$$\left(1+\frac{zm_P}{m_M}\right)^{1/2}=1+\frac{1}{2}\frac{zm_P}{m_M}-\frac{1}{8}\left(\frac{zm_P}{m_M}\right)^2+\cdots \tag{76}$$

When $m_M \gg m_P$, only the first two terms are retained; when $m_M > m_P$, the first three are used. Second, the relationship between molality and grams per liter units is given by the following in dilute solutions:

$$m_P=\frac{1000\,V_1^\circ}{M_1M_2}c_2 \tag{77}$$

All the forms presented in Table 4.4 are readily obtained from Eq. (75) by incorporating either (76) or (77) or both into (75). Now let us consider the physical significance of the resulting special cases.

The two extreme values of m_M, $m_M = 0$ and $m_M \gg m_P$, are especially interesting. The former corresponds to the case of no added salt (since the macroion is positive) and the latter to a large excess or "swamping" amount of added salt. Now suppose an osmotic pressure experiment were conducted on two solutions of the same colloid, assumed to have a fixed charge z, with the objective of determining the molecular weight of the colloid. Further suppose that the two determinations differ from one another in the sense that one corresponds to zero added salt, and the second to swamping electrolyte conditions. Finally, suppose the results are simply interpreted in terms of Eq. (49) to yield the molecular weight of the colloid. Comparing Eq. (49) with the results listed in Table 4.4 reveals that the correct molecular weight would be obtained for the charged colloid under swamping electrolyte conditions, but an apparent molecular weight less than the true weight by a factor $z+1$ is obtained in the absence of salt.

How are we to understand this odd result? The answer is easy when we remember that osmotic pressure counts solute particles. The macroion cannot pass through the semipermeable membrane. In the absence of added salt, its counterions will not pass through the membrane either since the electroneutrality of the solution must be maintained. Therefore, the equilibrium pressure is that associated with $z+1$ particles. Failure to consider the presence of the counterions will lead to the interpretation of a low molecular weight for the colloid. As we already saw, the presence of increasing amounts of salt leads to a leveling off of the ion concentrations on the two sides of the membrane. The effect of the charge on the macroion is essentially "swamped out" with increasing electrolyte.

One interesting aspect of the limiting case of swamping electrolyte is the fact that the conclusion is totally independent of the specific nature of the ions. This is a partial justification for an assumption that was implicitly made at an earlier point. In writing Eqs. (66) and (67), we assume that the only ions present are M^+, X^-, and P^{+z}. In aqueous solutions, however, H^+ and OH^- are always present also. The latter have clearly been assumed to be negligible in writing Eqs. (66) and (67). The swamping electrolyte concentration may always be chosen to justify neglecting these contributions.

Table 4.4 also includes an approximation for the case in which the concentration of the salt exceeds that of the colloid, but not to the swamping extent: $m_M > m_P$.

Comparison of that case with the result given in Eq. (48) suggests that the contribution of charge to the second virial coefficient of the solution is given by

$$B = \frac{1000z^2}{4M_2^2\rho_1 m_{MX}} \tag{78}$$

Strictly speaking, this contribution should be added to the excluded volume of the particles. The latter becomes more important as the concentration of the salt increases, a conclusion that may be seen in two ways:

1. Charged colloids behave as if they were uncharged under swamping electrolyte conditions.
2. The electrostatic contribution to B, Eq. (78), is inversely proportional to salt concentration.

This effect may be qualitatively understood as follows. In a charged system the colloid consists of the macroion *and* its low-molecular-weight counterions. The latter are, of course, distributed through a portion of the solution in the neighborhood of the macroion. Thus we visualize the colloidal ion as being surrounded by an ion atmosphere, the same sort of model that is invoked in the Debye–Hückel theory of electrolyte nonideality. The "excluded volume" that is required, therefore, includes both the volume of the colloidal particle and the volume of that part of the solution which contains the counterions. It is reasonable to expect the latter to be sufficiently larger than the former in dilute solutions so that the volume actually excluded by the colloidal particle can be neglected.

The precise distribution of counterions around a charged particle is the subject matter of Chap. 9 and, to a lesser extent, Chap. 11. In those chapters, we shall see that the extent of the domain over which the ion atmosphere extends decreases as the electrolyte concentration increases.

We have already seen that the second virial coefficient may be determined experimentally from a plot of the reduced osmotic pressure versus concentration. Since all other quantities in Eq. (78) are measurable, the charge of a macroion may be determined from the second virial coefficient of a solution with a known amount of salt.

Figure 4.5 shows some actual experimental data which illustrate how the charge of the dispersed particles affects the osmotic pressure of a protein solution. In this figure, the π/c ratio is plotted versus c for bovine serum albumin molecules in 0.15 molal NaCl. The curves are determined for pH values of 7.00 and 5.37, the latter pH corresponding to a state of no net charge as determined by electrophoresis (see Chap. 11). The slope and intercepts of these curves may be interpreted as follows:

1. The common intercept (0.268 mm Hg kg g^{-1}) of the two curves yields the molecular weight of the protein:

$$M = \frac{(0.082)(298)}{(0.268/760)} = 69{,}000 \text{ g mole}^{-1} \tag{79}$$

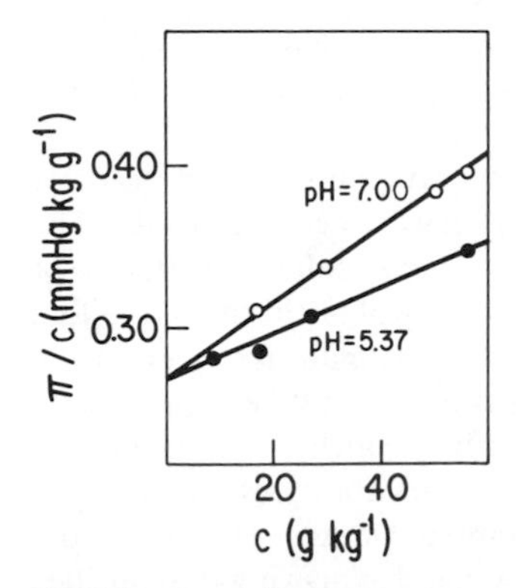

FIGURE 4.5 *π/c versus concentration for bovine serum albumin in 0.15 M NaCl at pH 7.00 and 5.37. [Taken from G. Scatchard, A. C. Batchelder, and A. Brown,* J. Am. Chem. Soc. **68**:2320 *(1946), copyright by the American Chemical Society.]*

2. The second virial coefficient for the protein solution at pH 5.37 (1.37×10^{-3} mm kg^2 g^{-2}) describes the excluded volume for this nearly spherical molecule:

$$B' = \frac{(1.37\times10^{-3})(2)(69{,}000)}{(0.268)(0.018)} \simeq 39{,}000 = 4\frac{\bar{V}_2}{\bar{V}_1} \tag{80}$$

$$\bar{V}_2 \simeq 176{,}000 \text{ cm}^3 \text{ mole}^{-1} \tag{81}$$

$$R_{\text{eq sph}} \simeq 4.1\times10^{-7} \text{ cm} \simeq 41 \text{ Å} \tag{82}$$

3. The difference between the second virial coefficients at pH 7.00 and 5.37 [$(2.28-1.37)10^{-3} = 9.1\times10^{-4}$ mm kg^2 g^{-2}] may be attributed to the charge of the particles at the higher pH:

$$z^2 = \frac{(9.1\times10^{-4})(4)(0.15)(69{,}000)}{(0.268)} \simeq 140 \tag{83}$$

$$z \simeq 12 \tag{84}$$

At pH levels above 5.37 the protein carries a negative charge. Titration results are also consistent with a value of -12 for the charge of the protein at pH 7.00.

4.10 DIALYSIS

Substances with particles in the colloidal size range are often obtained in a form which contains low-molecular-weight impurities. For example, enzymes are separated from homogenized tissue samples by extraction in a buffer solution. The enzyme preparation, therefore, is "contaminated" by the components of the buffer. Likewise, synthetic high polymers generally contain unreacted monomer, initiator, and catalyst. There are many experiments in which traces of low-molecular-weight impurities would have no effect, as, for example, in sedimentation. This chapter has shown clearly, however, that the presence of low-molecular-weight solutes may have large effects in an osmotic pressure experiment if these substances are retained by the membrane, either directly or through the electroneutrality condition.

Experiments with semipermeable membranes not only point out the need for purification but also suggest a means by which this may be accomplished. The procedure of dialysis is one technique for removing low-molecular-weight salts or nonelectrolytes from a colloid. The method consists merely of enclosing the colloid to be purified in a bag which is made of some semipermeable material. The sealed bag is then placed in a quantity of the solvent. The membrane must be permeable to the solvent and to any low-molecular-weight impurities which are present, but impermeable to the colloid. As a result of the semipermeability of the bag, the impurities will distribute themselves through both compartments. The outer portion is replaced often or even continuously, so the low-molecular-weight impurities are gradually flushed away.

Since the membrane is also permeable to the solvent, the latter simultaneously diffuses into the bag, diluting the colloid. Ample air space must be present in the bag to begin with, otherwise it will rupture due to the pressure developed by the solvent imbibed. There is also a danger that the porosity of the membrane will increase if the bag is stretched as a result of internal pressure buildup. Cellophane tubing is most commonly used as the membrane material. It is sold in rolls for this purpose and may be cut to length and tied at the ends to make the required bags.

Purification by dialysis is a slow process. Its rate is increased, however, by increasing the surface area of the membrane since that is where the exchange of solute between the two phases takes place. Stirring and frequent replacement of the solvent accelerate the process by maintaining the maximum gradient of concentration across the membrane. With ionic contaminants, the rate of dialysis may be enhanced by placing electrodes in the compartment surrounding the enclosed colloid and taking advantage of the migration of the ions in an electric field. This modification is known as electrodialysis.

Finally, it might be noted that colloids may be concentrated by a slight modification of the dialysis procedure. The liquid against which the colloid of interest is being dialyzed may itself be a concentrated colloid. With aqueous dispersions, for example, polyethylene oxide solutions may be used as the second colloid.

The second colloid is prepared at higher activity; therefore, the solvent is drawn toward the more concentrated phase. This increases the concentration of the colloid of interest. Alternatively, the concentration increase may be accomplished by allowing the solvent to evaporate from the outer surface of the bag.

Just as with osmotic pressure, the membranes in dialysis must be carefully selected to be compatible with the system under study. Specifically, this amounts to impermeability with respect to the colloid(s) involved and permeability with respect to low-molecular-weight components.

4.11 SOME APPLICATIONS OF OSMOMETRY

In this section, two applications of osmometry are presented which do not involve significant amounts of colloidal solutes. The topics have contemporary relevance, however, and are included primarily for that reason. It is not stretching the definition of colloid and surface chemistry, however, to view the membrane as a colloidal system in itself. In this case, the solid portion of the membrane would be the continuous phase and the pores, necessarily small if the membrane is to be effective, the "dispersed" phase. According to this point of view, numerous aspects of membrane technology become part of the interests of colloid and surface chemists.

The topics to be considered briefly in this section are reverse osmosis as a method of water desalination and the mixing of fresh and salt water as a source of energy.

As we saw in Sec. 4.3, samples of solution and solvent separated by a semipermeable membrane will be at equilibrium only when the solution is at a greater pressure than the solvent. This is the osmotic pressure. If the solution is under less pressure than the equilibrium osmotic pressure, solvent will flow from the pure phase into the solution. If, on the other hand, the solution is under a pressure greater than the equilibrium osmotic pressure, the pure solvent will flow in the reverse direction, from the solution to the solvent phase. In the latter case, the semipermeable membrane functions like a filter which separates solvent from solute molecules. In fact, the process is referred to in the literature by the names "hyperfiltration" and "ultrafiltration" as well as reverse osmosis. However, the latter name is enjoying common usage these days.

As we have seen already, the property of membrane semipermeability applies to all sorts of systems. Likewise, reverse osmosis may be applied to a wide variety of systems. An application that has attracted a great deal of interest in recent years is the production of potable water from saline water. Since no phase transitions are involved, as, for example, in distillation, the method offers some prospect of economic feasibility in coastal regions.

Cellulose acetate seems to be the most thoroughly investigated of many possible membrane materials. Cellulose acetate membranes are capable of yielding 96–98% retention of NaCl, for example, and of delivering about $0.2\ cm^3\ s^{-1}\ atm^{-1}\ m^{-2}$. This amounts to about $50\ gal\ day^{-1}\ ft^{-2}$ at 100 atm. At present, suitable membranes for reverse osmosis can be manufactured at a cost of less than $10¢\ ft^{-2}$. In all likelihood, future research will increase the rate of flow and lower the cost still further.

It is the rate of separation rather than the efficiency of salt retention that is the primary practical issue in the development of reverse osmosis desalination. In addition to a variety of other factors, the rate of reverse osmotic flow depends on the excess pressure across the membrane. Therefore, the problem of rapid flow is tied into the technology of developing membranes capable of withstanding high pressures. The osmotic pressure of seawater at 25°C is about 25 atm. This means that no reverse osmosis will occur until the applied pressure exceeds this value. This corresponds to a water column about 840 ft high at this temperature. Suppose a pipe were sealed at one end with a suitable membrane and that this end of the pipe were submerged more than 840 ft (or to an appropriate depth considering the temperature of the water) in seawater. Then fresh water would collect in the pipe: the arrangement would function like a freshwater well in the sea. As yet no such sea well as been developed, but numerous pieces of laboratory apparatus have been designed which operate on the basis of an applied mechanical pressure.

Note that a solution more concentrated than the original also results from the reverse osmosis process. This means that the method of reverse osmosis may also be

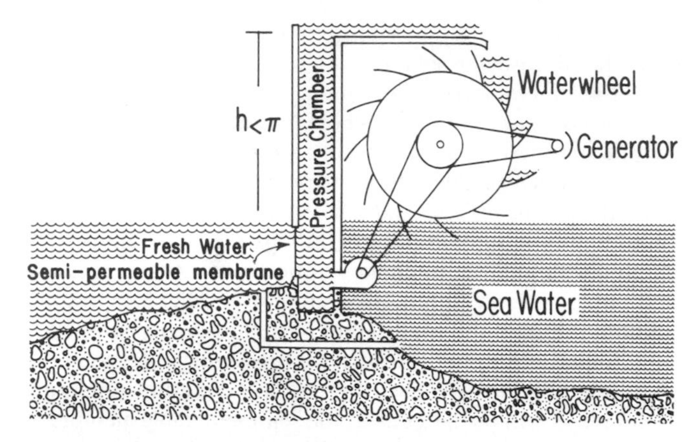

FIGURE 4.6 *Schematic of an energy converter which derives electrical energy from the free energy of mixing fresh water and salt water. [Redrawn with permission from R. S. Norman,* Science **186**:350 (1974), *copyright* 1974 *by the American Association for the Advancement of Science.*]

used as a method for concentrating solutions. Fruit juices and radioactive wastes, for example, have been concentrated by this method.

Figure 4.6 represents a less fully developed application of osmometry, but one that also illustrates the potential usefulness of the earth's resources in this regard. In essence, a tall chamber would be constructed at the mouth of a river at a point where fresh water and salt water mix. Below the liquid surface on the freshwater side, a semipermeable membrane like those used in reverse osmosis would be built into the wall of the chamber. The chamber itself would be filled with seawater as shown in the figure. As already noted, the osmotic pressure of seawater at 25°C is sufficient to support a pressure head over 800 ft high. In this application, a chamber is constructed that is less than the equilibrium column in height. Solvent will then flow through the membrane to move the system toward osmotic equilibrium, but the chamber will overflow before the equilibrium pressure is reached. The overflow is then used to run a generator to produce electrical energy.

It is essentially the free energy of mixing fresh water and salt water that is being tapped as a source of useful energy in this application. The method depends on the properties of semipermeable membranes to develop a waterfall in the mouth of a river. The enormous potential of this application may be appreciated by imagining the mouth of every stream and river in the world terminated by a waterfall over 800 ft high. It must also be conceded, however, that the effect of such developments on the estuarine environment would be catastrophic.

As with reverse osmosis, the critical consideration in evaluating the feasibility of this idea must be the rate at which the water is transported through the membrane. Semipermeability with respect to ionic solutes, per se, is not the limiting factor.

REFERENCES

1. A. W. Adamson, *A Textbook of Physical Chemistry*, Academic Press, New York, 1973.
2. R. U. Bonnar, M. Dimbat, and F. H. Stross, *Number Average Molecular Weights*, Wiley-Interscience, New York, 1958.
3. P. J. Flory, *Principles of Polymer Chemistry*, Cornell University Press, Ithaca, New York, 1953.
4. S. Souririjan, *Reverse Osmosis*, Academic Press, New York, 1970.
5. C. Tanford, *Physical Chemistry of Macromolecules*, Wiley, New York, 1961.

PROBLEMS

1. Criticize or defend the following proposition: "As proof that no low-molecular-weight fractions of polymer have passed through the membrane in an osmotic pressure experiment, the following test may be performed. A quantity of 'poor' solvent is added to an aliquot taken from the solvent side of the membrane. The absence of precipitate proves that no low-molecular-weight polymer passed through the membrane."
2. The osmotic pressure of solutions of a fractionated, atactic poly(isopropyl acrylate) solution was measured at 25°C with the following results*:

π (g cm^{-2})	1.39	2.46	4.20	6.52
$c \times 10^2$ (g cm^{-3})	0.47	0.69	1.05	1.36

Prepare a plot of π/c versus c for these results and evaluate $(\pi/c)_0$. Calculate M and B for this system. Define what is meant by an "atactic" polymer and compare with syndiotactic and isotactic polymers. List reference(s) consulted for these definitions.

3. The osmotic pressure of cellulose nitrate solutions in acetone was measured at 25°C in an apparatus like that shown in Fig. 4.2c.† For two different molecular weight fractions, the following data were obtained:

Sample BA	$\pi/c \times 10^4$ (liter atm g^{-1})	2.1	2.7	3.6	4.5
	c (g liter^{-1})	3.9	5.9	8.9	11.8
Sample C	$\pi/c \times 10^4$ (liter atm g^{-1})	4.0	4.3	5.1	5.5
	c (g liter^{-1})	2.7	3.9	7.9	9.9

Evaluate M and B for these two samples.

* J. E. Mark, R. A. Wessling, and R. E. Hughes, *J. Phys. Chem.* **70**:1895 (1966).
† A. M. Holtzer, H. Benoit, and P. Doty, *J. Phys. Chem.* **58**:624 (1954).

4. When the osmotic pressure of a fractionated polystyrene sample was studied at 25°C in dioxane and in chlorobenzene the following data were obtained*:

Chloro-benzene	c (g cm^{-3})	0.282	0.468	0.498	0.616	0.953	1.638	2.770
	π/c (cm)	2.90	3.20	3.44	3.49	4.05	5.28	7.75
Dioxane	c (g cm^{-3})	0.502	0.691	0.983	1.007	1.416	1.976	3.094
	π/c (cm)	3.08	3.13	3.42	3.33	3.89	4.24	7.29

Evaluate M for the polystyrene and B for the two polymer–solvent systems. Which of these solvents is the better solvent for this polymer? Are these results consistent with your expectations based on the structure of these molecules?

5. An important assumption made in truncating Eq. (48) is that the third virial coefficient is small. It is known (cf. Flory [3]) that the third virial coefficient depends strongly on the second so that it approaches zero in poor solvents even faster than B does. In fact, if Γ_2 is defined as the product BM_2, it is known that Γ_3 is approximately $0.25\Gamma_2{}^2$ in a good solvent. That is, Eq. (48) may be written with one additional term as

$$\frac{\pi}{RTc} = \frac{1}{M}(1 + \Gamma_2 c + 0.25\Gamma_2{}^2 c^2)$$

Describe how this result may be used to facilitate the evaluation of M and B in the event that a plot of reduced osmotic pressure versus c still contains too much curvature to permit a meaningful straight line to be drawn.

6. Osmotic pressures for aqueous solutions of n-dodecyl hexaoxyethylene monoether, $C_{12}H_{25}(OC_2H_4)_4OC_2H_5$, were measured at 25°C. At concentrations below 0.038 g liter^{-1} no osmotic pressure develops, indicating complete membrane permeability. Above this concentration a pressure develops, indicating the presence of impermeable species. In the following data† these pressures are reported for various $c - c_0$ values, where $c_0 = 0.038$ g liter^{-1}, the threshold for an osmotic effect:

π (cm)	4.90	6.53	7.62	10.58
$c - c_0$ (g liter^{-1})	29.72	38.12	43.90	58.46

Plot $\pi/(c - c_0)$ versus $c - c_0$ to evaluate the molecular weight and B for the species responsible for the osmotic pressure.

7. The data of the preceding problem may be interpreted by assuming the following model. Above 0.038 g liter^{-1}, solute molecules associate into aggregates. By comparing the M value obtained in Problem 6 with the molecular weight of the original ether, calculate the number of molecules in the aggregate

* J. Leonard and H. Daoust, *J. Phys. Chem.* **69**:1174 (1965).
† D. Attwood, P. H. Elworthy, and S. B. Kayne, *J. Phys. Chem.* **74**:3529 (1970).

according to this model. Assuming the colloidal particles are spherical, use the second virial coefficient evaluated in Problem 6 to estimate the molar volume and radius of the aggregates (see Table 4.2). Do the quantities calculated in this problem seem reasonably self-consistent?

8. Solutions of bovine serum albumin in 0.15 *M* NaCl were studied at other pH levels in addition to those shown in Fig. 4.5. The following data are examples of additional measurements (reference in legend of Fig. 4.5):

pH	m_P (g protein kg^{-1})	π (mm Hg)
6.19	57.71	21.48
6.64	56.17	21.40
4.23	9.63	2.65

Since the limiting value of π/m_P shown in Fig. 4.5 applies to these data also, it is possible to evaluate the second virial coefficient at these pH levels. Evaluate B and z, the effective protein charge, at the pH values shown. In what way does the value of z at pH 4.23 differ from the z values at other pH levels? (*Note*: Slight variation in NaCl concentrations at these different pH levels should be taken into account for more accurate determination of z.)

9. The osmotic pressure of salt-free (electrodialyzed) bovine serum albumin solutions was measured at pH = 5.37(±?) (reference listed in Fig. 4.5). At this pH, the net charge of the protein molecules is zero. The following data were obtained in different runs:

π (mm Hg)	4.26	7.44	8.14	12.97	11.31	19.62
m_P (g protein kg^{-1})	19.56	19.71	40.88	46.05	60.81	63.25

Discuss the reasons why it is so difficult to obtain meaningful osmotic pressure data in salt-free solutions. Consider specifically the reconciliation of the electrolyte-free aspect of the experiment with the accurate control of pH.

10. In reverse osmosis, both solvent and solute diffuse because of gradients in their chemical potentials. For the solvent, there is no gradient of chemical potential at an osmotic pressure of π; at applied pressures p greater than π, there is such a gradient which is proportional to the difference $p - \pi$. To a first approximation, the gradient of the solute chemical potential is independent of p and depends on the difference between concentrations on opposite sides of the membrane. This leads to the result that the fraction of solute retained varies as $[1 + \text{const}/(p - \pi)]^{-1}$. Verify that the following data* for a reverse osmosis experiment with 0.1 *M* NaCl and a cellulose acetate membrane follow this relationship.

Applied p (atm)	10	13	20	38	51	75
% salt retained	63	79	88	94	95	97

(π is about 2.6 atm for 0.1 *M* NaCl.)

* Data of J. E. Breton, Jr., cited by H. K. Lonsdale, in *Desalination by Reverse Osmosis* (U. Merton, ed.), MIT Press, 1966.

Light Scattering

5

... an interesting and oft-investigated question, "What is the origin of light?" and the solution of it has been repeatedly attempted, with no other result than to crowd our lunatic asylums with the would-be solvers. Hence, after fruitless attempts to suppress such investigations by making them liable to a heavy tax, the Legislature ... absolutely prohibited them. [From Abbott's *Flatland*]

5.1 INTRODUCTION

In Chap. 1 we described dark field microscopy in which particles too small for direct microscopic observation could be detected against a dark field by horizontal illumination. Airborne dust or smoke particles show up in a beam of light in an otherwise dark room in the same way. In both cases, the particles interact with the light which strikes them and deflect some of that light from its original direction. We speak of this light as being "scattered." A whole assortment of optical phenomena related to this are generally known as light scattering effects.

It turns out that the intensity of scattered light at any angle depends on the wavelength of the incident light, the size and shape of the scattering particles, and the optical properties of the scatterers as well as the angle of observation. Further, the functional relationship among these variables is known, at least for spherical particles and other geometries, under certain circumstances. By applying these relationships to light scattering experiments, information about the particles responsible for the scattering can be deduced. We shall see how the weight and a characteristic linear dimension of the particle may be determined for some systems from light scattering.

A general relationship for the intensity of light scattered by a spherical particle was derived by L. Lorenz in the latter part of the nineteenth century and applied to colloids by G. Mie in 1908. Some quotations from two of the references at the end of this chapter will give an indication of the historical development of light scattering since Mie's complicated theory was presented. In a book published in 1956 [4], Stacey remarks that scattering patterns "have been tabulated in only a few cases because the computation is so laborious" [4, p. 56]. In his 1969 book [2], Kerker writes of the same patterns that so many have been published "that these can hardly be coped with in the usual tabular form, much less published in the normal way" [2, p. 77]. Elsewhere, Kerker notes that "there were more than 400 publications dealing with molecular weight determination by light scattering during the two-year period 1964–65" [2, p. 509].

We might ask what happened in the period of slightly more than one decade between the Stacey and Kerker books to account for the differences they reveal. The answer is that during this period high-speed computers were applied extensively to the computational problems that arise in light scattering theory. Light scattering is a field which is still growing rapidly; another decade will probably reveal numerous additional applications of this technique.

Even though there was a lapse of more than 50 years between the Mie theory and its extensive application, this does not mean that light scattering was unknown or unused during this period. On the contrary, much effort was devoted to finding approximations to the rigorous theory that would be easier to use. Lord Rayleigh's work (which was published in 1871, historically preceding the general theory) consists of an approximation which applies only to very small particles whose refractive index is not too large. Similarly, Debye's approximation (1909) applies to slightly larger particles than the Rayliegh approximation, but to a narrower range of refractive indices. The Rayleigh and Debye approximations have proved to be extremely valuable in the characterization of certain colloids, especially polymeric materials whose refractive index is not too much different from that of the solvent.

In this chapter, we shall derive the Rayleigh approximation in considerable detail. The Debye approximation is described in enough detail to reveal its utility. The general theory itself is discussed primarily as it applies to several specific systems, rather than in any comprehensive way.

5.2 ELECTROMAGNETIC RADIATION

The phenomena with which we are concerned in this chapter are displayed by the entire spectrum of electromagnetic radiation. In applications to colloid chemistry, however, most work is done in the visible part of the spectrum; hence, the common designation: *light* scattering. Visible light shares a variety of parameters and descriptive relationships with other regions of the electromagnetic spectrum. The purpose of this section is to examine briefly some of the characteristics of electromagnetic radiation, particularly those that are needed for an understanding of light scattering.

Electromagnetic radiation consists of oscillating electrical (E) and magnetic (H) fields which are perpendicular to each other and are both perpendicular to the direction of propagation of the wave, as shown in Fig. 5.1. Under vacuum, the velocity of propagation of an electromagnetic wave c is about 3×10^{10} cm s^{-1} and is independent of the wavelength of the radiation. The frequency ν, wavelength λ_0, and velocity of the radiation are related through the familiar equation

$$c = \lambda_0 \nu \tag{1}$$

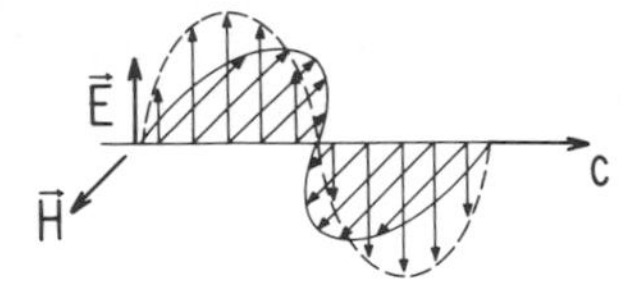

FIGURE 5.1 *The relationship between the electric and magnetic fields and the direction of propagation of electromagnetic radiation.*

If the radiation is passing through a medium other than vacuum, its velocity and wavelength are both diminished by a factor n, the refractive index of the medium. Then Eq. (1) becomes

$$\frac{c}{n}=\frac{\lambda_0}{n}\nu \tag{2}$$

or

$$v=\lambda\nu \tag{3}$$

We shall use the symbols c and λ_0 to refer to the velocity and wavelength under vacuum only. In this chapter, the symbol λ, without a subscript, always refers to the wavelength of the radiation in the medium.

The dominant characteristic of the electrical and magnetic fields that comprise this radiation is their periodically oscillating nature, a fact that enables us to describe them by the mathematics of waves. As far as light scattering is concerned, it is the electric field in which we are interested. The oscillating nature of an electric field propagating in the positive x direction is described by the equation

$$\mathbf{E}=\mathbf{E}_0\cos 2\pi\left(\nu t-\frac{x}{\lambda}\right) \tag{4}$$

in which $\mathbf{E}_0$ is the maximum amplitude of the field.

Since we have postulated the x direction as the direction of propagation, the electric field lies in the yz plane and may, in general, be resolved into y and z components, since $\mathbf{E}$ is a vector. Both the y and z components of the field are described by Eq. (4) when the latter is modified by the inclusion of phase angles. This is because the two components need not be in phase with each other. Accordingly, we write

$$\mathbf{E}_y=\mathbf{E}_{0y}\cos 2\pi\left(\nu t-\frac{x}{\lambda}+\delta_y\right) \tag{5}$$

and

$$\mathbf{E}_z=\mathbf{E}_{0z}\cos 2\pi\left(\nu t-\frac{x}{\lambda}+\delta_z\right) \tag{6}$$

in which the δ terms are the phase angles of the two components. In the most general case, these two equations mean that the electric field vector traces an ellipse in the yz plane. There are two special cases of note within this general situation. If the phase difference between the two components of the field, $\delta_y - \delta_z$, is zero or some integral multiple of π, the ellipse flattens to a line. If the phase difference is $\pi/2$ or any odd integral multiple of $\pi/2$ and the amplitudes of the two components are equal, the ellipse is rounded to a circle. In the former case, we speak of the radiation as being plane polarized; and in the latter case, as circularly polarized.

Ordinary light is said to be unpolarized. This last term is somewhat unfortunate because all light displays some form of polarization. In ordinary light, however, all forms of polarization are present so the individual effects cancel out. The use of various filters makes it possible to conduct experiments with radiation which shows a unique state of polarization. Polaroid filters, for example, transmit plane-polarized light only. In discussing light scattering, we are concerned primarily with unpolarized light and, occasionally, with plane-polarized light.

An interesting example of polarization arises in the study of light reflected from a surface. Suppose we consider a beam of light incident upon the planar surface of some material having a higher refractive index than the medium from which the beam approaches. At the surface, some of the light will be refracted, and some will be reflected. Figure 5.2 illustrates this for the case in which the reflected and the refracted beams are separated by an angle of 90°. In this situation, two very different results are obtained, depending on whether the incident light is linearly polarized in the plane of the figure or perpendicular to it. If the light is polarized in the plane of the figure, no light will be reflected at all. On the other hand, if it is initially polarized perpendicular to the plane of the figure, it will be reflected with the same polarization. If ordinary light—a mixture of the two—is used for the incident radiation, only one of the components contributes to the reflection. Furthermore, the reflected light is polarized perpendicular to the plane of the figure. Polaroid filters are used in photography and in sunglasses to reduced the glare of reflected light since this light is polarized. This behavior is not observed uniquely when the angle between the two beams is 90°; rather, the intensity of the reflected beam varies continuously with the angle. At 90°, however, the polarization effect is most pronounced. We shall see that some scattering phenomena also show an angular dependence. Included among these is the fact that scattered light displays maximum polarization at 90° also.

FIGURE 5.2 *The relationship between the incident, reflected, and refracted beams of radiation at a plane surface.*

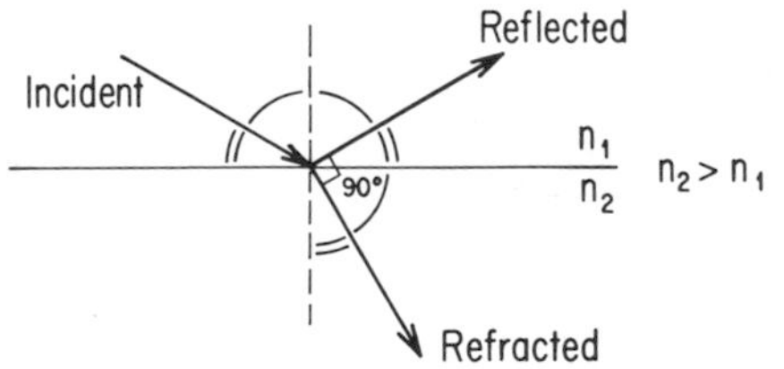

5.3 OSCILLATING ELECTRIC CHARGE

In this section we discuss the interaction between an electric field and a charge which is free to move with the field. Such a charge experiences a force which accelerates it with the field. If the field is oscillating, the acceleration of the charge will also oscillate. One of the basic results of classical electromagnetics is that the acceleration of a charge leads to the emission of radiation. It was the apparent violation of this requirement that led to the postulate of quantization in the Bohr theory of the hydrogen atom. However, we are concerned here with the classical result in which the charge does radiate. Our objective is to describe the emitted radiation some distance r from the emitter.

The radiation emitted by an oscillating charge may be described by its electric field vector which is given by

$$\mathbf{E} = \frac{q\mathbf{a}_p \sin \phi_z}{c^2 r} \tag{7}$$

according to electromagnetic theory. In this equation, q equals the magnitude of the charge, $\mathbf{a}_p$ is its periodic acceleration, and c is the velocity of light. The coordinate system defined by Fig. 5.3a will help describe this field. The origin of the coordinates is located at the emitting charge. The angles between the line of sight—along which r is measured—and the x, y, and z axes are designated ϕ_x, ϕ_y, and ϕ_z, respectively.

The oscillating charge behaves like an antenna and Eq. (7) describes the field of such an antenna as long as r is large compared to the wavelength of the radiation which induces the oscillation. It should also be noted that the antenna to which Eq. (7) applies is aligned vertically (z axis) and is therefore "driven" by vertically polarized radiation.

We are using Eq.. (7), presented without proof, as the point of departure for our discussion of light scattering. Therefore, it is important that we find its predictions reasonable. First, let us consider the plausibility of the $\sin \phi_z$ factor. This factor ranges between 0 and 1 as ϕ_z varies from 0 to $\pi/2$. This means that the maximum field will be observed at right angles to the oscillating charge and no field will be observed along the axis of the oscillation. It is the projection of the acceleration in the plane perpendicular to the line of sight that induces the field. The strength of the field is proportional to this projection at any location, as shown in Fig. 5.3b. This

FIGURE 5.3 *(a) The coordinates of an electric field* **E** *relative to an oscillating charge located at the origin. (b) Projection of the acceleration in the plane perpendicular to the line of sight.*

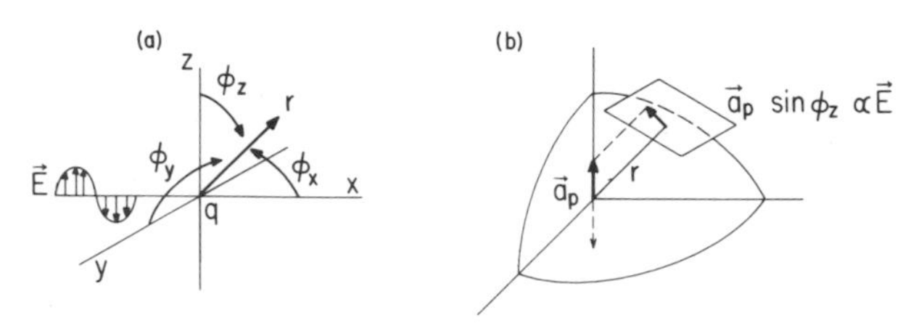

factor describes the entire angular dependence of the induced field produced by a vertical driving field. Since it depends on the angle ϕ_z alone, the induced field is seen to be symmetrical with respect to the z axis.

Next, we note that the induced field varies inversely with r. It makes sense that the field should decrease as we get farther from the antenna, but the inverse first power dependence may be unexpected since we are more familiar with inverse-square laws. However, it is the energy or intensity of the light that varies according to an inverse-square law. In the next section, we convert this expression for $\mathbf{E}$ to an expression for energy; the more familiar r^{-2} functionality will appear then.

Finally, it is sufficient for our purposes to think of the factor c^{-2} in Eq. (7) as supplying dimensional consistency to the equation. Note that $\mathbf{E}$ must have units of charge per square length.

Equation (7) describes the field emitted by an antenna which, in turn, is driven by another field. The oscillation of one field promotes the oscillation of a charge in the antenna, and this induces another electric field. The frequency is the same for all three.

This description of antennas may seem more appropriate to a discussion of radio or television waves. We must realize, however, that at the molecular level dipoles behave exactly like antennas. Since molecules are made up of charged parts, a dipole moment $\boldsymbol{\mu}$ is induced in any material through which radiation passes by the electric field of the radiation. We shall consider isotropic materials characterized by a polarizability α. In this case, the dipole moment and the field are related by the expression

$$\boldsymbol{\mu} = \alpha \mathbf{E} \tag{8}$$

If we imagine the molecule to lie at the origin of a coordinate system so that $x = 0$, we may substitute Eq. (4) into (8) to obtain

$$\boldsymbol{\mu} = \alpha \mathbf{E}_0 \cos 2\pi\nu t \tag{9}$$

A dipole moment may be regarded as the product of the distance ξ between two charges times the magnitude of the charge q. A useful way of looking at Eq. (9) is to identify the charge as

$$q = \alpha \mathbf{E}_0 \tag{10}$$

and the separation of the charges as

$$\xi = \cos 2\pi\nu t \tag{11}$$

Then the periodic acceleration of the charge is given by

$$\mathbf{a}_p = \frac{d^2\xi}{dt^2} = -4\pi^2\nu^2 \cos 2\pi\nu t \tag{12}$$

Equations (10) and (12) may now be substituted into (7) to give

$$\mathbf{E} = -\frac{\alpha \mathbf{E}_0 4\pi^2\nu^2 \cos(2\pi\nu t) \sin \phi_z}{c^2 r} \tag{13}$$

For maximum generality, we must remember that the field is periodic in space as well as in time; hence, the cosine factor in Eq. (13) is corrected by analogy with Eq. (4):

$$\mathbf{E} = -\frac{4\pi^2\nu^2\alpha\mathbf{E}_0 \cos 2\pi(\nu t - r/\lambda) \sin \phi_z}{c^2 r} \tag{14}$$

Equation (14) describes the induced field a distance r from the dipole.

5.4 RAYLEIGH SCATTERING

In this section, we discuss the first of several light scattering theories to be considered in this chapter, Rayleigh scattering. We shall see presently that Rayleigh scattering applies only when the scattering centers are small in dimension compared to the wavelength of the radiation. As such, it is severely limited in its applicability to colloidal particles, at least when visible light is the radiation involved. Rayleigh scattering is the easiest of the scattering theories to understand, however, so it is a logical place to begin. Furthermore, we shall extend its applicability to larger particles in later sections by introducing suitable correction factors.

When a beam of radiation is incident upon a molecule, a certain fraction of that radiation will undergo the process described in the preceding section and be emitted by the dipole. Any light that does not interact this way will continue past the molecule along the original path. This undeviated or transmitted light will be attenuated compared to the incident light since some of the original beam of light is scattered from its initial path. Note that this attenuation has nothing to do with absorption: The effect we are considering is a classical result and does not involve transitions between quantum states.

In order to evaluate the amount of light which is transmitted past a scattering center, it is necessary to add up the light scattered over all possible angles since this is the light responsible for the attenuation. In saying that the total light equals the sum of the transmitted and the scattered light we are merely stating that energy is conserved. In other words, the summation of which we speak must be carried out using the energy of the light rather than its electric field strength.

Light intensity is defined as the radiation energy falling on a unit of area in a unit of time. The ratio of intensities, therefore, is the same as the ratio of energies at a given location. Intensity is also proportional to the square of the electric field of the radiation. According to Eq. (14), therefore, the intensity i of light scattered at an angle ϕ_z and measured a distance r from the scattering is

$$i \propto \frac{16\pi^4\nu^4\alpha^2\mathbf{E}_0^2 \cos^2 2\pi(\nu t - r/\lambda) \sin^2 \phi_z}{c^4 r^2} \tag{15}$$

By analogy, the intensity of the incident light I_0 is

$$I_0 \propto \mathbf{E}_0^2 \cos^2 2\pi\left(\nu t - \frac{r}{\lambda}\right) \tag{16}$$

Since the proportionality factor is the same in both of these expressions, we may write the ratio of intensities as

$$\frac{i_v}{I_{0,v}} = \frac{16\pi^4\nu^4\alpha^2 \sin^2 \phi_z}{c^4r^2} \tag{17}$$

Note that both the numerator and the denominator of the ratio have the subscript v. This indicates vertical polarization and reminds us that the entire development of the argument since Eq. (7) has been based on the assumption that the original radiation was polarized in the vertical plane. The scattered radiation, therefore, also possesses this polarization.

Now it is relatively easy to consider the scattering pattern that arises when the original radiation is polarized horizontally. In this case, the electric field of the original radiation would lie in the xy plane of Fig. 5.3a. It is still the tangential projection of the dipole in which we are interested; this would be given by the factor $\sin \phi_y$, where ϕ_y is the angle between the y axis and r as shown in Fig. 5.3a. Therefore, we may directly write the expression for $i_h/I_{0,h}$, the scattered intensity relative to the incident intensity when the latter is horizontally polarized:

$$\frac{i_h}{I_{0,h}} = \frac{16\pi^4\nu^4\alpha^2 \sin^2 \phi_y}{c^4r^2} \tag{18}$$

The angular parts of these two scattered intensity functions are illustrated in Fig. 5.4. It should be recalled that the scatter pattern for i_v is symmetrical with respect to the z axis. The actual intensity envelope for the vertically polarized light is the figure of revolution that results from rotating Fig. 5.4a around the z axis. Likewise, the envelope for the horizontal component results from rotating Fig. 5.4b around the y axis.

FIGURE 5.4 *The intensity pattern for light scattered by a particle located at the origin. The incident and the scattered light are vertically polarized in (a) and horizontally polarized in (b). Length of radius vector proportional to scattered intensity at each angle.*

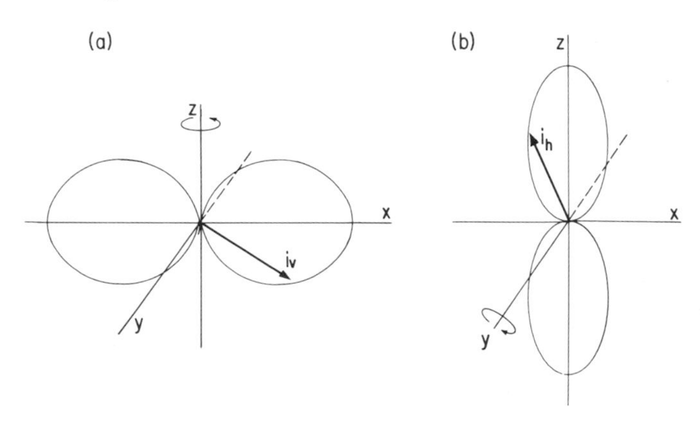

If the original light is unpolarized, we may think of it as consisting of equal amounts of horizontally and vertically polarized light. In this case the intensity ratio equals

$$\frac{i}{I_{0,u}} = \frac{\frac{1}{2}(i_v + i_h)}{I_{0,u}}$$

$$= \frac{1}{2} \frac{16\pi^4 \nu^4 \alpha^2}{c^4 r^2} (\sin^2 \phi_y + \sin^2 \phi_z) \tag{19}$$

where the subscript u reminds us that the incident light in unpolarized and the factor $\frac{1}{2}$ arises because two equal incident sources have been combined. In this case, the scattered light consists of a horizontally polarized component (proportional to $\sin^2 \phi_y$) and a vertically polarized component (proportional to $\sin^2 \phi_z$). It is inconvenient to use the two different angles to describe the intensity pattern for light scattered from an initially unpolarized source. Fortunately, it is not difficult to replace them by a single angle.

As shown in Fig. 5.3a, ϕ_x is the angle between the x axis and r. This means that $r \cos \phi$ is the projection of r onto the x, y, or z axis depending on which angle is considered. It follows, therefore, that

$$r^2(\cos^2 \phi_x + \cos^2 \phi_y + \cos^2 \phi_z) = r^2 \tag{20}$$

Replacing $\cos^2 \phi$ by $1 - \sin^2 \phi$ and rearranging leads to

$$\sin^2 \phi_y + \sin^2 \phi_z = 1 + \cos^2 \phi_x \tag{21}$$

This permits us to rewrite Eq. (19) as

$$\frac{i}{I_{0,u}} = \frac{8\pi^4 \nu^4 \alpha^2}{c^4 r^2} (1 + \cos^2 \phi_x) \tag{22}$$

A plot of Eq. (22) is shown in Fig. 5.5. The intensity is clearly a maximum for $\phi_x = 0$ and π, and falls to half this value at $\phi_x = \pi/2$. When $\phi_x = \pi/2$, the scattered light is totally polarized since only one of the components contributes to the intensity

FIGURE 5.5 *Projection in the xz plane of i_v, i_h, and their resultant i_{total} for light scattered by a molecule situated at the origin. The intensity of scattered unpolarized light at any angle ϕ_x is proportional to the length of the radius vector at this angle.*

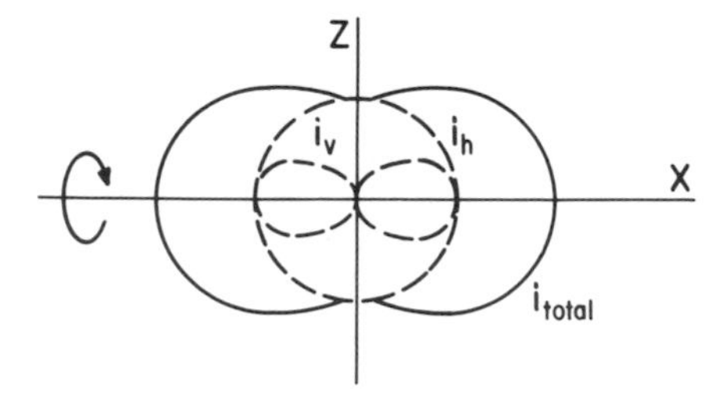

at this angle. The envelope for the total intensity of scattered unpolarized light is generated by rotating the contour for total intensity around the x axis. Figure 5.5 also includes the contributions of the vertically and horizontally polarized components in the xz plane. Equation (22) and its precursors (17) and (18) may be written in a slightly different form by incorporating Eq. (1) into them. Thus, Eq. (22) becomes

$$\frac{i}{I_{0,u}} = \frac{8\pi^4\alpha^2}{\lambda_0^4 r^2}(1+\cos^2\phi_x) \tag{23}$$

This equation gives the intensity of the light scattered by a single molecule to the point described by the coordinates r and ϕ_x when the incident light is unpolarized.

Next, we must consider how to evaluate the light scattered by an array of molecules. If the molecules in a scattering sample are far apart, as in a gas, then each may be regarded as an independent source of scattered light. Then we may multiply Eq. (23) by the number of molecules per unit volume to get an expression for the light scattered from that volume element. The number of molecules per unit volume is given by $N_A\rho/M$ where ρ and M are the density and molecular weight of the scattering gas and N_A is Avogadro's number. Then Eq. (23) becomes

$$\frac{i_s}{I_{0,u}} = \frac{8\pi^4 N_A\rho\alpha^2}{r^2\lambda_0^4 M}(1+\cos^2\phi_x) \tag{24}$$

where i_s is the intensity of light scattered *per unit volume* (subscript s).

It is shown in physical chemistry that the polarizability and refractive index of dielectric particles are related by the expression

$$\alpha = \frac{3M}{4\pi N_A\rho}\frac{n^2-1}{n^2+2} \tag{25}$$

A particularly useful form of this relationship applies to material for which n is not much larger than unity, as would be the case for a nonabsorbing gas. In this case,

$$\frac{n^2-1}{n^2+2} = \frac{(n+1)(n-1)}{n^2+2} \simeq \frac{2}{3}(n-1) \tag{26}$$

and

$$\alpha \simeq \frac{M}{2\pi\rho N_A}(n-1) \tag{27}$$

Therefore, Eq. (24) may be written

$$\frac{i_s}{I_{0,u}} = \frac{2\pi^2 M}{r^2\lambda_0^4 N_A\rho}(n-1)^2(1+\cos^2\phi_x) \tag{28}$$

The results we have considered in this section were derived by Lord Rayleigh in 1871. A rather interesting early application of Eq. (28) was the determination of Avogadro's number by measuring the light scattered by air. Equation (28) also explains why light scattered by the earth's atmosphere appears blue. This arises from the inverse fourth power dependence on λ_0 for i_s. Suppose, for example, that two radiations are compared whose wavelengths differ by a factor of 2. Then the scattered intensity of the shorter wavelength will be 16 times as great as that of the longer wavelength. Although red and blue light do not differ by quite this much in wavelength, the blue component of white light, say sunlight, is scattered very much more than the red. Accordingly, the sky overhead appears blue. At sunset, we see mostly transmitted light. Since the blue has been most extensively removed from this by scattering, the sky appears red at sunset. This simple picture does not apply when the scattering molecules are absorbing or when the atmosphere contains dust particles, water drops, or other particles whose dimensions are larger than ordinary gas molecules.

5.5 RAYLEIGH SCATTERING APPLIED TO SOLUTIONS: FLUCTUATIONS

A crucial aspect of the transition from Eq. (23) to Eq. (24) is the requirement that the individual molecules are far enough apart to be treated as independent sources. This assumption is justified for gases, but in liquids the molecules are close enough together so that interference occurs between the waves emitted from different centers. As a matter of fact, there would be complete destructive interference of all scattered light in liquids if the molecules were randomly arranged and were stationary. Nevertheless pure liquids do scatter light. It is not the individual molecules which are the scattering centers in this case, but rather the small domains of compression or rarefaction that arise from fluctuations.

We saw in Chap. 3 that molecular motion results in small fluctuations in density at the molecular level. Although the average density of a liquid is a constant equaling the experimental density, there will be small transient domains within it which have densities larger or smaller than the mean value.

Liquid solutions also scatter light by a similar mechanism. In the case of a solution, the scattering may be traced to two sources: fluctuations in solvent density and fluctuations in solute concentration. The former are most easily handled empirically by subtracting a solvent "blank" correction from measurements of the intensity of light scattered from solutions. What we are concerned with in this section, then, is the remaining scattering, which is due to fluctuations in the solute concentration in the solution.

A fluctuation in the concentration of a small volume element δV of solution will result in a change in the properties of that volume element. We begin the analysis of this situation by defining δc and $\delta \alpha$ as the fluctuations in solution concentration and polarizability, respectively, in this element. The first thing that we recognize about these quantities is that their average values are zero since both positive and negative fluctuations occur. Although their averages may be zero, the averages of their squares are not zero. Remember that similar situations were encountered in Secs. 3.11 and 3.12 where we discussed particle displacements due to diffusion and segment displacements in a random coil. In both of these cases, it was considering the average values of the square of the displacements that meaningful quantities could be obtained. Similarly, the average values of δc and $\delta \alpha$ are zero, but their mean square values will be different from zero. Equation (23) shows that the intensity of scattered light depends on the square of polarizability. We conclude, therefore, that the way to adapt this equation to the scattering by solutions is to replace α^2 in Eq. (23) by $\overline{\delta\alpha^2}$:

$$\frac{i}{I_0} = \frac{8\pi^4 \overline{\delta\alpha^2}}{r^2\lambda^4}(1 + \cos^2 \phi_x) \tag{29}$$

Note that the wavelength of light in the medium, λ, rather than the value under vacuum is used in this expression, since this is the light that reaches the scattering center.

Next, we must consider how to extend this result to a unit volume of solution and how to relate $\delta\alpha$ to δc. The steps involved in these extensions are not difficult but they are lengthy. Accordingly, we shall not develop the entire argument in detail. Instead, some of the key steps in the development along with a brief justification for each are summarized in Table 5.1. In this table, each major substitution is presented along with variations of Eq. (29) which reflect the cumulative effects of all the substitutions. The first entry in the table, for example, converts Eq. (29) to an expression for the relative light scattered by a unit of volume i_s/I_0 by multiplying (29) by the number of volume elements in 1 cm^3: $1/\delta V$.

Although the justifications in Table 5.1 are sketchy, they provide hints which will show the interested reader how to proceed in order to develop the required relationship in detail. Only the third entry in the table, the connection between $\overline{\delta c^2}$ and $\partial^2 G/\partial c^2$ requires a more elaborate proof than the qualitative justification supplied.

The cumulative result of these substitutions is to replace $\overline{\delta\alpha^2}$ in Eq. (29) by a number of other factors, all of which are experimentally measurable:

1. The refractive index gradient, dn/dc. This is simply the local slope of a plot of the refractive index of a solution versus its concentration.
2. The concentration of the solution, c. This is expressed in units of grams per volume.
3. The quantity $\partial\pi/\partial c$ is evaluated for an equilibrium solution. This is the significance of the subscript 0.

In Chap. 4, we developed expressions for the equilibrium osmotic pressure of a solution as a function of its concentration. In view of these substitutions, Eq. (29) becomes

$$\frac{i_s}{I_0} = \frac{2\pi^2 (n\, dn/dc)^2 kTc}{r^2\lambda^4 (\partial\pi/\partial c)_0}(1 + \cos^2 \phi_x) \tag{30}$$

Equation (4.48) may be written

$$\pi = RT\left(\frac{c}{M} + Bc^2\right) \tag{31}$$

by incorporating the value of the second virial coefficient supplied by (4.50). Since

TABLE 5.1 *Some key substitutions and their justifications for the transformation of Eq. (29) to Eq. (30)*

Substitution	Justification	Cumulative effect on Eq. (29)
$\frac{\overline{\delta\alpha^2}}{\text{volume}} = \frac{1}{\delta V} \frac{\overline{\delta\alpha^2}}{\text{fluctuation}}$	(1) $\frac{1}{\delta V}$ domains of volume δV can fit into 1 cm^3 of solution	$\frac{i_s}{I_0} = \frac{8\pi^4 \overline{\delta\alpha^2}}{r^2\lambda^4 \delta V}(1+\cos^2\phi_x)$
$\overline{\delta\alpha^2} = \left(\frac{\delta V}{4\pi}\right)^2 \left(2n\frac{dn}{dc}\right)^2 \delta c^2$	(1) δV replaces $\frac{M}{\rho N_A}$ in Eq. (25)	$\frac{i_s}{I_0} = \frac{2\pi^2 \delta V(n\,dn/dc)^2 \delta c^2}{r^2\lambda^4}(1+\cos^2\phi_x)$
	(2) For $n \simeq 1$, $\alpha = \frac{\delta V}{4\pi}(n^2-1)$	
	(3) $\delta\alpha = \frac{\delta V}{4\pi} 2n\,dn$ and $dn = \frac{dn}{dc}\delta c$	
$\overline{\delta c^2} = \frac{kT}{(\partial^2 G/\partial c^2)_0}$	(1) Taylor series (Appendix A) expansion of G around equilibrium value:	$\frac{i_s}{I_0} = \frac{2\pi^2 \delta V(n\,dn/dc)^2 kT}{r^2\lambda^4(\partial^2 G/\partial c^2)_0}(1+\cos^2\phi_x)$
Subscript 0: evaluated at equilibrium	$G = G_0 + \left(\frac{\partial G}{\partial c}\right)_0 \delta c + \frac{1}{2}\left(\frac{\partial^2 G}{\partial c^2}\right)_0 \delta c^2$	
	(2) $\left(\frac{\partial G}{\partial c}\right)_0 = 0$ at equilibrium	

	(3) $G - G_0 \simeq \frac{1}{2}kT$ since fluctuation is due to thermal energy ($\frac{1}{2}kT$ per degree of freedom)	
$\frac{\partial^2 G}{\partial c^2} = \frac{\delta V}{\bar{V}_1 c}\left(-\frac{\partial \mu_1}{\partial c}\right)_0$	(1) $dG = \mu_1\, dn_1 + \mu_2\, dn_2$ (2) Since $\bar{V}_1\, dn_1 = -\bar{V}_2\, dn_2$, $dG = \left[\mu_2 - \frac{\bar{V}_2}{\bar{V}_1}\mu_1\right] dn_2$ (3) Since $M\, dn_2 = dc\, \delta V$, $\frac{dG}{dc} = \left[\mu_2 - \frac{\bar{V}_2}{\bar{V}_1}\mu_1\right]\frac{\delta V}{M}$ (4) By the Gibbs–Duhem equation, $\frac{\partial^2 G}{\partial c^2} = -\frac{\delta V}{M}\left[\frac{\bar{V}_2}{\bar{V}_1} + \frac{n_1}{n_2}\right]$ (5) $c = \frac{n_2 M}{n_1 \bar{V}_1} + n_2 \bar{V}_2$	$\frac{i_s}{I_0} = \frac{2\pi^2 (n\, dn/dc)^2 kT \bar{V}_1 c}{r^2 \lambda^4 (-\partial \mu_1/\partial c)_0}(1 + \cos^2 \phi_x)$
$\left(\frac{\partial \mu_1}{\partial c}\right)_0 = -\bar{V}_1 \left(\frac{\partial \pi}{\partial c}\right)_0$	(1) By Eq. (4.34)	$\frac{i_s}{I_0} = \frac{2\pi^2 (n\, dn/dc)^2 kTc}{r^2 \lambda^4 (\partial \pi/\partial c)_0}(1 + \cos^2 \phi_x)$

Eq. (31) applies at equilibrium, we may evaluate $(\partial\pi/\partial c)_0$ from (31):

$$\left(\frac{\partial \pi}{\partial c}\right)_0 = RT\left[\frac{1}{M}+2Bc\right] \tag{32}$$

Combining Eqs. (30) and (32) yields

$$\frac{i_s}{I_0} = \frac{2\pi^2(n\,dn/dc)^2 c}{N_A r^2 \lambda^4 (1/M+2Bc)}(1+\cos^2\phi_x) \tag{33}$$

Before looking at the experimental aspects of light scattering, it is convenient to define several more quantities. First, a quantity known as the Rayleigh ratio R_θ is defined as

$$R_\theta = \frac{i_s r^2}{I_0(1+\cos^2\theta)} \tag{34}$$

where θ is the value of ϕ_x measured in the horizontal plane. According to Eq. (34), the Rayleigh ratio should be independent of both θ and r. An experimental verification of this is one way of testing the applicability of the Rayleigh theory to the experimental data. Next, it is convenient to identify the numerical and optical constants in Eq. (33) as follows:

$$K = \frac{2\pi^2 n^2 (dn/dc)^2}{N_A \lambda^4} \tag{35}$$

With these changes in notation, Eq. (33) becomes

$$R_\theta = \frac{Kc}{1/M+2Bc} \tag{36}$$

or

$$\frac{Kc}{R_\theta} = \frac{1}{M}+2Bc \tag{37}$$

This suggests that a plot of Kc/R_θ versus c should be a straight line for which the intercept and slope have the following significance:

$$\text{Intercept} = \frac{1}{M} \tag{38}$$

$$\text{Slope} = 2B \tag{39}$$

Comparing Eqs. (38) and (39) with Eqs. (4.49) and (4.50) reveals that plots of π/RTc versus c and Kc/R_θ versus c (a) have identical intercepts at least for monodisperse colloids (see Sec. 5.7 for a discussion of the average obtained for polydisperse systems), and (b) have slopes differing by a factor of 2, with the light scattering results having the larger slope.

The Rayleigh ratio as defined by Eq. (34) has a precise meaning, yet it is a quantity which is somewhat difficult to visualise physically. After we have discussed the experimental aspects of light scattering, we shall see that R_θ is directly proportional to the turbidity of the solution, where turbidity is the same as the absorbance determined spectrophotometrically.

5.6 EXPERIMENTAL ASPECTS OF LIGHT SCATTERING

In order to determine M and B by means of Eq. (37) it is clear that all the other quantities in the equation must be measured. It is convenient to group these factors into two categories for the purposes of our discussion: optical and concentration terms. To begin with, concentration enters the light scattering expressions primarily through the equations for osmotic pressure. Hence the same conditions apply in this application as in osmometry. Specifically, light scattering should be measured under isothermal conditions. Although concentration units other than weight per unit volume (the units of c) may be used, few of the alternatives are as useful as these. In consulting the literature, however, one should be attentive to the possibility that various workers may use slightly different units for c.

All the remaining variables in Eq. (37) are optical in origin. The factors to be provided with numerical values are the refractive index of the solution and the refractive index gradient, dn/dc in K [Eq. (35)], and the Rayleigh ratio [Eq. (35)]. All these optical parameters are wavelength dependent; therefore, each should be measured at the same wavelength. It is the value of this working wavelength that is used as the numerical substitution for λ in K.

The actual measurement of the refractive index of the solution poses no difficulty, but the evaluation of the refractive index gradient is more troublesome. The assumptions of the derivation of Eq. (33) restrict its applicability to dilute solutions. The refractive index of a dilute solution changes very gradually with concentration; hence, a plot of n versus c, the slope of which equals dn/dc, will be nearly horizontal. Since the intensity ratio depends on the square of dn/dc, it is clear that successful interpretation of Eq. (33) depends on the accuracy with which this small quantity is evaluated. Measuring the absolute refractive indices of various solutions and determining dn/dc by difference or graphically would introduce an unacceptable error. A more precise method must be used to measure this quantity.

A differential refractometer is a device which specifically measures *differences* in refractive indices. By means of a differential refractometer the difference between the refractive index of a solution and solvent may be measured directly with the necessary precision. We shall not concern ourselves with either the theory or the technique of the differential refractometer. It is sufficient to say that these intruments are commercially available and are capable of measuring differences in refractive index as large as 0.01 with a sensitivity of about 3 in the sixth decimal place. A differential refractometer, then, is an indispensible accessory in a laboratory conducting light scattering experiments.

Now let us consider the actual measurement of light intensity. A light scattering photometer differs from an ordinary spectrophotometer primarily in the fact that the photoelectric cell which measures the scattered light is mounted on an arm that permits it to be located at various angular positions relative to the sample. In

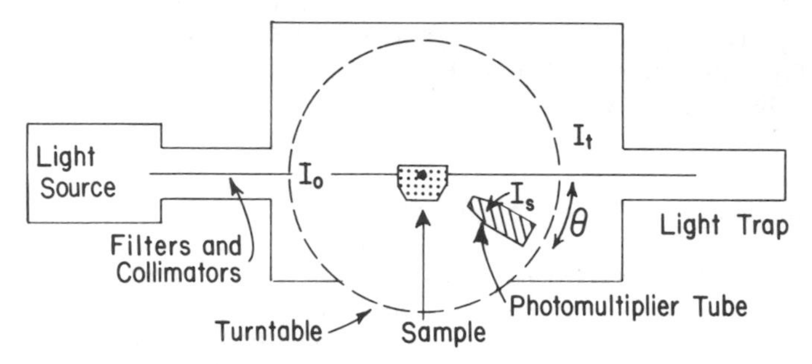

FIGURE 5.6 *Schematic top view of light scattering photometer, showing definition of θ.*

commercial light scattering devices, the detector rests on a turntable, the center of which coincides with the center of the sample. Thus, the angle ϕ_x is measured in the horizontal plane. From now on, we shall use the symbol θ to signify the angle of observation in the horizontal plane. The angle θ is measured from the direction of the transmitted beam where $\theta = 0°$. The incident beam, therefore, strikes the sample at $\theta = 180°$. A schematic top view of a light scattering photometer is shown in Fig. 5.6.

The light scattering photometer includes a number of distinct components. First, there is the light source, usually a mercury vapor lamp. Next, there are filters to provide monochromatic radiation. Most measurements are made at either the 436 or 546 nm line of mercury. The light must be carefully collimated so that the sample is, in fact, illuminated by a parallel beam as required by theory. Polarization filters are also included so that experiments with polarized light are possible. The intensity of the scattered light is low so that a sensitive photomultiplier is required to amplify the signal. The output from the latter may be connected to a high-sensitivity galvanometer or to a recorder.

The inside of the scattering compartment must be painted black to prevent light reflected from the apparatus from entering the photomultiplier. It is especially important to remove the transmitted light beam so a light trap is built into the scattering compartment at the 0° position.

The cells in which the scattering solutions are measured should have flat windows at the angle at which the scattering is measured. Cells with octagonal cross sections (actually, only half an octagon is used) are especially convenient since they present flat faces at 0, 45, 90, 135, and 180°. Cylindrical cells have been used, but they must be corrected for reflections from the walls. Regardless of the cell geometry, it is imperative that the cells be clean, otherwise the scattering from a fingerprint may exceed that from the solute! The solvent must also be purified of all extraneous matter. Filtration through sintered glass or centrifugation is usually employed to remove any dust particles which would also invalidate the measurement.

The easiest way to calibrate a light scattering photometer is to use a suitable standard as a reference. Although polymer solutions and dispersions of colloidal silica have been used for this purpose, commercial photometers are equipped with opal glass reference standards.

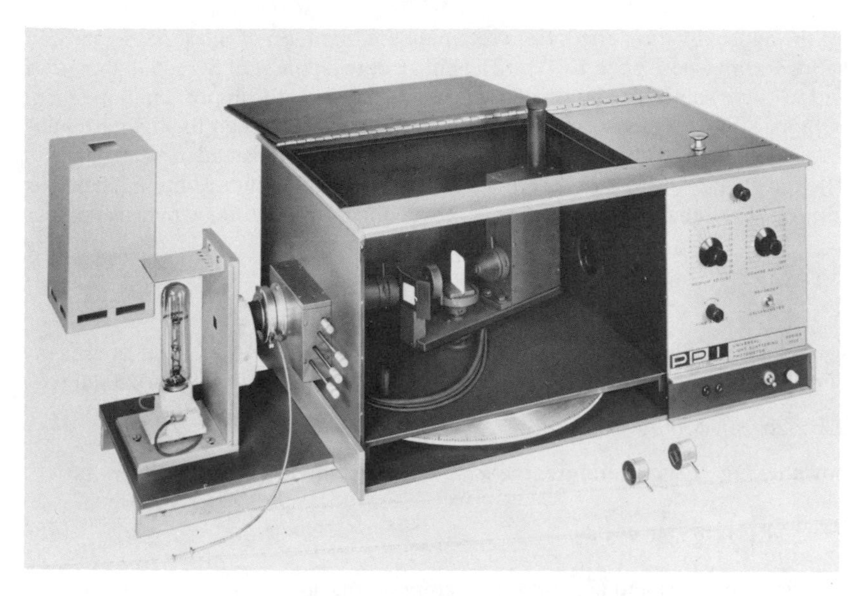

FIGURE 5.7 *Cutaway photograph of light scattering photometer, the Brice-Phoenix Universal Scattering Photometer. (From The Virtis Company, Gardiner, New York, used with permission.)*

Figure 5.7 is a photograph of a light scattering photometer with one side removed to show its inner workings. The apparatus shown is the Brice-Phoenix Universal Scattering Photometer, Series 2000 (The Virtis Company, Gardiner, New York 12525). Note that the arm holding the photomultiplier is located at about $\theta = -10°$.

Except for the movable photomultiplier tube, a light scattering photometer is very nearly identical to an ordinary spectrophotometer. The latter measures the ratio of the intensity of transmitted light to the intensity of incident light, I_t/I_0. The absorbance ε per unit optical path is defined in terms of this quantity as

$$\varepsilon = -\ln \frac{I_t}{I_0} \tag{40}$$

Now let us examine the relationship between absorbance and the intensity of scattered light. In a light scattering experiment with nonabsorbing materials, the intensity of the transmitted light equals the initial intensity minus the intensity of the light scattered *in all directions*, I_s:

$$I_t = I_0 - I_s \tag{41}$$

Combining Eqs. (40) and (41) leads to the result

$$\varepsilon = -\ln\left(\frac{I_0 - I_s}{I_0}\right) = -\ln\left(1 - \frac{I_s}{I_0}\right) \simeq \frac{I_s}{I_0} \tag{42}$$

where the approximation arises from retaining the first term of the series expansion of the logarithm (see Appendix A). The entire development of Secs. 5.4 and 5.5 is limited to dilute solutions and small n values. Therefore, the approximation in Eq. (42) is applicable to the systems we have been discussing. When the light attenuation is due to scattering, the ratio I_s/I_0 is called the turbitity τ instead of absorbance.

The quantity I_s in Eq. (42) is not the same as the light scattered to a particular point (r, ϕ_x), but equals the summation of these contributions, totaled over all angles:

$$\frac{I_s}{I_0} = \sum_{\substack{\text{all}\\ \text{angles}}} \frac{i_s}{I_0} \tag{43}$$

This summation may be replaced by an integral as follows. An element of area on the surface of a sphere of radius r and making an angle ϕ_x with the horizontal is

$$dA = 2\pi r \sin \phi_x (r\, d\phi_x) \tag{44}$$

as shown by Fig. 5.8. Therefore, the *total* scattered intensity ratio is given by

$$\frac{I_s}{I_0} = \int_0^\pi \left(\frac{i_s}{I_0}\right) 2\pi r^2 \sin \phi_x\, d\phi_x \tag{45}$$

Substituting Eqs. (33) and (35) into this expression yields

$$\frac{I_s}{I_0} = \int_0^\pi \frac{Kc(1+\cos^2 \phi_x) 2\pi r^2 \sin \phi_x\, d\phi_x}{r^2(1/M + 2Bc)} \tag{46}$$

The factor r^2 cancels out of Eq. (46) and the value of the integral over ϕ_x is $\frac{8}{3}$. Therefore,

$$\frac{I_s}{I_0} = \tau = \frac{16\pi Kc}{3(1/M + 2Bc)} \tag{47}$$

The parameter H is defined to equal the cluster of constants

$$H = \frac{16\pi K}{3} = \frac{32\pi^3 n^2 (dn/dc)^2}{3N_A \lambda^4} \tag{48}$$

FIGURE 5.8 *Definition of an element of area required for the summation over all angles of the intensity of scattered light.*

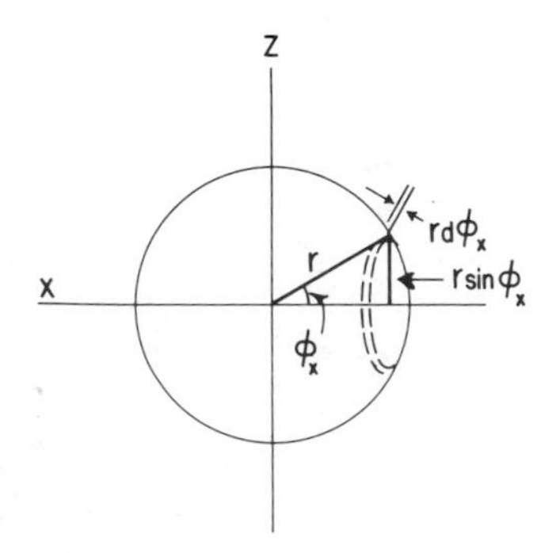

TABLE 5.2 *Summary of the cgs units used for various light scattering parameters*

Quantity	Units
c	$\mathrm{g\ cm^{-3}}$
$\dfrac{dn}{dc}$	$\mathrm{cm^3\ g^{-1}}$
$\dfrac{1}{M}$ and Bc	moles $\mathrm{g^{-1}}$
B	moles $\mathrm{cm^3\ g^{-2}}$
i_s/I_0 [Eq. (33)]	$\mathrm{cm^{-3}}$ (i.e., per unit volume)
K and H [Eqs. (35) and (48)]	moles $\mathrm{cm^2\ g^{-2}}$
$\dfrac{Hc}{\tau}=\dfrac{Kc}{R_\theta}$	moles $\mathrm{g^{-1}}$
R_θ and τ [Eqs. (34) and (47)]	$\mathrm{cm^{-1}}$ (i.e., per unit path)

and in terms of this quantity, Eq. (47) becomes

$$\frac{Hc}{\tau}=\frac{1}{M}+2Bc \tag{49}$$

The formal similarity between Eqs. (37) and (49) helps us understand somewhat better the physical significance of the Rayleigh ratio R_θ. It is directly proportional to the attenuation of the light per unit optical path, measured as absorbance, when the attenuation is due to scattering alone. In this case, absorbance is more properly called turbidity.

The dimensional consistency of the various quantities encountered in the presentation of light scattering results is shown in Table 5.2. Now let us consider some actual results from light scattering experiments on systems which satisfy the assumptions of the theory.

5.7 RESULTS FROM LIGHT SCATTERING EXPERIMENTS: WEIGHT AVERAGE MOLECULAR WEIGHTS

In the preceding section, we saw how turbidity measurements are made and how they may be analyzed to yield numerical values for some of the parameters of interest in colloid chemistry. Figure 5.9 shows a plot of Hc/τ versus c for three different fractions of polystyrene in methyl ethyl ketone. The measurements shown in the figure were made at 25°C and at a wavelength of 436 nm. The molecular weights for the three fractions as determined from the values of the intercepts of the lines according to Eq. (44) are included in the figure. It should be noted that all three lines

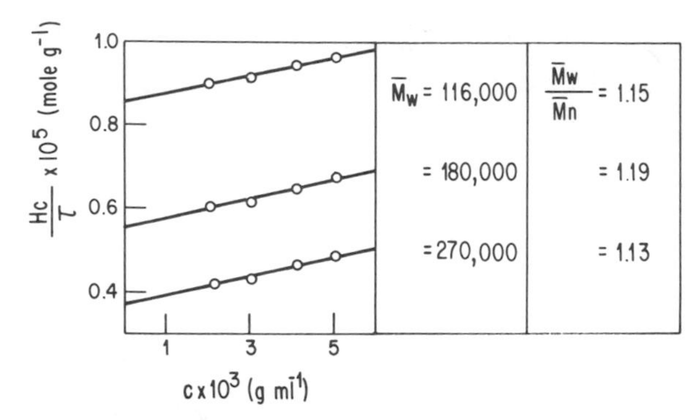

FIGURE 5.9 *Plots of* Hc/τ *versus c for three different fractions of polystyrene in methylethyl ketone. Curves are labeled with* $\bar{M}_w$ *and* $\bar{M}_w/\bar{M}_n$ *values of polymer. [B. A. Brice, M. Halwer, and R. Speiser,* J. Opt. Soc. Am. **40**: 768 (1950), *used with permission.]*

have essentially the same slope, a situation which closely resembles Fig. 4.4a for osmotic pressure data.

At first glance, it appears that light scattering experiments provide no information that is not already available from osmometry. Indeed, for monodisperse colloids this is true, at least for the experiments we have discussed until now. The apparent redundancy between osmotic pressure and light scattering results should not be interpreted to mean that the two procedures duplicate one another entirely. For one thing, light scattering is free from the limitations imposed on osmometry by the availability of a suitable semipermeable membrane. Further, turbidity measurements do not require time for equilibration, and hence they may be used for systems which change with time in a manner that is not possible with osmometry. In addition to these practical considerations, there are other ways in which light scattering and osmometry differ. Some of these will become apparent only in subsequent sections where additional characteristics of light scattering are developed. Another importance difference, however, arises in the type of average that is measured for polydisperse systems.

In Chap. 4, we saw that osmometry enables us to measure the number average molecular weight for a polydisperse colloid. In view of the way the osmotic pressure enters the development of Eqs. (37) and (49), it appears that the same type of average is obtained from turbidity experiments also. This is not the case. The following argument shows that light scattering measures the weight average molecular weight.

For a polydisperse system, Eq. (49) relates the experimental concentration, the experimental turbidity, and the average molecular weight:

$$\frac{Hc_{\text{exp}}}{\tau_{\text{exp}}} = \frac{1}{\bar{M}} \tag{50}$$

It is sufficient to consider only the leading term of Eq. (49) in writing (50) since the molecular weight is evaluated from the intercept at infinite dilution. Likewise, we expect that Eq. (49) will also apply to each molecular weight fraction in the polydisperse system:

$$\frac{Hc_i}{\tau_i} = \frac{1}{M_i} \tag{51}$$

It is the relationship between the value of $\bar{M}$ and the distribution of M_i values that we wish to determine. To accomplish this, we note that

$$c_{\text{exp}} = \sum_i c_i \tag{52}$$

and

$$\tau_{\text{exp}} = \sum_i \tau_i \tag{53}$$

Combining the last four equations gives

$$\bar{M} = \frac{\tau_{\text{exp}}}{Hc_{\text{exp}}} = \frac{\Sigma_i \tau_i}{H \Sigma_i c_i} = \frac{H \Sigma_i c_i M_i}{H \Sigma_i c_i} = \frac{\Sigma_i c_i M_i}{\Sigma_i c_i} \tag{54}$$

Now recalling that $c = n_i M_i / V$ enables us to write

$$\bar{M} = \frac{\Sigma_i n_i M_i^2}{\Sigma_i n_i M_i} = \bar{M}_w \tag{55}$$

Equation (55) corresponds to the weight average molecular weight $\bar{M}_w$ as defined by Eq. (1.21).

It will be recalled from Chap. 1 that the *number* of particles in a molecular weight class provides the weighting factor used to compute the number average molecular weight. The *weight* of particles in a class gives the weighting factor for the weight average molecular weight. For this reason, the weight average is especially influenced by the larger particles in a distribution. Therefore, the weight average molecular weight is always larger than the number average for a polydisperse system. As we saw in Chap. 1, the ratio of the two different molecular weights is a useful measure of the polydispersity of a sample. Figure 5.9 also gives the ratios between the two averages for the fractionated samples used in that study.

Thus we see that the redundancy between osmometry and light scattering is only an apparent effect for polydisperse systems. In fact, the combination of the two analyses provides additional information about the characteristics of the system.

Equation (49) shows that the slope of a light scattering plot is twice the value of the slope of a comparable plot from osmometry. In addition to the factor of 2, there is a more subtle difference between the slopes arising from a difference in the two values of the second virial coefficient. Examination of Eq. (4.50) reveals that the second virial coefficient is inversely proportional to the square of the molecular weight of the solute. For polydisperse systems, it is the average molecular weight that appears in this expression. Since the weight average molecular weight is larger than the number average, the second virial coefficient B will be somewhat smaller as determined by light scattering than by osmometry after the factor of 2 has been taken into account.

Aside from the difference just noted, the interpretation of the second virial coefficient in light scattering is exactly the same as that developed in Sec. 4.7. It should be noted, however, that Eq. (49) does not apply to charged systems. The reason for this lies in the fact that the charge of macroions is also a fluctuating quantity, and this must also be considered in developing a scattering theory for charged particles. The resulting analysis shows that it is a plot of Hc/τ versus $c^{1/2}$ which is linear in this case, with the limiting slope proportional to $\overline{z^2}$, the average value of the square of the charge. The slope is also predicted to be negative in this situation.

In concluding this section, it should be emphasized that the turbidity values which are plotted to interpret light scattering experiments are the *solution* turbidities, corrected for scattering by the solvent. Also, the entire theoretical development leading to Eq. (49) is based on the assumption that the scatterers are isotropic. In this case, unpolarized incident light will produce a scattered beam which is totally polarized at $\theta = 90°$ as may be seen from Fig. 5.5. When anisotropic particles are present, there is a depolarization of the light scattered at 90°. The ratio of the horizontally to vertically polarized scattered light may be determined by inserting a Polaroid filter between the sample and the photomultiplier. From the measured value of this depolarization ratio, a correction factor (called the Cabannes factor) may be introduced to allow for anisotropy. In the sample with $M = 116{,}000$ in Fig. 5.9, for example, the ratio of the two different polarizations at 90° has a value of 0.013 for which the Cabannes factor equals 0.98. The turbidity should be multiplied by this factor to correct for the fact that the anisotropy enhances the amount of scattered light.

5.8 EXTENSION TO LARGER PARTICLES

In the remainder of this chapter we shall see that a good deal more information about scattering particles may be deduced, at least under some circumstances, from the study of the light scattered by a sample. In developing the Rayleigh theory and applying it to solutions, a definite model was postulated and certain variables emerged as factors which affect the intensity of the scattered light. Before we extend light scattering theory to more complex systems, it will be convenient to review once again the assumptions of the Rayleigh model:

1. The scattering centers are isotropic, dielectric, and nonabsorbing.
2. The scatterers have a refractive index which is not too large [see Eq. (26)].
3. The particles are small in dimension compared to the wavelength of the light.

This last assumption is fundamental to the theory and originates as early as Eq. (14), where it is assumed that the field which drives the oscillating dipole is the same throughout the scattering center. It is generally held that particles must have no dimension larger than about $\lambda/20$ for this assumption to apply.

Equation (49) suggests that light scattering is a technique which is ideally suited to the study of particles in the colloidal size range since the turbidity increases with the molecular weight of the particle. However, the assumptions underlying the derivation of Eq. (49) impose a limitation. The Rayleigh theory shows that turbidity increases with molecular weight at least until the particle is large enough to have some dimension exceeding about $\lambda/20$. In terms of the theory presented so far, we have no way of interpreting the light scattered by larger particles. The Debye theory which we examine in the following section will show us how to overcome this limitation.

The Rayleigh approximation shows that the intensity of scattered light depends on the wavelength of the light, the refractive index of the system (subject to the limitation already cited), the angle of observation, and the concentration of the solution (which is also restricted to dilute solutions). In the Rayleigh theory the size and shape of the scatterers (M and B) enter the picture through thermodynamic rather than optical considerations.

A fully developed theory of light scattering which allows all the variables, including particle size and shape, to take on a full range of values is extremely complex. Because of this complexity, many treatments such as the Rayleigh theory are approximations which apply only to a narrow range of values of the parameters. In the Debye approximation most of the preceding restrictions will continue to apply except that the limitation on particle size will be relaxed considerably. At the same time, however, the stipulation of low values of the refractive index becomes even more stringent. As we shall see presently, the Debye approximation introduces some additional complexity to the theory of light scattering and trades off some range in refractive index for extra range in particle size. From this a positive dividend emerges: We shall be able to determine a characteristic linear dimension of the scattering particles *without any assumptions* about the shape of the particles. Aside from direct microscopic examination, this is the first place we have encountered a single technique which could give such information. In the cases to which it applies, this information is definitely worth the price of a little additional complexity.

We shall omit most of the mathematical details in developing the additional theory. Instead, we shall emphasize the major concepts of the theory, its range of applicability, and its plausibility in limiting cases.

5.9 THE DEBYE SCATTERING THEORY

The Rayleigh approximation is restricted to particles whose dimensions are small compared to the wavelength of light. Suppose we now relax this restriction in our model of a scattering particle, allowing the particles to take on dimensions comparable to λ. Under these circumstances, different regions of the same particle will behave as scattering centers. Because the distances between these various scattering centers are of the same magnitude as the wavelength of light, there will be interference between the waves of light scattered from different parts of the same particle.

Figure 5.10 shows how this comes about. The electric field of the incident light beam is out of phase when it strikes two different portions of the scatterer, designated A and B. The light which is scattered from each of these sites is characterized by the same field as that which induces the oscillation at these locations. Therefore, the

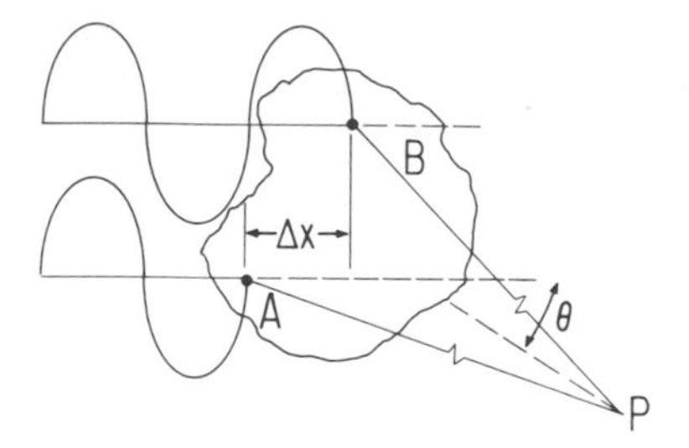

FIGURE 5.10 *Schematic illustration of light scattered from two different regions of the same particle when the particle size is comparable to λ.*

light which is scattered from regions A and B will be out of phase and will display interference when observed at a large distance (compared to Δx), P in Fig. 5.10. We propose, therefore, that the Rayleigh ratio defined by Eq. (34) must be multiplied by a correction factor $P(\theta)$ to correct for the interference effects which were not considered previously. Several things may be anticipated about this factor:

1. As the dimensions of the particle become negligible compared to the wavelength of light ($\Delta x \to 0$), $P(\theta) \to 1$—under these conditions, no correction is necessary.
2. The correction factor is a function of the angle of observation—as implied by the notation $P(\theta)$—since it is essentially an interference effect.
3. Multiplying the denominator of both sides of Eq. (37) by $P(\theta)$ corrects i_s (a theoretical quantity calculated without interference) for interference:

$$\frac{Kc}{i_s P(\theta) r^2/[I_0(1+\cos^2\theta)]} = \frac{1}{P(\theta)}\left[\frac{1}{M} + 2Bc\right] \tag{56}$$

4. The actual intensity of scattered light measured at P, i_P, equals the product i_s times $P(\theta)$. Therefore, the experimental Rayleigh ratio obeys Eq. (37) modified as follows:

$$\frac{Kc}{R_\theta} = \frac{1}{P(\theta)}\left[\frac{1}{M} + 2Bc\right] \tag{57}$$

In view of these considerations, it is apparent that the way to extend the previous theories to larger particles is to evaluate $P(\theta)$. To do this in a general way, two complications must be introduced into Fig. 5.10:

1. Two scattering centers in a large particle are not simply displaced from one another in the x direction by an amount Δx—rather, the coordinates of one relative to the other must be described by a radial distance and two angles, for example θ and ϕ.
2. A large particle does not consist of merely two scattering centers, but may be subdivided into several centers, the number of which increases with the size of the particle.

These two considerations must be incorporated into any general expression for $P(\theta)$.

We shall begin our summary of the derivation of $P(\theta)$, however, by considering only the pair of scattering centers shown in Fig. 5.10. To do this quantitatively, imagine that region A is at the origin of a coordinate system, so that the light which reaches B has to travel an additional distance Δx. The field at A and B is now represented by the following formulations of Eq. (4):

$$\mathbf{E}_A = \mathbf{E}_0 \cos 2\pi\nu t \tag{58}$$

and

$$\mathbf{E}_B = \mathbf{E}_0 \cos 2\pi\left(\nu t - \frac{\Delta x}{\lambda}\right) \tag{59}$$

The light scattered from each of these sites will be characterized by the same field as that which induces the oscillation. Therefore, the light scattered from A and B will be out of phase by an amount $2\pi\ \Delta x/\lambda$.

Let us now consider the net scattered light which reaches the point P, a (large) distance r from the scatterer. The field at P is the sum of the fields emerging from A and B

$$\mathbf{E}_P = \mathbf{E}_A + \mathbf{E}_B = \mathbf{E}_0\left[\cos 2\pi\nu t + \cos 2\pi\left(\nu t - \frac{\Delta x}{\lambda}\right)\right] \tag{60}$$

If we use the appropriate trigonometric formula for the sum of two cosines, this may be written

$$\mathbf{E}_P = \left(2\cos\frac{\pi\ \Delta x}{\lambda}\right)\mathbf{E}_0 \cos 2\pi\left(\nu t - \frac{\Delta x}{2\lambda}\right) \tag{61}$$

This equation shows that the electric field of light scattered to P is not altered in frequency or wavelength, but that the amplitude is modified by the factor $2\cos(\pi\ \Delta x/\lambda)$. The intensity of light depends on the square of the field amplitude; therefore, with interference:

$$i_P \propto \left(2\cos\frac{\pi\ \Delta x}{\lambda}\right)^2 \mathbf{E}_0^2 \tag{62}$$

without interference ($\Delta x \to 0$):

$$i_s \propto 2^2\mathbf{E}_0^2 \tag{63}$$

Combining Eqs. (62) and (63) leads to

$$\frac{i_P}{i_s} = \cos^2\frac{\pi\ \Delta x}{\lambda} \tag{64}$$

In view of Eq. (56), the factor $\cos^2(\pi\ \Delta x/\lambda)$ must equal $P(\theta)$ for this simple case.

A fair amount of straightforward but tedious trigonometry is necessary to establish the relationship between Δx and r, ϕ_x and θ for the general case of any orientation between A and B. The result of this analysis is the expression (given without proof)

$$\Delta x = 2r \cos \phi_x \sin \frac{\theta}{2} \tag{65}$$

where ϕ_x is defined the same as in Fig. 5.8 and θ continues to be the scattering angle measured in the horizontal plane.

It is not any specific value of ϕ_x in which we are interested, but in all possible values. This means that Eq. (65) is to be integrated over all values of ϕ_x, a procedure which leads to the result (given without proof)

$$\frac{i_P}{i_s} = 1 + \frac{\sin[(4\pi r/\lambda)\sin(\theta/2)]}{(4\pi r/\lambda)\sin(\theta/2)} \tag{66}$$

Equation (66) benefits considerably from some simplification in notation. Accordingly, we define

$$s = \frac{4\pi}{\lambda} \sin \frac{\theta}{2} \tag{67}$$

Using this notation permits us to rewrite Eq. (66) as

$$\frac{i_P}{i_s} = 1 + \frac{\sin sr}{sr} \tag{68}$$

The right-hand side of Eq. (68) correctly defines $P(\theta)$ for interference between *two* regions of a large particle. The next question we must consider is how this result may be applied to a particle which consists of N, not just two, scattering centers. If the composition of the particle is such that the scattering from one part of the particle has no effect (other than interference) on the scattering from another part, then the particle may be subdivided into N scattering elements and Eq. (68) can be applied to all possible pairs. If we define r_{ij} as the distance between the ith and jth members of the set of N scattering elements, then this argument leads to the result (given without proof)

$$\frac{i_P}{i_s} \propto \sum_i \sum_j \left(\frac{\sin sr_{ij}}{sr_{ij}} \right) \tag{69}$$

Expression (69) is written as a proportionality rather than an equation because the summation requires that a normalization factor be introduced. The ratio i_P/i_s must equal unity when r_{ij} is small since it explicitly corrects for interference effects which vanish under these circumstances. For small values of sr_{ij}, $(\sin sr_{ij})/sr_{ij}$ equals unity and $\Sigma_i \Sigma_j[(\sin sr_{ij})/sr_{ij}]$ equals N^2. Therefore, normalization requires that we write

$$\frac{i_P}{i_s} = \frac{1}{N^2} \sum_i \sum_j \frac{\sin sr_{ij}}{sr_{ij}} = P(\theta) \tag{70}$$

Note that the first term of Eq. (68) is implicitly present in (70) when r_{ij} equals zero. Equation (70) provides the general expression for $P(\theta)$ which we have sought. It is possible that the reader will recognize Eq. (70) from another context. It is exactly the same expression which describes the diffraction of x rays by polyatomic molecules, the situation for which it was derived by Debye (Nobel Prize, 1936). An important insight emerges from this realization. All interference phenomena between electromagnetic radiation and matter follow the same mathematical laws. Interference becomes important when there is some characteristic distance D in the material under consideration which is of the same magnitude as the wavelength of the available radiation. The interference phenomena will be identical for identical values of D/λ, regardless of the separate values of D and λ. Thus, x-ray diffraction is an experimental technique that uses x rays for which $\lambda \simeq 1$ Å to measure interatomic distances which are on the order of angstroms. Likewise, we may use visible light for which $\lambda \simeq 5000$ Å to measure particles whose dimensions are in the colloidal size range. The scattering of microwaves by atmospheric rain and snow particles is an example of still another situation in which interference phenomena may be observed by scaling the wavelength of the radiation to suit the particle sizes under investigation.

Before turning to the applications of the Debye approximation, we should elaborate more fully on a point that was glossed over. This is the assumption—made at the outset but explicated in going from Eq. (68) to (69)—that the scattering behavior of each scattering element is independent of what happens elsewhere in the particle. The approximation that the phase difference between scattered waves depends only on their location in the particle and is independent of any material property of the particle is valid as long as

$$\frac{2\pi D}{\lambda}(n-1) \ll 1 \tag{71}$$

It is the condition expressed by the inequality (71) that requires $n-1$ to become smaller and smaller as the theory is applied to progressively larger particles. For example, when $\theta = 10°$, the Debye approximation is good to within 10% for spheres whose radius is about 62, 37, and 25 times λ at $n = 1.1$, 1.2, and 1.3, respectively. As we shall see presently, the approximation applies to a somewhat wider range of R and n values at smaller angles of observation and to a narrower range at larger angles.

5.10 ZIMM PLOTS

Although Eq. (70) may correct (57) for the turbidity of systems in which the scattering centers are not negligible in size compared to the wavelength of light, it is not a very promising looking result. A little more mathematical manipulation will correct this impression. Specifically, suppose we examine Eq. (67) in the case in which $\theta/2$ is small. This will make s small regardless of the value of r_{ij} so that $\sin(sr_{ij})$ in Eq. (70) may be expressed as a power series (Appendix A):

$$P(\theta) = \frac{1}{N^2}\sum_i \sum_j \frac{\sin sr_{ij}}{sr_{ij}} = \frac{1}{N^2}\sum_i \sum_j \frac{sr_{ij} - (sr_{ij})^3/3! + \cdots}{sr_{ij}} \tag{72}$$

For small values of s, the expansion may be limited to the first two terms:

$$P(\theta) = \frac{1}{N^2}\sum_i\sum_j 1 - \frac{(sr_{ij})^2}{6} = 1 - \frac{s^2}{6N^2}\sum_i\sum_j r_{ij}^2 \tag{73}$$

It is actually $1/P(\theta)$ in which we are interested, so we may again take advantage of the fact that s is small to write Eq. (73) as

$$\frac{1}{P(\theta)} \simeq 1 + \frac{s^2}{6N^2}\sum_i\sum_j r_{ij}^2 \tag{74}$$

In the next section, we examine the definition of the radius of gyration for particles possessing a variety of geometries. For now, we note that the radius of gyration R_g is given by

$$R_g^{\,2} = \frac{1}{2N^2}\sum_i\sum_j r_{ij}^2 \tag{75}$$

This means that Eq. (74) may be written

$$\frac{1}{P(\theta)} = 1 + \tfrac{1}{3}R_g^{\,2}s^2 \tag{76}$$

Equation (76) is valid in the limit of small values of $\theta/2$. Of course, if the scattering particle is not too large, the expansions of Eqs. (73) and (74) will be valid at larger θ, assisted by the fact that the values of r_{ij} will be small. This explains why the range of parameters to which the Debye equation applies depends on the angle of observation, becoming narrower for larger angles and broader at small angles.

Substitution of Eq. (76) into (57) yields

$$\frac{Kc}{R_\theta} = \left(\frac{1}{M} + 2Bc\right)\left(1 + \frac{16\pi^2 R_g^{\,2}}{3\lambda^2}\sin^2\frac{\theta}{2}\right) \tag{77}$$

Let us consider Eq. (77) in three important limiting cases with the objective in mind of developing a graphical technique for using (77):

1. In the limit of $\theta = 0$, Eq. (77) reduces to (37). That is, there is no interference effect in the scattered light.
2. In the limit of $c \to 0$, Kc/R_θ is proportional to $\sin^2(\theta/2)$.
3. If both c and θ equal zero, Kc/R_θ equals $1/M$.

These limits suggest how experimental data might be collected, plotted, extrapolated, and interpreted. The resulting graph is known as the Zimm plot, after its originator.

Equation (77) shows clearly that i should be measured as a function of both concentration *and* angle of observation in order to take full advantage of the Debye theory. The light scattering photometer described in Sec. 5.6 is designed with this capability, so this requirement introduces no new experimental difficulties. The data collected then consist of an array of i/I_0 values (no subscript is needed on i since it now applies to small and large particles), measured over a range of c and θ values.

By means of Eq. (34), the i/I_0 ratios are converted to R_θ values. Then the results are plotted with Kc/R_θ as the ordinate and $\sin^2(\theta/2)+c$ as the abscissa. Figure 5.11a is a schematic of how such a plot might appear. Each of the points will correspond to a particular pair of c, θ values. When the points measured at the same values of c and those measured at the same values of θ are connected, a grid of lines is obtained like that sketched in the figure. If all the experimental c and θ values are small enough for the theories to hold exactly, then the grid would consist of two sets of parallel straight lines with different slopes. In general, however, the range of experimental c and θ values exceeds the range of validity of the theory, and the lines will show some curvature.

The next step in the treatment of the data is the extrapolation of the curves drawn at constant θ to $c=0$ and the extrapolation of those drawn at constant c to $\theta=0$. This is done by placing a mark (the triangles in Fig. 5.11a) on the smooth lines drawn through the experimental points at that value of the abscissa which corresponds to the value of that coordinate at the desired limit. For example, when the limit for $c=0$ of the θ_2 line is located, it will lie on the θ_2 line and the value of the abscissa will be $\sin^2(\theta_2/2)$. Similar arguments apply to the extrapolation to $\theta=0$, as shown in Fig. 5.11a. The triangles in that figure now define two straight lines as shown in Fig. 5.11b.

FIGURE 5.11 *Schematic Zimm plot*: (*a*) *The array of experimental points*, (*b*) *The extrapolated points, connected by the appropriate straight lines. The intercept and slopes of these lines are given by Eqs.* (78) *to* (80).

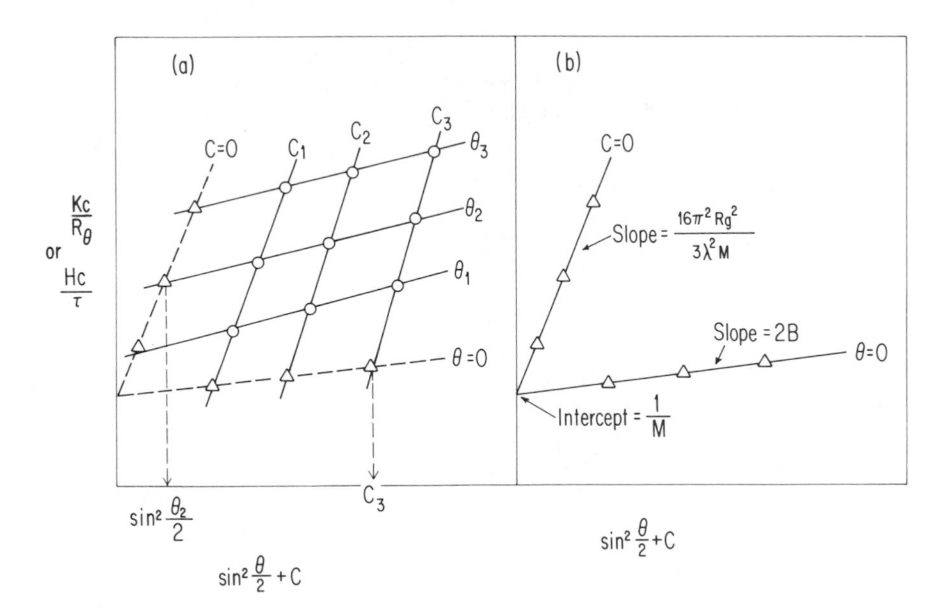

Let us now consider the interpretation of this graph. First, it should be remembered that interference effects vanish at $\theta = 0$. The line so labeled in Fig. 5.11b corresponds to values of Kc/R_θ at different values of c all expressed at $\theta = 0$ where eq. (37) is valid. According to that equation, the slope of this line equals $2B$ and the intercept equals $1/M$. This is the same result we obtained previously for small particles. By extrapolating to $\theta = 0$, the procedure now applies to larger particles as well. In formulas, then, we write for the $\theta = 0$ line

$$(\text{Slope})_{\theta=0} = 2B \tag{78}$$

and

$$(\text{Intercept})_{\theta=0} = \frac{1}{M} \tag{79}$$

The new feature of the Zimm plot is the second extrapolated line, corresponding to $c = 0$. This line connects values of Kc/R_θ measured at different values of θ and extrapolated to $c = 0$. Accordingly, it is described by Eq. (77) with $c = 0$. That is, theory predicts the $c = 0$ line to have an intercept of $1/M$ and a slope equal to $16\pi^2 R_g^2/3\lambda^2 M$. It is expected, then, that the two lines will extrapolate to a common intercept, $1/M$, and that the slope of the $c = 0$ line will be proportional to R_g^2. Summarizing in formulas, we write for the $c = 0$ line

$$(\text{Slope})_{c=0} = \frac{16\pi^2 R_g^2}{3\lambda^2 M} \tag{80}$$

and

$$(\text{Intercept})_{c=0} = \frac{1}{M} \tag{81}$$

FIGURE 5.12 *Experimental Zimm plot for cellulose nitrate in acetone. [Reprinted with permission from H. Benoit, A. M. Holtzer, and P. Doty,* J. Phys. Chem. **58**:635 (1954), *copyright by the American Chemical Society.]*

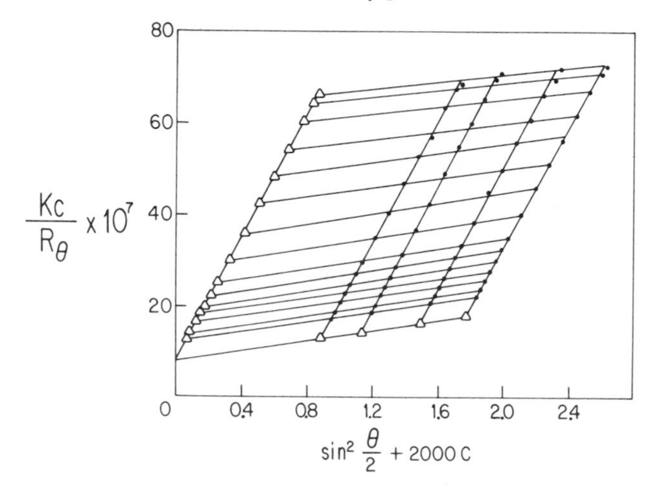

TABLE 5.3 *Characteristic parameters for the cellulose nitrate–acetone system as evaluated from Fig.* 5.12

Experimental quantities	Derived quantities
Intercept: 7.87×10^{-7} moles g^{-1}	Molecular weight: $M = 1.27 \times 10^{6}$ g $mole^{-1}$
Slope $c = 0$ line: 6.78×10^{-6} ($\lambda' = 321$ nm)	$R_g^2 = 1.69 \times 10^{-10}$ cm^2 $(R_g)_{rms} = 1300$ Å $(r)_{rms} = 3180$ Å
Slope $\theta = 0$ line: 5.7×10^{-7} moles cm^3 g^{-2} (slope $\times 2000 = 1.14 \times 10^{-3}$ moles cm^3 g^{-2})	$B = 5.7 \times 10^{-4}$ moles cm^3 g^{-2}

It follows, therefore, that the square of the radius of gyration equals

$$R_g^2 = \frac{3\lambda^2}{16\pi^2}\left(\frac{\text{slope}}{\text{intercept}}\right)_{c=0} \tag{82}$$

In the following section, we shall see how the radius of gyration is related to the actual physical dimensions of the scattering particles. For now, however, it is enough to note that this quantity is a size parameter which characterizes the scattering particle unambiguously. Note, too, that it is obtained by this method without any assumptions as to the shape of the particles.

Figure 5.12 is a graph of experimental data for cellulose nitrate in acetone at 25°C plotted according to Zimm's procedure. Note that the abscissa in Fig. 5.12 is different from that in Fig. 5.11 inasmuch as the concentrations have been multiplied by a constant quantity (2000) before adding them to $\sin^2(\theta/2)$. This produces a more useful array of points, and is readily eliminated from the calculated slope.

Table 5.3 summarizes the evaluation of particle parameters by Eqs. (78) through (82) for the cellulose nitrate–acetone system from Fig. 5.12.

5.11 THE RADIUS OF GYRATION

The typical student of chemistry has probably not heard much about the radius of gyration since studying general physics. Suddenly, however, this quantity appears in the study of light scattering phenomena, so a short review seems in order.

We assume that the particle whose radius of gyration is under discussion may be subdivided into a number of volume elements of mass m_i. Then the moment of inertia I about the axis of rotation of the body is given by

$$I = \sum_i m_i r_i^2 \tag{83}$$

where r_i is the distance of the ith volume element from the axis of rotation.

Regardless of the shape of the particle, there is a radial distance at which the entire mass of the particle could be located such that the moment of inertia would be the same as that of the actual distribution of the mass. This distance is the radius of gyration R_g. According to this definition, it is clear that

$$R_g^2 \sum_i m_i = I = \sum_i m_i r_i^2 \tag{84}$$

Therefore, we see that the value of R_g^2 for an array of volume elements is the weight average value of r_i^2 for the array

$$R_g^2 = \frac{\Sigma_i m_i r_i^2}{\Sigma_i m_i} \tag{85}$$

The relationship between the radius of gyration of a particle and its actual dimensions depends on the shape of the particle. These relationships are derived in most elementary physics texts for rigid bodies of various geometries; it has also been worked out for the random coil. We shall merely present the final results.

The radius of gyration of the following bodies about an axis through the center of the particle is given by these expressions:

1. For a sphere of radius R:

$$R_g^2 = \tfrac{3}{5} R^2 \tag{86}$$

2. For a thin rod of length L:

$$R_g^2 = \frac{L^2}{12} \tag{87}$$

3. For a random coil of mean square end-to-end distance $\overline{r^2}$:

$$R_g^2 = \frac{\overline{r^2}}{6} \tag{88}$$

Thus we see that light scattering gives us values of R_g without any assumptions as to particle shape; but to translate these to actual particle dimensions, some shape must be assumed. It should be emphasized, however, that the radius of gyration in itself is a perfectly legitimate way of describing the dimensions of a particle. The specification of a particular geometry is really optional. Any assumed particle geometry must also be consistent with the molecular weight as obtained from the same data, so the range of possible geometries for unsolvated particles (where the density is known) is quite limited. If the characterization of a particle by a linear dimension alone is sufficient, an analysis known as the dissymmetry method is a relatively easy way of obtaining this information. In this technique, one measures i/I_0 at $\theta = 45°$ and $\theta = 135°$ for dispersions at several different concentrations. It should be noted that the factor $1 + \cos^2 \theta$ in Eq. (34) has the same value for these two angles of observation. Therefore, any deviation of the ratio of the intensities, z, from unity must measure the ratio of the $P(\theta)$ values at these two angles:

$$z = \frac{i_{45°}}{i_{135°}} = \frac{P(45°)}{P(135°)} \tag{89}$$

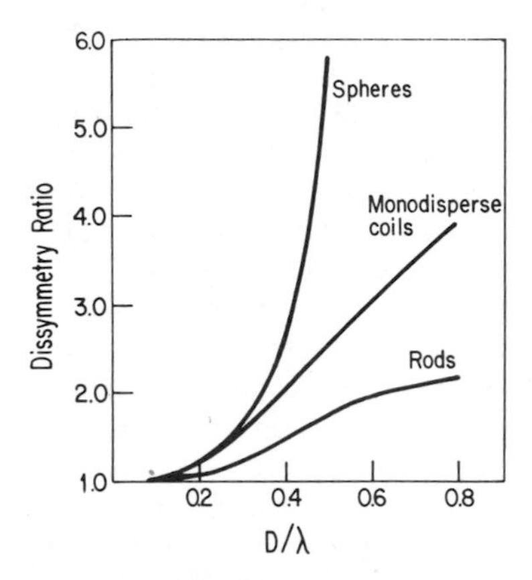

FIGURE 5.13 *Values of the dissymmetry ratio z versus the size parameter D/λ for spheres, random coils, and rods.* (*Data from Stacey* [4].)

The last ratio could be evaluated directly from Eq. (70) for different values of R_g. A more common procedure, however, is to plot the dissymmetry ratio versus a characteristic particle dimension D relative to λ using Eqs. (86), (87), and (88) for spheres, rods, and coils. The curves shown in Fig. 5.13 are the result of such calculations. The experimental value of the dissymmetry ratio is measured at several different concentrations and extrapolated to zero concentration. The extrapolated dissymetry ratio is then converted into a size parameter by reading the appropriate value off one of the curves of Fig. 5.13.

Again, this method involves measuring the turbidity at different c and θ values. This method of interpretation actually yields less information than one could obtain from the same results by the Zimm plot. The latter also permits evaluation of M and B.

It will be noted from the ordinate values in Fig. 5.13 that the scattering is always larger in the forward direction for particles showing interference effects. This is one reason why the presence of dust particles raises such havoc in a scattering experiment on particles that lie in the Rayleigh region.

5.12 LARGE, ABSORBING PARTICLES: SCATTERING AND ABSORPTION CROSS SECTIONS

All the applications of light scattering that have been discussed so far have been restricted to very small particles and fairly small indices of refraction or to fairly small particles and very small indices of refraction. We have finally reached the point where it seems appropriate to relax all these restrictions and consider the scattering by a particle of arbitrary size and index of refraction.

The general problem is to solve Maxwell's electromagnetic equations both inside and outside a particle, where the indices of refraction are different in the two regions.

The problem has been solved for both spherical and cylindrical particles; we shall limit our considerations to the former. In the most general case, some provision must be made for the possibility that the particle absorbs as well as scatters light. This contingency is introduced by defining the refractive index of an absorbing material as a complex number $n - ik$ where $i = \sqrt{-1}$. For nonabsorbing materials $k = 0$. Both n and k are wavelength-dependent characteristics of the material; k obviously increases as the wavelength of an absorption peak is approached.

The idea of representing the refractive index of an absorbing material by a complex number may seem strange so the following analysis will be helpful. Suppose we consider the passage of a beam of light through a layer (in the yz plane) of unspecified material of thickness Δx. If the layer contains a vacuum ($n = 1$), then the electric field transmitted through the layer will be given by Eq. (4):

$$\mathbf{E}_{n=1} = \mathbf{E}_0 \cos 2\pi\left(\nu t - \frac{x}{\lambda}\right) \tag{90}$$

On the other hand, suppose the layer consists of a material of refractive index n. The light will now take an increment of time Δt longer to pass through the layer due to the delaying effect of the medium. In this case the emerging field would be given by

$$\mathbf{E}_n = \mathbf{E}_0 \cos 2\pi\left[\nu(t + \Delta t) - \frac{x}{\lambda}\right] \tag{91}$$

The delay may be related to the thickness of the layer and the refractive index as follows:

$$\Delta t = t_n - t_0 = \frac{\Delta x}{c/n} - \frac{\Delta x}{c} = (n-1)\frac{\Delta x}{c} \tag{92}$$

This means Eq. (91) may be written

$$\mathbf{E}_n = \mathbf{E}_0 \cos 2\pi\left[\nu t + (n-1)\frac{\Delta x}{\lambda} - \frac{x}{\lambda}\right] \tag{93}$$

which shows that, compared with Eq. (90), the field experiences a phase shift in the medium.

A very important relationship involving complex numbers is

$$e^{i\theta} = \cos\theta + i\sin\theta \tag{94}$$

Cosine trigonometric functions, in other words, are given by the real part of the function $e^{i\theta}$. This means that Eqs. (90) and (93) may be written

$$\mathbf{E}_{n=1} = \mathbf{E}_0 \operatorname{Re} \exp\left[2\pi i\left(\nu t - \frac{x}{\lambda}\right)\right] \tag{95}$$

and

$$\mathbf{E}_n = \mathbf{E}_0 \operatorname{Re} \exp\left\{2\pi i\left[\nu t + (n-1)\frac{\Delta x}{\lambda} - \frac{x}{\lambda}\right]\right\} \tag{96}$$

where Re reminds us that it is only the real part of the complex number in which we are interested. Equation (96) may be written

$$\mathbf{E}_n = \mathbf{E}_0 \operatorname{Re} \exp\left[2\pi i\left(\nu t - \frac{x}{\lambda}\right)\right] \exp\left[2\pi i(n-1)\frac{\Delta x}{\lambda}\right]$$

$$= \mathbf{E}_{n=1} \exp\left[2\pi i(n-1)\frac{\Delta x}{\lambda}\right] \tag{97}$$

As we have seen repeatedly in this chapter, the intensity of light is proportional to the field amplitude squared; therefore, we write

$$\frac{I_n}{I_{n=1}} = \exp\left[4\pi i(n-1)\frac{\Delta x}{\lambda}\right] \tag{98}$$

Now let us consider the case in which the refractive index of the layer is a complex number. In that case, Eq. (98) becomes

$$\frac{I_n}{I_{n=1}} = \exp\left[4\pi i(n+ik-1)\frac{\Delta x}{\lambda}\right] = \exp\left[4\pi i(n-1)\frac{\Delta x}{\lambda}\right] \exp\left(-4\pi k\frac{\Delta x}{\lambda}\right) \tag{99}$$

The first factor of Eq. (99) is exactly what it would be if the material were nonabsorbing. It is modified, however, by a second term, $\exp(-4\pi k\, \Delta x/\lambda)$, which is real but contains the imaginary part of the index of refraction.

The Beer–Lambert equation is another formalism by which the arrangement we have been discussing might be described. In the Beer-Lambert equation, the intensity of the transmitted light I_t relative to the intensity of the incident light I_0 is given by Eq. (40), which may be written

$$\left(\frac{I_t}{I_0}\right)_{\text{abs}} = \exp(-\varepsilon\, \Delta x) \tag{100}$$

where ε is the absorbance of the material. Because of its usefulness in analytical chemistry, chemistry students are more likely to be familiar with this formalism than with that which leads to Eq. (99). Comparison of Eqs. (99) and (100) reveals that both describe the attenuation of light due to absorption in terms of a negative exponential which is proportional to the path length through the absorbing material. Since the two approaches describe the same situation in the same functional form, the two proportionality factors must also be equal. Therefore, the imaginary part of the complex refractive index and the absorbance must be related as follows:

$$\varepsilon = \frac{4\pi k}{\lambda} \tag{101}$$

We observed in Sec. 5.6 that, for nonabsorbing systems, the turbidity is a concept analogous to the absorbance; that is,

$$\left(\frac{I_t}{I_0}\right)_{\text{sca}} = \exp(-\tau\, \Delta x) \tag{102}$$

Examination of Eqs. (100) and (102) shows how the two complement each other: The first describes absorption without scattering, the second describes scattering without absorption. For a system which displays these two optical effects simultaneously, the following composite relationship applies:

$$\left(\frac{I_t}{I_o}\right) = \exp[-(\varepsilon + \tau)\,\Delta x] \tag{103}$$

The experimental extinction in a system which displays these two effects equals the sum of ε plus τ.

The literature on this topic often factors absorbance and turbidity into the product of several terms. Both the nomenclature and notation which are used in this area vary from author to author, but the principal breakdown of both ε and τ goes as follows:

$$\varepsilon = N\pi R^2 Q_{abs} \tag{104}$$

and

$$\tau = N\pi R^2 Q_{sca} \tag{105}$$

where N is the number of dispersed particles per unit volume, R is their radius, and the Q terms are known as the efficiency factors for absorption and scattering. The quantity πR^2 in these equations gives the geometrical cross section of the dispersed spheres. This area times the efficiency factor defines a quantity known as the cross section C for absorption and scattering:

$$C_{abs} = \pi R^2 Q_{abs} \tag{106}$$

$$C_{sca} = \pi R^2 Q_{sca} \tag{107}$$

Dimensionally, ε and τ describe the attenuation of light per unit optical path and have the cgs units reciprocal centimeters. The area πR^2 has units of square centimeters per particle and N has units of particles per cubic centimeter. The efficiency factors are dimensionless. The cross sections have units of area per particle. These should not be taken literally as cross-sectional areas, but rather as the "blocking power" of a particle as far as the transmission of incident light is concerned.

As stated previously, Maxwell's equations have been solved for spherical particles of arbitrary size and arbitrary refractive index. Furthermore, the refractive index may be complex, so the general theory applies to both nonabsorbing and absorbing particles. The solutions to Maxwell's equations are generally given in terms of the efficiency factors Q where the magnitude of the two efficiencies depends on the wavelength of light, the size of the particles, the real and imaginary parts of the refractive index, and the angle of observation:

$$Q = f(\lambda, R, n, k, \theta) \tag{108}$$

For nonscattering particles, Q_{sca} is zero, and for nonabsorbing particles, Q_{abs} is zero. The results of such an analysis are generally reported by giving values of Q_{sca} and Q_{abs} as a function of a size parameter α, where α is defined to be

$$\alpha = 2\pi\frac{R}{\lambda} \tag{109}$$

The calculations involved in computing these results are formidable and the results which were available prior to the widespread utilization of computers were very limited. As noted in the introduction to this chapter, the advent of the computer broadened the applicability of light scattering enormously.

It is difficult to make any generalizations about the variation of the efficiencies with α since the functions are so complicated and vary greatly with the refractive index. About the best that can be done along these lines is the following: (a) At any given angle of observation, Q tends to be an oscillating function of α; (b) for nonabsorbing particles (n real), the amount of oscillation in the Q_{sca} versus α curve is more pronounced the larger n is—if the particles absorb, the amount of oscillation in the curves decreases with increasing k; (c) for any given value of α, Q varies with the angle of observation θ. The number of oscillations in this curve is greater for larger values of α and n. These generalizations are consistent with the approximations discussed earlier in the chapter: Small particles with low values of n are the simplest to describe.

In the following two sections, we consider two specific systems which have been studied extensively, aqueous dispersions of colloidal gold (Sec. 5.13) and aqueous dispersions of colloidal sulfur (Sec. 5.14). These will illustrate some of the statements made previously as well as show the sort of information that may be obtained from light scattering experiments in this very general case. No attempt has been made to be comprehensive in this presentation. Much more has been done with the systems to be discussed, and cross sections for many other systems have been calculated. The book by Kerker [2] contains a good bibliography of published scattering functions.

5.13 THE MIE THEORY: GOLD SOLS

The preceding section indicated some of the complications that arise when the optical properties of dispersions are calculated without placing narrow limitations on the size and refractive index of the particles. The difficulties are still large, but somewhat more manageable, if only the refractive index is given full range, the particle dimensions being somewhat restricted. This is the situation that was treated by Mie in 1908.

Mie wrote the scattering and absorption cross sections as power series in the size parameter α, restricting the series to the first few terms. This truncation of the series restricts the Mie theory to particles with dimensions less than the wavelength of light but, unlike the Rayleigh and Debye approximations, applies to absorbing and nonabsorbing particles. The following equations give some indication of the nature of these expansions:

$$Q_{abs} = A\alpha + B\alpha^3 + C\alpha^4 + \cdots \tag{110}$$

and

$$Q_{sca} = D\alpha^4 + \cdots \tag{111}$$

where the values of the coefficients A, B, C, and D are listed in Table 5.4. In this

TABLE 5.4 *Values for the constants A to D in Eqs.* (110) *and* (111)[a]

Coefficient	General case	Special case of $k=0$
A	$\dfrac{24nk}{(n^2+k^2)^2+4(n^2-k^2)+4}$	0
B	$\dfrac{4nk}{15}+\dfrac{20nk}{3[4(n^2+k^2)^2+12(n^2-k^2)+9]}$	0
	$+\dfrac{4.8nk[7(n^2+k^2)^2+4(n^2-k^2-5)]^2}{[(n^2+k^2)^2+4(n^2-k^2)+4]^2}$	
C	$\dfrac{-192n^2k^2}{[(n^2+k^2)^2+4(n^2-k^2)+4]^2}$	0
D	$\dfrac{\frac{8}{3}[(n^2+k^2)^2+n^2-k^2-2]^2+36n^2k^2}{[(n^2+k^2)^2+4(n^2-k^2)+4]^2}$	$\dfrac{\frac{8}{3}(n^2-1)^2}{(n^2+2)^2}$

[a] From R. B. Penndorf, *J. Opt. Soc. Am.* **52**:896 (1062).

presentation, we have intentionally limited the series to include no terms higher than fourth order in α. Thus only the leading term in Q_{sca} is represented, even though the first three terms in Q_{abs} are included. The neglect of higher order terms permits us to apply this discussion rigorously only to particles which are sufficiently small that terms in α^5 or higher would make negligible contribution. Even with only these terms retained, it is possible to draw several informative conclusions about Q_{abs} and Q_{sca}: (a) The absorption and scattering efficiencies do not show the same dependence on the particle size parameter α; (b) both numerical coefficients of the complex refractive index, n and k, appear in both Q_{abs} and Q_{sca} (Table 5.4); (c) the coefficients A, B, and C equal zero and D reduces to Rayleigh's law if $k=0$; (d) the efficiencies are functions of the dimensionless variable α alone; (e) for dispersions of uniform spheres, the entire wavelength dependence of the extinction is given by Eqs. (110) and (111); (f) the "wavelength dependence of the extinction" is simply the spectrum of the dispersion, which is, therefore, predicted theoretically by the general equations.

In the remainder of this section, we shall see how the theoretical calculations of Mie account for the observed spectrum of colloidal gold. In the next section, we shall consider the inverse problem for a simpler system: how to interpret the experimental spectrum of sulfur sols in terms of the size and concentration of the particles. Both of these example systems consist of relatively monodisperse particles. Polydispersity complicates the spectrum of a colloid since the same Q value will occur at different λ-values for spheres of different radii according to Eqs. (109), (110), and (111).

Colloidal gold is of considerable historic importance in colloid chemistry since many of the scientists who led the early development of the field conducted experiments on this system. Mie set out specifically to account for the brilliant colors displayed by sols of gold and other metals.

Chloroauric acid, $HAuCl_4$, is easily reduced to metallic gold by a wide variety of reducing agents. However, characteristics of the resulting gold are widely different for different reducing agents. Thus, if phosphorus is used, a polydisperse system containing very small particles forms rapidly. If the resulting colloid is used to seed a reaction in which hydrogen peroxide is the reducing agent, the following reduction takes place slowly without additional nucleation:

$$2AuCl_4^- + 3H_2O_2 \rightarrow 2Au + 8Cl^- + 6H^+ + 3O_2$$

The gold particles grow to a larger size by this process with considerable sharpening of the particle size distribution. The particle size may be regulated to some extent by varying the amount of reagents used. Thus, approximately monodisperse colloids of several different particle sizes may be prepared and compared. They are found to display different colors depending on the particle size: The smaller particles produce a red dispersion, somewhat larger particles impart a blue color to the dispersion.

In calculating efficiency factors for absorption and scattering, the wavelength dependence of both the real and imaginary parts of the refractive index must be considered.

Figure 5.14 shows the real and imaginary parts of the complex refractive index of gold plotted against the wavelength of light in air and in water. These values of the refractive index were used to calculate the values of C_{abs} and C_{sca} for three different

FIGURE 5.14 *The real and imaginary parts of the complex refractive index of gold versus wavelength in air and in water. (Data from van de Hulst* [6].)

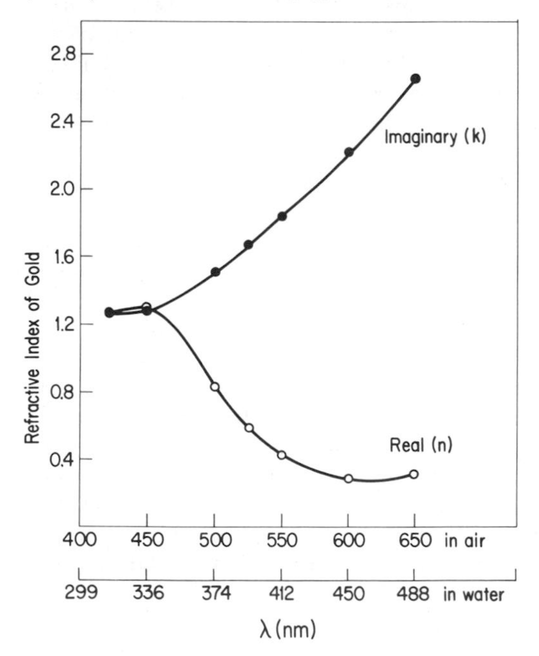

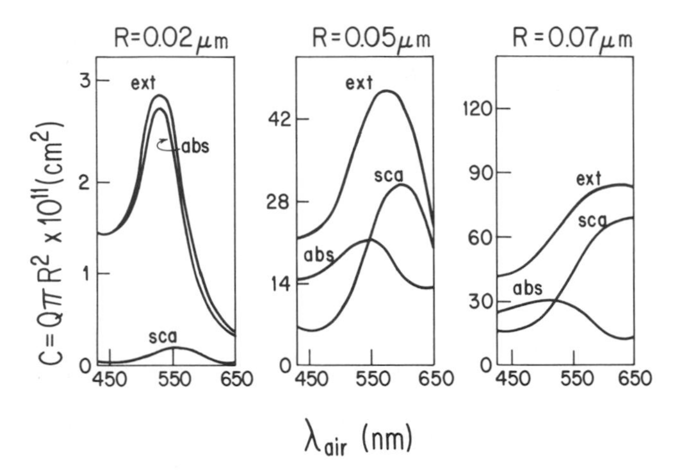

FIGURE 5.15 *Scattering coefficients versus wavelength for spheres of colloidal gold having three different radii. Note different scales for the ordinates in each figure.* (*Data from van de Hulst* [6].)

size spherical gold particles. The results are shown in Fig. 5.15 as a function of the wavelength in air. The figure also includes the sum of the two cross sections, C_{ext}, which describes the total extinction.

The magnitude and location of the maximum for each of the cross section curves are informative. It will be observed that the height and the location of the absorption curve in Fig. 5.15 change relatively little with particle size (note different scales for ordinates) as we might expect. The cross section for scattering, on the other hand, is practically negligible for the smallest of the particles and increases to be roughly 50 and 100% larger than absorption for the larger particles. Also noteworthy is the fact that the wavelength location of the scattering maximum shifts to longer wavelengths as the particle size increases. This is a consequence of the fact that the efficiency depends on α, rather than the separate values of R and λ. The wavelength at which maximum total extinction occurs lies in the green, yellow, and red portions of the spectrum, respectively, for the particles with $R = 0.02$, 0.05, and 0.07 μm. On the basis of color complementarity, these would appear red, violet, and blue, respectively. These are the hues displayed by actual dispersions. Thus, the Mie theory not only accounts for the color of the dispersions but shows how the color displayed may be used to characterize the particle size of the dispersed phase.

Note that the index of refraction of the continuous phase as well as that of the dispersed particles enter the evaluation of the various efficiencies. The light which actually strikes the particles is used in the determination of Q. This light differs from that under vacuum by the refractive index of the medium. This effect enters the calculation of the Q values in that it is the ratio of the refractive index of the particle relative to that of the medium which determines the extinction.

5.14 MONODISPERSE SULFUR SOLS: HIGHER ORDER TYNDALL SPECTRA

Another colloidal system whose light scattering characteristics have been widely studied is the so-called monodisperse sulfur sol. Although not actually monodisperse, the particle size distribution in this preparation is narrow enough to make it an ideal system for the study of optical phenomena.

The colloid is prepared by rapidly mixing dilute solutions of sodium thiosulfate and hydrochloric acid so that the final concentration of each is about 0.002 *M*. The following reaction then occurs so slowly that the sulfur precipitates only on those particles which nucleate first:

$$H^+ + S_2O_3^{2-} \rightarrow HSO_3^- + S$$

Slow growth on the original nuclei is how the narrow distribution of particle sizes is obtained just as with the colloidal gold described in the preceding section. The formation of sulfur may be terminated at any time by adding I_2 to react with the remaining thiosulfate. These monodisperse sulfur sols have been studied extensively, notably by V. K. LaMer and co-workers.

Since these particles are nonabsorbing in the visible spectrum, the range of particle sizes which may be conveniently dealt with is broader than for absorbing particles such as gold.

Using the Mie theory, the scattering efficiency as a function of α is evaluated for particles having a refractive index relative to the medium of 1.50 which describes the sulfur-water system.

Figure 5.16 shows a plot of Q versus α (on a double logarithmic scale) for this ratio of refractive indices. The same graph also shows the experimental spectrum (total extinction versus λ^{-1}, also double logarithmic) for a preparation which was allowed to grow for about 4.75 h before quenching. It will be observed that the theoretical and experimental curves agree very well in shape, but are displaced with respect to one another. The displacement may be analyzed with respect to R and N as follows. A prominent feature which appears on both curves is identified and the values of the abscissa for the two points are treated as matching values. For example, the minimum on the theoretical curve in Fig. 5.16 occurs at $\alpha = 7.5$ whereas the corresponding minimum on the experimental curve occurs at $\lambda = 336$ nm. Substituting these values into the definition for α [Eq. (109)] gives

$$R = \frac{(7.5)(336)}{2\pi} = 400 \text{ nm} = 4 \times 10^{-5} \text{ cm} \tag{112}$$

Similarly, the extinction at this point is 0.256 for a cell in which $\Delta x = 44.4$ cm while the corresponding value of Q_{sca} is 1.77. Next, Eq. (105) can be used with the value of R given by (112) to yield a value for N:

$$N = \frac{(2.303)(0.256)/44.4}{\pi(4 \times 10^{-5})^2(1.77)} = 1.5 \times 10^6 \text{ cm}^{-3} \tag{113}$$

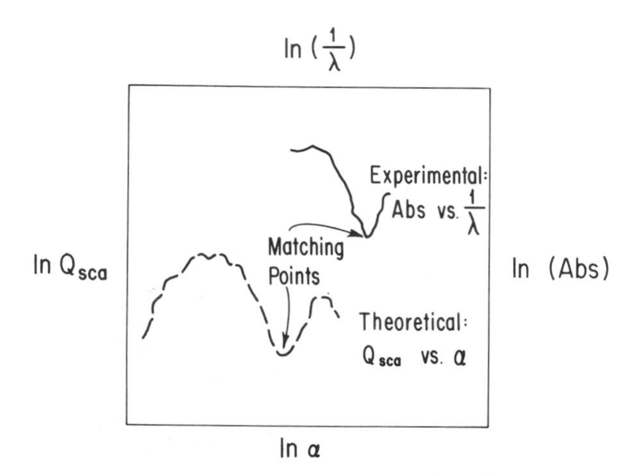

FIGURE 5.16 *Analysis of scattering spectrum for sulfur sols. Broken line: Q versus α from theory; solid line: extinction versus* $\ln(1/\lambda)$ *from experiment, both on double logarithmic scale.* [*Data from V. K. LaMer and M. D. Barnes,* J. Colloid Sci.**1**:71, 79 (1946).]

By this type of analysis, it has been possible to show that the number of particles per unit volume does not change appreciably as R increases, thus substantiating the mechanism described to account for the monodispersity of sulfur sols.

This procedure or related variations have been applied to systems other than the sulfur sols described here. The success of the method clearly depends on having a system for which some outstanding feature of the spectrum (e.g., the minimum of Fig. 5.16) lies at a value of R/λ which is within the range of α values studied. Methods have been developed for generating additional functions from a measured spectrum which are systematically displaced from the original spectrum to assist the matching of theoretical and experimental curves.

Another interesting application of scattering theory consists of combining scattering functions with expressions for particle size distribution to generate theoretical spectra for polydisperse systems. Such spectra have been calculated for different values of the relative refractive index (particle/medium). In this manner, the spectrum of a dispersion may be interpreted to yield estimates of both a mean and a standard deviation for the size of the dispersed particles. Excellent agreement with electron microscopic results has been obtained by this method.

The monodisperse sulfur sols we have discussed herein are good examples of another light scattering phenomenon: the higher order Tyndall spectrum. We observed in Sec. 5.12 that the scattering cross section is an irregularly oscillating function of θ, at least above a certain threshold value of α. Here it should be recalled that the complete theory reduces to the Rayleigh approximation for very small particles and to the Debye approximation for somewhat larger particles, provided the refractive index values are in the proper range. The full solution of the Mie theory provides quantitive information about the dependence of the efficiency factors on θ and λ. For uniform spheres over some range of refractive index and size, different colors of light will be scattered in different directions. The sulfur sols described here have the required properties to display this effect. Therefore, if a beam of white light is shown through a sample of the dispersion, various colors will be seen at different θ values. The resulting array of colors is known as the higher order Tyndall spectrum (HOTS). Red and green bands are most evident in the sulfur sols, and the number of times these bands repeat increases with the size of the sulfur particles. Therefore, the number and angular positions of the colored bands provide a unique characterization of the particle size. In the monodisperse sulfur sols, for example, particles having a radius of 0.30 μm are expected to show red bands at about 60, 100, and 140°. Particles with a radius of 0.40 μm, on the other hand, show red bands at about 42, 66, 105, 132, and 160°. Particle size determinations based on observations of this sort agree well with those determined from electron microscopy. These sulfur sols are quite easy to prepare, and it is interesting to observe the development of higher orders in the Tyndall spectrum as the thiosulfate decomposition reaction progresses.

It was once thought that the appearance of HOTS was evidence in itself for the presence of a monodisperse system. The argument was that one particle size would scatter, say, red light at a particular angle whereas another particle size would have the same α value and therefore the same scattering behavior for light of a complementary color. The resultant would be the obliteration of any distinct color: The scattered light would appear white. Although there may indeed be fortuitous cancellations of this sort at certain angles, it is also possible for certain bands to reinforce. In general, then, it is best to say that polydisperse systems may show HOTS, but in this case the angular distribution of bands is a characteristic of the particle size *distribution*. The angular location and number of bands as determined theoretically for uniform particles may not be used to interpret the HOTS of a disperse system correctly.

5.15 WRAP-UP: CHAPTERS 1 THROUGH 5

Although they dealt with a variety of different phenomena, Chaps. 1 through 5 have shared certain common features. In each chapter, at least one of the major topics was the development of a theoretical framework and an experimental technique which would permit the determination of the molecular weight of particles in the colloidal size range. In most cases, other facts concerning the nature of the system were also involved—either as part of the assumed model or as other parameters to be evaluated. Among the latter, we have considered particle shape, linear dimension, solvation, and polydispersity.

It should be evident that not all the methods we have discussed are equally useful for all systems, nor are they equally convenient when more than one is applicable. Often it is only by combining the results from more than one technique that ambiguity about particle characterization can be removed, so the availability of more than one experimental method is a valuable asset.

Direct observation, whether by optical or electron microscopy, permits the straightforward measurement of particle size from which molecular weights may be calculated. The method is tedious, however, and is subject to statistical complications and may be confused by artifacts of the preparative procedure.

Viscosity measurements are fast, and through the Staudinger equation permit the evaluation of molecular weight once suitable calibration data have been determined. For particles which combine unknown amounts of ellipticity and solvation, the viscosity cannot be interpreted unambiguously.

Sedimentation studies are versatile inasmuch as they may be applied to particles as small as individual molecules (in an ultracentrifuge) and as large as grains of sand (in gravitational settling). In the latter case, the assumption of spherical shape is needed to convert the results to molecular weights. For smaller particles, however, the combination of sedimentation and diffusion studies permits the unambiguous determination of particle mass.

Osmotic pressure gives the number average molecular weight of all particles restrained by a semipermeable membrane. If these are limited to the solute of interest, the molecular weight of that solute is measured with no assumptions about its shape. The procedure depends on the existence of a suitable membrane, however, and this may not always be as selective as might be desired.

Light scattering is theoretically complicated, but not too difficult in practice. Rigorous solutions of scattering functions are becoming more and more readily available, so the range of this technique and the amount of information that may be obtained from it continue to expand.

In the next three chapters, we shall be less concerned with the size and shape of dispersed particles as with the characteristics of their surfaces. As a matter of fact, the "dispersed phase" idea is scarcely mentioned. We are interested in the interfaces between various phases. Whether or not one of these phases is finely subdivided is of little consequence to us in these chapters. In summary, we are temporarily leaving colloid chemistry as such, and are entering the realm of surface chemistry. Clearly, the two disciplines are closely related since the specific area of a dispersed phase varies inversely with the size of the dispersed particle. We need to know more about the surfaces of these particles, particularly with respect to adsorption, before we can proceed with a study of colloidal phenomena. This information is presented in Chaps. 6 to 8, and Chap. 9 through 11 will again be concerned with dispersed particles.

REFERENCES

1. M. B. Huglin (ed.), *Light Scattering from Polymer Solutions*, Academic Press, New York, 1972.
2. M. Kerker, *The Scattering of Light and Other Electromagnetic Radiation*, Academic Press, New York, 1969.
3. D. McIntyre and F. Gormick (eds.), *Light Scattering from Dilute Polymer Solutions*, Gordon & Breach, New York, 1964.

4. K. A. Stacey, *Light Scattering in Physical Chemistry*, Butterworth, London, 1956.
5. C. Tanford, *Physical Chemistry of Macromolecules*, Wiley, New York, 1961.
6. H. C. Van de Hulst, *Light Scattering by Small Particles*, Wiley, New York, 1957.

PROBLEMS

1. The turbidity of solutions of sodium silicate containing SiO_2 and Na_2O in 3.75 ratio has been studied as function of time.* When the total solute content is 0.02 g cm^{-3}, the following data are obtained:

Time (days)	0	1	3	15	45	88	166	199	351	455
$\tau \times 10^4$ (cm^{-1})	0.81	2.32	3.80	5.47	6.70	8.06	9.53	10.47	13.00	14.43

(During the same time, the pH of the solution changes from 10.85 to 11.02.) Suggest a qualitative explanation for these observations in terms of the chemical behavior of silicates (be sure to include references to whatever sources you consult). Why is it important to the study of light scattering to realize that solutions such as these show a variation of turbidity with time?

2. Equation (3.66) suggest that the quantity $\overline{\delta c^2}$ in Table 5.1 is evaluated by solving

$$\overline{\delta c^2} = \int_0^\infty (\delta c)^2 P(\delta c)\, d(\delta c)$$

Argue that the appropriate form for $P(\delta c)$ is

$$P(\delta c) = A \exp(-\frac{G - G_0}{kT})$$

Use the value of $G - G_0$ for a fluctuation from Table 5.1 $[G - G_0 = \frac{1}{2}(\partial^2 G/\partial c^2)(\delta c)^2]$ to verify the value for $\overline{\delta c^2}$.

3. The refractive indices of NaCl solutions have been measured at 20°C as a function of concentration by means of a differential refractometer† with the following results:

$c \times 10^3$ (g NaCl cm^{-3})	3.749	5.468	7.498	7.920
$(n - n_0) \times 10^4$	6.73	9.83	15.18	14.06

* P. Debye and R. V. Hauman, *J. Phys. Chem.* **65**:5 (1961).
† P. Debye and R. V. Hauman, *J. Phys. Chem.* **65**:8 (1961).

Use these data to evaluate the quantity dn/dc for NaCl solutions in this concentration range. On the basis of these data, what is the apparent uncertainty introduced in a light scattering experiment through this quantity?

4. Above a certain concentration (the turbidity and concentration of which are taken to characterize the medium), the turbidity of sodium dodecyl sulfate (molecular weight = 288) solutions increases with concentration as if particles in the colloidal size range were present. Use the following data* to evaluate the apparent molecular weight of the species responsible for the scattering. For this system $H = 3.99 \times 10^{-6}$.

$c \times 10^3$ (g cm^{-3})	2.7	4.2	7.7	9.7	13.2	17.7	22.2
$\tau \times 10^4$ (cm^{-1})	1.10	1.29	1.71	1.98	2.02	2.14	2.33

Assuming the scattering centers to be aggregates of sodium dodecyl sulfate molecules, estimate the number of these units in the aggregate.

5. The turbidity of "Ludox" (a colloidal silica manufactured by DuPont) has been studied as a function of concentration with the following results†:

$c \times 10^2$ (g cm^{-3})	0.57	1.14	1.70	2.30
$\tau \times 10^2$ (cm^{-1})	1.56	2.97	4.25	5.36

Evaluate the molecular weight of the Ludox particles, using a value of $H = 4.08 \times 10^{-7}$ for the system. Calculate the characteristic diameter for these particles, assuming them to be uniform spheres of density 2.2 g cm^{-3}.

6. Criticize or defend the following proposition: "The accompanying data for the turbidity of dodecylamine hydrochloride solutions‡ suggest that at concentrations exceeding about 0.003 g cm^{-3} the solute associates into aggregates of colloidal dimensions."

$c \times 10^3$(g cm^{-3})	0.77	1.73	3.10	3.31	6.15	8.31
$\tau \times 10^4$(cm^{-1})	0	0	0	0.95	1.91	2.55

H is approximately 7.7×10^{-6} for this system.

7. A. I. Krasna§ has measured the turbidity of calf thymus DNA in aqueous solutions. The accompanying table gives $R_\theta \times 10^5$ (in cm^{-1}) for this system,

* H. V. Tartar and A. L. M. Lelong, *J. Phys. Chem.* **59**:1185 (1956).
† G. Deželic and J. P. Kratohvil, *J. Phys. Chem.* **66**:1377 (1962).
‡ P. Debye, *J. Phys. Chem.* **53**:1 (1949).
§ A. I. Krasna, *J. Colloid Interface Sci.* **39**:632 (1972).

measured at 546 nm:

θ (deg) \ $c \times 10^6$ (g cm^{-3})	20.6	41.4	62.0	82.5
26	2.47	4.80	6.94	8.75
30	2.06	4.02	5.83	7.67
34	1.77	3.39	4.84	6.44
38	1.75	2.98	4.28	5.61
42	1.38	2.57	3.84	4.87
50	1.01	1.90	2.91	3.71
60	0.76	1.45	2.23	2.89

Prepare a Zimm plot of these results (using $K = 3.63 \times 10^{-7}$) and evaluate M, B, and the radius of gyration of the DNA in this preparation.

8. The following table gives $R_\theta/K \times 10^{-3}$ for different values of c and θ in the system polystyrene–decalin at 30°C, measured with the mercury 435.8 nm line*:

θ (deg) \ $c \times 10^3$ (g cm^{-3})	0.50	0.99	1.49
30	0.735	1.34	—
45	0.685	1.27	2.04
60	0.625	1.17	1.62
75	0.562	1.05	1.49
90	0.510	0.96	1.37
105	0.467	0.88	1.25

Prepare a Zimm plot of these data and evaluate M, B, and the radius of gyration of the polymer under these conditions.

9. The following table shows the angular location of the green bands in the HOTS of various size spheres of relative refractive index 1.46 [2, p. 409]:

Number of green bands \ Radius (μm)	0.2	0.3	0.4	0.5	0.6	0.8
First	25°	7.5°				
Second	140°	77.5°	57.5°	42.5°	32.5°	17.5°
Third		150°	95°	72.5°	60°	40°
Fourth			140°	117.5°	85°	62.5°
Fifth				115°	110°	87.5°
Sixth				(blue)	130°	110°
Seventh					157.5°	127.5°
Eighth						142.5°

* M. D. Lechner and G. V. Schulz, *J. Colloid Interface Sci.* **39**:469 (1972).

Formulate a generalization correlating the observed number of green bands in the HOTS for this system with the approximate particle size. Briefly describe how this information could be used to "grow" a monodisperse sulfur sol whose dimension corresponds approximately to a predetermined size.

10. The spectra of monodisperse sulfur sols such as that shown in Fig. 5.16 have been observed after allowing the thiosulfate reduction to proceed for various lengths of time.* Each spectrum contains a recognizable minimum which may be identified with the minimum observed in the theoretical plot of K versus α. The coordinates of these minima for curves measured at different times are as follows:

Age (h)	λ_{min} (nm)	$(\text{Absorbance})_{min}$
7.77	358	0.313
15.90	442	0.484
22.68	526	0.421

Evaluate the radius and concentration of the sulfur particles at each of these times and comment on the correlation between these results and the mechanism proposed for monodispersity in Sec. 5.14.

11. General solutions of light scattering equations generate oscillating curves when Q is plotted versus α for a particular value of the relative refractive index of the dispersed phase compared to the continuous phase. Such a curve can be described by the expression $Q = K\alpha^{-n}$ where $-n$ is the local value of the slope of the $\ln Q$ versus $\ln \alpha$ curve at specific values of α. Suppose this parameter n is known as a function of α for an experimental system. Describe the kind of experimental data and the analysis required thereof to yield a size parameter for the dispersed particles. What are the limitations of this method? How does this method differ from the curve-matching techniques shown in Fig. 5.16?

12. Suppose your employer intends to develop a new laboratory to characterize particles in the colloidal size range. Your assignment is to prepare a list of the equipment which should be purchased for such a facility. A brief justification for each major item should be included along with a priority ranking based on the versatility of the method. Assume that your laboratory is already well stocked with such nonspecialized items as laboratory glassware, balances, and the like.

* M. D. Barnes and V. K. LaMer, *J. Colloid Sci.* **1**:79 (1946).

The Surface Tension of Pure Substances

. . . I don't think I said anything about the Third Dimension; and I am sure I did not say one word about "Upward, not Northward," for that would be nonsense, you know. How could a thing move Upward, and not Northward? Upward and not Northward! . . . How silly it is! [From Abbott's *Flatland*]

6.1 INTRODUCTION

The primary theme of this chapter is the analysis of situations which permit the surface tension of an interface to be evaluated. Surface tension γ is the contractile force which always exists in the boundary between two phases at equilibrium. Even though the evaluation of surface tension is the unifying topic of the chapter, it is actually the analysis of the physical phenomena involving surface tension which interests us. In the course of these analyses we shall consider:

1. surface tension as a force (Sec. 6.2, 6.8, and 6.10),
2. surface tension as surface free energy (Sec. 6.3 and 6.9),
3. surface tension and the shape of mobile interfaces (Secs. 6.4 and 6.5),
4. surface tension and capillarity (Sec. 6.6), and
5. surface tension and intermolecular forces (Sec. 6.12).

Several aspects of the opening statement require elaboration. First, the boundary between any pair of phases possesses a surface tension. This property is not limited to liquids although it is certainly most evident there. In this chapter we shall examine the tension in the interface between four different pairs of phases: liquid–vapor (LV), solid–vapor (SV), liquid–solid (LS), and liquid–liquid (L_1L_2 or AB). These parenthetical symbols are used as subscripts to identify the surfaces involved in the relationships of this chapter.

In both this chapter and Chap. 7, we are primarily concerned with equilibrium surfaces and equilibrium values of γ. The difference between the two chapters lies in the nature of the equilibrium phases which meet at the surface. In this chapter, we are concerned with the interface between two "pure" substances; in Chap. 7, we shall consider the effects of solutes of variable concentration. At equilibrium, of course, even "pure" phases are mutually saturated, but we are concentrating on situations in which the solubility of substances in the adjoining phases is low. For the most part, those phenomena which permit the evaluation of surface tension for pure liquids also apply to solutions. The major developments of this chapter will, therefore, apply to the following chapter also. In Chap. 7, the adsorption of added solutes is the new feature considered.

In this chapter, we shall discuss relatively few practical applications, except insofar as they refer to the measurement of γ. Applications frequently involve the effects of deliberately added solutes, so we shall postpone discussing them until Chap. 7.

A semiempirical interpretation of the origin of surface tension at the molecular level is also considered in this chapter. An especially important concept is the contribution of long-range van der Waals forces to γ. The so-called dispersion component of the surface tension γ^d quantitatively measures this contribution. It will appear again in our discussion of the stability of dispersions in Chap. 10.

6.2 SURFACE TENSION AS A FORCE: THE WILHELMY PLATE

The surface of a liquid appears to be stretched by the liquid it encloses. The beading of water drops on certain surfaces and the climbing of most liquids in glass capillaries are two phenomena which we can readily associate with this stretching of liquid surfaces. This effect is quantitatively measured by the surface tension of the liquid, γ. As the name implies, this is a force which acts on the surface and operates perpendicular and inward from the boundaries of the surface, tending to decrease the area of the interface.

Figure 6.1a shows a simple apparatus based on the notion just mentioned which would permit the measurement of surface tension. In practice, other experimental techniques are actually used for this measurement as we shall see presently. The arrangement of Fig. 6.1a has the advantage of simplicity, however, and serves to illustrate how the tension in the liquid surface is indeed measured by γ. The figure represents a loop of wire with one movable side upon which a film could be formed by dipping the frame into a liquid. The surface tension of a stretched film in the loop will cause the slide wire to move in the direction of decreasing film area unless an opposing force F is applied. In an actual apparatus, the fraction of the slide wire might be sufficient for this. In an idealized, frictionless apparatus like that in Fig. 6.1a the force opposing γ could be measured. The force evidently operates along the entire edge of the film, and will vary with the length l of the slide wire. Therefore, it is the force per unit length of edge which is the intrinsic property of the liquid surface. Since the film in the figure has two sides, the surface tension as measured by this apparatus equals

$$\gamma = \frac{F}{2l} \tag{1}$$

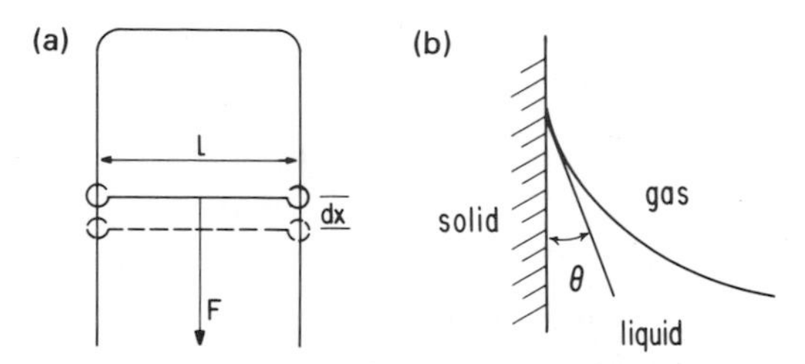

FIGURE 6.1 *(a) A wire loop with slide wire upon which a soap film might be formed and stretched by an applied force F. (b) Profile of a three-phase (solid, liquid, gas) boundary which defines the contact angle θ.*

Several points should be noted before proceeding any further:

1. Equation (1) defines the units of surface tension to be those of force per length or dynes per centimeter in the cgs system. We shall see presently that these are not the only units used for γ.
2. The apparatus shown in Fig. 6.1a resembles a two-dimensional cylinder/piston arrangement. With this similarity in mind, the suggestion that surface tension is analogous to a two-dimensional pressure seems plausible. With certain refinements, this notion will prove very useful in Chap. 7.
3. A gas in the frictionless, three-dimensional equivalent to the apparatus of the figure would tend to expand spontaneously. For a film, however, the direction of spontaneous change is contraction.

A quantity that is closely related to surface tension is the contact angle. The contact angle θ is defined as the angle (measured in the liquid) that is formed at the junction of three phases, for example, at the solid–liquid–gas junction as shown in Fig. 6.1b. Although the surface tension is a proporty of the two phases which form the interface, θ requires that three phases be specified for its characterization. We shall discuss contact angles in some detail in Sec. 6.8. At this time, however, it will be sufficient to consider the experimental arrangement illustrated in Fig. 6.2a.

Figure 6.2a represents a thin vertical plate suspended at a liquid surface from the arm of a tared balance. For simplicity, the plate is positioned so that the lower edge is in the same plane as the horizontal surface of the liquid away from the plate as shown in the figure. The manifestation of surface tension and contact angle in this situation is the entrainment of a meniscus around the perimeter of the suspended plate. Assuming the apparatus is balanced before the liquid surface is raised to the contact position, the imbalance that occurs on contact is due to the weight of the entrained meniscus. Since the meniscus is held up by the tension on the liquid surface, the weight measured by the apparatus can be analyzed to yield a value for γ.

The observed weight of the meniscus, w, must equal the upward force provided by the surface. This, in turn, equals the vertical component of γ—$\gamma \cos \theta$ where θ is the contact angle—times the perimeter of the plate. If the plate has a rectangular

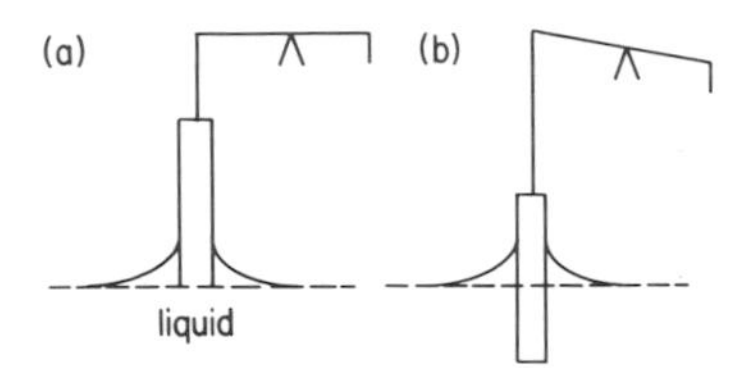

FIGURE 6.2 *The Wilhelmy plate method for measuring γ. In (a) the base of the plate does not extend below the horizontal liquid surface. In (b) the plate is partially submerged so buoyancy must be considered.*

cross section of length l and thickness t, the following equation applies:

$$w = 2(l + t)\gamma \cos \theta \tag{2}$$

In principle, all the quantities in Eq. (2) other than γ are independently measurable. Therefore, this simple experiment provides an excellent method for the determination of γ.

The technique we have described is called the Wilhelmy plate method for surface tension determination. It is one of the most straightforward procedures we shall find for this purpose. The method obviously depends on a knowledge of the contact angle. This quantity is not always easy to determine unambiguously so Eq. (2) does suffer from this limitation. Aside from this, however, the Wilhelmy plate technique requires less instrumentation and less computation than any other method for the precise evaluation of γ. Some of the sources of error which are common to this and all other procedures whereby γ is measured are mentioned in Sec. 6.7.

Because of the difficulties in measuring θ, the Wilhelmy plate method is most frequently used for systems in which $\theta = 0$. In that case, Eq. (2) becomes

$$w = 2(l + t)\gamma \tag{3}$$

Since the thickness of the plates used in the Wilhelmy method is generally negligible compared to their length ($t \ll l$), Eq. (2) may be approximated

$$w \simeq 2l\gamma \tag{4}$$

In practice, an end correction may be determined by working with a set of plates of different length but uniform thickness. The contribution of the force exerted by the meniscus on the ends is obtained by extrapolating the weights to zero length.

From the point of view of measuring γ for liquids, this discussion of the Wilhelmy plate is a valuable contribution. However, it does relatively little to relate surface tension and contact angle to the numerous other phenomena we associate with these properties in liquids, not to mention the determination of γ for solids. To proceed further, we must express γ in energetic terms so that it can be interpreted by the powerful techniques of thermodynamics.

6.3 SURFACE TENSION AS SURFACE EXCESS FREE ENERGY

Application of a force infinitesimally larger than the equilibrium force to the slide wire in Fig. 6.1a will displace the wire through a distance dx. The product of force times distance equals energy, in this case the energy spent in increasing the area of the film by the amount $dA = 2l\,dx$. Therefore, the work done on the system is given by

$$\text{Work} = F\,dx = \gamma 2l\,dx = \gamma\,dA \tag{5}$$

This supplies a second definition of surface tension: It equals the work per unit area required to produce new surface. In terms of this definition, the units of γ are energy per area or ergs per square centimeter in the cgs system.

We see, therefore, that there are two equivalent interpretations of γ: force per unit length of boundary of the surface and energy per unit area of the surface. The dimensional equivalency of the two is evident from the following manipulation of units:

$$\frac{\text{dynes}}{\text{cm}} = \frac{\text{dynes}}{\text{cm}} \cdot \frac{\text{cm}}{\text{cm}} = \frac{\text{ergs}}{\text{cm}^2}$$

In dealing with surface tension, we shall find it convenient to use both of these viewpoints because there are contexts in which one concept will be more helpful than the other. For example, Eq. (2) is an expression which represents force balancing. In this context, it is clearly convenient to use the concept of force to relate γ to other variables. On the other hand, when discussing the surfaces of solids, the notion of "tension" on the surface of the solid is harder to visualize, and the energy viewpoint is more convenient.

Equation (5) relates γ to the work required to increase the area of a surface. From thermodynamics, it will be recalled that work is a path-dependent process: How much work is done depends on how it is done. Based on this realization, then, it seems desirable to examine Eq. (5) a little more fully. As already noted, there is a tendency for mobile surfaces to decrease spontaneously in area. Therefore, it is convenient to shift our emphasis from work done *on* the system to work done *by* the system in such a reduction of area. If the quantity $\delta w'$ is defined to be the work done by the system when its area is changed, then Eq. (5) becomes

$$\delta w' = \gamma\,dA \tag{6}$$

According to Eq. (6), a decrease in area (dA negative) corresponds to work done *by* the system, whereas an increase in area requires work to be done on the system (dA positive and $\delta w'$ negative). This sign convention is consistent with the ideas expressed in Chap. 1 that energy is stored in surfaces.

We are now in a position to relate the quantity $\delta w'$ to other thermodynamic variables. To do this, a brief review of some basic thermodynamics is useful.

According to the first law, the change in the energy E of a system equals

$$dE = \delta q - \delta w \tag{7}$$

in which δw is the work done *by* the system and δq is the heat absorbed *by* the system. The quantity δw is conveniently divided into a pressure/volume term and a non-pressure/volume term:

$$\delta w = \delta w_{\text{pV}} + \delta w_{\text{non-pV}} = p\,dV + \delta w_{\text{non-pV}} \tag{8}$$

It will be recalled from physical chemistry that chemical work is the usual substitution for $\delta w_{\text{non-pV}}$. However, the work defined by Eq. (6) may also be classified as non-pressure/volume work.

The second law tells us that for reversible processes

$$\delta q_{\text{rev}} = T\,dS \tag{9}$$

Substituting Eqs. (8) and (9) into (7), with the stipulation of reversibility as required by (9), enables us to write

$$dE_{\text{rev}} = T\,dS - p\,dV - \delta w_{\text{non-pV}} \tag{10}$$

Next, we recall the definition of the Gibbs free energy G:

$$G = H - TS = E + pV - TS \tag{11}$$

which may be differentiated to give

$$dG = dE + p\,dV + V\,dp - T\,dS - S\,dT \tag{12}$$

Substituting Eq. (10) into (12) gives

$$dG_{\text{rev}} = T\,dS - p\,dV - \delta w_{\text{non-pV}} + p\,dV + V\,dp - T\,dS - S\,dT \tag{13}$$

This is a fundamental equation of physical chemistry because it enables us to assign a physical significance to G as defined by Eq. (11). Equation (13) shows that for a constant temperature, constant pressure, and reversible process

$$dG = -\delta w_{\text{non-pV}} \tag{14}$$

That is, dG equals the maximum non-pressure/volume work derivable from such a process since maximum work is associated with reversible processes.

We have already seen by Eq. (6) that changes in surface area entail non-pressure/volume work. Therefore, we identify $\delta w'$ from (6) with $\delta w_{\text{non-pV}}$ in (14) and write

$$dG = \gamma\,dA \tag{15}$$

Even better, in view of the stipulations made going from Eq. (13) to (14), we write

$$\gamma = \left(\frac{\partial G}{\partial A}\right)_{T,p} \tag{16}$$

Several things should be noted about Eq. (16). This relationship identifies the surface tension as the increment in Gibbs free energy per unit increment in area. The path-dependent variable $\delta w'$ is replaced by a state variable as a result of this analysis. Another notation that is often encountered which emphasizes the fact that γ is identical to the excess Gibbs free energy per unit area arising from the surface is to write it as G^s. The energy interpretation of γ, then, has been carried to the point where it has been identified with a specific thermodynamic function. Many of the general relationships which apply to G apply equally to γ. For example,

$$G^s = \gamma = H^s - TS^s \tag{17}$$

TABLE 6.1 *Several representative values of γ, S^s, and H^s for a variety of liquids near room temperature*[a]

Substance	γ at 20°C (ergs cm^{-2})	$d\gamma/dT = -S^s$ (ergs cm^{-2} deg^{-1})	$H^s = \gamma - T\,(d\gamma/dT)$ (ergs cm^{-2})
n-Hexane	18.4	−0.105	49.2
Ethyl ether	17.0	−0.116	51.0
n-Octane	21.8	−0.096	49.9
Carbon tetrachloride	26.9	−0.092	53.9
m-Xylene	28.9	−0.077	51.4
Toluene	28.5	−0.081	52.2
Benzene	29.0	−0.099	58.0
Chloroform	28.5	−0.135	68.3
1,2-Dichloroethane	32.2	−0.139	72.9
Carbon disulfide	32.3	−0.138	72.7
Water	72.8	−0.152	117.3
Mercury	484	−0.220	548

[a] Data from Kaelbe [6].

and

$$\left(\frac{\partial G^s}{\partial T}\right)_p = \left(\frac{\partial \gamma}{\partial T}\right)_p = -S^s \tag{18}$$

Equations (17) and (18) may be combined to give

$$\gamma = H^s + T\left(\frac{\partial \gamma}{\partial T}\right)_p \tag{19}$$

In this book we shall not distinguish between H^s and E^s; therefore, we may also write

$$E^s = H^s = \gamma - T\left(\frac{\partial \gamma}{\partial T}\right)_p \tag{20}$$

For water at 20°C, γ is about 72.8 ergs cm^{-2} and $d\gamma/dT$ is about −0.152 ergs cm^{-2} deg^{-1}. Therefore, Eq. (20) gives

$$E^s = 72.8 - (293)(-0.152) = 117 \text{ ergs cm}^{-2} \tag{21}$$

Additional values of G^s, S^s, and H^s for various substances are listed in Table 6.1.

6.4 THE LAPLACE EQUATION AND THE SHAPE OF LIQUID SURFACES

The interface between two phases will be curved when there is a pressure difference across the interface. The difference is such that the greater pressure is on the concave side. The relationship between the pressure difference and the curvature may be seen by the following application of thermodynamic concepts to surfaces.

Figure 6.3 shows a portion $ABCD$ of a curved surface. The surface has been cut by two planes which are perpendicular to one another. Each of the planes, therefore, contains a portion of arc where it intersects the curved surface. In the figure the radii of curvature are designated R_1 and R_2, and the lengths are designated x and y, respectively, for these two intercepted arcs. Now suppose the curved surface is moved outward by a small amount dz to a new position $A'B'C'D'$. Since the corners of the surface continue to lie along extensions of the diverging radial lines, this move increases the arc lengths to $x+dx$ and $y+dy$. Obviously the area of surface must also increase. The work required to accomplish this must be supplied by a pressure difference Δp across the element of surface area.

The increase in area when the surface is displaced is given by

$$dA = (x+dx)(y+dy) - xy = x\,dy + y\,dx + dx\,dy \simeq x\,dy + y\,dx \tag{22}$$

where the approximation arises from neglecting second-order differential quantities. The increase in free energy associated with this increase in area is given by $\gamma\,dA$:

$$dG = \gamma(x\,dy + y\,dx) \tag{23}$$

If ordinary pressure/volume work is responsible for the expansion of this surface, then the work equals $\Delta p\,dV$ where dV is the volume swept by the moving surface. In terms of Fig. 6.3 this equals

$$dw = \Delta pxy\,dz \tag{24}$$

Setting Eqs. (23) and (24) equal to one another gives

$$\gamma(x\,dy + y\,dx) = \Delta pxy\,dz \tag{25}$$

By using similar triangles, we may set up the following proportions from Fig. 6.3:

$$\frac{x+dx}{R_1+dz} = \frac{x}{R_1} \tag{26}$$

FIGURE 6.3 *Definition of coordinates describing the displacement of an element of curved surface ABCD to A'B'C'D'.*

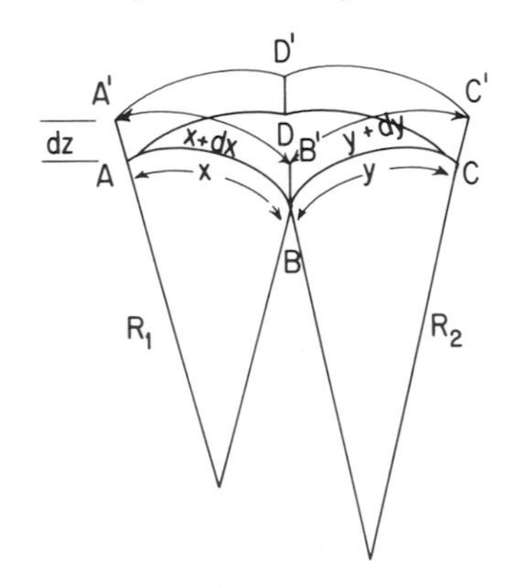

and

$$\frac{y+dy}{R_2+dz}=\frac{y}{R_2} \tag{27}$$

which simplify to

$$\frac{dx}{x\,dz}=\frac{1}{R_1} \tag{28}$$

and

$$\frac{dy}{y\,dz}=\frac{1}{R_2} \tag{29}$$

Substituting Eqs. (28) and (29) into (25) enables us to write the relationship of Δp to R_1, R_2, and γ:

$$\Delta p=\gamma\left(\frac{1}{R_1}+\frac{1}{R_2}\right) \tag{30}$$

This expression is known as the Laplace equation, and was derived in 1805.

Several special cases of Eq. (30) are worthy of note:

1. For a spherical surface: $R_1=R_2=R$, therefore,

$$\Delta p=\frac{2\gamma}{R} \tag{31}$$

2. For a cylindrical surface: $R_1=\infty$, therefore,

$$\Delta p=\frac{\gamma}{R_2} \tag{32}$$

3. For a planar surface: $R_1=R_2=\infty$, therefore,

$$\Delta p=0 \tag{33}$$

However, planar surfaces are not the only ones for which the pressure differential is equal to zero. Consider, for example, a soap film stretched over a curved but open framework. At first glance, this appears to produce a contradictory situation. Pressure differences are associated with curved surfaces, but how can a pressure difference be maintained in an openly connected space? It cannot; therefore, we conclude that in this case

$$\frac{1}{R_1}=-\frac{1}{R_2} \tag{34}$$

Physically, this means that the two radii of curvature lie on opposite sides of the surface, generating a saddle-shaped surface.

Until now, our discussion has been limited to small elements of area which could be described by constant radii of curvature. For macroscopic surfaces, however, these radii vary from position to position. This means that Δp across the surface will also vary with location, and may be expressed as a function of the height of the point under consideration above or below some reference plane. With these ideas in mind, the Laplace equation becomes

$$\Delta p(z) = \gamma[R_1^{-1}(x, y, z) + R_2^{-1}(x, y, z)] \tag{35}$$

In Eq. (35), we have added the notion that Δp, R_1^{-1}, and R_2^{-1} may be functions of location in space for any given surface.

The following expressions from analytical geometry are general functions for R_1^{-1} and R_2^{-1} for surfaces with an axis of symmetry:

$$R_1^{-1} = \frac{d^2z/dx^2}{[1+(dz/dx)^2]^{3/2}} \tag{36}$$

and

$$R_2^{-1} = \frac{dz/dx}{x[1+(dz/dx)^2]^{1/2}} \tag{37}$$

Axial symmetry is encountered in many systems of interest, so we shall focus attention on this case, accepting the loss of generality as the price for some mathematical simplification. Substitution of Eqs. (36) and (37) into (35) generates a complicated differential equation, the solution to which relates the shape of an axially symmetrical interface to γ. In principle then, Eq. (35) permits us to understand the shapes assumed by mobile interfaces and suggests that γ might be measurable through a study of these shapes.

We shall not attempt any general discussion of this problem. Instead, our attention will focus on those specific surfaces which most readily allow the experimental determination of γ. The shape assumed by a meniscus in a cylindrical capillary and the shape assumed by a drop resting on a planar surface (called a sessile drop) are most useful in this regard. Fortunately, in both of these cases, the surface contains an axis of symmetry which permits the use of Eqs. (36) and (37) also.

Both the sessile drop and the meniscus may be treated by the same mathematical formalism at the outset. Figure 6.4 may be regarded as a portion of the surface of either of these cases. As drawn, the curve represents the profile of a sessile drop; inverted, the solid portion represents the profile of a meniscus. The actual surfaces are generated by rotating these profiles around the axis of symmetry. In order to solve Eq. (35) for these interfaces, it is necessary to introduce a suitable coordinate system to describe these cylindrically symmetrical shapes.

In Fig. 6.4, the origin of the coordinate system O is situated at the apex of the surface. The two radii of curvature at point S are defined as follows. The one designated R_1 lies in the plane of the figure and describes the curvature of the profile shown. The radius of rotation of point S around the z axis equals x. The relationship between x and R_2 is given by

$$x = R_2 \sin\phi \tag{38}$$

since R_2 is defined as a vector originating along the z axis at P and making an angle ϕ with the axis of symmetry as shown in the figure.

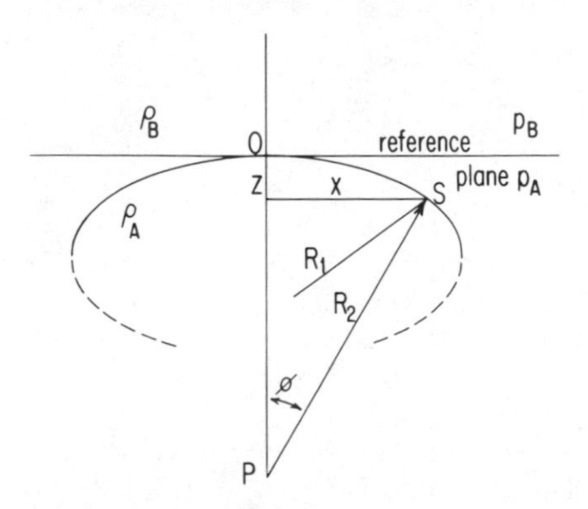

FIGURE 6.4 *Definition of coordinates for describing surfaces with an axis of symmetry* (OP).

Because of the symmetry of the surface, both values of R must be equal at the apex of the drop. The value of the radius of curvature at this location is symbolized b. Therefore, at the apex (subscript 0)

$$(\Delta p)_0 = \frac{2\gamma}{b} \tag{39}$$

according to Eq. (31).

Next, let us calculate the pressure at point S. At S the value of Δp equals the difference between the pressure at S in each of the phases. These may be expressed relative to the pressure at the reference plane through the apex (subscript 0) as follows:

1. In phase A:

$$p_A = (p_A)_0 + \rho_A gz \tag{40}$$

2. In phase B:

$$p_B = (p_B)_0 + \rho_B gz \tag{41}$$

Therefore, Δp at S equals

$$(\Delta p)_S = p_A - p_B = (p_A)_0 - (p_B)_0 + (\rho_A - \rho_B)gz = (\Delta p)_0 + \Delta\rho gz \tag{42}$$

where $\Delta\rho = \rho_A - \rho_B$. Now Eq. (39) may be substituted for Δp at the apex to give

$$(\Delta p)_S = \frac{2\gamma}{b} + \Delta\rho gz \tag{43}$$

The general form of the Laplace equation may be expressed in terms of the coordinates of Fig. 6.4 by combining Eqs. (35), (38), and (43):

$$\gamma\left(\frac{\sin\phi}{x} + \frac{1}{R_1}\right) = \frac{2\gamma}{b} + \Delta\rho gz \tag{44}$$

TABLE 6.2 *x/b and z/b for β = 25 and* $0° < \phi \leq 180°$[a]

	$\beta = 25$	
ϕ (deg)	x/b	z/b
5	0.08521	0.00368
10	0.16035	0.01348
15	0.22230	0.02712
20	0.27250	0.04288
25	0.31333	0.05974
30	0.34684	0.07713
35	0.37455	0.09475
40	0.39755	0.11236
45	0.41666	0.12985
50	0.43249	0.14711
55	0.44551	0.16405
60	0.45609	0.18063
65	0.46451	0.19678
70	0.47101	0.21246
75	0.47579	0.22761
80	0.47905	0.24221
85	0.48089	0.25626
90	0.48148	0.26966
95	0.48092	0.28243
100	0.47934	0.29435
105	0.47682	0.30594
110	0.47345	0.31665
115	0.46931	0.32662
120	0.46452	0.33585
125	0.45911	0.34433
130	0.45319	0.35204
135	0.44682	0.35901
140	0.44008	0.36519
145	0.43302	0.37061
150	0.42571	0.37526
155	0.41823	0.37915
160	0.41062	0.38231
165	0.40296	0.38472
170	0.39528	0.38643
175	0.38766	0.38744
180	0.38014	0.38776

[a] From F. Bashforth and J. C. Adams, *An Attempt to Test the Theory of Capillary Action*, Cambridge University Press, London and New York, 1883.

In Eq. (44), R_1^{-1} is given by (36). Expression (44) is known as the Bashforth and Adams equation. It is conventional to express this equation in dimensionless form by expressing all distances relative to the radius at the apex b:

$$\frac{\sin\phi}{x/b}+\frac{1}{R_1/b}=2+\frac{\Delta\rho g b^2}{\gamma}\frac{z}{b}=2+\beta\frac{z}{b} \tag{45}$$

The cluster of constants in Eq. (45) is defined by the symbol β:

$$\beta=\frac{\Delta\rho g b^2}{\gamma} \tag{46}$$

Surfaces of different shape are characterized by different values of β. For example, if $\rho_A > \rho_B$, β will be positive and the drop will be oblate in shape since the weight of the fluid tends to flatten the surface. A profile with positive β is shown in Fig. 6.4, assuming that gravity operates in the negative z direction. If $\rho_A < \rho_B$, a prolate drop is formed since the larger buoyant force leads to a surface with much greater vertical elongation. In this case, β is negative. A β value of zero corresponds to a spherical drop and, in a gravitational field, is expected only when $\Delta\rho = 0$. Positive values of β correspond to sessile drops of liquid in a gaseous environment. Negative β values correspond to sessile bubbles extending into a liquid. These statements imply that the drop is resting *on* a supporting surface. If, instead, the drop is suspended *from* a support (called pendant drops or bubbles), g becomes negative, and it is the liquid drop that will have the prolate ($\beta < 0$) shape, and the gas bubble the oblate ($\beta > 0$) shape.

The Bashforth and Adams equation—the composite of Eqs. (36) and (45)—is a differential equation which may be solved numerically with β and ϕ as parameters. Bashforth and Adams solved this equation for a large number of β values between 0.125 and 100 by compiling values of x/b and z/b for $0° < \phi < 180°$. Their tabular results, calculated by hand before the days of computers, were published in 1883. Other workers subsequently extended these tables. A very useful compilation of these results is found in the chapter by Padday [7]. Table 6.2 shows a typical result from these tables for $\beta = 25$. The surface profiles sketched in Fig. 6.5 were also drawn from these tabulated results for (a) $\beta = +10.0$ and (b) $\beta = -0.45$.

6.5 MEASURING SURFACE TENSION: SESSILE DROPS

We have already seen how the Wilhelmy plate method can be used to evaluate γ. That method depends on a knowledge of θ, however, and the latter is sometimes difficult to obtain. The Bashforth and Adams tables provide an alternate way of evaluating γ by observing the profile of a sessile drop of the liquid under investigation. If, after all, the drop profiles of Fig. 6.5 can be drawn using β as a parameter, then it should also be possible to match an experimental drop profile with the β value that characterizes it. Equation (46) then relates γ to β and other measurable quantities. This method is claimed to have an accuracy of 0.1%, but it is slow and tedious, and hence not often the method of choice in practice.

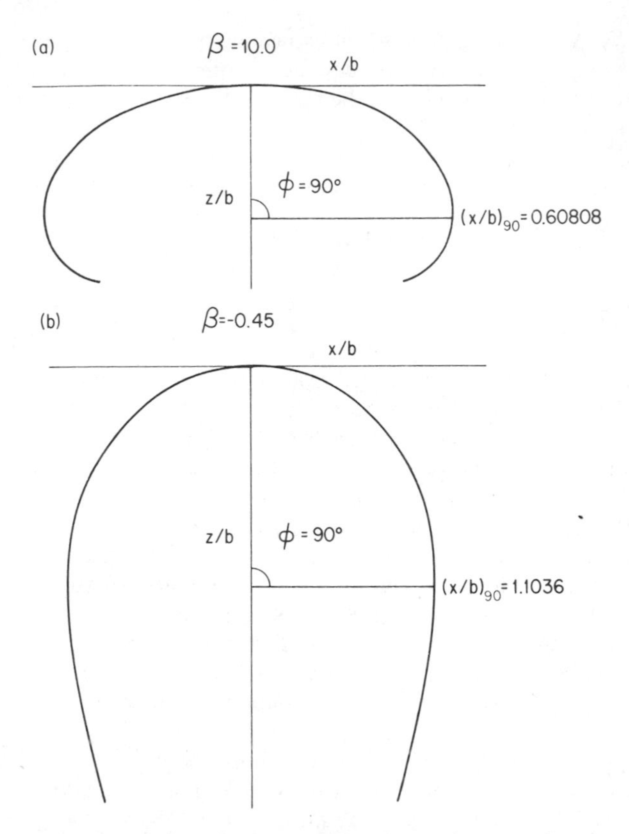

FIGURE 6.5 *Sessile drop profiles drawn from results of Bashforth and Adams tables. (a)* $\beta = 10.0$; *(b)* $\beta = -0.45$. *(Data from Padday* [7].)

The Bashforth and Adams tables describe sessile (or pendant) drops (or bubbles) resting against a horizontal surface. In order to obtain the profile of one of these surfaces experimentally, it is best to photograph the silhouette of the surface. Since surface tension is temperature dependent, the drop/bubble must be enclosed in a thermostated chamber. The support must be vibration-free to guarantee an equilibrium shape, and the geometry of the photographic arrangement must be distortion-free to allow magnification to be determined.

Once the profile of an experimental surface is obtained it is compared with theoretical profiles. To make the comparison, the experimental and all the theoretical profiles must have the same maximum diameter. With care, it is possible to interpolate between theoretical profiles and arrive at the β value which describes the surface under consideration. Equation (46) shows that knowledge of β alone is not sufficient to permit the evaluation of γ; b and $\Delta\rho$ must also be known. Evaluating the density difference poses no special difficulty. Let us next consider how b is measured.

Once β is known for a particular profile, the Bashforth and Adams tables may be used further to evaluate b:

1. For the appropriate β value, the value of x/b at $\phi = 90°$ is read from the tables. This gives the maximum radius of the drop in units of b.
2. From the photographic image of the drop, this radius may be measured since the magnification of the photograph is known.
3. Comparing the actual maximum radius with the value of $(x/b)_{90°}$ permits the evaluation of b.

Figure 6.5a may be used as a numerical example to illustrate this procedure. Suppose an actual experimental drop profile is matched with theoretical profiles and is shown to correspond to a β value of 10.0. Then b is evaluated as follows:

1. The value of $(x/b)_{90°}$ for $\beta = 10$ is found to be 0.60808 from the tables.
2. Assume the radius of the actual drop is 0.500 cm at its widest point.
3. Items (1) and (2) describe the same point; therefore, $b = 0.500/0.608 = 0.822$ cm. This would be the radius at the apex of the drop shown in Fig. 6.5a if the maximum radius were 0.5 cm.
4. This numerical example may be completed by using the definition of β to complete the evaluation of γ for the liquid of Fig. 6.5a. Assuming $\Delta\rho$ to be $+1.00$ g cm^{-3} and taking $g = 980$ cm s^{-2} gives for γ:

$$\gamma = \frac{\Delta\rho g b^2}{\beta} = \frac{(1.00)(980)(0.822)}{(10.0)} = 66.3 \text{ ergs cm}^{-2} \tag{47}$$

Several additional points might be noted about the use of the Bashforth and Adams tables to evaluate γ. If the interpolation is necessary to arrive at the proper β value, then interpolation will also be necessary to determine $(x/b)_{90°}$. This results in some loss of accuracy. With pendant drops or sessile bubbles (i.e., negative β values), it is difficult to measure the maximum radius since the curvature is least along the equator of such drops (see Fig. 6.5b). The Bashforth and Adams tables have been rearranged to facilitate using them with pendant drops. The interested reader will find tables adapted for pendant drops in the chapter by Padday [7]. The pendant drop method utilizes an equilibrium drop attached to a support and should not be confused with the drop–weight method (to be described later) which involves drop detachment.

6.6 MEASURING SURFACE TENSION: CAPILLARY RISE

The rise of liquids in a capillary is one of the most familiar manifestations of surface tension. Figure 6.6 shows schematically how such an experiment is carried out. Conventionally, the height of a liquid column in a capillary above the reference level in a large dish is measured. It is important that the dish be large enough in diameter so that the reference level has a well-defined horizontal surface. In the capillary, the liquid will have a curved meniscus and, as usual, it is the height of the bottom of the meniscus above the horizontal that is measured. It should be noted that capillary depression is also observed, as with mercury for example. In this case, the capillary "rise" is a negative quantity. The objective of this section is to relate the equilibrium liquid column height h to the surface tension of the liquid.

A simple—but incorrect—relationship between the height of capillary rise, capillary radius, contact angle, and surface tension is easily derived. At equilibrium, the vertical component of the surface tension $2\pi R\gamma \cos\theta$ equals the weight of the liquid column, approximated as the weight of a cylinder of height h and radius R. This leads to the approximation

$$2\pi R\gamma \cos\theta \simeq \pi R^2 h \, \Delta\rho g \tag{48}$$

It is extremely difficult to obtain reproducible results unless $\theta = 0°$, in which case Eq. (48) simplifies to

$$Rh \simeq \frac{2\gamma}{\Delta\rho g} \tag{49}$$

The cluster of constants $2\gamma/(\Delta\rho g)$ is defined as the capillary constant (some authors define it without the factor 2) and is given the symbol a^2:

$$a^2 = \frac{2\gamma}{\Delta\rho g} \tag{50}$$

It should be noted that a has the units of length, centimeters in the cgs system. The approximation in expression (48) arises from neglecting the weight of liquid in the "crown" of the curved meniscus.

The Laplace equation must be used to obtain a rigorous solution to the problem of capillary rise. Since the meniscus is curved, there is a pressure difference across it, and it is this pressure difference that is responsible for the capillary rise. Thus, if the meniscus is concave as in Fig. 6.6, the pressure just under the surface is less than it would be for a planar surface at the same level. The pressure difference in this case pushes the liquid up the tube until an equilibrium height is reached. In the case of a convex meniscus, the pressure difference has the opposite sign and leads to capillary depression.

The apex of the curved surface is identified as the point from which h is measured. As we have seen before, both radii of curvature are equal to b at this point. At the

FIGURE 6.6 *Schematic illustration of capillary rise in a cylindrical tube of radius R.*

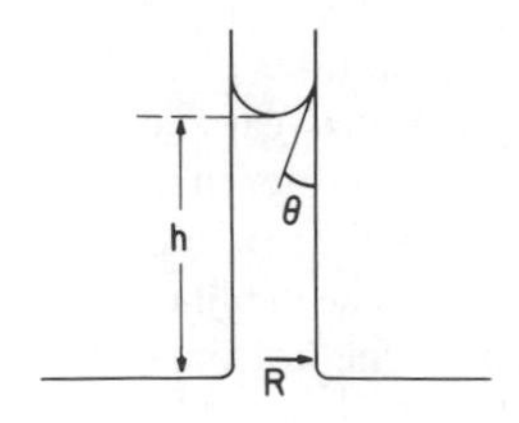

apex of the meniscus, the equilibrium force balance leads to the result

$$\Delta p = \frac{2\gamma}{b} = \Delta \rho g h \tag{51}$$

Combining Eqs. (50) and (51) yields

$$bh = a^2 \tag{52}$$

Unlike Eq. (49), Eq. (52) is exact since both the right- and left-hand sides refer to the same point, the apex of the meniscus. Equation (48) is valid only when $R = b$, that is, for a hemispherical meniscus. In general this is not the case and b is not readily measured so we have not yet arrived at a practical method of evaluating γ from the height of capillary rise. Again, the tables of Bashforth and Adams provide the necessary information.

For a liquid to make an angle of 0° (the usual situation in glass capillaries) with the supporting walls, the walls must be tangent to the profile of the surface at its widest point. Accordingly, $(x/b)_{90°}$ in the Bashforth and Adams tables must correspond to R/b. Since the radius of the capillary is measurable, this information permits the determination of b for a meniscus in which $\theta = 0$. However, there is a catch. Use of the Bashforth and Adams tables depends on knowing the shape factor β. It is not feasible to match the profile of a meniscus with theoretical contours, so we must find a way of circumventing this problem.

The procedure calls for using successive approximations to evaluate β. Like any iterative procedure, some initial values are fed into a computational loop and recycled until no further change results from additional cycles of calculation. In this instance, initial estimates of a and b (a_1 and b_1) are combined with Eqs. (46) and (50) to yield a first approximation to β (β_1). The value of $(x/b)_{90°}$ for β_1 is read or interpolated from the tables. This value and R are used to generate a second approximation to b (b_2). By Eq. (52) a second approximation of a (a_2) is also obtained and—starting from a_2 and b_2—a second round of calculations is conducted. Table 6.3 shows an example of this procedure. It will be noted that the third and fourth rounds of approximation show negligible differences, so the β value which

TABLE 6.3 *Use of the Bashforth and Adams tables to solve the capillary rise problem by successive approximations. β values are obtained by linear interpolation of the tabulated quantities*[a]

Parameter	Approximation number			
	First	Second	Third	Fourth
b (cm)	$\sim R = 0.25$	0.288	0.295	0.295
a (cm)	$\sim\sqrt{Rh} = 0.32$	$\sqrt{b_2 h} = 0.339$	0.344	0.344
$\beta = 2(b/a)^2$	1.25	1.44	1.47	1.47
$(x/b)_{90°} = R/b_{i+1}$	0.867	0.846	0.847	—
R/a	0.78	0.736	0.727	0.728

[a] Data from Padday [7].

characterizes the meniscus is readily established. It is sometimes troublesome to find a starting point for these iterative calculations. The following estimates are helpful for the capillary rise problem:

1. From Eqs. (49) and (50):

$$a_1 \simeq \sqrt{Rh} \tag{53}$$

2. Treating the meniscus as a hemisphere:

$$b_1 \simeq R \tag{54}$$

The initial values of a and b used in the calculations of Table 6.3 are generated by these approximations, assuming $R = 0.25$ cm and $h = 0.40$ cm.

We began this discussion by noting that we needed a value of β to characterize the shape of the meniscus in order to correct Eq. (49). Although we have obtained such a value by these iterations, we have also arrived at the a value which characterizes the interface. The surface tension may be evaluated directly from this value. For the example in Table 6.3, for instance, $a = 0.344$. If we assume $\Delta\rho$ to be 1.00 g cm^{-3} in this case, then $\gamma = 57.9$ ergs cm^{-2} according to Eq. (50). Although the procedure just outlined shows the application of the Bashforth and Adams equation to the phenomenon of capillary rise, supplementary tables—compiled by Sugden and reproduced by Padday [7]—are somewhat more convenient to work with. The latter tabulate pairs of R/a, R/b ratios which satisfy the Bashforth and Adams equation. The values $R/a = 0.73$ and $R/b = 0.85$ in Table 6.3 are examples of a pair of points meeting this condition.

6.7 ADDITIONAL METHODS FOR DETERMINING SURFACE TENSION

Besides the Wilhelmy plate, sessile drops, and capillary rise, there are several additional methods for the determination of γ which are well known and widely used. We shall mention only two: the drop–weight method and the DuNouy ring. In the former, a counted number of liquid drops are allowed to fall into a container. The total amount of liquid collected is determined by weighing, so the weight per drop may be calculated. It is assumed (incorrectly) that the drop breaks away when its weight exceeds the upward pull arising from the surface tension. If R is the radius of the tip from which the drop is formed (the inside radius is used if the liquid does not spread over the tip, otherwise the outside radius is used), then the balance of forces gives

$$2\pi R\gamma \simeq w_d \tag{55}$$

where w_d is the weight per drop. This is only an approximate relation for several reasons. First, it is clear from the Bashforth and Adams profiles that a pendant drop narrows near its base, the tip in this case. This means that the force of surface tension is directed toward the axis of symmetry and only the vertical component opposes the weight of the drop. Second, only part of a drop that has reached the point of instability actually breaks away. The process of drop formation is quite complex. The narrow portion elongates to form a thin neck. Part of the latter forms a satellite drop that falls with the main drop, and the rest recedes back to the tip. Correction tables for Eq. (55) have been prepared which allow for these deviations from the simple model for drop formation [7].

The DuNouy ring method somewhat resembles the Wilhelmy plate technique except that a horizontal ring rather than a vertical plate is pulled from the surface. The force to pull a ring of radius R from the surface may be measured by a balance and may be (incorrectly) related to γ by the equation

$$F = 2(2\pi R)\gamma \tag{56}$$

where the extra factor 2 arises from the fact that the surface is stretched at both the inner and outer edges of the ring. As with the drop method, a small residue of liquid clings to the ring as it pulls away from the surface, making Eq. (56) inexact. Tables of correction factors for the DuNouy ring are also available [7].

In contrast to the Wilhelmy plate, sessile drop, and capillary rise methods, the drop–weight and DuNouy methods involve the detachment of a liquid from some type of support. At the moment of detachment, new surface is forming rapidly and the surface may not satisfy all equilibrium requirements. This is not a problem for pure liquids; but for solutions, where concentration effects at the surface are common, it may be quite a serious objection against these detachment methods.

All the methods we have described for measuring γ are clearly applicable to the liquid–vapor or liquid–air interface. In principle, they should also apply to the interface between two liquids; but, in practice, there are complications. Contact angles of zero are very rare at the junction of a liquid–liquid interface with a solid surface. Therefore, the expressions for capillary rise, the Wilhelmy plate, and the DuNouy ring may not apply to liquid–liquid interfaces. This objection does not apply to the drop–weight or the sessile drop methods. Since the first of these is subject to the objection of fresh surface being formed at the moment of detachment, the drop profile method seems best for liquid–liquid interfaces.

It is impossible to complete a discussion of the measurement of surface tension without saying something about the need for extreme cleanliness in any determination of γ. Any precision chemical measurement requires attention to this consideration, but surface tension is exceptionally sensitive to impurities. In Sec. 6.10 we shall see how the presence of a surface film only one molecule thick can dramatically lower the surface tension of liquids with large values of γ, such as water and mercury. It is often noted that touching the surface of 100 cm^2 of water with a fingertip deposits enough contamination on the water to introduce a 10% error in the value of γ. Not only must all pieces of equipment be clean, but also the experiments must be performed within enclosures or in very clean environments to prevent outside contamination.

6.8 MEASURING CONTACT ANGLES

The contact angle ranks with surface tension as an important parameter in the characterization of liquid surfaces. Like γ, θ must be measured with very clean samples and careful temperature control. Even when these requirements are met, the contact angle turns out to be quite difficult to evaluate uniquely.

The easiest method for the evaluation of contact angles involves the formation and observation of sessile or pendant drops or bubbles. If a photograph of a sessile drop has been prepared, there is no difficulty in measuring the angle it makes with the support from the silhouette. Alternatively, a telescope with a goniometer eyepiece may be used on the sample directly without recording the entire profile. No knowledge of $\Delta\rho$ or γ is required by this method. Ordinarily, an accuracy of about 1° is good enough, so the measurement is not difficult to make. Small quantities of liquid are ordinarily sufficient. However, obtaining a suitable solid is more troublesome. We shall say more about the role of the solid presently. For now, it is assumed to be planar, smooth, and chemically homogeneous.

Another method for measuring contact angles, known as the tilted plate method, is illustrated in Fig. 6.7. In this method, the angle of inclination of a smooth solid cutting the surface of a liquid is varied until an angle is found at which the liquid forms a horizontal contact with the solid. This technique requires a larger volume of liquid than is needed in the sessile drop method.

In addition to these procedures, the Wilhelmy plate may also be used to evaluate nonzero contact angles [see Eq. (2)] if γ is known from an independent experiment.

The experimental methods used to evaluate θ are not particularly difficult, but the results obtained may be quite confusing. The situation is best introduced by referring to Fig. 6.8 which shows a sessile drop on a tilted plane. "Teardrop" shapes such as this are familiar to everyone: Just look at a raindrop on a window pane. The problem, of course, is that the contact angle is different at different points of contact with the support. It is conventional to call the larger value the advancing angle θ_a and the smaller one, the receding angle θ_r. The two may be quite different. The presence of contamination will definitely contribute to this effect, but it is by no means the only cause. Therefore, even with carefully purified materials, both advancing and receding contact angles should be measured.

All the techniques described here are easily conducted so that both θ_a and θ_r may be observed. With the sessile drop, the advancing angle is observed when the drop is emerging from a syringe or pipet at the solid surface. The receding angle is obtained by removing liquid from the drop. Vibrations must be carefully avoided in these manipulations. When the tilted plate method is used to evaluate the contact angle, θ_r values are obtained if the plate has been pulled out (emersion) from the liquid; θ_a results if the plate is pushed into the liquid (immersion). Likewise, both values of θ may be obtained from the Wilhelmy method, depending on whether the liquid is making an initial contact (θ_a) with the plate or is draining from it (θ_r).

A rather interesting example of the difference between advancing and receding contact angles is obtained from the Wilhelmy plate method when the contact angle

FIGURE 6.7 *The tilted plate method for measuring contact angles.*

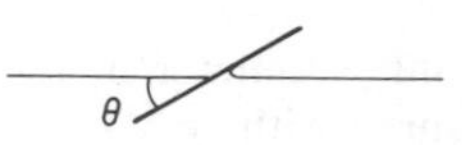

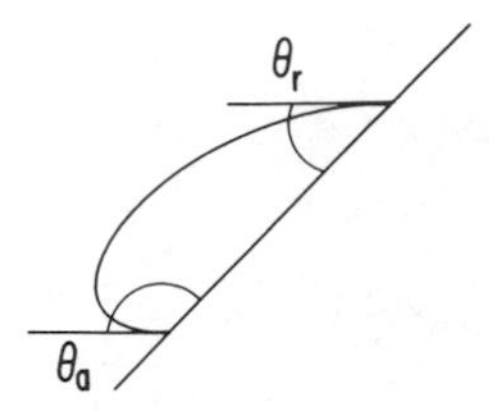

FIGURE 6.8 *A sessile drop on a tilted plane, showing advancing and receding contact angles. (From Johnson and Dettree [5], used with permission.)*

has a nonzero value. Suppose the Wilhelmy plate is allowed to dip beneath the horizontal liquid surface, as shown in Fig. 6.2b. In this case the weight of the meniscus as given by Eq. (2) will be decreased by a term w', the buoyant force on the submerged plate. The latter will clearly be proportional to the depth of immersion, d. Therefore, we write

$$w = 2\gamma(l+t)\cos\theta - w' = 2\gamma(l+t)\cos\theta - kd \tag{57}$$

where k is a suitable proportionality constant. This equation shows that the apparent weight of the meniscus, w, should give a straight line when plotted again the depth of immersion and that the intercept should be proportional to $\cos\theta$. If a single value of θ applies to both the immersion and emersion steps, then the line shown in Fig. 6.9a would result. Because of the difference between θ_a and θ_r, a curve like that shown in Fig. 6.9b is obtained instead. When the immersion–emersion cycle is repeated, a hysteresis loop is obtained which may be as reproducible as the analogous loops observed in magnetization–demagnetization cycles. The difference $\theta_a - \theta_r$ is called the hysteresis of a contact angle.

The general requirement for hysteresis is the existence of a large number of metastable states which differ slightly in energy and are separated from each other by small energy barriers. The situation is shown schematically in Fig. 6.10. The equilibrium contact angle corresponds to the free energy minimum. However, systems may be "frozen" in metastable states of somewhat higher energy by lacking

FIGURE 6.9 *Weight of a meniscus in a Wilhelmy plate experiment versus depth of immersion of the plate. In (a), both advancing and receding contact angles are equal. In (b) $\theta_a > \theta_r$.*

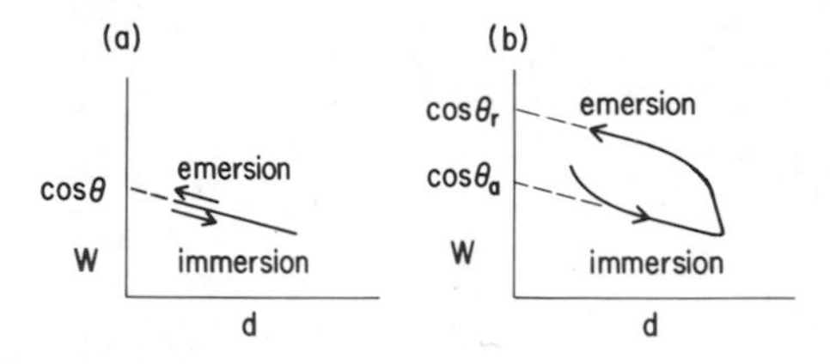

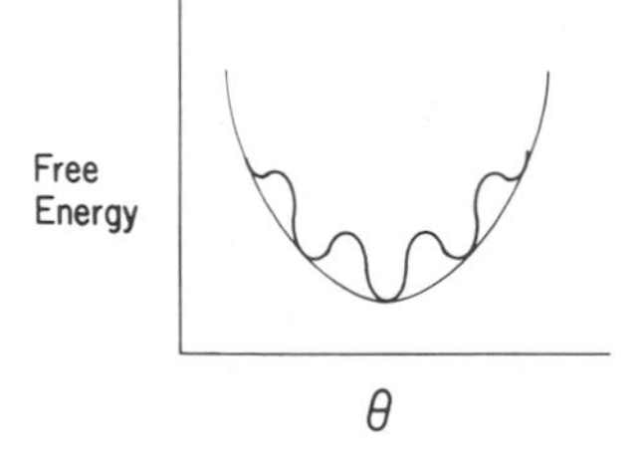

FIGURE 6.10 *Schematic energy diagram for metastable states corresponding to different contact angles.*

sufficient energy to overcome the energy barrier separating them from equilibrium. An interesting experimental observation is that advancing and receding values of θ converge to a common value when the surface is vibrated. Presumably, the mechanical energy imparted to the liquid by the vibration assists it in passing over the energy barriers and reaching equilibrium. In this sense, maximum vibration rather than vibration-free conditions may appear to be the ideal conditions for measuring θ, but "vibration" is a difficult parameter to control reproducibly so this concept is of little help practically.

Now let us briefly consider the origin of these metastable states. Excluding the effect of impurities, the metastable states are generally attributed to either the roughness of the solid surface or its chemical heterogeneity, or both. Of course, a well-prepared laboratory sample, will be fabricated and cleaned as effectively as possible to eliminate gross roughness and chemical heterogeneity. What we are talking about are microscopic irregularities that cannot be eliminated. In size and distribution, these will follow a random pattern on actual surfaces.

An informative model for contact angle hysteresis is obtained by postulating the surface to contain a set of concentric grooves upon which a sessile drop rests. Figure 6.11 represents the profile of two different drops on such a surface. In both of the profiles shown, the angles of contact between the liquid–vapor interface and the solid are identical, θ_0. With respect to the horizontal, however, two very different

FIGURE 6.11 *Cross section of a sessile drop resting on a surface containing a set of concentric grooves. For both profiles, the contact angle is identical microscopically, although macroscopically different.* (*From Johnson and Dettree* [5], *used with permission.*)

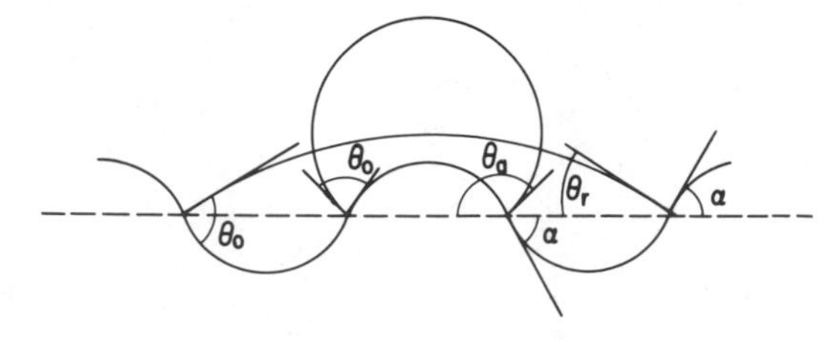

apparent contact angles are observed. The two extremes are identified as θ_a and θ_r in this model.

The two drop configurations in Fig. 6.11 differ in surface area and in the elevation of their centers of gravity. Thus, they possess different energies. The change from one configuration to the other involves the disortion of the shape of the drop, which accounts for the energy barrier between the two configurations. Thus, the model qualitatively accounts for the kind of metastable states shown in Fig. 6.10 which are required for hysteresis. According to this model, contact angle hysteresis arises when a three-phase boundary gets trapped in transit, lacking sufficient energy to surmount the energy barrier to a lower energy state. The teardrop profile of Fig. 6.8 corresponds to the situation in which one edge of the drop has one configuration while the other edge has the second configuration.

This model can also be applied to hysteresis which arises from chemical heterogeneity. However, this time the surface is assumed to be smooth and to contain concentric rings of different chemical composition and hence different θ. Actual heterogeneity may arise from impurities concentrated at the surface, from crystal imperfections, or from differences in the properties of different crystal faces. The distribution of such heterogeneities on an actual surface will obviously be more complex than the model considers, but the qualitative features of hysteresis are explained by the model nevertheless. Johnson and Dettree [5] present additional details of model experiments of this sort.

In general, surfaces are both rough *and* heterogeneous. Ordinarily, heterogeneity is the principal cause of hysteresis unless the roughness of the surface is fairly large. For example, the irregularities introduced by typical machining of metals has little effect on θ. In summary, then, it appears that unique values of θ are expected only for a smooth, homogeneous surface which—for solids at least—is often closer to being an ideal than an attainable fact. Experimental contact angles, then, must be regarded as some sort of average value for the surface. In fact, the average of θ_a and θ_r is often used as the value of θ in calculations.

6.9 THE KELVIN EQUATION

In Sec. 6.4 we saw how the pressure difference across a curved interface contributes to the shape of a mobile surface. Another extremely important result of this pressure difference is the effect it has on the free energy of the material possessing the curved surface. To evaluate this free energy effect, suppose we consider the process of transferring molecules of a liquid from a bulk phase with a vast horizontal surface to a small spherical drop of radius r.

According to Eq. (33), no pressure difference exists across a plane surface: the pressure is simply p_0, the normal vapor pressure. However, a pressure difference given by Eq. (31) exists across a spherical surface. Therefore, for liquid–vapor equilibrium at a spherical surface both the liquid and the vapor must be brought to the same pressure $p_0 + \Delta p$. Assuming the liquid to be incompressible and the vapor to be ideal, ΔG for the process of increasing the pressure from p_0 to $p_0 + \Delta p$ is as follows:

1. For the liquid:

$$\Delta G = \int_{p_0}^{p_0+\Delta p} V_L \, dp = V_L \, \Delta p = \frac{2V_L\gamma}{r} \tag{58}$$

where V_L is the molar volume of the liquid.
2. For the vapor:

$$\Delta G = RT \ln \frac{p_0 + \Delta p}{p_0} = RT \ln \frac{p}{p_0} \tag{59}$$

When liquid and vapor are at equilibrium, these two values of ΔG are equal:

$$RT \ln \frac{p}{p_0} = \frac{2V_L\gamma}{r} = \frac{2M\gamma}{\rho r} \tag{60}$$

since the volume per mole equals M/ρ where M and ρ are the molecular weight and density of the liquid, respectively. In either of these forms, this expression is known as the Kelvin equation. The Kelvin equation enables us to evaluate the *actual* pressure above a spherical surface and not just the pressure difference across the interface, as was the case with the Laplace equation. Using the surface tension of water at 20°C, 72.8 ergs cm^{-2}, the ratio p/p_0 is seen to be

$$\frac{p}{p_0} = \exp\left[\frac{(2)(18.0)(72.8)}{(0.998)(8.31\times 10^7)(293)r}\right] = \exp \frac{1.08\times 10^{-7}}{r} \tag{61}$$

or 1.0011, 1.0184, 1.1139, and 2.9404 for drops of radius 10^{-4}, 10^{-5}, 10^{-6}, and 10^{-7} cm, respectively. Thus, for small drops the vapor pressure may be considerably larger than for flat surfaces. For very small drops there is the usual question of whether the conventional value of γ applies to very small groups of molecules.

It might be mentioned that the Kelvin equation also applies to gas bubbles in a liquid. In this case, the vapor pressure is less inside the bubble because of surface curvature.

The Kelvin equation may also be applied to the equilibrium solubility of a solid in a liquid. In this case the ratio p/p_0 in Eq. (60) is replaced by the ratio a/a_0 where a_0 is the activity of dissolved solute in equilibrium with a flat surface and a is the analogous quantity for a spherical surface. For an ionic compound having the general formula M_mX_n, the activity of a dilute solution is related to the molar solubility S as follows:

$$a = (mS)^m (nS)^n \tag{62}$$

Therefore, for a solid sphere

$$\frac{2M\gamma}{\rho r} = RT \ln \frac{a}{a_0} = (m+n)RT \ln \frac{S}{S_0} \tag{63}$$

where S and S_0 are the solubilities of the spherical and flat particles, respectively. In principle, Eq. (63) provides a thermodynamically valid way to determine γ_{SL}. For example, the value of γ_{SL} for the $SrSO_4$–water surface has been found to be 85 ergs cm^{-2} and for the NaCl–alcohol surface to be 171 ergs cm^{-2} by this method.

Although the increase in solubility of small particles is unquestionably a real effect, using it quantitatively as a means of evaluating γ_{SL} is fraught with difficulties:

1. The difference in solubility between a small particle and a larger one will probably differ by less than 10%. Since a phase boundary exists at all, the solubility is probably low to begin with so there may be some difficulty in determining the experimental solubilities accurately.
2. Solid particles are not likely to be uniform spheres even if the sample is carefully fractionated. Rather, they will be irregularly shaped and polydisperse, although the particle size distribution may be narrow. The smallest particles will have the largest effect on the solubility, but they may be the hardest to measure.
3. The radius of curvature of sharp points or protuberances on the particles has a larger effect on the solubility of irregular particles than the equivalent radius of the particles themselves.

The Kelvin equation helps explain an assortment of supersaturation phenomena. All of these—supercooled vapors, supersaturated solutions, supercooled melts—involve the onset of phase separation. In each case, the difficulty is the nucleation of the new phase: Chemists are familiar with the use of seed crystals and the effectiveness of foreign nuclei to initiate the formation of the second phase.

The Kelvin equation shows that the ratio S/S_0 or p/p_0 increases rapidly as r decreases toward zero. Applying Eq. (63) rigorously (and incorrectly) down to $r = 0$ would imply infinite supersaturation, and make the appearance of a new phase impossible. Equation (60) is derived on the basis of two phases already in existence. To arrive at an understanding of the emergence of a new phase, we must consider what is going on at the molecular level at the threshold of phase separation. A highly purified vapor, for example, may remain entirely as a gas even thought its pressure exceeds the normal vapor pressure of the liquid for the temperature in question. At such a point, the vapor state is thermodynamically unstable with respect to the formation of a liquid phase with flat surfaces. Whatever stability the vapor has is kinetic stability, arising from the high activation energy required to start the formation of the second phase. A crude kinetic picture of the processes occurring at the molecular level may be informative.

Although a supersaturated vapor is still a gas, it is a very nonideal gas indeed! Clusters of molecules are continually forming and disintegrating: embryonic nuclei of the new phase. Some of these clusters will be dimers, some trimers, in general n-mers. Each will have its own characteristic radius r_n. An abbreviated derivation of the rate of formation of n-mers proceeds as follows. The rate law will contain both a frequency factor and a Boltzmann factor. The energy term in the latter may be estimated by Eq. (58) with V_L taken to be $\frac{4}{3}\pi r_n^3$. The frequency factor will involve the probability of additional molecules adding by collision. That is, it will depend on the surface area of the cluster ($4\pi r_n^2$) and the frequency of collisions with a wall as given by kinetic molecular theory. Therefore, the rate law may be approximated

$$\text{Rate} \simeq Z(4\pi r_n^2) \exp\left(-\frac{2\gamma}{r_n}\frac{4\pi r_n^3}{3kT}\right) = c_1 p r_n^2 \exp(-c_2 r_n^3) \tag{64}$$

where Z is the collision frequency. Since the latter certainly increases with p, the pressure has been factored out of the second expression in (64) where c_1 and c_2 are constants.

Two aspects of Eq. (64) are especially informative. First, we observe that the pre-exponential term increases with increasing r_n whereas the exponential term decreases. This means that there exists some critical radius for which the rate law shows a maximum. Clusters with this critical radius may be compared with reaction intermediates or transition states in ordinary chemical reactions. Those clusters which manage to overcome the energy barrier associated with this critical size are capable of further growth leading to the appearance of the new phase, and smaller clusters disintegrate. In addition, Eq. (64) also shows that the rate of cluster growth increases as the pressure increases. All clusters, including those of the critical size, form more rapidly at higher pressures.

The preceding considerations suggest that a point is ultimately reached in the course of increasing supersaturation where the liquefaction process becomes kinetically, as well as thermodynamically, favorable. It must be remembered here that the initial state of the system is one of instability or, more correctly, metastability. Once liquefaction begins, the pressure drops until the radius–pressure combination that satisfies Eq. (60) is reached.

A great deal of work has been done on the kinetics of phase formation; the arguments presented here are intended merely to suggest the direction taken in more detailed treatments. Many aspects of nucleation are of extreme interest in colloid and surface chemistry. The monodisperse colloids described in Chap. 5 are formed by carefully controlling the formation of the solid phase. In the case of the monodisperse sulfur sols, for example, the decomposition of $S_2O_3^{2-}$ (Sec. 5.14) proceeds slowly to quite a high level of supersaturation (the solution must be free of foreign nuclei). Ultimately, clusters of sulfur atoms exceeding the critical size form, nucleate precipitation, and grow until the supersaturation is relaxed. Any additional sulfur that is formed beyond this point will deposit on these particles without forming a new "crop" of nuclei. Relatively monodisperse colloids formed by such condensation processes depend on the fact that supersaturation and, therefore, nucleation occur only once during the formation of the colloid. If the rate of crystallization were too slow, then a second stage of supersaturation and nucleation could be reached, and a polydisperse colloid would result.

Of course, nucleation may also be accomplished by seeding or adding externally formed nuclei. The monodisperse gold sols described in Sec. 5.13 are prepared by this method. The use of AgI crystals and other materials as nuclei in cloud seeding has also been studied extensively. This is especially interesting in view of possible applications to weather modification.

6.10 THE YOUNG EQUATION

In the preceding sections, a variety of important related concepts have been introduced. Among these are the equivalence of surface tension and surface free energy, the applicability of these concepts to solids as well as liquids, and the notion of the contact angle. All these concepts are brought together in discussing the physical situation sketched in Fig. 6.12. Suppose a drop of liquid is placed on a perfectly smooth solid surface and these phases are allowed to come to equilibrium with the surrounding vapor phase. Viewing the surface tensions as forces acting along the perimeter of the drop enables us to write immediately an equation which describes the equilibrium force balance:

$$\gamma_{LV} \cos \theta = \gamma_{SV} - \gamma_{SL} \tag{65}$$

This result was qualitatively proposed by Thomas Young in 1805 and is generally known as Young's equation.

Young's equation is a plausible, widely used result; but its apparent simplicity is highly deceptive. The two terms which involve the interface between the solid and other phases cannot be measured independently, so it is debatable whether the equation has ever been verified experimentally, although a variety of experiments have been directed along these lines.

There are also a number of objections to Young's equation. These objections may be classified into two categories: those based on the noncompliance of the experimental system to the postulates of the derivation and those critical of the assumption of thermodynamic equilibrium in the solid. We shall discuss these two classes of objections separately.

Real solid surfaces may be quite different from the idealized one in this derivation. As we saw in Sec. 6.8, real solid surfaces are apt to be rough and even chemically heterogeneous. In principle, both of these considerations can be incorporated into Young's equation in the form of empirical corrections. For example, if a surface is rough, a correction factor r is traditionally introduced as a weighting factor for $\cos \theta$, where $r > 1$. The logic underlying this correction goes as follows. The factor $\cos \theta$ enters Eq. (65) by projecting γ_{LV} onto the solid surface. If the solid surface is rough, a larger area will be "overshadowed" by the projection than if the surface were smooth. The roughness factor measures this effect. With the empirical

FIGURE 6.12 *Components of interfacial tension needed to derive Young's equation.*

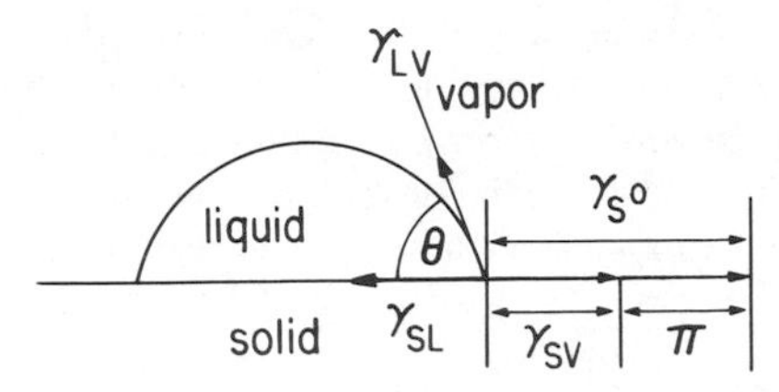

correction factor for roughness included, Young's equation becomes

$$r\gamma_{LV} \cos \theta = \gamma_{SV} - \gamma_{SL} \tag{66}$$

A surface may also be chemically heterogeneous. Assuming, for simplicity, that the surface is divided into fractions f_1 and f_2 of chemical types 1 and 2, we may write

$$\gamma_{LV} \cos \theta = f_1(\gamma_{S_1V} - \gamma_{S_1L}) + f_2(\gamma_{S_2V} - \gamma_{S_2L}) \tag{67}$$

where $f_1 + f_2 = 1$.

Both roughness and heterogeneity may be present in real surfaces. In such a case, the correction factors defined by Eqs. (66) and (67) are both present. Although such modification adapts Young's equation to nonideal surfaces, it introduces additional terms which are difficult to evaluate independently. Therefore, the validity of Eq. (65) continues to be questioned.

More fundamental objections to Young's equation center on the issue of whether the surface is in a true state of thermodynamic equilibrium. In short, it may be argued that the liquid surface exerts a force perpendicular to the solid surface, $\gamma_{LV} \sin \theta$. On deformable solids, a ridge is produced at the perimeter of a drop; on harder solids, the stress is not sufficient to cause deformation of the surface. This is the heart of the objection. Is it correct to assume that a surface under this stress is thermodynamically the same as the idealized surface which is free from stress? Clearly, the troublesome stress component is absent only when $\theta = 0$, in which case the liquid spreads freely over the surface and the concept of the sessile drop becomes meaningless.

In answer to this the following argument has been suggested, based on the fact that it is the difference $\gamma_{SV} - \gamma_{SL}$ which appears in Young's equation. Since the same solid is common to both terms it is really only the local difference at the surface between an adjacent phase which is liquid and one which is vapor that is being measured. According to this point of view, the nonequilibrium state of the solid is immaterial: The same solid is involved at both interfaces.

Another approach to resolving this theoretical objection to Young's equation is to eliminate the difference $\gamma_{SV} - \gamma_{SL}$ from the equation entirely, replacing it by some equivalent quantity, thereby shifting attention away from the notion of solid surface tension. It is an experimental fact that small quantities of heat (order of magnitude 10–10^2 ergs cm^{-2}) are liberated when a solid is immersed in a liquid. It is not particularly difficult to show how the heat of immersion of a solid in a liquid must depend on γ_{LV}, θ, and the temperature variation of each if Young's equation is valid. In this way Eq. (65) could be tested, sidestepping the cloudy questions about γ for the solid. We have already seen, however, that measurements of θ are complicated by difficulties with surface roughness or heterogeneity so this method is subject to exactly the same criticisms as those attempts to verify Young's equation directly.

In summary, then, Young's equation is still controversial despite the fact that it has been in existence since the beginning of the nineteenth century. The reader will appreciate that any relationship that has been around so long and has eluded definitive empirical verification has been the center of much research. Accordingly, the relationship is very widely encountered in the literature of surface chemistry.

Before concluding this section, it is worthwhile to consider one additional aspect of Young's equation. Suppose, instead of Eq. (65), we write (incorrectly)

$$\gamma_{LV} \cos \theta = \gamma_{S^0} - \gamma_{SL} \tag{68}$$

where γ_{S^0} is the surface energy of the solid interface with its own vapor and γ_{SV} refers to the interface between the solid and the gaseous state of the liquid component. We must assume that γ_{SV} and γ_{S^0} may be different, although we have no factual basis, as yet, for distinguishing between the two.

Let us consider what occurs when the vapor of a volatile liquid is added to an evacuated sample of a nonvolatile solid. Experience shows that the vapor is adsorbed on the solid surface, the extent of adsorption depending on the gas pressure, the temperature, the nature of the solid, and the nature of the vapor. Gas adsorption is discussed in detail in Chap. 8. This is closely related to the observation that the interface between a solution and another phase will differ from the corresponding interface for the pure solvent due to the adsorption of solute from solution. Adsorption from solution is discussed in Chap. 7.

For now, we may anticipate a result from these chapters to note that adsorption always leads to a decrease in γ. In the present context, therefore, we write

$$\gamma_{S^0} \geqslant \gamma_{SV} \tag{69}$$

We shall use the symbol π_e to signify the difference

$$\gamma_{S^0} - \gamma_{SV} = \pi_e \tag{70}$$

and call this quantity the equilibrium film pressure. The word "equilibrium" in this designation refers explicitly to the fact that the adsorbed molecules are in equilibrium with a sessile drop of bulk liquid. The molecules adsorbed at an interface may be regarded as repelling one another or as rebounding off one another, thereby relieving some of the tension in the surface. This interpretation makes it sensible to call the reduction of γ due to adsorption, π_e, a "pressure." Note that π_e is a *two-dimensional* pressure, measuring the force exerted per unit length of perimeter (dynes per centimeter) by the adsorbed molecules. We shall have a good deal more to say about this quantity in the following chapter.

With these ideas in mind, why Eq. (68) is incorrect and how to correct it become clear. The state described by Eq. (68) cannot be an equilibrium situation; it must be corrected in the light of (70) to give Young's equation:

$$\gamma_{LV} \cos \theta = \gamma_{S^0} - \pi - \gamma_{SL} \tag{71}$$

Figure 6.12 shows the relationship between γ_{S^0}, γ_{SV}, and π_e in terms of the sessile drop. The figure suggests that the shape of the drop might be quite different in equilibrium and nonequilibrium situations depending on the magnitude of π_e.

There are several concepts which will assist us in anticipating the range of π_e values:

1. Spontaneously occurring processes are characterized by negative values of ΔG.
2. Surface tension is the surface excess free energy; therefore, the lowering of γ with adsorption is consistent with the fact that adsorption occurs spontaneously.

3. Surfaces which initially possess the higher free energies have the most to gain in terms of decreasing the free energy of their surfaces by adsorption.
4. A surface energy value in the neighborhood of 100 ergs cm^{-2} is generally considered the cutoff value between "high energy" and "low energy" surfaces. For the latter, the adsorption is negligible, so π_e is taken to be zero in these cases. For high-energy surfaces, $\pi_e > 0$.

6.11 ADHESION AND COHESION

In this section, we shall consider two processes—adhesion and cohesion—which will eventually lead us to an understanding of the origin of surface tension at the molecular level. Figure 6.13 illustrates these processes schematically. In parts a and b of the figure, a column of liquid of unit cross section is portrayed. It is to be understood that each of these is a volume element in an infinite volume of liquid. Now by some hypothetical process, each of the columns is separated into two portions. We wish to consider the value of ΔG for each of these processes.

In Fig. 6.13a—which applies to a pure liquid—the process consists of producing two new interfaces, each of unit cross section. Therefore, for the separation process

$$\Delta G = 2\gamma_A = W_{AA} \tag{72}$$

The quantity W_{AA} is known as the work of cohesion since it equals the work required to pull a column of liquid A apart. It measures the attraction between the molecules of the two portions. We shall return to a consideration of the work of cohesion as a measure of van der Waals forces presently.

Now let us consider the value of ΔG for the separation of the immiscible layers shows in Fig. 6.13b. Setting up the difference between the final and the initial free

FIGURE 6.13 *Schematic illustrations of the processes for which ΔG equals (a) the work of cohesion and (b) the work of adhesion.*

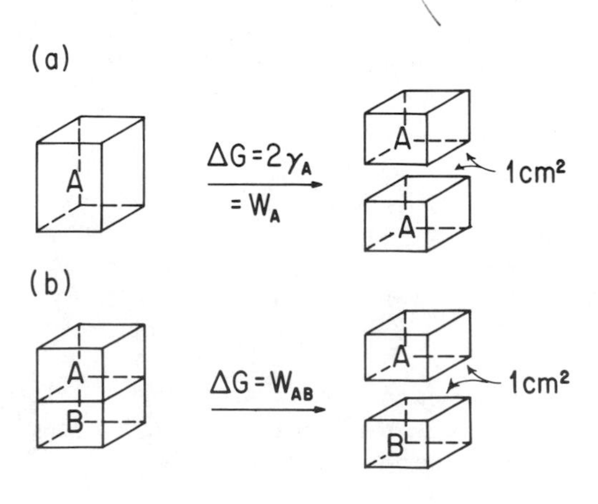

energies for this process yields

$$\Delta G = W_{AB} = \gamma_{final} - \gamma_{initial} = \gamma_A + \gamma_B - \gamma_{AB} \quad (73)$$

This quantity is known as the work of adhesion and measures the attraction between the two different phases.

The work of adhesion between a solid and a liquid phase may be defined by analogy with Eq. (73):

$$W_{SL} = \gamma_{S^0} + \gamma_{LV} - \gamma_{SL} \quad (74)$$

By means of Eq. (70), γ_{S^0} may be eliminated from this expression to give

$$W_{SL} = \gamma_{SV} + \pi_e + \gamma_{LV} - \gamma_{SL} \quad (75)$$

Finally, Young's equation may be used to eliminate the difference $\gamma_{SV} - \gamma_{SL}$:

$$W_{SL} = \gamma_{LV}(1 + \cos\theta) + \pi_e \quad (76)$$

As we have noted previously, $\pi_e \geqslant 0$ where the equality holds in the absence of adsorption. High-energy surfaces bind enough adsorbed molecules to make π_e significant. On the other hand, π_e is negligible for a solid which possesses a low-energy surface. Metals, metal oxides, metal sulfides, and other inorganic salts, silica, and glass are examples of high-energy surfaces. Most solid organic compounds, including organic polymers, have low-energy surfaces. Because of the simplification arising from $\pi_e = 0$ for low-energy surfaces, the latter are often chosen as model systems in fundamental research.

It is informative to apply Eq. (76) to low-energy surfaces for two extreme values of θ, 0° and 180°, for which $\cos\theta$ is 1 and -1, respectively. For $\theta = 0°$, $W_{SL} = 2\gamma_{LV} = W_{AA}$, the work of solid–liquid adhesion is identical to the work of cohesion for the liquid. In this case, interactions between solid–solid, liquid–liquid, and solid–liquid molecules are all equivalent. At the other extreme where $\theta = 180°$, $W_{SL} = 0$. In this case, the sessile drop is totally tangent to the solid; there is no interaction between the phases.

The difference between the work of adhesion and the work of cohesion of two substances defines a quantity known as the spreading coefficient of B on A, $S_{B/A}$:

$$S_{B/A} = W_{AB} - W_{BB} \quad (77)$$

If $W_{AB} > W_{BB}$, the A–B interaction is sufficiently strong to promote the wetting of A by B. This is the significance of a positive spreading coefficient. Conversely, no wetting occurs if $W_{BB} > W_{AB}$ since the work required to overcome the attraction between two B molecules is not compensated by the attraction between A and B. Thus a negative spreading coefficient means that B will not spread over A.

A geometric interpretation of the spreading coefficient for A on B is seen as follows. Substituting Eqs. (72) and (73) into (77) leads to

$$S_{B/A} = \gamma_A - \gamma_B - \gamma_{AB} = \gamma_A - (\gamma_B + \gamma_{AB}) \quad (78)$$

Although it fails to project the forces into a common plane, it is clear that the spreading coefficient measures the difference between the force which tends to expand the perimeter of the drop (γ_A) and those which tend to contract it ($\gamma_B + \gamma_{AB}$). Again we see that positive and negative values of S correspond to situations in which liquid B spreads freely over (positive) or does not wet (negative) the surface of A.

6.12 THE DISPERSION COMPONENT OF SURFACE TENSION

Throughout this chapter we have dealt with surface tension from a phenomenological point of view almost exclusively. To chemists, however, descriptions from a molecular perspective are often more illuminating than descriptions of phenomena alone. In condensed phases, where interactions involve many molecules, rigorous derivations based on the cumulative behavior of individual molecules are extremely difficult. We shall not attempt to review any of the efforts directed along these lines for surface tension. Instead we use a semiempirical approach to isolate one component of surface tension, the so-called dispersion component, and discuss this in molecular terms. In doing this, we neglect many specific molecular interactions which are responsible for the unique properties of various substances and focus attention on a type of interaction which exists between all molecules. It turns out that the dispersion component of γ shows up again in the discussion of colloidal stability in Chap. 10.

The molecules that comprise a condensed phase are certainly not separated by fixed distances. Nevertheless, an equilibrium spacing is a meaningful quantity in both liquids and solids. The equilibrium spacing R^0 between molecular centers would occur at the minimum in a plot of energy versus distance of separation such as those shown in Fig. 6.14. Qualitatively, we may think of such a curve as the resultant of several contributions. For simplicity, we shall limit our consideration to three contributions. At small separations, a very steep component of the curve describes

FIGURE 6.14 *Interaction energy versus distance of separation between molecules (schematic). Curve* 1: *repulsion*; *curve* 2: *dispersion component of attraction*; *curve* 3: *specific interactions*; *curve* 4: *resultant* 1+2+3.

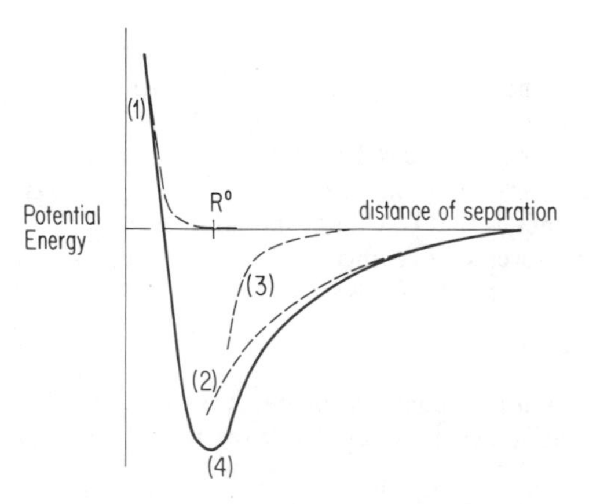

repulsion between molecules. The repulsion, which originates from the overlap of electron domains of neighboring molecules, drops off very rapidly with separation. Curve 1 in Fig. 6.14 represents this repulsion. In the figure, positive interaction energies correspond to repulsion, and negative values are attractions. At larger separations, two components of attraction are shown. Let us first consider the contribution of dispersion forces to attraction. Dispersion is one of the dominant contributions to van der Waals forces.

For now, all that is necessary is to recognize that dispersion forces arise from the interaction between the dipoles induced in a pair of atoms or molecules by the fluctuating charge density in each. There are three consequences of this which interest us: (a) Dispersion forces are always present; (b) they are always attractive; and (c) they are relatively more long range than many other types of molecular interactions. Curve 2 in Fig. 6.14 represents the variation with distance of intermolecular attraction due to dispersion forces. We shall discuss dispersion forces more quantitatively in Chap. 10.

Both repulsion and the dispersion component of attraction are always present, regardless of the specific chemical characteristics of the molecules under consideration. Curve 3 in Fig. 6.14 also represents a component of attraction. It is purposely drawn with a more pronounced curvature than the dispersion curve to emphasize the fact that it is associated with some specific and, therefore, highly localized interaction such as hydrogen bonding or metallic bonding. We shall not identify this component any further than to label it a "specific interaction." By nature, the "specific" component operates only *within* the phase to which it applies. On the other hand, the dispersion component operates across phase boundaries. This difference in behavior is inherent in the distinction between specific and nonspecific interactions. The two attraction curves and one repulsion curve generate the resultant, curve 4, when added together. The minimum in this net interaction curve identifies the equilibrium separation of the molecules of the material, R^0.

Next, we apply these concepts to a molecule of A in two different environments: in the interior of a bulk phase and near an interface between two phases.

A molecule in the interior of a bulk phase is surrounded on all sides by a homogeneous molecular environment of A. Any movement that would increase its separation from some neighbors would automatically decrease the separation from others, so large deviations from the equilibrium value of the separation on curve 4 are improbable.

For a molecule at an interface between a condensed phase and the gas phase, the environment is quite unsymmetrical. Movement toward the bulk phase is impeded by the excluded volume of the A molecules. Movement away from the bulk phase meets no resistance of this sort, although the prevailing attraction between molecules in the condensed phase opposes the A molecule from escaping the condensed phase altogether. Because of this, the equilibrium separation between the molecules at the surface will be larger than between those in the interior. The intermolecular separation has been "stretched" in bringing a molecule to the surface. The contractile force in the interface of the substance is simply the restoring force attempting to return the molecules to their bulk spacing. From an energetic point of view, the difference between the energy of the bulk and surface minimum separations gives the work needed to bring a molecule from the interior to the surface.

Next, let us consider the situation in which a second condensed phase B adjoins the reference phase A. The new consideration in this case is the dispersion component of attraction of the molecules in condensed phase B for the A molecules in the interface. This A–B attraction partially overcomes the A–A attraction which opposes the movement of an A molecule to any interface. As a consequence, there is a difference in the energy which must be expended to bring an A molecule to an interface with a gas and to an interface with another condensed phase. We call this difference ΔE^s. Recall that ΔE^s arises from the dispersion forces between the phases.

F. Fowkes has proposed that any interfacial tension may be written as the summation of contributions arising from the various types of interactions which would operate in the materials under consideration. In general then,

$$\gamma = \gamma^d + \gamma^h + \gamma^m + \gamma^\pi + \gamma^i = \gamma^d + \gamma^{sp} \tag{79}$$

where the superscripts refer to dispersion forces (d), hydrogen bonds (h), metallic bonds (m), electron interactions (π), and ionic interactions (i). Clearly not all of these are present in all materials. In fact, the dispersion component is the only component which does operate between all molecules. The other contributions (γ^{sp}) are specific for individual substances. It is argued that only the dispersion component of γ operates across an interface to decrease the work that is required to bring a molecule to a surface. Fowkes has assumed that this reduction in work, ΔE^s, equals the geometric mean of the dispersion component of the surface tensions of the two condensed phases:

$$\Delta E^s = \sqrt{\gamma_A{}^d \gamma_B{}^d} \tag{80}$$

The assignment of a geometric mean rather than an arithmetic mean or some other function of the two γ terms is justified primarily on the basis of the successful use of this type of averaging in the theory of nonelectrolyte solubility. Only the dispersion component of γ is used since it is the part of γ which crosses phase boundaries.

According to these ideas, the work required to bring an A molecule to the AB interface is given by

$$(\text{Work})_A = \gamma_A - (\Delta E^s)_A = \gamma_A - \sqrt{\gamma_A{}^d \gamma_B{}^d} \tag{81}$$

when A and B are both in condensed phases. A similar expression applies to the work required to bring a B molecule to the interface. The total work of forming the AB interface is the sum of these contributions:

$$\gamma_{AB} = (\gamma_A - \sqrt{\gamma_A{}^d \gamma_B{}^d}) + (\gamma_B - \sqrt{\gamma_A{}^d \gamma_B{}^d}) = \gamma_A + \gamma_B - 2\sqrt{\gamma_A{}^d \gamma_B{}^d} \tag{82}$$

Comparing Eqs. (73) and (82) reveals that

$$W_{AB} = 2(\gamma_A{}^d \gamma_B{}^d)^{1/2} \tag{83}$$

This is consistent with the concepts:

1. that W_{AB} measures the attraction between the columns of separated material as illustrated in Fig. 6.13, and
2. that it is the dispersion component of surface tension that operates across phase boundaries.

By judiciously selecting systems and making some reasonable assumptions about the kinds of interactions involved therein, Fowkes was able to arrive at some estimates of γ^d. A summary of the method is presented here. In the following liquids, Eq. (79) is assumed to become:

1. For hydrocarbons (H):

$$\gamma_H = \gamma^d \tag{84}$$

2. For water (W):

$$\gamma_W = \gamma^d + \gamma^h \tag{85}$$

3. For mercury (Hg):

$$\gamma_{Hg} = \gamma^d + \gamma^m \tag{86}$$

Therefore, for the mercury–hydrocarbon interface, Eq. (82) becomes

$$\gamma_{HgH} = \gamma_{Hg} + \gamma_H - 2\sqrt{\gamma_{Hg}{}^d \gamma_H{}^d} \tag{87}$$

All the quantities in this expression except $\gamma_{Hg}{}^d$ are directly measurable, so Eq. (87) provides a method of evaluating this quantity. Table 6.4 lists values of γ_H and γ_{HgH} for a number of liquid hydrocarbons layered on mercury, along with values of $\gamma_{Hg}{}^d$ calculated by Eq. (87). The latter are seen to be relatively constant, averaging 200 ± 7 ergs cm^{-2}.

The same procedure may be used to evaluate γ^d for water. For the water–hydrocarbon interface, Eq. (82) becomes

$$\gamma_{WH} = \gamma_W + \gamma_H - 2\sqrt{\gamma_W{}^d \gamma_H{}^d} \tag{88}$$

Table 6.4 also lists experimental values for γ_H and γ_{WH} for several liquid hydrocarbons, as well as values of γ^d for water as calculated by Eq. (88). Here again the values are seen to be fairly constant, averaging 21.8 ± 0.7 ergs cm^{-2}.

From the preceding examples, it is evident that the value of γ for the water–mercury interface can be calculated for comparison with experiment. In this case, Eq. (82) becomes

$$\gamma_{WHg} = \gamma_W + \gamma_{Hg} - 2\sqrt{\gamma_W{}^d \gamma_{Hg}{}^d} \tag{89}$$

Substitution of the numerical values for $\gamma_W{}^d$ and $\gamma_{Hg}{}^d$ just obtained gives:

$$\gamma_{WHg} = 72.8 + 284 - 2\sqrt{(21.8)(200)} = 425 \text{ ergs cm}^{-2} \tag{90}$$

TABLE 6.4 *Experimental values for* γ_H, γ_{HW}, *and* γ_{HHg}, *with* H = *hydrocarbon, at* 20°C. $\gamma_{Hg}{}^d$ *calculated from Eq.* (87); $\gamma_W{}^d$ *from* (88). *All values in* ergs cm^{-2}[a]

Hydrocarbon	γ	Mercury (γ_{Hg} = 484) γ_{HHg}	Mercury (γ_{Hg} = 484) $\gamma_{Hg}{}^d$	Water (γ_W = 72.8) γ_{HW}	Water (γ_W = 72.8) $\gamma_W{}^d$
n-Hexane	18.4	378	210	51.1	21.8
n-Heptane	18.4	—	—	50.2	22.6
n-Octane	21.8	375	199	50.8	22.0
n-Nonane	22.8	372	199	—	—
n-Decane	23.9	—	—	51.2	21.6
n-Tetradecane	25.6	—	—	52.2	20.8
Cyclohexane	25.5	—	—	50.2	22.7
Decalin	29.9	—	—	51.4	22.0
Benzene	28.85	363	194	—	—
Toluene	28.5	359	208	—	—
o-Xylene	30.1	359	200	—	—
m-Xylene	28.9	357	211	—	—
p-Xylene	28.4	361	203	—	—
n-Propylbenzene	29.0	363	194	—	—
n-Butylbenzene	29.2	363	193	—	—
Average			200 ± 7		21.8 ± 0.7

[a] Data from F. M. Fowkes, *Ind. Eng. Chem.* **56**:40 (1964).

which compares quite favorably with the experimental value for this quantity: 426 ergs cm^{-2}.

The average values of γ^d for water and mercury are clearly successful in their ability to calculate γ correctly for the water–mercury interface. Individually, however, they are the averages of slightly divergent values measured for several different interfaces with hydrocarbons. In this sense, the values of γ^d which we have considered are analogous to mean bond energies in physical chemistry. The latter, too, are averages obtained from a variety of compounds. Although mean bond energies are very useful, they are, by nature, insensitive to unique, specific effects. With both mean bond energies and values of γ^d, the user must be careful that no such special interactions are present, otherwise quite serious errors could arise. Further, it is important to realize that any errors are perpetuated and compounded by this scheme for evaluating γ^d.

It is not difficult to apply the concept of the dispersion component of γ to solid surfaces. In doing this, it is necessary to treat high- and low-energy surfaces differently. We shall not consider solid interfaces in detail; our treatment is limited to the following observations:

1. For low-energy surfaces, $\pi_e \simeq 0$. Manipulation of Young's equation [Eq. (65)] generates a relationship that expresses $\gamma_S{}^d$ in terms of θ and other experimental quantities.

2. For high-energy surfaces, $\pi_e > 0$ due to adsorption. Relationships have been derived which express $\gamma_S{}^d$ in terms of gas adsorption.

Values of $\gamma_S{}^d$ which have been determined for high- and low-energy surfaces by these two methods are listed in Table 6.5.

Examination of the values of γ^d for high- and low-energy solids in Table 6.5 reveals several interesting points. To begin with, there is a slight overlap in the data inasmuch as polypropylene was studied by the gas adsorption technique even though it might reasonably be expected to resemble paraffin wax or polyethylene in γ^d value. As the table shows, the value of γ^d for this substance lies in the same neighborhood as the values for these other compounds, even though very widely different procedures are used to arrive at the different values. The multiple values of γ^d listed for some high-energy solids correspond to evaluations based on the adsorption of different gases at widely different temperatures. For example, the γ^d values for TiO_2 are obtained from adsorption studies conducted with butane at 0°C, heptane at 25°C, and N_2 at −195°C. With these ideas in mind, the observed variation between ostensibly duplicate values becomes more acceptable. The evaluation of $\gamma_S{}^d$ also depends on the accuracy of the value of γ^d for a substance other than the solid. Again

TABLE 6.5 *Values of $\gamma_S{}^d$ (in ergs cm^{-2}) for a variety of solids as determined from contact angles and from gas adsorption*[a]

	$\gamma_S{}^d$ values as determined from measurements of:	
Material	θ	π
Dodecanoic acid on Pt	10.4, 13.1	—
Polyhexafluoropropylene	11.7,* 18.0	—
Polytetrafluoropropylene	19.5	—
n-$C_{36}H_{74}$	21.0	—
n-Octadecylamine on Pt	22.0,* 22.1	—
Paraffin wax	23.2,* 25.5	—
Polypropylene	—	26, 28.5
Polytrifluoromonochloroethylene (Kel F)	30.8	—
Nylon-6,6	33.6*	—
Polyethylene	31.3,* 35.0	—
Polyethyleneterephthalate	36.6*	—
Polystyrene	38.4,* 44.0	—
$BaSO_4$	—	76
Silica	—	78
Anatase (TiO_2)	—	89, 92, 141
Iron	—	89, 106, 108
Graphite	—	115, 120, 123, 132

[a] Most data from F. M. Fowkes, *Ind. Eng. Chem.* **56**:40 (1964). Those data labeled with an asterisk are from Kaelbe [6].

we see that errors may be propagated in the analyses that led to the results presented in Table 6.5.

Once γ and γ^d are known for a particular surface, their difference may also be evaluated. As Eq. (79) shows, the difference equals the contribution of specific interactions to γ. Generally speaking, the dispersion component is found to be the largest contributor to γ; however, when hydrogen bonding is present, it is found to make a contribution to γ approximately equal to that of the dispersion forces. We shall not examine the other contributions to γ in any more detail; the dispersion component of surface free energy is our major interest. As we shall see in Chap. 10, dispersion is the source of the long-range interaction between colloidal particles and is responsible for flocculation. In Chap. 9 we shall examine the electrical interaction between charged particles. It turns out, in many cases, that electrical and dispersion forces are comparable in range and magnitude. This means that the stability of a dispersion with respect to flocculation is quite sensitive to the parameters which determine these forces. This is why the Fowkes theory was singled out of many different theories for γ since it allows us to estimate quantitatively the dispersion component of van der Waals forces between macroscopic phases.

REFERENCES

1. N. K. Adam, *The Physics and Chemistry of Surfaces*, Dover, New York, 1968.
2. A. W. Adamson, *Physical Chemistry of Surfaces* (2nd ed.), Wiley-Interscience, New York, 1967.
3. J. J. Bikerman, *Physical Surfaces*, Academic Press, New York, 1970.
4. J. T. Davies and E. K. Rideal, *Interfacial Phenomena*, Academic Press, New York, 1961.
5. R. E. Johnson, Jr., and R. H. Dettree, in *Surface and Colloid Science*, Vol. 2 (E. Matijević, ed.), Wiley-Interscience, New York, 1969.
6. D. H. Kaelbe, *Physical Chemistry of Adhesion*, Wiley-Interscience, New York, 1971.
7. J. F. Padday, in *Surface and Colloid Science*, Vol. 1 (E. Matijević, ed.), Wiley-Interscience, New York, 1969.
8. A. C. Zettlemoyer, in *Hydrophobic Surfaces* (F. M. Fowkes, ed.), Academic Press, New York, 1969.

PROBLEMS

1. The frictional force was measured on a strand of viscous rayon fiber moving through a wad of identical fibers as a function of the water content of the wadded fiber. It was found that the friction increased from 59 to 133 mN cm^{-1} when the water content decreased to the point at which capillary "necking" between the fibers occurred.* A model for this situation may be visualized by considering two parallel, tangent cylinders connected by a "neck" of water held

* J. Skelton, *Science* **190**:15 (1975).

in the neighborhood of the contact by capillary forces. Sketch the situation represented by this model and explain why the force of fiber–fiber attraction increases with decreasing water content. Use this model to discuss: (a) the behavior of a wet paintbrush, (b) the practice of wetting the tip of a thread before threading a needle, and (c) the dewatering of cellulose fibers to form paper.

2. Use the data of Table 6.2 to plot the profile of a drop with $\beta = 25$. Measure (in cm) the radius of the drop you have drawn at its widest point. By comparing this value with the value of $(x/b)_{90°}$ from the table, evaluate b (in cm) for the drop as you have drawn it. Suppose an actual drop is characterized by this value of β. If the actual radius at the widest point is 0.25 cm and $\Delta\rho =$ 0.50 g cm^{-3}, what is γ for the interface of the drop?

3. Suppose the drop profile shown in Fig. 6.5b describes an actual drop for which the radius at the widest point equals 0.135 cm. Use the value of $(x/b)_{90°}$ from the figure to calculate γ for each of the following situations:

System	$\Delta\rho$ (g cm^{-3})
(a) Oil in water	0.20
(b) Water in oil	0.20
(c) Oil in air	0.80
(d) Air in water	1.00
(e) Water in air	1.00

State whether the drop is pendant or sessile in each case.

4. The accompanying data give experimental values of the capillary rise for various liquids*:

Liquid	$\Delta\rho$ (g cm^{-3})	h (cm)	R (cm)
Water	0.9972	1.4343	0.10099
Benzene	0.8775	1.5425	0.043135
$CHCl_3$	1.4869	1.921	0.1932

Use values from the following table (interpolated from Padday [7]) to evaluate a^2 (and γ) by the successive approximation technique illustrated in Table 6.3:

β	0	0.02	0.04	0.06	0.08	0.10	0.12	0.14	0.16	0.18	0.20
$(x/b)_{90°}$	1.00	0.997	0.994	0.991	0.987	0.984	0.981	0.978	0.975	0.972	0.970

Compare the values of γ calculated by this procedure with those obtained by the approximation given by Eq. (49).

5. A cylindrical rod may be used instead of a rectangular plate in a slight variation of the Wilhelmy method. Derive an expression equivalent to Eq. (57) for a

* T. W. Richards and E. K. Carver, *J. Am. Chem. Soc.* **43**:827 (1921).

suspended solid of cylindrical geometry. Prepare semiquantitative plots analogous to Fig. 6.9b based on the following data, assuming cylindrical rods 1.0 mm in diameter at the air–water interface*:

Metal	ρ_S (g cm^{-3})	θ_a (deg)	θ_r (deg)
Au	19.3	70	40
Pt	21.5	63	28

In both cases the metal surfaces were carefully polished, washed, steamed, and then heated in an oven for 1 h at 100°C.

6. The vertical rod method of the preceding problem was used to study the contact angle of water at the gold–water–air junction at 25°C. The following data show how the value of $\theta(\theta_a)$ depends on the prior history of the metal surface.† For a gold surface polished, washed, and heat-treated for 1 h at T°C, then allowed to stand in air for t h:

T (°C)	100	200	300	400	500	600	600	600	600	600	600
t (h)	$<\frac{1}{4}$	$<\frac{1}{4}$	$<\frac{1}{4}$	$<\frac{1}{4}$	$<\frac{1}{4}$	$\frac{1}{4}$	1	5	10	24	120
θ_a (deg)	68	57	45	36	25	13	22	38	47	53	55

Calculate the work of adhesion between water and gold for each of these cases on the assumption that $\pi_e = 0$. Is the variation in W_{SL} consistent (qualitatively? quantitatively?) with the expected validity of the assumption concerning π_e?

7. Finely dispersed sodium chloride particles were prepared, their specific area was measured, and their solubility in ethanol at 25°C was studied‡. It was found that a preparation with a specific area of 4.25×10^5 cm^2 g^{-1} showed a supersaturation of 6.71%. Estimate the radius of the NaCl ($\rho = 2.17$ g cm^{-3}) particles, assuming uniform spheres. Calculate γ for the NaCl–alcohol interface from the solubility behavior of this sample.

8. Enüstün and Turkevich§ prepared $SrSO_4$ ($\rho = 3.96$ g cm^{-3}) precipitates under conditions which resulted in different particle sizes. Particle sizes were characterized by electron microscopy and solubilities were determined at 25°C by a radiotracer technique. In the following data the supersaturation ratios are presented for different preparations, each of which is characterized by an

* F. E. Bartell, J. A. Culbertson, and M. A. Miller, *J. Phys. Chem.* **40**:881 (1936).
† F. E. Bartell and M. A. Miller, *J. Phys. Chem.* **40**:889 (1936).
‡ F. Van Zeggeren and G. C. Benson, *Can. J. Chem.* **35**:1150 (1957).
§ B. V. Enüstün and J. Turkevich, *J. Am. Chem. Soc.* **82**:4502 (1960).

average particle width and a minimum particle width:

x_{mean} (Å)	x_{min} (Å)	a/a_0
247	96	1.43
269	130	1.35
388	155	1.28
541	168	1.29
629	252	1.16
1260	378	1.10
1660	500	1.07

Which size parameter gives the best agreement with the Kelvin equation? Explain. Use the best fitting data to evaluate γ for the $SrSO_4$–H_2O interface.

9. Water drops were formed at the mercury–benzene interface by means of a syringe, and the contact angle (measured in the water) was recorded as a function of time.* For the interface between Hg and benzene saturated with water, γ was measured independently as a function of time. The following table summarizes these data (all measured at 25°C):

Time (h)	$\gamma_{Hg\text{-}benzene}$ (ergs cm^{-2})	θ_{obs} (deg)
0.10	363.0	118
0.42	359.5	119
1.0	358.0	122
2.5	354.5	138
5.0	350.0	144
13	336.0	—
23	—	180

Using 379.5 and 34.0 ergs cm^{-2}, respectively, as the values of γ for the mercury–water and benzene–water interfaces, compare the observed contact angles with the predictions of Young's equation. Comment on the fact that constant values are used for γ_{HgW} and γ_{BW}.

10. Bartell and Osterhof† describe an experimental procedure for measuring the work of adhesion between liquids and solids. With carbon (lampblack) as the solid, the following values for the work of adhesion were obtained:

Liquid	Benzene	Toluene	CCl_4	CS_2	Ethyl ether	H_2O
W_{AB} (ergs cm^{-2})	109.4	110.2	112.4	122.1	76.4	126.8

Use these data together with the surface tensions of the pure liquids from Table 6.1 to calculate the spreading coefficients for the various liquids on carbon black. Use your results to interpret the authors' observations: "About equal

* F. E. Bartell and C. W. Bjorkland, *J. Phys. Chem.* **56**:435 (1952).
† F. E. Bartell and H. J. Osterhof, *J. Phys. Chem.* **37**:543 (1933).

quantities of water and organic liquid were put into a test tube with a small amount of the finely divided solid and shaken. It was noted that the carbon went exclusively to the organic liquid phase . . . ."

11. The effect of mutual saturation on the L–V and L–L interfacial tensions is effectively illustrated by considering the spreading coefficient of one liquid on another using both the initial (unsaturated) and equilibrium values of γ. Use the following data to calculate $S_{B/A}$ (equilibrium) and $S'_{B/A}$ (nonequilibrium)*:

	H_2O/air	H_2O/IAA	IAA/air	H_2O/air	H_2O/CS_2	CS_2/air
γ' (ergs cm^{-2})	25.9	23.6	5.0	70.3	31.8	48.4
γ (ergs cm^{-2})	72.8	23.7	5.0	72.8	32.4	48.4

Describe what happens when a drop of pure isoamyl alcohol (IAA) is placed on the surface of pure water. What happens with the passage of time? Repeat this description for the case of pure CS_2 on pure water.

12. The tendency of spilled mercury to disperse as small drops which roll freely on most surfaces is a well-known characteristic of this liquid. Discuss this behavior in terms of (a) the work of adhesion and the spreading coefficient for mercury on various substrates, (b) surface tension versus contact angle (measured in Hg) as causes of this behavior, and (c) the implications of the Kelvin equation on the health hazards associated with mercury spills.

13. By a suitable combination of Eqs. (71) and (82) show that

$$\gamma_S{}^d = \frac{\gamma_{LV}^2}{4\gamma_L{}^d}(1+\cos\theta)^2$$

for low-energy surfaces. Use the following data (see Table 6.4 for reference) to evaluate either $\gamma_S{}^d$ or $\gamma_L{}^d$ as appropriate:

S	L	γ_{LV} (ergs cm^{-2})	γ^d (ergs cm^{-2})	θ (deg), in liquid
Dodecanoic acid on Pt	α-Bromonaphthalene	44.6	10.4 for S	92
Kel F	α-Bromonaphthalene	44.6	30.8 for S	48
Paraffin wax	Glycerol	63.4	36 for L	97
Paraffin wax	Fluorolube	20.2	13.5 for L	31

14. The equation derived in the preceding problem suggests that a plot of $\cos\theta$ (as ordinate) versus $\sqrt{\gamma_L{}^d}/\gamma_L$ (as abscissa) should be linear with a slope of $2\sqrt{\gamma_S{}^d}$

* W. D. Harkins, *The Physical Chemistry of Surface Films*, Van Nostrand-Reinhold, Princeton, New Jersey, 1952.

and an intercept of -1. Describe how this result can be used to evaluate $\gamma_S{}^d$ when contact angle measurements are made on a particular solid with a variety of liquids for which γ_L and $\gamma_L{}^d$ are known. Describe how the same graphing procedure can be used to evaluate $\sqrt{\gamma_L{}^d/\gamma_L}$ when contact angles are measured in a liquid of unknown $\gamma_L{}^d$ on different solids of known $\gamma_S{}^d$. Use the data of the preceding problem to illustrate these two graphical interpretations. Include the additional datum that α-bromonaphthalene forms a contact angle of 58.5° with paraffin (see Table 6.4 for reference).

Adsorption from Solution

7

You are living on a plane. What you style Flatland is the vast level surface of what I may call a fluid, on, or in, the top of which you and your countrymen move about, without rising above it or falling below it.

[From Abbott's *Flatland*]

7.1 INTRODUCTION

Until now, we have intentionally excluded solutes of variable concentration from our consideration of surfaces. In this chapter, the effects of such solutes are our specific interest. We shall be especially concerned with a particular class of solutes which show dramatic effects on surface tension. These are said to be surface active, and are often simply called surfactants. The primary emphasis is on the relationship between adsorption phenomena and surface tension–surface energy. In the process of describing experimental methods, results, and interpretations, however, a variety of related concepts will enter the picture.

In studying the material of this chapter, it may be helpful to realize that the topics covered may be grouped in several different ways. Let us enumerate what these various ways of looking at the material are.

First, we may focus our attention on the solubility of the adsorbed species in one or both of the adjacent phases. In this way there is generated the study of two broad categories of phenomena: insoluble and soluble surface layers.

A second way of classifying the material of this chapter is on the basis of the experimental methods involved. For mobile interfaces, surface tension is easily measured. For these, it is easiest to examine the surface tension–adsorption relationship starting with surface tension data. When insoluble surface films are involved, we shall see how the difference in γ between a clean surface and one with an adsorbed film may be measured directly. For solid surfaces, surface tension is not readily available from experiments. In this case, adsorption may be measurable directly, and the relationship between adsorption and surface tension may be examined from the reverse perspective.

Third, the material of this chapter may be considered as either theoretical or descriptive. Only two major theoretical results are presented, the equations of Gibbs and Langmuir. Both are extremely important in surface chemistry. This chapter includes considerably more descriptive material than most other chapters in this book. In a certain sense, it is like chapters in general physical chemistry texts which deal with phase diagrams. That is, the emphasis is on principles, but the method for presenting them involves consideration of specific systems.

Finally, the material of this chapter may be regarded as a mixture of fundamentals and applications. Although the entire book stresses principles, applications are considered from time to time as examples of more abstract ideas. This is also the intent of the sections on applications in this chapter. In addition, however, many applications of adsorption phenomena are the basis of large and important areas of technology. To omit mention of them would lead to a very incomplete picture of these fields. As it is, many important applications must be omitted for lack of space, and those which are mentioned are sketched in only a superficial way.

7.2 SPREAD MONOLAYERS

Suppose a dilute solution is prepared from an aliphatic solvent and an organic solute RX where R is a long-chain alkyl group and X is a polar group. Then a small amount of this solution is placed on a large volume of water with a horizontal surface. The components of this system were chosen because they are assumed to meet the following experimental criteria:

1. The solubility in water of both components of the organic phase is negligible at room temperature.
2. The likelihood of any complex being formed between the organic solvent and solute is exceedingly low.
3. The volatility of the organic solvent is high and that of the solute is low.

With these facts in mind, let us examine the fate of the drop of solution placed on the surface of water. The initial spreading coefficient [Eq. (6.78)] for the organic layer on water $S_{O/W}$ is positive. This is primarily because γ_{OW} is unusually low and γ_W is high, even with an adsorbed layer of the organic solvent. After spreading, we allow sufficient time to elapse for all the solvent to evaporate from the spread layer. At this point, the surface will contain a layer of the organic solute similar to that which would result from the spreading of a sessile drop of pure liquid solute or from the adsorption of vapors of the solute component from the gas phase. Using a solution with a volatile solvent to form such a layer is a very common technique, and has the advantage of permitting very small amounts of solute to be quantitatively deposited on a surface.

After the solvent has evaporated, the nature of the remaining layer depends on the amount of solute deposited and the area available to it. It is convenient to distinguish between three situations in this regard. If the amount of added material and the area are such that the water surface is covered uniformly to a depth of one molecule with the solute, the resulting film is called a monolayer. On the other hand, submonolayer coverage and multilayer coverage result when the amount of added material per area is less or more, respectively, than that which produces the monolayer. In this chapter, we shall be concerned mostly with degrees of coverage up to and including the monolayer. If a large excess of spread material (beyond the amount needed for monolayer coverage) is used, the excess collects into droplets of a bulk phase. The equilibrium situation is then identical to what would be produced by the spreading of a sessile drop of the solute material.

Films of the sort described here are called either spread monolayers (when the method of their preparation is stressed) or insoluble monolayers (when the chemical nature of the solute is emphasized). We shall use these terms interchangeably. Now, let us examine some of the properties of the spread monolayer that we have described.

It was seen in the preceding chapter [e.g., Eq. (6.70)] that the presence of an adsorbed layer lowers the surface tension of an interface. The phenomenon is quite general, so we shall redefine π in the following symbols:

$$\pi = \gamma_0 - \gamma \tag{1}$$

where γ_0 refers to the surface of any phase in the absence of adsorption and γ refers to the same surface with an adsorbed layer. Specific subscripts are used only when the problem clearly involves more than one interface.

The spread monolayer just described may be discussed from two points of view. First, there are those aspects of the film that pertain explicitly to the chemical nature of the components: water and the organic solute. Second, there are certain properties of the monolayer that depend on physical variables such as temperature, area of the water surface, and number of molecules of RX present. Let us briefly discuss both of these viewpoints.

The organic solute RX is a prototype of an important array of surface-active materials. Many surface-active substances are composed of what are known as amphipathic molecules. This term simply means that the molecule consists of two parts, each of which has an affinity for a different phase. We shall be concerned mostly with surfaces in which one of the phases is aqueous, so the surfactants that we consider will contain polar or ionic groups or "heads" and nonpolar organic residues or "tails." In the compound RX, for example, R is an alkyl group, generally containing 10 or more carbon atoms. The literature of surface chemistry contains many references to these organic groups by both their IUPAC and their common names. Table 7.1 lists some of the more commonly encountered examples. In RX, the polar X group may be —OH, —COOH, —CN, $—CONH_2$, or —COOR′ or an ionic group such as $—SO_3^-$, $—OSO_3^-$, or $—NR_3^+$.

With the foregoing ideas in mind, one characteristic of the adsorbed monolayer becomes apparent: molecular orientation at surfaces. For a film of RX on water, the picture that emerges is one in which the polar groups are incorporated into the aqueous phase with the hydrocarbon part of the molecule oriented away from the water. Such details as the depth of immersion of the tail and the configuration of the alkyl group will be best approached by considering how the properties of the monolayer depend on the physical variables.

TABLE 7.1 *IUPAC and common names for a variety of normal saturated and unsaturated surface-active compounds*

	Normal, saturated compounds		
	Carboxylic acids		Alcohols, amines, sulfates, etc.
n (No. C atoms)	IUPAC name	Common name	Common name
12	Dodecanoic	Lauric	Lauryl
14	Tetradecanoic	Myristic	Myristyl
16	Hexadecanoic	Palmitic	Cetyl
17	Heptadecanoic	Magaric	Heptadecyl
18	Octadecanoic	Stearic	Stearyl
20	Eicosanoic	Arachidic	Eicosyl, arachic
22	Docosanoic	Behenic	Docosyl

	Normal, unsaturated carboxylic acids	
	IUPAC name	Common name
18	*cis*-9-Octadecenoic	Oleic
18	*cis,cis*-6,9-Octadecenoic	Linoleic
18	*cis,cis,cis*-3,6,9-Octadecenoic	Linolenic
18	*trans*-9-Octadecenoic	Elaidic
22	*cis*-9-Docosenoic	Erucic
22	*trans*-9-Docosenoic	Brassidic

Next, let us consider some of the physical properties of the spread monolayer we have described. Equation (1) states that the surface tension of the covered surface will be less than that of pure water. It is quite clear, however, that the magnitude of γ must depend on both the amount of material adsorbed and the area over which it is distributed. The spreading technique already described enables us to control the quantity of solute added, but so far we have been vague about the area over which it spreads. Fortunately, once the material is deposited on the surface, it stays there—it has been specified to be insoluble and nonvolatile for precisely this reason. This means that some sort of barrier resting on the surface of the water may be used to "corral" the adsorbed molecules. Furthermore, moving such a barrier permits the area accessible to the surface film to be varied systematically. In the laboratory this adjustment of area is quite easy to do in principle. As we shall see later, the actual experiments must be performed with great care to prevent contamination.

Suppose that the initial film is spread on water which fills to the brim a shallow tray made of some inert material. Rods with low-energy surfaces may then be drawn across the water to adjust the area accessible to the molecules of the monolayer. Figure 7.1a indicates schematically how such an arrangement might appear. In practice, several barriers would be used; first to sweep the surface free of insoluble contaminants, then to confine the monolayer.

Next, an experiment such as that shown in Fig. 7.1b could be conducted. The apparatus consists of a pair of Wilhelmy plates, attached to two arms of a balance. One plate contacts the clean surface, the other the surface with the monolayer. Note that the barrier separates the two portions of surface. The surface tension will be different in the two regions, and the weight (and volume) of the meniscus entrained by the plate will be larger for the clean surface because of its higher surface tension [Eq. (6.2)]. Ideally, the two plates are identical in weight, perimeter, and contact angle although the last may be difficult to achieve in practice. If these conditions are met, however, the additional weight needed to bring the apparatus to balance measures the difference in γ for the two surfaces. By Eq. (6.2), this is given by

$$w_{\text{clean}} - w_{\text{film}} = 2(l + t) \cos \theta (\gamma_0 - \gamma) \tag{2}$$

Of course, there is no necessity to measure both γ_0 and γ on the same apparatus; they may be determined independently by any of the nondetachment methods of the preceding chapter. The experiment represented by the figure is intended mainly to emphasize that the surface tension of the two areas will be different. Furthermore, as the barrier is moved in such a way as to compress the area of the spread monolayer, the difference in π will increase.

Although π and the area A of the surface vary inversely, the precise functional form by which they are related is more difficult to describe. For very large areas, π and A show a simple inverse proportionality such as pressure and volume for an ideal gas. As the area is decreased, a more complex relationship is needed to connect these variables, just as the equation of state becomes more complex for nonideal gases and condensed phases. This analogy of π and A with p and V turns out to be a very profitable way of thinking about insoluble monolayers. For one thing, it suggests an alternative to the difference between two values of surface tension as a means of measuring π. In addition, the analogy to three-dimensional states suggests models for understanding monolayers. In succeeding sections each of these points is developed in greater detail.

Identification of area as the two-dimensional equivalent of volume is a straightforward geometrical concept. That π should be interpreted as the two-

FIGURE 7.1 *(a) Schematic illustration of a barrier delineating the area of a monolayer. (b) A Wilhelmy plate arrangement for measuring the difference in γ on opposite sides of barrier.*

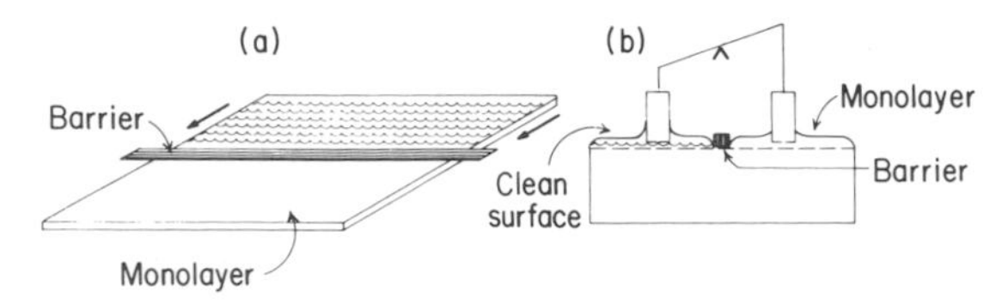

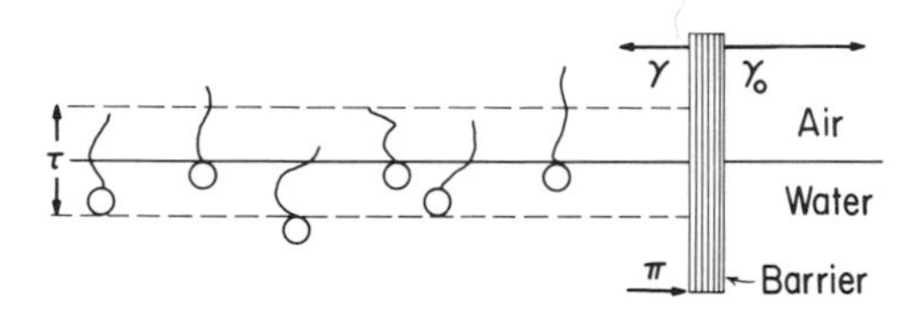

FIGURE 7.2 *Schematic profile of the air–water interface at a barrier which separates a monolayer from the clean surface.*

dimensional equivalent of pressure is not so evident, however, even though the notion was introduced without discussion in Sec. 6.10. Figure 7.2 will help to clarify this equivalency as well as suggest how to quantitatively compare two- and three-dimensional pressures. The figure sketches a possible profile of the air–water surface with an adsorbed layer of amphipathic molecules present. In general, we must allow for the fact that different configurations might exist among the adsorbed molecules; nevertheless, the surface layer has some mean thickness τ.

If the barrier represents the limit of the monolayer, then it is clear that the contractile force exerted by the surface is different on opposite sides of the barrier. Since γ is less than γ_0, it is as if the film were exerting a force on the barrier along the perimeter of the film equal to π. Force per unit length—the units of γ—is the two-dimensional equivalent of force per unit area, the units of pressure.

The surface layer does not have zero thickness, of course, even though it is conceptually convenient to think of it as two-dimensional matter. If we assume that the film pressure π extends over the entire thickness of the film, then it is an easy problem to convert the two-dimensional pressure to its three-dimensional equivalent. Taking 10 dynes cm^{-1} as a typical value for π and 10 Å as a typical value for τ, enables us to write

$$p = \frac{\pi}{\tau} = \frac{10 \text{ dynes cm}^{-1}}{10^{-7} \text{ cm}} = 10^8 \text{ dynes cm}^{-2} \tag{3}$$

or

$$p = 10^8 \text{ dynes cm}^{-2} \times \frac{1 \text{ atm}}{1.013 \text{ dynes cm}^{-2}} \simeq 100 \text{ atm} \tag{4}$$

In view of this calculation, it is not too surprising that insoluble monolayers do not usually display a simple inverse proportionality between π and A. At pressures this high, three-dimensional matter is not likely to obey the ideal gas law either.

7.3 THE LANGMUIR FILM BALANCE

The considerations of the preceding section suggest a second way to study spread monolayers. This technique involves measuring the film pressure directly, rather than calculating it from surface tension differences by Eq. (1). Figure 7.3 is a schematic representation of an apparatus called the Langmuir film balance after I. Langmuir (Nobel Prize, 1932), a pioneering worker in this field. Its base is a shallow

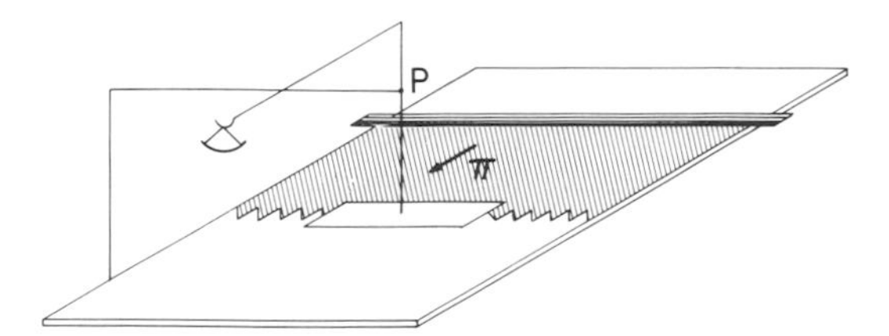

FIGURE 7.3 *Schematic representation of a Langmuir film balance.*

tray or trough of some inert material. As was the case in Fig. 7.1a, the surface must be swept by barriers both to clean the surface and to compress monolayers. In the Langmuir balance, however, one of the barriers is attached to a pivoted arm, arranged in such a way that a torque balance around point *P* can be measured, for example by adding weights to the pan. As the figure shows, an insoluble monolayer may be confined to a portion of the surface adjacent to the pressure-sensing float. The shaded area in the figure corresponds to the area of the film. It is obviously adjustable by moving the other barrier. To prevent the film from leaking past the edges of the float, flexible barriers connect the ends of the float to the edges of the tray.

By means of this apparatus, it is possible to vary the area of a spread monolayer and measure the corresponding film pressure directly. Many different variations of the film balance have been built. Figure 7.4 is a photograph of a commercial apparatus in which the float is attached to a torsion wire which may be calibrated to

FIGURE 7.4 *Photograph of a commercial film balance. (Courtesy of Central Scientific Company, Illinois, used with permission.)*

read film pressures directly. When the torsion wire is 0.010-in steel, a 1° rotation of the wire corresponds to about 0.3 dynes cm^{-1} pressure.

Although the Langmuir balance is quite simple conceptually, obtaining unambiguous results by this technique is far from simple. As we have done in discussing other experimental techniques in this book, we shall only touch upon those aspects of the method which will somehow contribute to our fundamental understanding of insoluble monolayers. Anyone considering experiments of this sort should consult more detailed discussions, such as the book by Gaines [3].

A convenient way to discuss the Langmuir balance is to examine the difficulties involved in measuring each of the two-dimensional state variables: π, A, T, and, of course, the number of moles of material in the insoluble layer, n.

The float–torsion wire assembly is the pressure-sensing system in the film balance. The flexible barriers that connect the float to the edges of the tray must be considered part of this mechanism. As with gas pressure determinations, it is essential that the system be leakproof. For this reason the float and flexible barriers must always be hydrophobic, that is, not wetted by the aqueous substrate. If these surfaces were wet by water, the possibility of surfactant transferring to them—that is, a leak in the system—would be enhanced. Thin pieces of mica are commonly used as float material, and platinum ribbons or threads of silk or nylon are often used for flexible barriers. All of these are waxed to give them suitably hydrophobic characteristics. The sweeping barriers and the tray must also be hydrophobic for the same reasons. These are usually waxed metal although Teflon is also quite popular because of its inertness. The barriers must make intimate contact with the edges of the tray, also to prevent leaks. Therefore, both tray and barriers must be carefully machined to assure good contact. Plastics are generally unsuitable as barrier materials because they are too light to make good contacts.

To convert the measured torque into a two-dimensional pressure, it is necessary to know both the length of the float and the distance between the float and the torsion wire. The latter requires that the water level be controlled quite accurately. As far as the length of the float is concerned, the flexible connectors must be included in this figure. Since they are anchored at one end, only part of their length may be considered a part of the pressure-sensing system. Some approximations are required here, but if the length of the connector is small compared to the total length of the float, the error is negligible.

The float is effectively a two-dimensional manometer, and, like its open-ended counterpart, it measures the film pressure difference between the two sides of the float. This is another reason why it is imperative that no leakage occurs past the float assembly: Leakage would increase the pressure on the reference side of the float. For the same reason, the side of the float opposite the monolayer must be carefully checked for any possible source of contamination, not just misplaced surfactant. One way of doing this is to slide a barrier toward the float from that side to verify that no displacement of the float occurs. In all aspects of film pressure measurement, the torque must be measured with sufficient sensitivity to yield meaningful results.

Measuring the area of the film is less troublesome. If the edges of the tray are parallel and the barriers perpendicular to them, the area of the rectangular surface is easily determined. The curvature of the surface at the hydrophobic boundaries introduces a small error, but—since the total area is of the order of magnitude of 10^2 cm^2—this is generally negligible.

The results obtained in π–A experiments may be sensitive to the rate at which the film area is changed. We shall not discuss the factors responsible for this but shall merely note that the same film pressures should be obtained on compression and expansion if true equilibrium values are being measured.

Temperature is an important variable in any equation of state. The experiments we are describing are isothermal; therefore, it is important that both the water substrate and the adjoining vapor be thermostated.

Next, let us consider those difficulties associated with the determination of the amount of material deposited on the surface. We have already noted that the method of depositing insoluble monolayers by spreading permits the accurate determination of n. Since the spreading technique requires solvent volatility, care must be exercised to prevent the stock solutions from changing concentration due to evaporation prior to their application to the surface. Also, precise microvolumetric methods must be used to dispense the solution on the aqueous surface, since the quantity used is small. The solvent (as well as the solute) must be free from contaminants. There is also the possibility that the solvent will extract spreadable contaminants from the waxed surfaces of the float, barriers, and tray. Some workers advocate addition and evaporation of one drop at a time to minimize this. Oily contaminants may also reach the water surface from the fingers and from the atmosphere. These last sources are particularly hard to control: Tests for reproducibility and blank compressions (i.e., move barrier toward float on "clean" surface) are the best evidence of their absence.

Not all solvents are equally suitable for spreading monolayers. The requirement that the solvent evaporate completely is self-evident. It has been suggested that if the organic solvent dissolves much water, the properties of the monolayer will be different from those in which no water is trapped. Verification that no artifacts are entering the observations from the solvent may be accomplished by conducting duplicate experiments with different solvents.

Until now, we have concentrated on those difficulties in using the Langmuir balance which arise from the determination of π, A, T, and n. A few remarks are also in order about considerations which may affect the monolayer and which originate in the adjacent phases. We have already discussed contaminants originating in the gaseous phase. For some monolayer materials, air oxidation may also be a problem. The aqueous substrate is the source of a wide assortment of contaminants in addition to spreadable oily matter. Ionic impurities, including those which affect the pH, are quite troublesome. The charge state of many amphipathic molecules, for example amines and carboxylic acids, is obviously pH dependent. Salts or complexes formed between amphipathic molecules and ions in the aqueous phase will have different monolayer properties than the unreacted surfactant molecule.

7.4 RESULTS OF FILM BALANCE STUDIES

The preceding section shows that it is possible to determine π–A isotherms for surfaces just as p–V isotherms may be measured for bulk matter. The results that

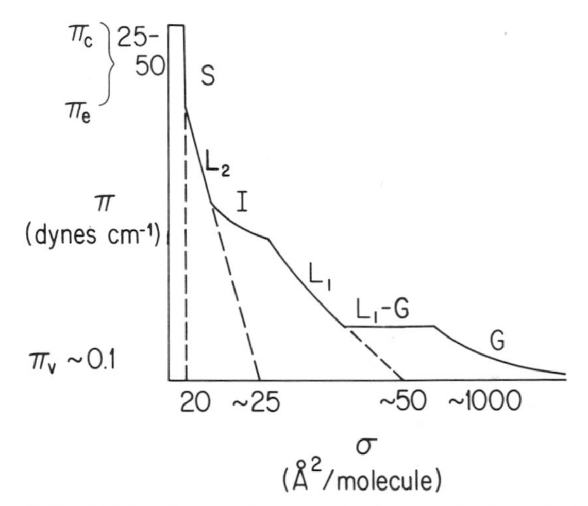

FIGURE 7.5 *Composite π–σ isotherm which includes a wide assortment of monolayer phenomena. Note that the scale of the figure is not uniform so that all features may be included on one set of coordinates.*

are obtained for surfaces are analogous to bulk observations also, although some caution must be expressed about an overly literal correlation between bulk and surface phenomena. We shall return to a discussion of these reservations later. There can be no doubt, however, that analogies with bulk behavior supply a familiar framework within which to consider π–A isotherms.

The curve sketched in Fig. 7.5, which is drawn with grossly distorted coordinates to encompass all features, contains several similarities to p–V isotherms. Not all the features shown here are always observed, nor are all known idiosyncrasies of π–A isotherms represented. The presence or absence of various features and their π–A coordinates vary with temperature for a particular amphipathic molecule and from one amphipathic substance to another. Lastly, there is some diversity in the terminology used to describe various monolayer phenomena. In short, Fig. 7.5 is a composite isotherm which will introduce and summarize a variety of observations.

In this section, we shall discuss in turn the various two-dimensional phases and phase equilibria represented in Fig. 7.5, progressing from low values of π to high. The existence of these two-dimensional states and the properties they possess are presumably unfamiliar to most readers. Therefore, it is important to keep the following ideas in mind in reading this section:

1. We are concerned with *two-dimensional matter* situated at the boundary between two bulk phases.
2. The properties of the two-dimensional phases are relatively independent of the properties of the bulk phases of the same material.
3. Many surface states are two-dimensional analogs of three-dimensional states. As with any analogy, however, there are points of similarity and points of difference between the surface and bulk states. For most of the states we discuss, we shall

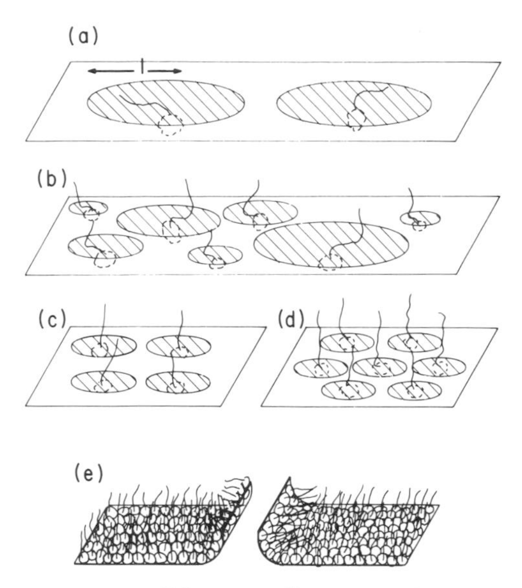

FIGURE 7.6 *Schematic illustration showing by shading the effective area per molecule at various stages of monolayer compression: (a) gaseous state, (b) liquid expanded state, (c) liquid condensed state, and (d) solid state. In (e) the collapse of the film is illustrated.*

consider the phenomenological behavior as represented by Fig. 7.5 and also suggest a molecular interpretation in terms of Fig. 7.6.

If measurements can be made at sufficiently low pressures, all monolayers will display gaseous behavior, represented by region G in Fig. 7.5. The gaseous region is characterized by an asymptotic limit as $\pi \rightarrow 0$. In the limit of very low film pressures, a two-dimensional equivalent to the ideal gas law applies:

$$\pi A = nRT \tag{5}$$

where R is the gas constant, usually in cgs units. This is a convenient place to define another quantity, the area occupied per molecule in the interface, σ. Since R equals Avogadro's number times the Boltzmann constant and nN_A equals the total number of surface molecules, we may write

$$\pi\left(\frac{A}{nN_A}\right) = \pi\sigma = kT \tag{6}$$

As a model for this highly expanded state we may choose a situation like that shown in Fig. 7.6a in which the hydrocarbon chain lies flat on the surface, blocking an area πl^2, where l is the length of the "tail." We may then use Eq. (6) to calculate the value of π corresponding to this area per molecule. At 25°C and with $l = 10$ Å, we obtain

$$\pi = \frac{kT}{\sigma} = \frac{(1.38 \times 10^{-16})(298)}{(3.14)(10^{-7})^2} = 1.31 \text{ dynes cm}^{-1} \tag{7}$$

Allowing only this area per molecule is equivalent to defining a distance of closest approach. Therefore, ideal behavior is not expected until areas per molecule which are larger than this, on the order of 10^3 Å^2 perhaps. Correspondingly lower film pressures will be involved also. If we recall that the pressure of this example is equivalent to a bulk gas pressure of about 13 atm, it is less surprising that such low film pressures are needed to observe gaseous monolayer behavior. The repulsion between particles in a charged monolayer increases the effective area these molecules occupy at the surface. The effect of this is to increase the pressure of charged films, making them more accessible to measurement. Since ionic surfactants are soluble, techniques other than the film balance must be used to study their π–σ isotherms.

Like its three-dimensional counterpart, Eq. (5) is a limiting law which means that deviations may be expected at higher pressures, lower temperatures, or with more strongly interacting molecules. Figure 7.7 is a plot of $\pi\sigma/kT$ versus π for several members of the carboxylic acid homologous series. The film balance was used to collect the data only for the C_{12} acid. Shorter chain compounds are soluble and were investigated by surface tension measurements and interpreted by the Gibbs equation, which we discuss in Sec. 7.8. The main point to note about this figure is the strong resemblance it bears to similar plots for three-dimensional gases. Negative deviations occur at low pressures, becoming more pronounced as the length of the alkyl chain increases. As with gases, this may be attributed to attraction between molecules, an effect that increases with chain length. At higher pressures, deviations

FIGURE 7.7 *Plots of $\pi\sigma/kT$ versus π for n-alkyl carboxylic acids. Curve number* $1 = C_4$, $2 = C_5$, $3 = C_6$, $4 = C_8$, $5 = C_{10}$, and $6 = C_{12}$. [*Data from N. K. Adam,* Chem. Rev. **3**:172 (1926).]

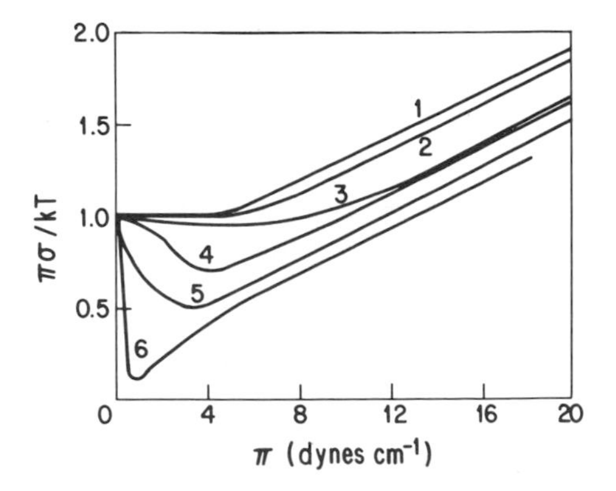

tend to be positive. This is analogous to the excluded volume effect for gases, except that it becomes an excluded area in two dimensions.

Next, let us return to Fig. 7.5, discussing the features labeled L_1 and L_1–G. The region L_1 is called the liquid state or, more commonly, the liquid expanded state to distinguish it from L_2, the liquid condensed state. The region L_1–G is a two-phase region in which the gaseous and liquid expanded state coexist in equilibrium.

The horizontal line of the L_1–G region in Fig. 7.5 is analogous in every way to the corresponding feature in bulk matter. At a given temperature, there is a constant-pressure region over which a significant compression occurs. The film pressures at which L_1–G equilibrium occurs are known as film vapor pressures π_v. Like the gaseous state itself, the L_1–G equilibrium occurs at very low pressures. Tetradecanol, for example, has a two-dimensional vapor pressure of 0.11 dynes cm^{-1} at 15°C.

It is important to remember the significance of π_V. It refers specifically to the equilibrium between two *surface* states. There is a danger of confusing π_v with the equilibrium spreading pressure π_e, introduced in Chap. 6. The latter is the pressure of the equilibrium film that exists in the presence of excess *bulk* material on the surface. It is the equilibrium-spreading pressure that is involved in the modification of Young's equation [Eq. (6.71)], where a bulk phase is present on the substrate. For tetradecanol at 15°C, the equilibrium-spreading pressure is about 45 dynes cm^{-1}, so π_e and π_v are very different from one another.

The temperature variation of π_v may be analyzed by a relationship analogous to the Clapeyron equation to yield the two-dimensional equivalent to the heat of vaporization. The numerical values obtained for this quantity more nearly resemble the bulk values for hydrocarbons than those for polar molecules. This suggests that most of the change in the surface transition involves the hydrocarbon tail of the molecule rather than the polar head.

Finally, note that the L_1–G equilibrium region disappears above a certain temperature which is the two-dimensional equivalent of the critical temperature for liquid–vapor equilibrium.

Because of the very low pressures at which they are observed, the surface states we have discussed so far are difficult to observe and relatively unimportant experimentally. They are relatively easy to understand, however, so they serve well to introduce the idea of surface states. The liquid expanded state, L_1 in Fig. 7.5, is the first of several condensed states we shall discuss. Because it is bound on the low-pressure side by a two-phase region with a critical temperature, the L_1 state is easily compared to a bulk liquid state. Since gaseous behavior is observed only at very low pressures, it is easy to extrapolate the isotherm for the L_1 state to the $\pi = 0$ value as the dashed line in the figure shows. For amphipathic molecules with saturated unbranched R groups, this intercept is in the range 45 to 55 $Å^2$. We shall identify the limiting area per molecule (superscript zero) for this state (subscript) by the symbol $\sigma^0_{L_1}$. The presence of branched chains or double bonds—particularly in the cis configuration—increases the value of this limiting area.

The precise structural details of the liquid expanded state at the molecular level are not fully understood, but several generalizations do appear to be justified. The value of $\sigma^0_{L_1}$ is several times the actual cross-sectional area of the amphipathic molecule, with the latter oriented perpendicular to the surface. At the same time, the area per molecule is considerably less than could be permitted if the entire tail were free to move in the surface. Figure 7.6b represents a model of the surface in the L_1 state at the molecular level. Here part of the hydrocarbon chain lies in the surface and some has been lifted out of the surface plane. That portion of the tail in the surface defines the effective area per molecule. Neither this area nor the length of the chain which is out of the surface will be the same for all molecules, so experimental values of $\sigma^0_{L_1}$ correspond to average values. Further, there will be considerable lateral interaction between those segments which are not in contact with the substrate.

The compressibility of the L_1 state is expected to be far less than that of the gaseous state, but not yet incompressible because the average area per molecule may be altered by squeezing additional CH_2 groups out of contact with the water.

With sufficient compression, the isotherm of the L_1 state shows a sharp break to enter a situation variously known as the intermediate or transition state, indicated by I in Fig. 7.5. From the sharpness of the break in the isotherm, this situation might be initially regarded as another two-phase equilibrium. This is especially tempting in view of the fact that one finds, at still greater compression, one or more additional states which are far less compressible than the liquid expanded state. Region I is definitely not horizontal, however, as would be required for a first-order phase transition. As a matter of fact, there is scarcely a discontinuity in the isotherm at the most compressed extreme of the intermediate state. Figure 7.8 shows a set of experimental isotherms for tetradecanoic acid on a substrate of 0.01 *M* HCl which

FIGURE 7.8 *Plots of π versus σ for myristic acid on* 0.01 *M HCl at various temperatures. The L_1, I, and L_2 states are shown.* [*N. K. Adam and G. Jessop,* Proc. Roy. Soc. **A112**:364 (1926).]

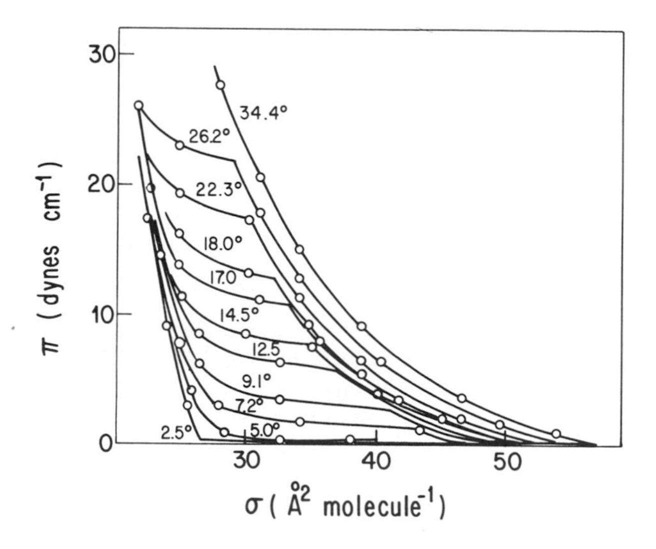

keeps the C_{14} acid un-ionized. These isotherms show region I particularly clearly and reveal that its characteristics vary quite regularly.

At present, no entirely satisfactory molecular interpretation of the state exists. Two proposed explanations center around the following ideas. One is that close-packed clusters of vertically oriented chains begin forming at the onset of the intermediate region. Another idea is that the number of configurational degrees of freedom undergoes a transition in this region over a narrow range of temperatures.

If the area of an insoluble monolayer is isothermally reduced still further, the compressibility eventually becomes very low. Because of the low compressibility, the states observed at these low values of σ are called condensed states. In general, the isotherm is essentially linear, although it may display a well-defined change in slope as π is increased as shown in Fig. 7.5. The (relatively) more expanded of these two linear portions is called the liquid condensed state L_2 and the less expanded is called the solid state S. It is clear from the low compressibility of these states that both the L_2 and the S state are held together by strong intermolecular forces so as to be relatively independent of the film pressure. Because of the near linearity of these portions of the isotherm, it is easy to extrapolate both regions to their value at $\pi = 0$. The intercepts for the solid and liquid condensed regions, $\sigma_S{}^0$ and $\sigma_{L_2}{}^0$, respectively, differ only slightly. Values of $\sigma_{L_2}{}^0$ for alcohols are about 22 Å^2 and for carboxylic acids about 25 Å^2, more or less independent of the length of the hydrocarbon chain. The intercept $\sigma_S{}^0$ has a value of about 20 Å^2, independent of both the length of the chain and the nature of the head. The film pressures in the condensed states (L_2 or S) are of the same magnitude as the equilibrium-spreading pressure for amphipathic molecules.

For the condensed L_2 and S states, a molecular interpretation is again possible. In both, the values of σ^0 are close to actual molecular cross sections when the molecules are oriented perpendicular to the surface. The difference between these two regions seems to involve the polar part of the molecule more than the hydrocarbon chain, which was more important for the more expanded states. The difference between $\sigma_S{}^0$ and $\sigma_{L_2}{}^0$ may involve a more efficient packing of the heads or the formation of fairly specific lateral interactions through hydrogen bonds, for example. Figures 7.6c and 7.6d present models of how the L_2 and S states, respectively, may differ in packing efficiency. The values of σ^0 which are observed for monolayers of saturated, n-alkyl compounds are only slightly larger than the close-packed cross sections obtained for these compounds in the bulk solid state by x-ray diffraction.

Additional compression eventually leads to the collapse of the film. The pressure at which this occurs, π_c, is somewhere in the vicinity of the equilibrium-spreading pressure. Figure 7.6e represents schematically how this film collapse may occur.

This figure represents the squeezing out of the surface of highly ordered aggregates which may quite plausibly be regarded as nuclei to bulk phase particles. If collapse marked the appearance of the amphipathic material in a bulk phase, then the collapse pressure and the equilibrium-spreading pressure should be identical. Here a significant complication appears. The collapse pressure is highly sensitive to the rate at which the film is compressed. This indicates nonequilibrium conditions, showing that the two-dimensional solid phase resembles three-dimensional solids in this respect also: difficulty in attainment of thermodynamic equilibrium.

In summary, it must be emphasized again that there are wide variations in the properties of insoluble monolayers. Some of the phenomena reported in the literature are probably artifacts due to impurities or nonequilibrium conditions. Others are probably unique effects which apply only to a very specific system. In the descriptive material of this section, both phenomenological and modelistic information were provided for various stages along the isotherm. At the very least, the models serve the pedagogical function of assisting the student in remembering an assortment of probably unfamiliar facts. At best, the models provide the basis for quantitatively understanding these phenomena. In the next section, we shall take a more quantitative look at the model for the gaseous state.

7.5 MODELS FOR THE GASEOUS STATE OF MONOLAYERS

It is not difficult to propose and develop a model for the gaseous state of insoluble monolayers. The arguments parallel those developed in kinetic molecular theory for three-dimensional gases and lead to equally appealing results. The problem, however, is that many assumptions of the model are far less plausible for monolayers than for bulk gases. To see this, a brief review of the derivation seems necessary.

Suppose we imagine a single molecule bouncing back and forth across a surface between two restraining barriers. If we define the direction of this motion to be the x direction and the velocity of the molecule to be v_x, then the change in momentum at each collision (assuming them to be elastic) is

$$\frac{\Delta\,(\text{Momentum})}{\text{Collision}} = mv_x - (-mv_x) = 2mv_x \tag{8}$$

where m is the mass of the molecule. The time interval between two successive collisions at the *same* wall is given by

$$\frac{\text{Elapsed time}}{\text{Collision}} = \frac{2l}{v_x} \tag{9}$$

if the distance between barriers is l, since the molecule must cross the distance between the barriers twice before returning to the same spot. The force exerted by the molecule on impact equals the rate of change in momentum which, in turn, equals the ratio of Eq. (8) to Eq. (9):

$$\frac{\Delta\,(\text{Momentum})}{\Delta\,(\text{Time})} = F_x = \frac{mv_x^{\;2}}{l} \tag{10}$$

This force is converted to two-dimensional pressure by dividing it by the length of the edge to which the force is applied. Assuming the accessible surface area to be a square means that the length of this edge is also l; the pressure contribution of this one collision equals

$$\pi = \frac{F_x}{l} = \frac{mv_x^{\;2}}{l^2} \tag{11}$$

The quantity l^2 in this equation clearly describes the area accessible to the molecule.

Since pressure is isotropic, we assume that the forces on the perpendicular barriers are identical. Therefore, if the surface contains N molecules, they behave as if $N/2$ were exerting a pressure given by Eq. (11) on the barriers perpendicular to the x direction, with the other $N/2$ exerting an identical pressure in the other direction. That is, for a surface containing N molecules

$$\pi A = \frac{N}{2} m\overline{v^2} = N(\overline{\text{KE}}) \tag{12}$$

The average value of the squared velocity has been used in Eq. (12) to allow for the fact that a distribution of molecular velocities exists. The nature of the averaging procedure to be used in this case is well established from physical chemistry. We also know from physical chemistry that the average kinetic energy per molecule $(\overline{\text{KE}})$ per degree of freedom is

$$\frac{\overline{\text{KE}}}{\text{Degree of freedom}} = \tfrac{1}{2}kT \tag{13}$$

Since the molecules on the surface have two translational degrees of freedom, Eqs. (12) and (13) may be combined to give

$$\pi \frac{A}{N} = \pi\sigma = kT \tag{14}$$

which is identical to Eq. (6).

The foregoing derivation is a straightforward two-dimensional analog of the three-dimensional case and leads to a result which describes the experimental facts. From a pragmatic point of view, it is a great success. One of the theoretical assumptions underlying Eq. (13), however, is that translational quantum states are sufficiently close together to justify treating them as continuous rather than discrete. This is unquestionably true for gases. For an amphipathic molecule, the polar head of which contacts—and interacts with—the aqueous substrate, it is somewhat harder to justify. We shall see presently that there is a totally different way of looking at Eq. (14) that is free from this objection.

We noted previously that the applicability of Eq. (14) to insoluble monolayers is severely restricted to very low values of π. Figure 7.7 shows that the deviations from Eq. (14) with increases in π are very similar to what is observed for nonideal gases. Specifically, the positive deviations associated with excluded volume effects in bulk gases and the negative deviations associated with intermolecular attractions are observed. It is tempting to try to correct Eq. (14) for these two causes of nonideality in a manner analogous to that used in the van der Waals equation:

$$\left(\pi + \frac{a}{\sigma^2}\right)(\sigma - b) = kT \tag{15}$$

where a and b are the two-dimensional analogs of the van der Waals constants. Note that b, the excluded area per molecule, is conceptually equivalent to σ^0, although which of the values ($\sigma_{L_1}{}^0$, $\sigma_{L_2}{}^0$, or $\sigma_S{}^0$) best fits the data cannot be predicted a priori.

The temperature at which the van der Waals equation goes from one having three real roots to one having one real root is generally identified with the critical temperature. In the same way, Eq. (15) may be considered to connect both the gaseous (G) and liquid expanded (L_1) states in monolayers, the transition between which also displays a critical point. Statistical mechanics shows, however, that the van der Waals "a" constant explicitly ignores orientation effects as contributing anything to the energy in gases; it is hard to imagine the properties of insoluble monolayers as being independent of orientation. Therefore, any attempt to correct Eq. (14) in such a way as to extend its range encounters difficulties. Ultimately, all objections to two-dimensional equations of state seem to center on their neglect or unsatisfactory inclusion of the substrate.

An alternative way of looking at monolayers is to consider them as two-dimensional binary solutions, rather than two-dimensional phases of a single component. The advantage of this approach is that it does acknowledge the presence of the substrate and the fact that the latter plays a role in the overall properties of the monolayer. Although quite an extensive body of thermodynamics applied to two-dimensional solutions has been developed, we shall consider only one aspect of this. We shall examine the film pressure as the two-dimensional equivalent of osmotic pressure. It will be recalled that, at least for low osmotic pressures, the relationship between π, V, n, and T is identical to the ideal gas law [Eq. (4.37)]. Perhaps the interpretation of film pressure in these terms is not too far fetched after all!

As we saw in Chap. 4, the heart of any osmotic pressure experiment is a semipermeable membrane which allows the solvent but not the solute to pass. The float of a Langmuir balance accomplishes this. That portion of surface with the monolayer is considered to be the two-dimensional solution; the clean surface is the two-dimensional solvent. The solvent can certainly pass from one region to another (remember that the mechanism of the partitioning has nothing to do with the equilibrium osmotic pressure) through the bulk substrate. However, the insoluble solute is restrained by the float.

For osmotic equilibrium, the chemical potential of the solvent must be the same on both sides of the membrane. In the two-dimensional analog, μ_1 must also be the same for the water on both sides of the float. The presence of the solute lowers the chemical potential of the solvent but the excess pressure compensates for this. Therefore, by analogy with Eq. (4.32), we write

$$\mu_{1s}^{0} = \mu_{1s}^{0} + RT \ln a_{1s} + \int_{0}^{\pi} \bar{A}_1 \, dp \qquad (16)$$

where the subscript s indicates the surface and the subscript 1 identifies the solvent. If the partial molal area of the solvent, $\bar{A}_1$, is assumed to be independent of π, Eq. (16) may be integrated to give

$$-RT \ln a_{1s} = \pi \bar{A}_1 \qquad (17)$$

For ideal (dilute) solutions, the activity is replaced by the mole fraction x_{1s} which, for a two-component surface solution, equals $1 - x_{2s}$. With the customary expansion of the logarithm as a power series (Appendix A), these substitutions yield

$$RTx_{2s} = \pi \bar{A}_1 \tag{18}$$

the two-dimensional equivalent of Eq. (4.36). Finally, it is necessary to relate the surface mole fraction and the molar area of the solvent to more familiar variables.

The surface mole fraction is entirely analogous to the bulk value of this quantity

$$x_{2s} = \frac{n_{2s}}{n_{1s} + n_{2s}} = \frac{N_{2s}}{N_{1s} + N_{2s}} \tag{19}$$

where the n terms are the numbers of moles and the N terms are the numbers of molecules. For dilute surface solutions—that is, expanded monolayers—$N_{1s} \gg N_{2s}$; therefore,

$$x_{2s} \simeq \frac{N_{2s}}{N_{1s}} \tag{20}$$

The total area of the surface may be written

$$A_T = n_{1s}\bar{A}_1 + n_{2s}\bar{A}_2 = N_{1s}\sigma_1{}^0 + N_{2s}\sigma_2{}^0 \tag{21}$$

Applying these various relationships to Eq. (18) leads to the following result for low film osmotic pressures

$$\pi(A_T - N_{2s}\sigma_2{}^0) = n_{2s}RT = N_{2s}kT \tag{22}$$

Dividing through by N_{2s} to express the total area as area per solute molecule gives

$$\pi(\sigma - \sigma_2{}^0) = kT \tag{23}$$

Equation (23) obviously gives the two-dimensional ideal gas law when $\sigma > \sigma_2{}^0$, and with the $\sigma_2{}^0$ term included represents part of the correction included in (15). This model for surfaces is, of course, no more successful than the one-component gas model used in the kinetic approach; however, it does call attention to the role of the substrate as part of the entire picture of monolayers. We saw in Chap. 4 that solution nonideality may also be considered in osmotic equilibrium. Pursuing this approach still further results in the concept of phase separation to form two immiscible surface solutions.

In summary, we see that insoluble monolayers may be viewed either as examples of two-dimensional phases of one component or as two-dimensional solutions with two components. The former model is somewhat simpler and is often adequate. The latter, although more complex, is more realistic. In spite of our interest in the monolayer, we must not neglect the fact that none of the monolayer phenomena would exist without the aqueous phase as the substrate.

7.6 APPLICATIONS OF INSOLUBLE MONOLAYERS

In this section, we shall briefly mention two areas in which insoluble monolayers have found applications: as model systems for cell membrane studies and as evaporation retardants.

Cell walls not only delineate the boundaries of the cell but they also perform a variety of functions. A considerable amount of interest in molecular biology is centered around one aspect or another of these and other biological membranes. The chemical components of biological membranes are known to include phospholipids, sterols, and proteins. Cell membrane lipids are derivatives of phosphatidic acid and have the general structural formula:

$$
\begin{array}{l}
\mathrm{CH_2OCOR_1} \\
| \\
\mathrm{CHOCOR_2} \\
| \qquad\qquad \mathrm{OH} \\
| \qquad\qquad\ \ | \\
\mathrm{CH_2{-}O{-}P{-}X} \\
\qquad\qquad\ \ | \\
\qquad\qquad \mathrm{OH}
\end{array}
$$

where R_1 and R_2 are C_{14} to C_{20} carboxylic acids, sometimes unsaturated. Oleic and palmitic acids are the most common substituents in these positions. Common substituents for X are ethanolamine $-O-CH_2-CH_2-NH_2$ and choline $-O-CH_2-CH_2-\overset{+}{N}(CH_3)_3$; phosphatidylcholines are also called lecithins. These substances form insoluble monolayers that include many of the general features described in Sec. 7.4.

It is now generally agreed that the basic structure of biological membranes is a bimolecular sheet consisting of two oriented monolayers stacked in a tail-to-tail arrangement, but otherwise analogous to the extruded portion of the collapsed monolayer sketched in Fig. 7.6e. The arrangement of membrane proteins, many of which play enzymatic functions, is not so well understood. One model holds that the proteins are randomly embedded in the lipid matrix, as shown in Fig. 7.9. Another view is that the protein molecules are layered on both sides of the bimolecular lipid film in a sandwich arrangement. Regardless of which model is correct, it is clear that the proper functioning of these membranes is related to the chemistry of these essentially two-dimensional solutions.

These systems are clearly much more complicated than the simple amphipathic monolayers just discussed, but the latter are ideally suited as models for membranes in living systems. Consider the following example which illustrates this principle.

It is known that normal mammalian cells grown in culture display contact inhibition; that is, they stop dividing when they have covered the culture surface. Malignant cells, on the other hand, lack this inhibition and divide beyond the normal limit. There is obviously considerable interest in the mechanism underlying this change. In terms of the membrane model shown in Fig. 7.9 and various monolayer concepts, a possible mechanism might go as follows. In a normal cell wall certain molecules which somehow play the role of "sensors" and supply feedback to the dividing cell concerning cell density may be randomly distributed over the surface of a cell. A phase change in the two-dimensional solution which comprises the membrane may lead to the aggregation of these sensors in more concentrated patches, thereby eliminating or at least reducing the efficiency of the inhibition mechanism. Malignant cells are also known to display enhanced agglutination by saccharide-binding plant agglutinins. The clustering of saccharide-binding sites due

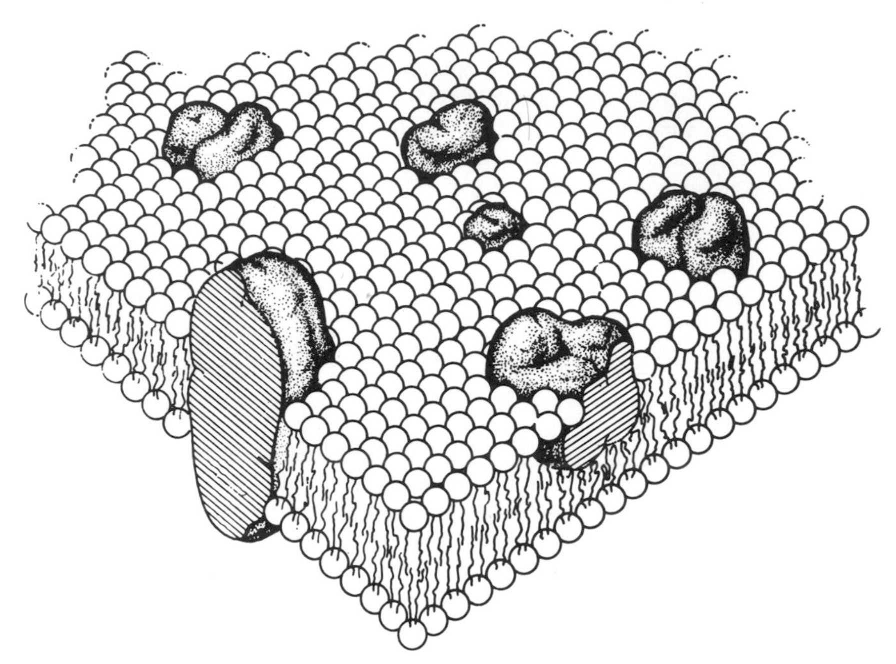

FIGURE 7.9 *Schematic representation of a biological membrane. The amphipathic phospholipid molecules form a bimolecular sheet with protein molecules embedded therein. [Reprinted with permission from S. J. Singer and G. L. Nicolson,* Science **175**:720 (1972), *copyright* 1972 *by the American Association for the Advancement of Science.*]

to two-dimensional phase transitions could also explain this observation. We have seen from monolayer studies that such two-dimensional phase changes do occur. Relatively small changes in the chemistry of a membrane may lead to quite different phases with significant biological consequences.

The foregoing is purely speculative, but is intended to show that the physical and reaction chemistry of monolayers—which supplied the original information suggesting the bimolecular layer structure in the cell membranes—continues to provide insights as to the structure and function of membranes.

A more concrete application of insoluble monolayers is their use in the retardation of evaporation. Particularly in arid regions of the world, evaporation of water from lakes and reservoirs constitutes an enormous loss of a vital resource. Under some conditions, the water level of such bodies may change as much as one foot per month due to evaporation. The usual unit for water reserves is the acre-foot, a volume of water covering an acre of surface to the depth of one foot. It equals about $\frac{1}{3}$ million gallons for each acre of water surface.

Considerable research has been conducted both in the laboratory and in the field on the effectiveness of insoluble monolayers in reducing evaporation. An American

Chemical Society Symposium in 1960 dealt exclusively with this topic; the proceedings of that symposium are given by LaMer [6].

Laboratory research in this area is conducted by suspending a porous box of desiccant very close to the surface of a film balance. The rate of water uptake is determined by weighing at various times. This way, the retardation of evaporation may be measured as a function of film pressure and correlated with other properties of the monolayer determined by the same method. As might be expected, the resistance to evaporation which a monolayer provides is enhanced by those conditions which promote the most coherent films, most notably high film pressures and straight chain compounds.

To be acceptable for use in the field, the monolayer material must have the following properties:

1. It must spread easily, probably as bulk material so a high value of π_e is desirable.
2. It must be self-healing since surface ripples will disrupt the monolayer. This implies viscous rather than rigid monolayers.
3. It must be inexpensive which means, effectively, capable of forming good films from naturally occurring mixtures.
4. It must be nontoxic and free from other deleterious effects on aquatic life.

Hexadecyl and octadecyl alochol have been extensively studied and shown to be highly effective in evaporation retardation. Scattering powdered samples of commercial grade alcohols by boat on lake surfaces or continuous addition of alcohol slurries from floating dispensers are two of the methods that have been employed to apply these monolayers. Wind conditions and the activity of aquatic birds have a

FIGURE 7.10 *Comparison of the water level in two adjacent lakes during the summer,* 1957. *The ordinate shows the level in the lake with the monolayer; the abscissa is the untreated lake.* (*W. J. Roberts in LaMer* [6], *redrawn with permission.*)

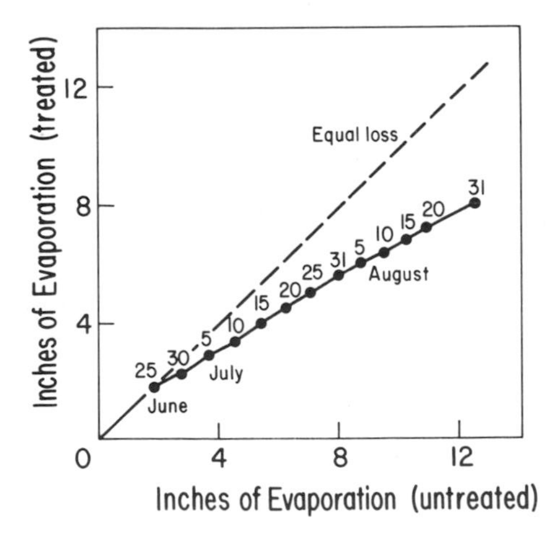

considerable effect on the stability of the monolayer and, therefore, on the rate at which the monolayer chemicals must be reapplied. Rates of application rarely exceed 0.5 lb $acre^{-1}$ day^{-1}, however, so that cost of the materials used is not excessive.

An indication of the effectiveness of such field treatment is seen in Fig. 7.10 which compares the amount of water lost by evaporation from two small adjacent lakes in Illinois. Treatment of North Lake (ordinate) with commercial hexadecanol was begun in late June 1957; untreated South Lake was the control (abscissa). Prior to treatment, the evaporation losses from the two lakes were identical, as shown by the 45° line in the figure. After treatment was begun, however, the loss of water from North Lake fell considerably behind that from South Lake. By the end of the summer, a difference of about 40% in the water loss was observed. This was equivalent to about 7600 gallons of water saved per pound of hexadecanol used. For areas where water is scarce—the southwestern United States, Israel, and western Australia, for example—such conservation of water is highly valued, and research continues for methods to improve the efficiency of this technique.

7.7 ADSORPTION FROM SOLUTION AND THE GIBBS EQUATION

Until now, we have discussed only insoluble monolayers. Although the behavior these display is complex, they have the conceptual simplicity of being localized in the interface. It has been noted, however, that even in the case of insoluble monolayers, the substrate should not be overlooked. The importance of the adjoining bulk phases is thrust into even more prominent view when soluble monolayers are discussed. In this case, the adsorbed material has appreciable solubility in one or both of the bulk phases which define the interface.

Gibbs treated this situation as part of his investigations on phase equilibria. Suppose we consider two phases α and β in equilibrium with a surface s dividing them. For the system so constituted, we may write

$$G = G^{\alpha} + G^{\beta} + G^{s} \tag{24}$$

where the superscripts indicate the contribution from each category. For the bulk phases,

$$G = E + pV - TS + \sum_i \mu_i n_i \tag{25}$$

where the chemical potential terms are summed for all components i. The volume term is replaced by an area term in the corresponding expression for G^s:

$$G^s = E^s + \gamma A - TS^s + \sum_i \mu_i n_i \tag{26}$$

Substituting Eqs. (25) and (26) into (24) and taking the total derivative yields

$$dG = \sum_{\alpha,\beta,s} \left(dE + p\,dV + V\,dp - T\,dS - S\,dT + \sum_i \mu_i\,dn_i + \sum_i n_i\,d\mu_i \right) + A\,d\gamma + \gamma\,dA \tag{27}$$

For a reversible process

$$dE = \delta q - \delta w = \sum_{\alpha,\beta,s} dE = \sum_{\alpha,\beta,s} [T\,dS - (p\,dV + \delta w_{\text{non-}pV})] \tag{28}$$

Substituting this result into Eq. (27) gives

$$dG = \sum_{\alpha,\beta,s} \left(V\,dp - S\,dT + \sum_i \mu_i\,dn_i + \sum_i n_i\,d\mu_i - \delta w_{\text{non-}pV} \right) + A\,d\gamma + \gamma\,dA \tag{29}$$

As we saw in Chap. 6 [Eq. (6.15)], the quantity $\gamma\,dA$ may be equated to non-pressure/volume work when surface energy is being considered. With this consideration, Eq. (29) simplifies still further to become

$$dG = \sum_{\alpha,\beta,s} \left(V\,dp - S\,dT + \sum_i \mu_i\,dn_i + \sum_i n_i\,d\mu_i \right) + A\,d\gamma \tag{30}$$

Another well-known relationship from physical chemistry may be introduced at this point:

$$dG = V\,dp - S\,dT - \sum_i \mu_i\,dn_i \tag{31}$$

Applying Eq. (31) to the bulk phases and the surface and subtracting the result from (30) gives

$$\sum_i n_i^{\alpha}\,d\mu_i + \sum_i n_i^{\beta}\,d\mu_i + \sum_i n_i^{s}\,d\mu_i + A\,d\gamma = 0 \tag{32}$$

When only one phase is under consideration, only one of the terms in Eq. (24) is required, and only one of the bulk phase summations in (32) survives. The result in this case is the famous Gibbs–Duhem equation:

$$\sum_i n_i\,d\mu_i = 0 \tag{33}$$

It will be recalled from physical chemistry that this relationship permits the evaluation of the activity of one component from measurements made on the other in binary solutions.

By means of the Gibbs–Duhem equation, we may eliminate the terms in Eq. (32) which apply to bulk phases and write

$$\sum_i n_i^{s}\,d\mu_i + A\,d\gamma = 0 \tag{34}$$

This is the Gibbs adsorption equation which relates γ to the number of moles and the chemical potentials of the components in the interface. In subsequent developments, we shall consider only two-component systems and shall identify the solvent (usually water) as component 1 and the solute as component 2. In terms of this stipulation, Eq. (34) becomes

$$n_1^{s}\,d\mu_1 + n_2^{s}\,d\mu_2 + A\,d\gamma = 0 \tag{35}$$

It is conventional to divide Eq. (35) through by A to give

$$d\gamma = -\frac{n_1^s}{A}\,d\mu_1 - \frac{n_2^s}{A}\,d\mu_2 \tag{36}$$

The quantity n_i^s/A is called the surface excess of component i and is given the symbol Γ_i:

$$\Gamma_i = \frac{n_i^s}{A} \tag{37}$$

In this notation, Eq. (36) becomes

$$-d\gamma = \Gamma_1\,d\mu_1 + \Gamma_2\,d\mu_2 \tag{38}$$

Before proceeding any further, it is necessary to examine just what the concept of a surface excess means. To do this, it is convenient to consider the changes which occur in some general property P as we move from phase α to phase β. The situation is represented schematically by Fig. 7.11 where x is the distance measured perpendicular to the interface. The scale of this figure is such that variations at the molecular level are shown. The interface is not a surface in the mathematical sense but rather a zone of thickness τ across which the properties of the system vary from values which characterize phase α to those characteristic of β. In spite of this, we generally do not assign any volume to the surface, but treat it as if the properties of α and β applied right up to some dividing plane situated at some specific value of x. What is this position x_0 at which we draw such a boundary?

Suppose the solid line in Fig. 7.11 represents the actual variation of property P. The squared-off extensions of the bulk values of this property represent the approximation made in assuming the surface to have zero thickness. Then the shaded area to the left of x_0 shows the amount by which the value of P for the system as a whole has been overestimated by extending P_α. Likewise, the shaded area to the right of x_0 shows how the extension of P_β leads to an underestimation of P for the system as a whole. In principle, the "surface" may be located at an x value such that these two areas compensate for one another. That is, x_0 may be chosen so that the two shaded areas in the figure are equal.

This is where the trouble begins! Generally speaking, the kind of profile sketched in Fig. 7.11 will be different for each property considered. Therefore, we

FIGURE 7.11 *Variation of some general property P with perpendicular distance from the surface in the vicinity of an interface between two phases, α and β.*

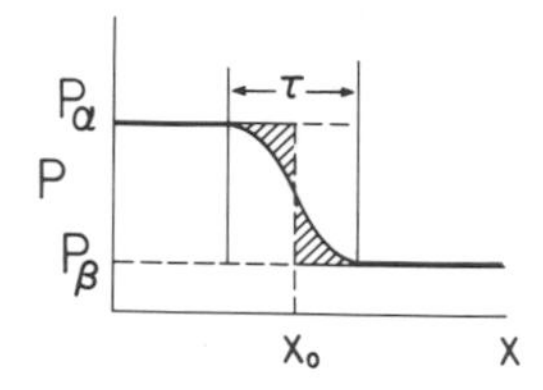

may choose x_0 to accomplish the compensation discussed herein for one property, but this same line will divide the profiles of other properties differently. The difference between the "overestimated" property and the "underestimated" one accounts for the "surface excess" of this property.

From the point of view of thermodynamics—which is oblivious to details at the molecular level—the dividing boundary may be placed at any value of x in the range τ. The actual placement of x_0 is governed by consideration of which properties of the system are most amenable to thermodynamic evaluation. More accurately, that property which is least convenient to handle mathematically may be eliminated by choosing x_0 so that the difficult quantity has a surface excess of zero.

For example, if the property in Fig. 7.11 were G and the dividing surface were placed so that the two shaded regions would be equal, then there would be no surface excess G: The last term in Eq. (24) would be zero. The Gibbs free energy is convenient to work with, however, so such a choice for x_0 would not be particularly helpful. Until now, we have not had any reason to identify the surface of physical phases with any specific mathematical surface. We had not, that is, until Eq. (38) was reached. Now things are somewhat different.

Suppose the property represented in Fig. 7.11 is the number of moles of solvent per unit area in a slice of solution at some value of x. This quantity will clearly undergo a transition in the vicinity of an interface. We elect to choose x_0 so that the shaded areas are equal when this is the quantity of interest. This placement of the dividing surface means

$$\Gamma_1 = 0 \tag{39}$$

With this situation, Eq. (38) becomes

$$d\gamma = -\Gamma_2 \, d\mu_2 \tag{40}$$

The physical significance of Γ_2 is now determined by the arbitrary placement of the mathematical surface which made $\Gamma_1 = 0$. That is, Γ_2 equals the algebraic difference between the "overestimated" and "underestimated" areas of the curve describing moles of solute, when this curve is divided at a location x_0 which makes the surface excess of the solvent zero.

It is important to realize that the mathematical dividing surface just discussed is a *reference level* rather than an actual physical boundary. What is physically represented by this situation may be summarized as follows. Two portions of solution containing an identical number of moles of solvent are compared. One is from the surface region, the other from the bulk solution. The number of moles of solute in the sample from the surface minus the number of moles of solute in the sample from the bulk gives the surface excess number of moles of solute, according to this convention. This quantity, divided by the area of the surface, equals Γ_2. To emphasize that the surface excess of component 1 has been chosen to be zero in this determination, the notation Γ_2^1 is generally used.

It should be evident from the foregoing discussion that the property defined to have zero surface excess may be chosen at will, the choice being governed by the experimental or mathematical features of the problem at hand. Choosing the surface excess number of moles of one component to be zero clearly simplifies Eq.

(38). The same simplification could have been accomplished by defining the mathematical surface so that Γ_2 would be zero, a choice which would obviously de-emphasize the solute. If the total number of moles N, the total volume V, or the total weight W had been the property chosen to show a zero surface excess, then in each case both Γ_1 and Γ_2 (which would be identified as Γ^N, Γ^V, or Γ^W for these three conventions) would have nonzero values. Lastly, note that the surface "excess" is an algebraic quantity which may be either positive or negative, depending on the convention chosen for Γ. A variety of different experimental methods are encountered in the literature to measure "surface excess" quantities; one must be careful to understand clearly what conventions are used in the definition of these quantities.

Equation (40), one form of the Gibbs equation, is an important result because it supplies the connection between the surface excess of solute and the surface tension of an interface. For systems in which γ can be determined, this measurement provides a method for evaluating the surface excess. It might be noted that the finite time required to establish equilibrium adsorption is why dynamic methods (e.g., drop detachment) are not favored for the determination of γ for solutions. At solid interfaces, γ is not directly measurable. However, if the amount of adsorbed material can be determined, this may be related to the reduction of surface free energy through Eq. (40). To understand and apply this equation, therefore, it is imperative that the significance of Γ_2 be appreciated.

Now let us return to the development of Eq. (40). The chemical potential depends on the activity according to the equation

$$\mu_2 = \mu_2^0 + RT \ln a_2 \tag{41}$$

In applying these results to adsorption from solution, the activity equals the pressure or concentration multiplied by the activity coefficient f. Differentiation of Eq. (41) at constant temperature yields

$$d\mu_2 = RT\frac{da_2}{a_2} = RT\, d\ln(fc) \tag{42}$$

This relationship may also be applied to the adsorption of gases by replacing concentration by gas pressure and continuing to use the appropriate activity coefficient. We shall return to the application of this result to the adsorption of gases in the following chapter.

For adsorption from dilute solutions, the activity coefficient approaches unity, in which case the combination of Eqs. (40) and (42) leads to the result

$$\Gamma_2^{\,1} = -\frac{c}{RT}\left(\frac{d\gamma}{dc}\right)_T = -\frac{1}{RT}\left(\frac{d\gamma}{d\ln c}\right)_T = -\frac{1}{2.303RT}\left(\frac{d\gamma}{d\log c}\right)_T \tag{43}$$

This form of the Gibbs equation shows that the slope of a plot of γ versus the logarithm of concentration (or activity if the solution is nonideal) measures the surface excess of the solute. It might also be noted that the choice of units for concentration is immaterial at this point.

7.8 THE GIBBS EQUATION: EXPERIMENTAL RESULTS

Surface tensions for the interface between air and aqueous solutions generally display one of the three forms indicated schematically in Fig. 7.12. The type of behavior indicated by curves 1 and 3 indicates positive adsorption of the solute. Since $d\gamma/dc$ and, therefore, $d\gamma/d \ln c$ are negative, $\Gamma_2{}^1$ must be positive. On the other hand, the positive slope for curve 2 indicates a negative surface excess or a surface depletion of the solute. Note that the magnitude of negative adsorption is also less than that of positive adsorption.

Curve 1 in Fig. 7.12 is the type of behavior characteristic of most un-ionized organic compounds. Curve 2 is typical of inorganic electrolytes and highly hydrated organic compounds. The type of behavior indicated by curve 3 is shown by soluble amphipathic species, especially ionic ones. The break in curve 3 is typical of these compounds; however, this degree of sharpness is observed only for highly purified compounds. If impurities are present, the curve will display a slight dip at this point. All three of these curves correspond to relatively dilute solutions. At higher concentrations, effects other than adsorption may lead to departures from these basic forms. We shall say a bit more about adsorption from binary solutions over the full range of compositions in Sec. 7.11.

For the limit as $c \to 0$, curve 1 may be presented by the equation of a straight line:

$$\gamma = \gamma_0 - mc \tag{44}$$

where m is the initial slope of the line. This is the same as

$$\pi = mc \tag{45}$$

From Eq. (44), $d\gamma/dc = -m$ and from (45) $c = \pi/m$; therefore (43) may be written

$$\Gamma_2{}^1 = \frac{\pi}{RT} \tag{46}$$

Recalling the definition of $\Gamma_2{}^1$ provided by Eq. (37), we see that (46) may also be written

$$\pi A = n_2{}^s RT \tag{47}$$

FIGURE 7.12 *Three types of variation of γ with c for aqueous solutions. Curve 1, simple organic solutes; curve 2, simple electrolytes; and curve 3, amphipathic solutes.*

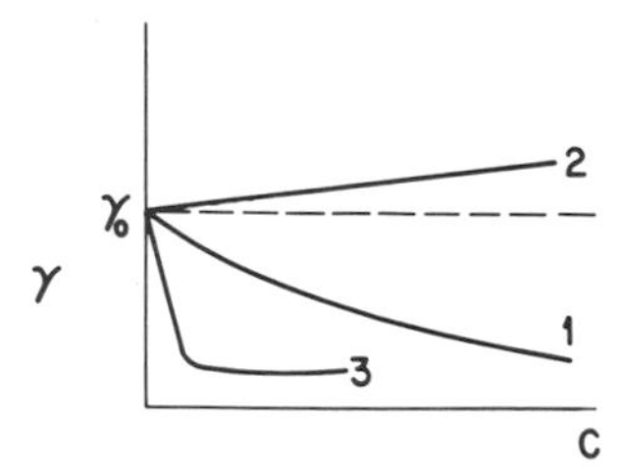

the two-dimensional ideal gas law again! Those carboxylic acids containing less than 12 carbons in the alkyl chain for which results were presented in Fig. 7.7 were investigated by this method. This same analysis also applies to the branch of curve 3 in Fig. 7.12 as $c \to 0$.

Curve 2 in Fig. 7.12 indicates a negative surface excess of simple electrolytes. This means that portions of solution from both the surface and bulk regions which contain the same number of moles of solvent will have more solute in the bulk region than at the surface. Obviously, the surface is enriched over the bulk in solvent, a fact that is easily understood when the hydration of the ions is considered. Water molecules interact extensively with ions, a fact that accounts in part for the excellent solvent properties of water for ionic compounds. To move an ion directly to the air–water interface would require considerable energy to partially dehydrate the ion. Accordingly, the first couple of molecular diameters into the solution will be a layer of essentially pure water, the ions being effectively excluded from this region. The surface tension is not that of pure water, but is increased slightly due to the small surface deficiency of solute. Other highly solvated solutes such as sucrose also show this effect.

Curve 3 in Fig. 7.12 applies primarily to amphipathic species. Most long-chain amphipathic molecules are insoluble unless the hydrophobic alkyl part of the molecule is offset by an ionic head or some other suitably polar head such as a polyethylene oxide chain, $-(CH_2CH_2O)_n-$. Like their insoluble counterparts, these substances form an oriented monolayer even at low concentrations. On the low concentration side of the break in the curve, the slope of a semilogarithmic plot of γ versus c is quite linear, indicating a constant surface excess of the solute in this concentration region. Figure 7.13 shows some actual experimental plots of type 3 for the ether which consists of a dodecyl chain and a hexaethylene oxide chain ($n = 6$) in the general formula just given. The slope of the 25°C plot is about -16.7 dynes cm^{-1}, which means that

$$\Gamma_2{}^1 = -\frac{-16.7}{(2.303)(8.31\times 10^7)(298)} = 2.93\times 10^{-10} \text{ mole cm}^{-2} \tag{48}$$

FIGURE 7.13 *γ versus $\log_{10} c$ for the dodecyl ether of hexaethylene oxide at* (1) 15°C, (2) 25°C, *and* (3) 35°C. [*J. M. Corkill, J. F. Goodman, and R. H. Ottewill,* Trans. Faraday Soc. **57**:1927 (1961).]

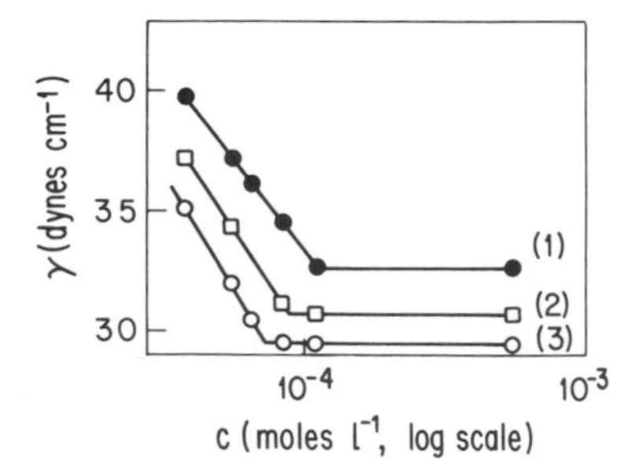

or

$$\sigma = \frac{1 \text{ cm}^2}{2.93 \times 10^{-10} \text{ mole}} \times \frac{1 \text{ mole}}{6.02 \times 10^{23} \text{ molecules}} \times \frac{10^{16} \text{ Å}^2}{1 \text{ cm}^2}$$

$$= 56.7 \text{ Å}^2 \text{ molecule}^{-1} \tag{49}$$

The polar heads of these solute molecules are much bulkier than those of the simple amphipathic molecules whose insoluble monolayers we discussed earlier. This is especially true when the hydration of the ether oxygens is considered. Therefore, this area probably represents a highly condensed surface layer.

The break in curve 3 is characteristic of these plots for soluble amphipathic molecules. We shall discuss the significance of this feature in the next section. First, however, it is instructive to consider the effect of dissociation on the adsorption of amphipathic substances, since many of the compounds which behave according to curve 3 are electrolytes. We shall consider only the case of strong 1 : 1 electrolytes; for weak electrolytes the equilibrium constant for dissociation must be considered.

If an ionic solute is totally dissociated into positive and negative ions, then its activity is given by

$$a_{\text{MR}} = a_{\text{M}} a_{\text{R}} \simeq c_{\text{M}} c_{\text{R}} \tag{50}$$

where the subscripts M and R refer to the cation and amphipathic anion, respectively. Analogous results would be obtained if the cation were the amphipathic species. The approximation included in Eq. (50) applies to the case in which the activity coefficient equals unity. Substituting this result into Eq. (43) gives

$$\Gamma_{\text{MR}}^1 = -\frac{1}{2RT}\left(\frac{d\gamma}{d \ln c}\right)_T \tag{51}$$

The assumption that no other electrolyte is present is implicit in this result. Now let us consider what happens when the system also contains a nonamphipathic electrolyte with a common ion to the surface-active electrolyte.

If a second electrolyte MX is present in addition to MR, then Eq. (38) must be written

$$-d\gamma = \Gamma_{\text{M}}^1 \, d\mu_{\text{M}} + \Gamma_{\text{R}}^1 \, d\mu_{\text{R}} + \Gamma_{\text{X}}^1 \, d\mu_{\text{X}} \tag{52}$$

Now the condition of surface neutrality becomes

$$\Gamma_{\text{M}}^1 = \Gamma_{\text{R}}^1 + \Gamma_{\text{X}}^1 \tag{53}$$

so Eq. (52) may be written

$$-d\gamma = \Gamma_{\text{R}}^1(d\mu_{\text{M}} + d\mu_{\text{R}}) + \Gamma_{\text{X}}^1(d\mu_{\text{M}} + d\mu_{\text{X}}) \tag{54}$$

This result may now be simplified by invoking some previous results. Recalling curve 2 from Fig. 7.12, we know that the surface excess of the X^- ion is likely to be a small negative number which we shall set equal to zero as a first approximation. With this approximation, Eq. (54) becomes

$$-d\gamma \simeq \Gamma_{\text{R}}^1(d\mu_{\text{M}} + d\mu_{\text{R}}) = \Gamma_{\text{R}}^1 RT\left(\frac{dc_{\text{M}}}{c_{\text{M}}} + \frac{dc_{\text{R}}}{c_{\text{R}}}\right) \tag{55}$$

Now let us consider a small change in the concentration of MR while the concentration of MX remains constant and considerably greater than the total MR concentration. Under these conditions, $dc_M = dc_R$ and $c_M \gg c_R$; therefore, $d \ln c_M \ll d \ln c_R$, and Eq. (55) becomes

$$-d\gamma = \Gamma_R^1(RT\,d \ln c) \tag{56}$$

Equations (51) and (56) are thus seen to describe the adsorption of MR in the absence of electrolyte and in the presence of swamping amounts of electrolyte, respectively. It is clear from the difference between these two results that extreme care must be taken in the study of charged monolayers if the effect of the charge on the state of the monolayer is to be properly considered in the interpretation of experimental results.

The difference between Eqs. (51) and (56) may be qualitatively understood by comparing the results with the Donnan equilibrium discussed in Chap. 4. The amphipathic ions may be regarded as restrained at the interface by a hypothetical membrane which is, of course, permeable to simple ions. Both the Donnan equilibrium [Eq. (4.64)] and the electroneutrality condition [Eq. (4.66)] may be combined to give the distribution of simple ions between the bulk and surface regions. As we saw in Chap. 4 (e.g., see Table 4.3), the restrained species behaves more and more as if it were uncharged as the concentration of the simple electrolyte is increased. In Chap. 9, we shall examine the distribution of ions near a charged surface from a statistical, rather than phenomenological, point of view.

We have noted previously that measuring γ as a function of concentration is a convenient means of determining the surface excess of a substance at a mobile interface. In view of the complications arising from charge considerations, the need for an independent method for measuring surface excesses becomes apparent. Some elaborate techniques have been developed which involve skimming a thin layer off the surface of a solution and comparing its concentration with that of the bulk solution. A simpler method for verifying the Gibbs equation involves the use of isotopically labeled surfactants. If the isotope emits a low-energy β particle, the range of the β in water will be very low. Thus, a detector placed just above the surface will count primarily those emissions originating from the surface region. Tritium (^{3}H), for example, emits a 0.0186 Mev β particle with a range in water of only about 1.7×10^{-3} cm, which means that only a negligible fraction of the β particles can travel farther than this in water. In fact, most are absorbed in an even shorter distance, so any ^{3}H β particles detected above an aqueous solution of tritiated surfactant probably originate within approximately 3 μm of the surface. The contribution of the bulk solution to the "background" of the former measurement is made using the same isotope in a compound which is known not to be adsorbed. By such studies, the kinds of effect just described have been investigated and verified.

The surface-active substances we have discussed have been purified, research-quality materials. In practical situations, the cost of synthesizing and purifying such surfactants is prohibitive. The materials commercially used, therefore, are inevitably mixtures. Commercial surfactants originate, for example, from the esterification of sugars or the sulfonation of alkylaryl mixtures. Such mixtures are marketed under a bewildering variety of trade names, and often as members of number- or

TABLE 7.2 *Some families of commercial surfactants showing specific examples from each*[a]

Name of series	General chemical nature	Specific designation ("——") and chemical nature of example	Example
Igepon "——"	Fatty acid amide of methyltaurine	"TN" R = palmityl	$RCON(CH_3)C_2H_4SO_3^-\ Na^+$
Aerosol "——"	Alkyl ester of sulfosuccinic acid	"OT" R = octyl	$CH_2{-}COOR$ | $CH_2{-}COOR$ | $SO_3^-\ Na^+$
Span "——"	Fatty acid esters of anhydrosorbitols	"60" R = stearyl	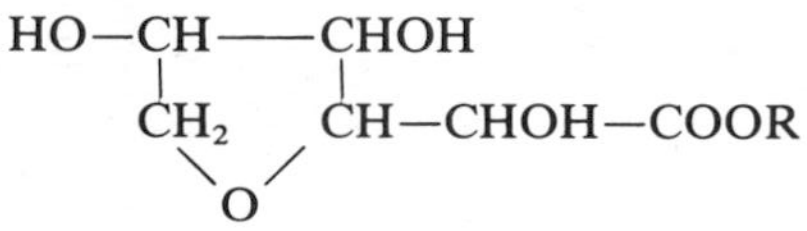 HO—CH——CHOH CH_2 CH—CHOH—COOR (ring closed through O)
Tween "——"	Fatty acid ester and ethylene oxide esters of anhydrosorbitols	"21" $n = 4$, R = lauryl	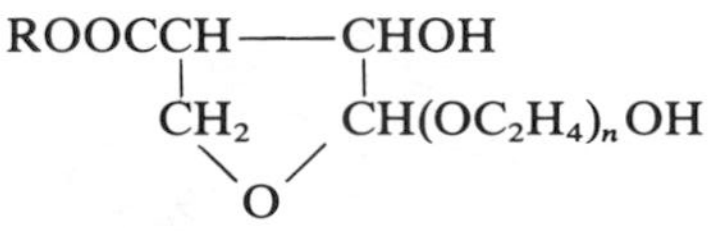 ROOCCH——CHOH CH_2 $CH(OC_2H_4)_nOH$ (ring closed through O)
Triton "——"	Ethylene oxide ethers of alkyl benzene	"X-45" $n = 5$, R = octyl	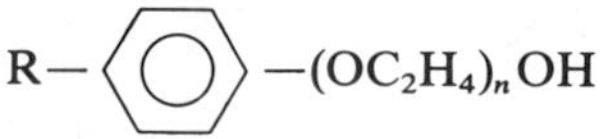 $R{-}C_6H_4{-}(OC_2H_4)_nOH$
Hyamine "——"	Alkylbenzyl dimethyl ammonium salts	"3500" $R = C_{12}{-}C_{16}$	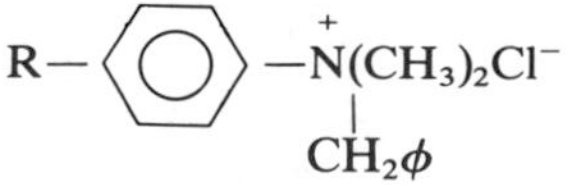 $R{-}C_6H_4{-}\overset{+}{N}(CH_3)_2Cl^-$ | $CH_2\phi$

[a] Data from "Surfactants," in *Encyclopedia of Chemical Technology*, Vol. 19, Wiley–Interscience, New York, 1969.

letter-designated series which correspond, roughly, to a set of homologs. Table 7.2 lists examples of several specific members of such series, along with a brief description of the general nature of the family to which they belong.

7.9 MICELLES AND THE CRITICAL MICELLE CONCENTRATION

One feature of curve 3 in Fig. 7.12 has not yet been discussed: the sharp change in the slope which occurs at low concentrations. Figure 7.13 not only shows this same feature, but also shows how it varies with temperature for a particular compound. Surface tension is not the only property of these solutions which displays such a discontinuity.

It is widely observed that soluble amphipathic substances, both ionic and nonionic, show sharp changes in a variety of properties at a well-defined concentration which is a characteristic of the solute in question. These phenomena are attributed to the association of solute molecules into clusters known as micelles. The concentration at which micellization occurs is known as the critical micelle concentration, generally abbreviated cmc. Figure 7.14 shows the superpositioned curves for a variety of properties versus concentration for sodium dodecyl sulfate solutions. A few remarks for each of the properties will serve to summarize the evidence for micellization, which occurs at a cmc of about 0.008 M at 25°C for this compound.

1. *Osmotic pressure* To a first approximation, we know that π/c is proportional to $1/M$ [Eq. (4.48)]. This is roughly equivalent to saying that the slope of a plot of π versus c is proportional to $1/M$. The decrease in the slope of the osmotic pressure plot at the cmc indicates an increase in the average molecular weight of solute at this point.

FIGURE 7.14 *Schematic illustration of a variety of properties* (κ *= conductivity,* τ *= turbidity,* Λ *= equivalent conductivity,* γ *= surface tension, and* π *= osmotic pressure) of sodium dodecyl sulfate solutions versus concentration.*

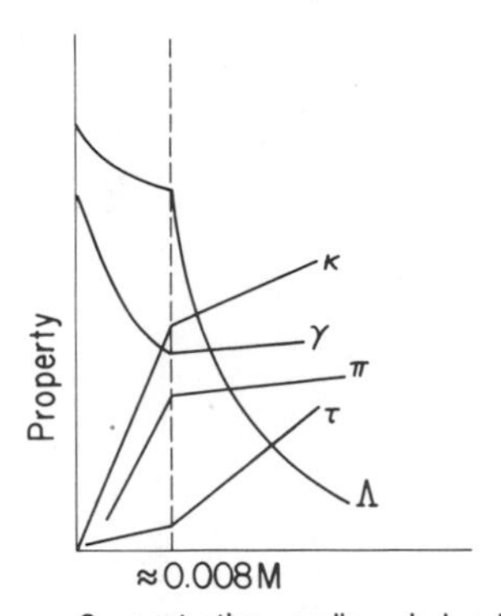

2. *Turbidity* Hc/τ is proportional to $1/M$ to a first approximation [Eq. (5.49)]. This means that the slope of a plot of τ versus c is roughly proportional to M. The break in the curve again corresponds to an increase in the molecular weight of the solute.
3. *Conductivity* κ *and specific conductivity* Λ The conductivity of the solution decreases at the cmc due to the lower mobility of the larger micelles. Dividing by concentration to convert to equivalent conductivity leads to a sharp reduction in the latter quantity at the threshold of micelle formation.

The association of amphipathic ions or molecules to form micelles is an area of considerable interest in both surface and colloid chemistry. Some of the same compounds which display surface activity below their cmc associate to form particles in the colloidal size range above this concentration. Most of the techniques discussed in Chaps. 2 through 5 have been utilized for the characterization of micelles. We shall not consider these results in detail except to note briefly how light scattering has been used in this connection. Equation (5.49) may be modified as follows to apply to the solution which contains micelles:

$$\frac{H(c-c_0)}{\tau-\tau_0}=\frac{1}{M}+2B(c-c_0) \tag{57}$$

where c_0 and τ_0 are the concentration and turbidity, respectively, of the solution at the cmc. The molecular weight of the micelle may be evaluated graphically in the usual way. For charged micelles, the situation is complicated by the fact that the micelle binds a certain number of counterions with the remainder required for electroneutrality distributed in an ion atmosphere surrounding the micelle. The theory has been worked out which relates the degree of aggregation n and the effective charge of the micelle z to the concentration of the amphipathic species and the concentration of the simple electrolyte.

Table 7.3 summarizes some of the results that have been obtained by such studies on systems containing charged micelles. The data show that as the salt concentration increases the cmc decreases, the degree of aggregation increases, and the effective percent ionization decreases. It may also be seen that an increase in the length of the alkyl chain lowers the cmc, increases the degree of aggregation, and lowers the effective percent ionization.

For nonionic micelles such as those formed by alkyl ethers of polyethylene oxide a few generalizations are also possible. Increasing the length of the alkyl chain increases the degree of aggregation and lowers the cmc. Increasing the length of the ethylene oxide chain raises the cmc, but the degree of aggregation appears to be relatively independent of this quantity.

The process of forming a charged micelle from a solution containing 1:1 electrolyte only may be represented by the equation

$$n\mathrm{R}^- + m\mathrm{M}^+ \rightarrow (\mathrm{M}_m\mathrm{R}_n)^{-z} \tag{58}$$

where the anion R^- is the amphipathic species. From this, it follows that $z=n-m$ and the effective fraction ionized equals z/n. The same equation may be used to

TABLE 7.3 *The critical micelle concentration, degree of aggregation, and effective fraction ionization for several surfactants with and without added salt*[a]

Surfactant	Solution	Critical micelle concentration (moles liter^{-1})	Aggregation number, n	Ratio of charge to aggregation number, z/n
Sodium dodecyl sulfate	Water	0.0081	80	0.18
	0.02 M NaCl	0.00382	94	0.14
	0.03 M NaCl	0.00309	100	0.13
	0.10 M NaCl	0.00139	112	0.12
	0.20 M NaCl	0.00083	118	0.14
	0.40 M NaCl	0.00052	126	0.13
Dodecylamine hydrochloride	Water	0.0131	56	0.14
	0.0157 M NaCl	0.0104	93	0.13
	0.0237 M NaCl	0.00925	101	0.12
	0.0460 M NaCl	0.00723	142	0.09
Decyl trimethyl ammonium bromide	Water	0.0680	36	0.25
	0.013 M NaCl	0.0634	38	0.26
Dodecyl trimethyl ammonium bromide	Water	0.0153	50	0.21
	0.013 M NaCl	0.0107	56	0.17
Tetradecyl trimethyl ammonium bromide	Water	0.00302	75	0.14
	0.013 M NaCl	0.00180	96	0.13

[a] J. N. Phillips, *Trans. Faraday Soc.* **51**:561 (1955).

describe the formation of nonionic micelles by letting m and z equal zero. It is instructive to write the equilibrium constant for the reaction described by Eq. (58) for both the ionic and nonionic cases. For ionic micelles,

$$K_{\text{ionic}} = \frac{[M_m R_n^{-z}]}{[M^+]^m [R^-]^n} \tag{59}$$

whereas for nonionic micelles

$$K_{\text{nonionic}} = \frac{[R_n]}{[R]^n} \tag{60}$$

Now let us compare these two equilibrium constants with respect to their order of magnitude. At the cmc, the concentration of both nonionic and ionic micelles may be equated:

$$[M_m X_n^{-z}]_{\text{cmc}} = [R_n]_{\text{cmc}} \tag{61}$$

and Eqs. (59) and (60) may be used as substitutions in (61) to give

$$K_{\text{ionic}}([M^+]^m [R^-]^n)_{\text{cmc}} = K_{\text{nonionic}}[R]^n_{\text{cmc}} \tag{62}$$

If we substitute into this result some typical values for the cmc of ionic and nonionic detergents, the relative magnitudes of the two K values can be determined. Typical values for the cmc of a C_{12} nonionic surfactant are of the order of magnitude of 10^{-4} M, whereas the values for C_{12} ionic substances are closer to 10^{-2} M. Substituting the former value for [R] and the latter for $[R^-]$ and $[M^+]$ in Eq. (62) yields

$$K_{\text{ionic}}(10^{-2})^n(10^{-2})^m = K_{\text{nonionic}}(10^{-4})^n \tag{63}$$

If m and n are taken to be approximately equal (Table 7.3 shows that m is about 85% n) and if the degree of aggregation n is roughly the same for both ionic and nonionic micelles, then Eq. (63) becomes

$$K_{\text{ionic}} \simeq K_{\text{nonionic}} \tag{64}$$

This is a very rough comparison of the micellization process for the two types of surface-active materials. Nevertheless, the difference in the cmc for ionic and nonionic substances is seen to follow primarily from the law of mass action applied to the two processes, whereas the two values of K and, therefore, ΔG^0 are not too different. This last quantity has been shown to be about $1.1kT$ per CH_2 per molecule for both ionic and nonionic compounds.

The fact that—to a first approximation—ΔG^0 is the same for the micellization of both ionic and nonionic surfactants does not mean that the values of ΔH^0 and ΔS^0 for the two processes are approximately equal as well. The heat of micellization may be calculated from the equation

$$\frac{d \ln(\text{cmc})}{dT} = -\frac{\Delta H^0}{RT^2} \tag{65}$$

The data in Fig. 7.13 show that $d \ln(\text{cmc})/dT$ for a nonionic surfactant is negative, which means, according to Eq. (65), that ΔH^0 is positive. For ionic materials, on the other hand, $d \ln(\text{cmc})/dT$ is positive with ΔH^0 negative.

Another interesting property of micelles is their ability to solubilize materials which would otherwise be insoluble in aqueous solutions. Insoluble organic matter, for example, may dissolve in the interior of the micelle even though it shows minimal solubility in water. Certain oil-soluble dyes barely color water, but give vividly colored solutions above the cmc. This solubilization of organic molecules in micelles is known to play an important part in the process of emulsion polymerization.

In discussing micellization, we have intentionally restricted attention to concentrations near the cmc where the micelles are fairly symmetrical in shape and far enough from each other to be treated as independent entities. At higher concentrations, neither of these conditions is met and more complex phase equilibria must be considered. Numerous studies have been published showing the effects of both organic and inorganic additives on the water–surfactant phase diagram. The phase diagrams for multicomponent systems may be quite important in any application where surface-active substances are mixed with several other components.

7.10 THE LANGMUIR EQUATION: THEORY

Throughout most of this chapter, we have been concerned with adsorption at mobile surfaces. In these systems, the surface excess may be determined directly from the experimentally accessible surface tension. At solid surfaces, this experimental advantage is missing. All we can obtain from the Gibbs equation in reference to adsorption at solid surfaces is a thermodynamic explanation for the driving force underlying adsorption. Whatever information we require about the surface excess must be obtained from other sources.

If a dilute solution of a surface-active substance is brought in contact with a large adsorbing surface, then extensive adsorption will occur with an attendant reduction in the concentration of the solution. To meet the requirement of a large surface available for adsorption, the solid—which is called the adsorbent—must be finely subdivided. From the analytical data describing the concentration change in the solution as well as a knowledge of the total amount of solid and solution equilibrated, it is possible to determine the amount of solute adsorbed—which is called the adsorbate—per unit weight of adsorbing solid. If the specific area of the latter is known, then the results may be expressed as amount adsorbed per unit area. These studies are generally conducted at constant temperature, and the results—which relate the amount of material adsorbed to the equilibrium concentration of the solution—describe what is known as the adsorption isotherm.

One isotherm that is both easy to understand theoretically and widely applicable to experimental data is due to Langmuir and is known as the Langmuir isotherm. In the following chapter, we shall see that the same function often describes the adsorption of gases at low pressures, with pressure substituted for concentration as the independent variable. We shall discuss the derivation of Langmuir's equation again in Chap. 8 specifically as it applies to gas adsorption. Now, however, adsorption from solution is our concern. In this section we consider only adsorption from dilute solutions. In the following section adsorption over the full range of binary solution concentrations is also mentioned.

Suppose we imagine a dilute solution in which both the solvent (component 1) and the solute (component 2) have molecules which occupy the same area when they are adsorbed on a surface. The adsorption of solute may then be schematically represented by the equation

$$\begin{aligned}&\text{adsorbed solvent} + \text{solute in solution}\\&\qquad \rightarrow \text{adsorbed solute} + \text{solvent in solution}\end{aligned} \tag{66}$$

The equilibrium constant for this reaction may be written as

$$K' = \frac{a_2{}^s a_1{}^b}{a_1{}^s a_2{}^b} \tag{67}$$

where a stands for the activity of the species and the superscripts s and b signify surface and bulk values, respectively. Next, let us assume that the two-dimensional surface solution is ideal, an assumption which enables us to replace the activity at the surface by the mole fraction at the surface x^s:

$$K' = \frac{x_2{}^s a_1{}^b}{x_1{}^s a_2{}^b} \tag{68}$$

Since the surface contains only two components, $x_1^s + x_2^s = 1$ and Eq. (68) becomes

$$K' = \frac{x_2^s a_1^b}{(1 - x_2^s) a_2^b} \quad (69)$$

Equation (69) may be rearranged to give

$$x_2^s = \frac{K' a_2^b / a_1^b}{K' a_2^b / a_1^b + 1} \quad (70)$$

In dilute solutions, the activity of the solvent is essentially constant so the ratio K'/a_1^b may be defined to equal a new constant K, in terms of which Eq. (70) becomes

$$x_2^s = \frac{K a_2^b}{K a_2^b + 1} \quad (71)$$

This is one form of the Langmuir adsorption isotherm.

An equivalent form of the Langmuir equation expressed in slightly different variables is obtained by multiplying both x_1^s and x_2^s in Eq. (68) by the total area of the surface A. We have already postulated that both the solvent and solute occupy equal areas on the surface. Therefore, $x_1^s \cdot A$ equals the fraction of the surface occupied by solvent, θ; since $\theta_1 + \theta_2 = 1$, the ratio θ_2/θ_1 rearranges the same as the ratio x_2^s/x_1^s to give

$$\theta_2 = \frac{K a_2^b}{K a_2^b + 1} \quad (72)$$

In this form, the Langmuir equation shows how the fraction of surface adsorption sites occupied by solute increases as the solute activity in solution increases. From now on, we shall drop the subscript 2 and the superscript b. Since Eq. (72) is written solely in terms of the solute, these designations are redundant.

Two limiting cases are of special interest:

1. At infinite dilution $a \to 0$ and Eq. (72) becomes

$$\theta = Ka \quad (73)$$

2. If $Ka \gg 1$, Eq. (72) becomes

$$\theta = 1 \quad (74)$$

Equation (73) shows that θ increases linearly with an initial slope that equals K. This slope will be larger the farther to the right the equilibrium represented by Eq. (66) lies. At higher concentrations, Eq. (74) indicates that saturation of the surface with adsorbed solute is achieved. Figure 7.15a shows how these two limiting conditions affect the appearance of the isotherm.

Experimentally, one does not measure the fraction of sites containing adsorbed solute directly. Instead, it is either the number of moles of solute adsorbed per unit weight of adsorbent n_2^s/w or the number of moles per unit area of adsorbent n_2^s/A

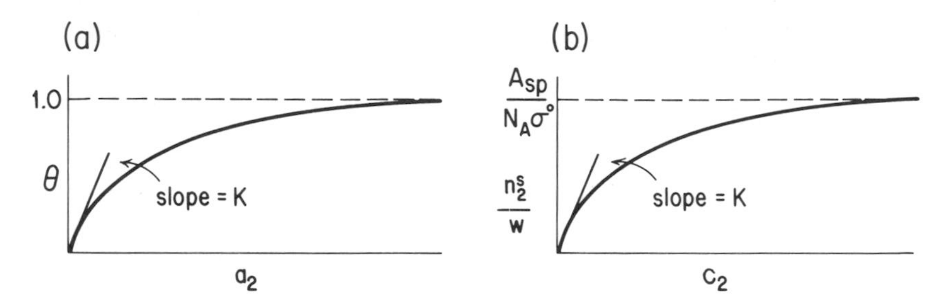

FIGURE 7.15 *Schematic plots of the Langmuir equation showing the significance of the initial slope and the saturation value of the ordinate. (a) The fraction covered is plotted versus solute activity. (b) The number of moles of solute adsorbed per unit weight of adsorbent is plotted versus concentration.*

which are measured. These quantities are related by the equation

$$\frac{n_2^s}{A} A_{sp} = \frac{n_2^s}{w} \tag{75}$$

where A_{sp} is the specific area of the adsorbent (see Sec. 1.2). The fraction covered is related to these quantities as follows:

$$\theta = \frac{n_2^s}{A} N_A \sigma^0 = \frac{n_2^s N_A \sigma^0}{w A_{sp}} \tag{76}$$

where N_A is Avogadro's number and σ^0 is the area occupied per molecule. The level at which saturation adsorption occurs may be identified with $\theta = 1$. Therefore, Eq. (76) shows the saturation values of the usual ordinates to be either

$$\left(\frac{n_2^s}{w}\right)_{sat} = \frac{A_{sp}}{N_A \sigma^0} \tag{77}$$

or

$$\left(\frac{n_2^s}{A}\right)_{sat} = \frac{1}{N_A \sigma^0} \tag{78}$$

Since the entire derivation of the Langmuir isotherm assumes dilute solutions, the concentration c of the solute rather than the activity is generally used in presenting experimental results. Figure 7.15b shows how actual experimental data might appear.

7.11 THE LANGMUIR EQUATION: APPLICATION TO RESULTS

Many systems which definitely do not conform to the Langmuir assumptions—the adsorption of polymers, for example—nevertheless display experimental isotherms which resemble Fig. 7.15. Although these can be fitted to Eq. (72), the significance of the constants is dubious. Therefore, the Langmuir equation is often written as

$$m\left(\frac{n_2^s}{w}\right) = \frac{(m/b)c}{(m/b)c+1} \tag{79}$$

where m and m/b are regarded simply as empirical constants. A method for obtaining the numerical values for these constants from experimental data is easily seen by rearranging Eq. (79) to the form

$$\frac{c}{n_2^s/w} = mc + b \tag{80}$$

This form suggests that a plot of $c/(n_2^s/w)$ versus c will be a straight line of slope m and intercept b.

If the experimental system matches the model, then the values of m and b can be assigned a physical significance by comparing Eq. (79) with (72) and (76):

$$m = \frac{N_A \sigma^0}{A_{sp}} \tag{81}$$

and

$$\frac{m}{b} = K \tag{82}$$

If the model does not apply, these constants are treated merely as empirical parameters which describe the adsorption isotherm.

When the model does apply, the experimental value of m permits A_{sp} to be evaluated if σ^0 is known, or σ^0 to be evaluated if A_{sp} is known. It is often difficult to decide what value of σ^0 best characterizes the adsorbed molecules at a solid surface. Sometimes, therefore, this method for determining A_{sp} is calibrated by measuring σ^0 for the adsorbed molecules on a solid of known area, rather than relying on some assumed model for molecular orientation and cross section.

It might also be noted that K' [Eq. (67)] may be related to ΔG^0 for the adsorption process if the model applies to the experimental system. Therefore, from studies of adsorption at different temperatures, values of ΔH^0 and ΔS^0 may be determined for the process described by Eq. (66). It must be emphasized that compliance with the form predicted by the Langmuir isotherm is not a sensitive test of the model; therefore, interpretations of this kind must be used cautiously.

Figure 7.16a is a plot of some actual experimental data for the adsorption of benzene from heptane on carbon black. This system suggests an approach toward a saturation value, so the data are replotted according to Eq. (80) in Fig. 7.16b. The slope of the initial linear portion of this figure was drawn to be 7.78 g C (mole benzene)$^{-1}$. This sample of carbon black was known to have a specific surface area of 325 m^2 g^{-1}. Therefore, if Eq. (81) is applied to the slope, a value of 42 Å^2 per molecule is obtained for the area occupied by the benzene. This is a very reasonable figure for the area of the benzene ring. Other studies suggest about 46 Å^2 for the area occupied by a heptane molecule, so the requirement of similar size molecules is met approximately by this system. Similar studies conducted with naphthalene and

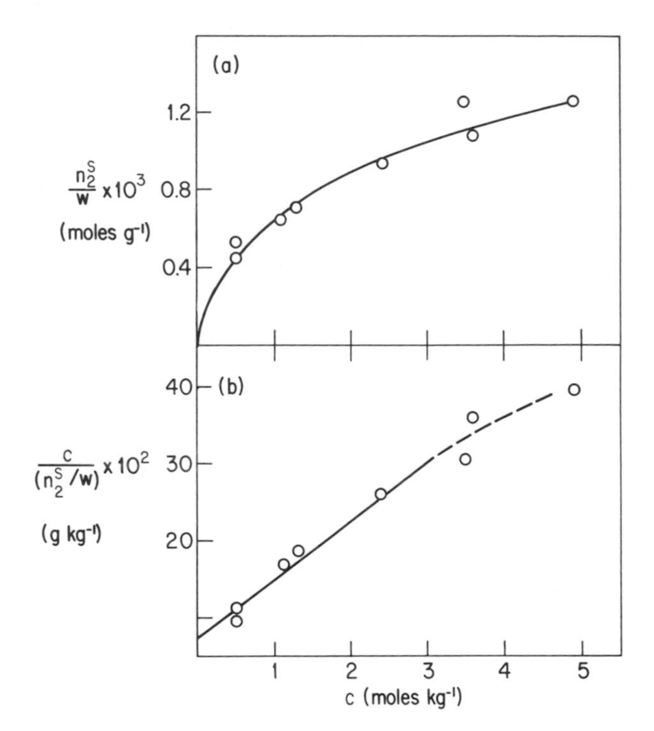

FIGURE 7.16 *The adsorption isotherm for benzene on carbon from heptane solution. (a) The data are plotted as n_2^s/w versus c; (b) plotted according to Eq.* (80). *[Data from M. van der Waarden,* J. Colloid Sci. **6**:443 (1951).]

anthracene indicate areas of about 67 and 83 Å², respectively, for these two substances. These values for σ^0 are not unreasonable even though the assumption that the areas of adsorbed solute and solvent be comparable is clearly getting worse and worse.

In summary, adsorption from dilute solutions frequently displays the qualitative form required by the Langmuir equation. If this form is observed, it may be quantitatively described by Eq. (80) in which m and b are empirical constants. Sometimes there may be a justification for further interpretation of these parameters in terms of the theoretical model.

We should not be too surprised that the Langmuir equation often yields only an empirical isotherm. There are several reasons why real systems are likely to deviate from the theoretical model:

1. The adsorption process described by Eq. (66) is a complex one, involving several different kinds of interactions: solvent–solute, solvent–adsorbent, and solute–adsorbent.
2. Few solid surfaces are homogeneous at the molecular level.

3. Few monolayers are ideal.
4. Our interest often extends beyond the region of dilute concentrations.

We shall briefly comment on each of these limitations.

In discussing adsorption from solution, there is nothing that can be done about the multiplicity of possible interactions, except possibly to avoid systems in which highly specific interactions are to be expected. In the next chapter, we shall again discuss the Langmuir isotherm as it applies to the adsorption of gases. In that case, there are considerably fewer interactions involved in the adsorption process, making the latter more amenable to analysis.

The assumption of surface homogeneity is one that was not explicitly stated in deriving the Langmuir equation. It is essential, however, otherwise a different value of K would apply to Eq. (66) at various places on the surface. Attempts to deal with surface heterogeneity have been undertaken, but this enterprise seems more likely to be successful for gas adsorption rather than for adsorption from solution because the variety of interactions which must be considered is less in the former than in the latter. There is an equation—known as the Freundlich isotherm—that may be derived by assuming a certain distribution function for sites having different ΔG^0 values for the process represented by Eq. (66) and assuming Langmuir adsorption at each type of site. The Freundlich isotherm is given by the expression

$$\theta = ac^{1/n} \tag{83}$$

in which a and n are constants with $n > 1$. This equation was in use long before the interpretation of a certain distribution of sites was assigned to it. Therefore, it is best regarded as an empirical isotherm, the constants for which may be evaluated from the slope and intercept of a log–log plot of θ versus c. The Freundlich isotherm is no cure-all for surface heterogeneity: Its theoretical derivation depends on a highly specific distribution of site energies. In addition, the Langmuir equation gives adequate results in many cases where surface heterogeneity is known to be present.

In the Langmuir derivation, the adsorbed molecules are allowed to interact with the adsorbent but not with each other: The adsorbed layer is assumed to be ideal. This necessarily limits adsorption to a monolayer. Once the surface is covered with adsorbed molecules, it has no further influence on the system. The assumption that adsorption is limited to monolayer formation was explicitly made in writing Eqs. (77) and (78) for the saturation value of the ordinate. It is an experimental fact, however, that adsorption frequently proceeds to an extent that exceeds the monolayer capacity of the surface for any plausible molecular orientation at the surface. That is, if monolayer coverage is postulated, the apparent area per molecule is only a small fraction of any likely projected area of the actual molecules. In this case, the assumption that adsorption is limited to the monolayer fails to apply. A model based on multilayer adsorption is indicated in this situation. This is easier to handle in the case of gas adsorption, so we shall defer until the following chapter a discussion of multilayer adsorption.

Next, let us consider adsorption from solutions which are not infinitely dilute. Suppose, for example, that adsorption is studied over the full range of binary liquid concentrations. Figure 7.17 is an example of such results for the benzene–ethanol

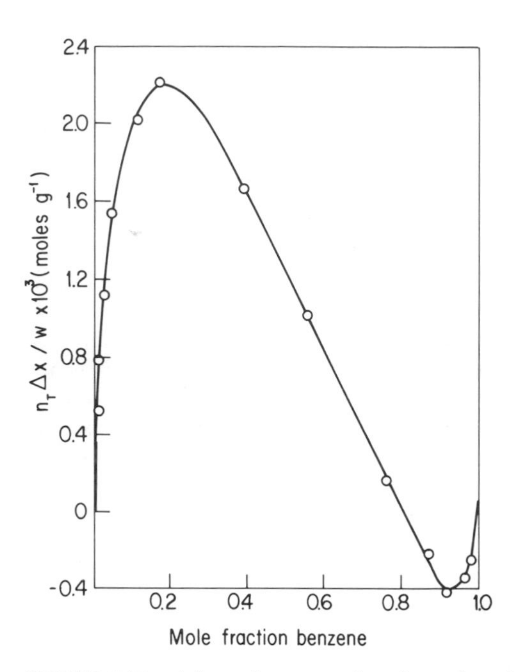

FIGURE 7.17 *Adsorption on carbon from the ethanol–benzene system. The ordinate equals the total number of moles of solution times the change in solution mole fraction per unit weight of carbon. [Data from F. E. Bartell and C. K. Sloan,* J. Am. Chem. Soc. **51**:1643 (1929).]

system adsorbed on carbon. At first, these results appear quite bewildering, displaying maximum, minimum, and negative adsorption. Recall, however, that what is actually measured is an isotherm of concentration change. The observed change in concentration is then expressed as moles of solute adsorbed. In a totally different range of solution concentrations, the solvent rather than the solute may adsorb. The associated change in the solution would then be an increase in solute concentration or the apparent negative adsorption of solute. A curve like that shown in Fig. 7.17 should, therefore, be understood as a composite of two distinctly different isotherms. A good deal of work has been done with composite isotherms, particularly toward separating them into individual isotherms. A summary of this kind of research can be found in the book by Kipling [4].

7.12 APPLICATIONS OF ADSORPTION FROM SOLUTION

No discussion of adsorption from solution is anywhere near complete unless it includes some indication of the enormous practical applicability of this topic. As a matter of fact, the examples we shall briefly consider—detergency and flotation—encompass a wide variety of concepts from almost all areas of surface and colloid chemistry. We have chosen to stress principles rather than applications, however, so these subjects will receive an amount of attention which belies their actual importance.

It is impossible to do justice to the complex phenomena of detergency and flotation in a few paragraphs. All we can do is point out some of the ways in which the principles of colloid and surface chemistry apply in these areas. There are several ways in which detergency and flotation phenomena resemble one another:

1. Both terms give simple names to processes involving many different steps. The more familiar of the two, detergency, may be defined as the process whereby some unwanted foreign matter is removed from a substrate by a combination of chemical treatment, temperature, and mechanical agitation. Flotation is the process by which a specific mineral component of an ore mixture is separated from other components (called "gangue") by being concentrated in the froth of an aerated slurry. Chemical additives and mechanical forces are involved here also.
2. In actual practice, both detergency and flotation deal with systems which are terribly difficult to idealize by any sort of model. In a laundering operation, for example, there will be present a variety of different fabric surfaces (cotton, polyester, etc.), different kinds of foreign matter (particulate, oily, etc.), and different chemical additives (detergent, inorganic phosphate, fluorescent whitening agents as well as the solvent, water). In flotation, all three states of matter—solid, liquid, and gas—are involved and each of these involves several chemical components. The ore is a complex mixture of minerals (assumed to be crushed to such an extent that each particle is a different phase), the air is a mixture of gases (including chemically reactive oxygen), and the liquid contains at least three deliberately added reagents (known as regulator, collector, and frother) in addition to whatever dissolved minerals are present in the water.
3. A third point of resemblance between detergency and flotation (perhaps redundant in view of what has already been said) is that both have developed largely by empirical research with (partially satisfactory) explanations trailing far behind the actual practice.

With this much general background, let us now consider these two processes separately. In discussing detergency, we must first examine the availability of the surfactant. Weak acid soaps form insoluble compounds with Ca^{2+}, for example, and are converted to insoluble molecular acids at low pH levels. One of the reasons for the addition of inorganic phosphate to laundry products is to prevent or minimize these reactions. In this discussion we shall assume that the impurity has not been imbibed into the interior of the fiber (soaking might help if it has) and that it is a semiliquid soiled spot rather than a solid contaminant with which we are dealing. One advantage of washing this type of soiled material at high temperatures is that the viscosity of the oily spot is lowered so that the shape of these drops is more readily altered.

The process of removing an oily drop from a solid substrate may be described in terms of the work of adhesion, given by Eq. (6.73). Applying this idea to the separation of oil (O) from solid (S) gives

$$W = \gamma_{OW} + \gamma_{SW} - \gamma_{OS} \tag{84}$$

where the subscript W describes the aqueous solution. For the separation to be spontaneous in the thermodynamic sense, this quantity (ΔG) must be negative. Positive surface excesses of surfactant molecules at the interface between the aqueous phase and the oil and/or the solid will lower γ for these surfaces. This change is a favorable one for the process of removing foreign matter.

In addition, the contact angle between an oil spot and a solid surface to be cleaned may be a contributing factor in detergency. For example, Fig. 7.18b and 7.18d illustrate schematically two different situations for an oily drop being lifted off a substrate by currents in the adjacent phase. The contact angles θ_1 between the drop and the substrate are assumed to be the same at "lift off" (b and d) as in the quiescent state (a and c). It is evident from the figure, however, that the necking of the drop for $\theta_1 < 90°$ is likely to leave a residue whereas $\theta_1 > 90°$ would lead to a clean detachment. Young's equation [Eq. (6.65)] may be applied to this situation:

$$\gamma_{OW} \cos \theta_1 = \gamma_{SW} - \gamma_{OS} \tag{85}$$

where θ_1 is measured in the oil drop as shown in Fig. 7.18. Equation (85) shows that $\theta_1 > 90°$ and $\theta_1 < 90°$ correspond to $\gamma_{SW} < \gamma_{OS}$ and $\gamma_{SW} > \gamma_{OS}$, respectively. Any adsorption at the solid–water interface will lower γ_{SW} and, therefore, be conducive to a contact angle which favors the complete "rollback" of the oily spot.

Once the dirty spot is removed from the substrate being laundered, it is important that it not be redeposited. Solubilization of the detached material in micelles of surfactant has been proposed as one mechanism which contributes to preventing the redeposition of foreign matter. Any process which promotes the stability of the detached dirt particles in the dispersed form will also facilitate this. We shall see in Chap. 9 how electrostatic effects promote colloidal stability. The adsorption of ions—especially amphipathic surfactant ions—onto the detached matter assists in blocking redeposition by stabilizing the dispersed particles. Materials such as carboxymethylcellulose are often added to washing preparations since these molecules also adsorb on the detached dirt particles and interfere with their redeposition.

FIGURE 7.18 *Schematic illustration of several configurations of three phases useful in the discussion of detergency and flotation. The shaded region represents the soiled spot in detergency and θ_1 is the relevant contact angle; the shaded region is an air bubble in flotation and θ_2 is the appropriate contact angle.*

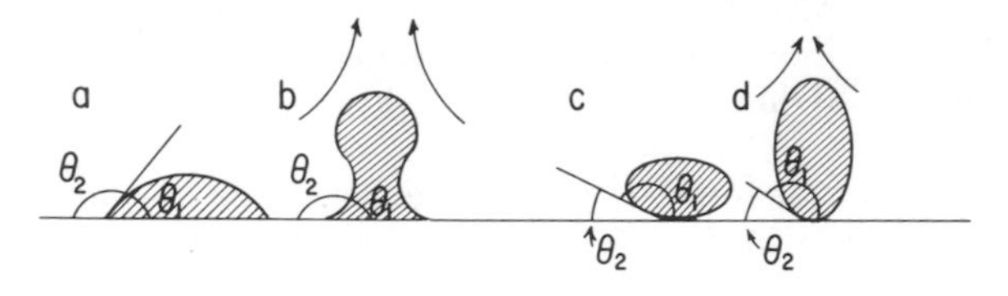

Now let us turn to a brief examination of flotation. Virtually all nonferrous metallic ores are concentrated by the flotation process. Sulfide ores have been studied particularly extensively, although the method has been used with oxides and carbonates as well as such nonmetallic materials as coal, graphite, sulfur, silica, and clay. Something of the order of a billion tons of ore a year are processed in this way.

We shall assume that the ore has been pulverized and mixed with water so that our involvement begins with a slurry known as "pulp." We shall consider, in turn, the chemical nature and the effects of each of the three broad classes of chemicals added in the flotation process.

The first class of chemical additive to be considered is the regulator, a compound which affects the adsorption of the collectors. Regulators, like catalysts, may be positive or negative in their role. For the case in which the collector adsorption is enhanced, the regulator is called an activator; when the effect is negative, it is called a depressant. Regulators are frequently compounds which control the pH and sequester metallic cations which would otherwise compete with the mineral particle surfaces for the surface-active collectors. The pH affects not only the availability of certain collectors but also the charge of the mineral particles (see Sec. 9.2 for a discussion of potential-determining ions). This last consideration plays a role in determining whether the mineral particles will be dispersed as small units (easier to lift by flotation) or whether they will be flocculated. Ammonia, lime, and sources of CN^- and HS^- are commonly used as regulating agents.

Collectors are surface-active additives which adsorb onto the mineral surface and prepare the surface for attachment to an air bubble so that it will float to the surface. Therefore, collectors must adsorb selectively if flotation is to result in any fractionation of the crude ore. In addition, the adsorbed collector must impart a hydrophobic character to the particle surface so that an air bubble will attach to the mineral or vice versa.

Amphipathic substances such as we have discussed throughout this chapter are used as collectors. Alkyl compounds with C_8 to C_{18} chains are widely used with carboxylate, sulfate, or amine polar heads. For sulfide minerals, sulfur-containing compounds such as mercaptans, monothiocarbonates, and dithiophosphates are used as collectors. The most important collectors for sulfides are xanthates, the general formula for which is

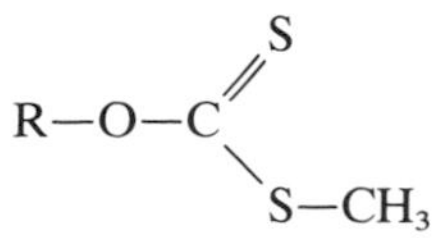

$$\mathrm{R{-}O{-}C}\begin{matrix} \nearrow\!\!\!\!\!\!/\ \mathrm{S} \\ \searrow\ \mathrm{S{-}CH_3} \end{matrix}$$

In the collectors used, R is generally in the C_2 to C_6 range. Xanthates are readily oxidized to dixanthogens, and the extent of this reaction may have a large effect on the efficiency of the collector.

The fundamental role of the collector is to produce a solid surface which is sufficiently hydrophobic that it will attach to an air bubble when the pulp is aerated. Figure 7.18 may also be used to represent this situation, except that for flotation the shaded region is an air bubble. Since contact angles are measured in the liquid phase, the contact angle in the flotation case will be θ_2. For good bubble adhesion, contact angles greater than 90° are preferred. Unlike the parallel situation in detergency, the adhesion of the bubble rather than its detachment is required for the success of the process.

Once again, we may use Young's equation to decide what adsorption situation is most conducive to values of $\theta_2 > 90°$:

$$\gamma_{WA} \cos \theta_2 = \gamma_{AS} - \gamma_{SW} \tag{86}$$

From this, we see the optimum condition corresponds to $\gamma_{SW} \gg \gamma_{AS}$, or extensive adsorption at the air–solid surface and minimum adsorption at the solid–water interface. The hydrophobic nature of the collectors and their chemical affinity for specific solids promote this situation.

The formation of a large bubble which facilitates flotation requires a large area of attachment, or more specifically, a large perimeter of attachment since the three-phase contact boundary occurs along the perimeter. Increasing this perimeter is favored by a positive value of the spreading coefficient [Eq. (6.78)]. In the notation of this problem (air spreading on solid in water), the spreading coefficient equals

$$S_{A/S} = \gamma_{SW} - (\gamma_{AW} + \gamma_{SA}) \tag{87}$$

The collector lowers γ_{SA}, an effect favorable to a positive spreading coefficient.

Finally, the frothing agents are intended to stabilize the mineral-laden foam at the surface of the aeration tank until it can be scooped off. Alkyl or aryl alcohols in the C_5 to C_{12} range are typical frothers. We have already seen how the long-chain members of this series form monolayers at the air–water surface. This lowers γ_{AW}, which is beneficial to the stability of the foam, and also favors a large contact angle (if $\theta_2 > 90°$), positive spreading, and large bubbles for flotation. Neither the collector nor the frother is adsorbed exclusively at the solid–air or the water–air surface where their respective effects would be greatest. To a certain extent these two classes of additives compete with each other for adsorption sites. Therefore, conditions under which each produces the maximum effect are difficult to achieve, so compromise conditions in which the net effect is optimized are sought.

Another aspect of the frothers used is the fact that they form fairly condensed and, therefore, relatively viscous slow-draining films. In addition to thermodynamic considerations, then, kinetic factors are also important in stabilizing the froth.

A number of additional applications of the ideas of this chapter could be profitably considered if space permitted. Included among these are adhesives, lubricants, foams, emulsions, waterproofing, and the recovery of oil from the pores of rocks. Like detergency and flotation, these topics involve a variety of surface and colloid phenomena. The interested reader will find an introduction to these fields in some of the references listed for this chapter, especially Adamson [1], Davies and Rideal [2], and Osipow [7].

REFERENCES

1. A. W. Adamson, *Physical Chemistry of Surfaces* (2nd ed.), Wiley–Interscience, New York, 1967.

2. J. T. Davies and E. K. Rideal, *Interfacial Phenomena*, Academic Press, New York, 1961.
3. G. L. Gaines, *Insoluble Monolayers at Liquid–Gas Interface*, Wiley–Interscience, New York, 1966.
4. J. J. Kipling, *Adsorption from Solutions of Nonelectrolytes*, Academic Press, New York, 1965.
5. V. I. Klassen and V. A. Mokrousov, *An Introduction to the Theory of Flotation*, Butterworth, London, 1963.
6. V. K. LaMer (ed.), *Retardation of Evaporation by Monolayers*, Academic Press, New York, 1962.
7. L. I. Osipow, *Surface Chemistry*, Van Nostrand–Reinhold, Princeton, New Jersey, 1962.
8. A. M. Schwartz, "The Physical Chemistry of Detergency," in *Surface and Colloid Science*, Vol. 5 (E. Matijević, ed.), Wiley–Interscience, New York, 1972.

PROBLEMS

1. By analogy with the behavior of gases, monolayers are expected to expand to cover an entire surface. When the underlying liquid is flowing, the motion supplies a natural barrier to spreading at the edge of the monolayer. Use this concept to interpret the following bucolic scene: "On a calm day an observer who finds the right place on a stream or a river will see an unobtrusive yet startling phenomenon, a line on the surface of the water. The line may lie still, or it may contort itself, one way and another, in response to eddies. Very likely he will think a spider thread has fallen onto the water and try to cut it with his canoe paddle. As the disturbance caused by the cutting fades, the line reappears, mended and whole."* Use the concepts of this chapter and Chap. 2 to discuss the existence of the bulge or line at the edge of the monolayer.
2. In some general chemistry laboratory courses, the following experiment is done to evaluate Avogadro's number. A watch glass is filled to the brim with water and then a solution of stearic acid in benzene is slowly deposited dropwise on the surface until such time that a drop is added which "will not spread out, but will instead form a thick, lens-shaped layer."† What is the name of the film pressure at the "end point" of this experiment? What would be a reasonable estimate for σ at this pressure? Estimate the number of 0.005-ml drops of stearic acid solution ($c = 0.200$ g liter^{-1}) needed to form a monolayer on a watch glass with a 14-cm diameter. Outline how data such as these could be interpreted to lead to a value for N_A. Discuss some of the sources of error in this experiment.
3. Isotherms of π versus σ at both 15 and 25°C were studied for oleic acid on a substrate of pH 2.0 using a high sensitivity film balance.‡ The following results

* C. W. McCutchen, *Science* **170**:61 (1970).

† J. B. Ifft and J. L. Roberts, Jr., *Frantz/Malm's Essentials of Chemistry in the Laboratory* (3rd ed.), Freeman, San Francisco, California, 1975.

‡ R. E. Pagano and N. L. Gershfeld, *J. Colloid Interface Sci.* **41**:311 (1972).

were obtained:

σ (Å^2 molecule^{-1})	π (dynes cm^{-1})	
	at 15°C	at 25°C
10,000	0.028	0.037
8,000	0.035	0.040
6,000	0.044	0.052
4,000	0.055	0.071
3,000	0.058	0.086
2,000	0.076	0.095
1,000	0.075	0.095
500	0.074	0.095

Prepare a plot of π versus σ from these data. What is the apparent significance of the discontinuity in the curves? What quantity can be evaluated from the temperature variation of this discontinuity? Estimate this quantity from the available data.

4. The accompanying data give σ values corresponding to different film pressures for monolayers of various lecithins spread on 0.1 M NaCl at 22°C*:

π (dynes cm^{-1})	σ (Å^2 molecule^{-1})				
Lecithin:	Dibehenoyl	Distearoyl	Dipalmitoyl	Dimyristoyl	Dicapryl
2	51.7	53.3	96.7	96.7	99.2
4	50.0	52.5	88.3	90.0	93.8
6	49.2	50.8	82.2	85.0	86.7
8	48.3	50.0	76.7	81.7	82.5
10	47.9	49.5	66.7	77.5	78.3
15	46.7	48.0	53.3	70.8	71.7
20	45.5	46.7	50.0	65.8	65.8
30	45.0	45.0	46.3	58.3	58.3
40	44.7	44.7	45.0	53.8	53.8

Plot π versus σ for these isotherms, and label the apparent two-dimensional phase present for various parts of the curves. Write the structural formulas for each of the lecithins and discuss the features of the curves in terms of the structure of the molecules.

5. If gas densities can be measured as a function of pressure, the molecular weight of a gas may be calculated from the expression

$$M = RT\left(\frac{d}{p}\right)_{\lim p \to 0}$$

Likewise, if surface concentrations (in weight per area) are measured as a function of π, the molecular weight of a solute which forms an insoluble

* M. C. Phillips and D. Chapman, *Biochim. Biophys. Acta* **163**:301 (1968).

monolayer may be determined. D. Romeo and H. L. Rosano* obtained the following data for a monolayer of acetyl lipopolysaccharide on 0.2 M NaCl at 20°C:

c (mg m^{-2})	0.06	0.09	0.11	0.14	0.17	0.23
π (millidynes cm^{-1})	10.3	16.4	20.4	25.9	34.3	50.0

What is the molecular weight of the acetyl lipopolysaccharide?

6. E. G. Cockbain† measured the interfacial tension of the water–decane surface at various concentrations of sodium dodecyl sulfate (NaDS). The experiments were done at 20°C both in the presence and absence of NaCl. Use the suitable form of the Gibbs equation in each case to calculate $\Gamma_R{}^1$ and σ at γ values of 10 and 20 dynes cm^{-1} from the following data:

Pure H_2O		0.1 M NaCl (swamping)	
c_{NaDS} (moles liter^{-1})	γ (dynes cm^{-1})	c_{NaDS} (moles liter^{-1})	γ (dynes cm^{-1})
0.0079	8.5	0.0014	5.2
0.00694	10.8	0.000694	11.7
0.00521	15.3	0.00347	17.4
0.00347	20.8	0.000173	22.7
0.001735	28.3	0.0000867	27.5

Is the variation of σ with interfacial film pressure qualitatively consistent with the expected behavior of monolayers in general? Of charged monolayers in particular?

7. The measurements of the preceding problem were extended to higher concentrations of NaDS. Above a certain concentration the interfacial tension levels off like the data shown in Fig. 7.13. The concentration at which the discontinuity in the γ versus c plot occurs varies with the NaCl content of the system as follows (reference in Problem 6):

c_{NaCl} (moles liter^{-1})	c_{NaDS} (moles liter^{-1})
0	0.0079
0.1	0.0014
0.2	0.001

Discuss the consistency of these observations (a) with other observations described in this chapter and (b) with the physical model for what occurs at this leveling off.

8. M. Blank (in LaMer [6]) has reported the permeability (in cm^3 of gas s^{-1} cm^{-2} surface) of various spread monolayers to water vapor at 25°C. For several

* D. Romeo and H. L. Rosano, *J. Colloid Interface Sci.* **33**:84 (1970).
† E. G. Cockbain, *Trans. Faraday Soc.* **50**:874 (1954).

different RX-type compounds at different π values, the permeabilities are as follows:

R	X	π(dynes cm^{-1})	Permeability $\times 10^3$ (cm^3 s^{-1} cm^{-2})
C_{16}	OH	44	380
C_{18}	OH	44	300
C_{17}	COOH	24	430

Discuss the observed differences in permeability (a) between the two alcohols at the same film pressure, (b) between the two 18-carbon surfactants at different pressures. In your comments, include comparisons of the molecular structure of the surfactants and the efficiencies of these monolayers in retarding evaporation.

9. The pendant drop method has been used to measure the interfacial tension at the surface between mercury and cyclohexane solutions of stearic acid at 30 and 50°C.* Interpret the accompanying data by means of the Gibbs equation to evaluate Γ^1 and σ for stearic acid at these two temperatures when the equilibrium bulk concentrations are 10^{-3}, 10^{-4}, and 10^{-5} M.

$T = 30°C$		$T = 50°C$	
γ (dynes cm^{-1})	c_{eq} (moles liter^{-1})	γ (dynes cm^{-1})	c_{eq} (moles liter^{-1})
362	4.8×10^{-6}	364	4.8×10^{-6}
355	8.5×10^{-6}	362	9.6×10^{-6}
334	6.6×10^{-5}	354	7.4×10^{-5}
307	2.7×10^{-4}	334	4.4×10^{-4}
286	1.0×10^{-3}	314	1.6×10^{-3}

Estimate the concentrations at which the stearic acid film at the interface reaches a condensed packing at the two temperatures.

10. A scintillation counter is used to measure tritium β particles adjacent to the surfaces of tritiated sodium dodecyl sulfate in 0.115 M aqueous NaCl solution and tritiated dodecanol in dodecanol. The former system is surface active and the latter is not, so the difference between the measured radioactivity above the two indicates the surface excess of sodium dodecyl sulfate. The number of counts per minute arising from the surface excess A_s is related to the surface excess in moles per square centimeter Γ^1 by the relationship $A_s = 4.7 \times 10^{12}\,\Gamma^1$.† Use the following data (25°C) to construct the adsorption isotherm for sodium dodecyl sulfate on 0.115 M NaCl.

* D. S. Ambwani, R. A. Jao, and T. Fort, Jr., *J. Colloid Interface Sci.* **42**:8 (1973).
† M. Muramatsu, K. Tajima, M. Iwahashi, and K. Nukina, *J. Colloid Interface Sci.* **43**:499 (1973).

$\mu C\ ^3H$ (g solution)$^{-1}$	Activity $\times 10^{-3}$ (cpm)		Surfactant conc. $\times 10^3$ (moles kg^{-1})
	Tritiated nonsurfactant	Tritiated surfactant	
2	—	1.9	0.17
4	—	2.1	0.34
6	—	2.3	0.50
8	0.50	2.5	0.67
10	—	2.6	0.84
15	0.95	2.9	1.20
20	—	3.2	1.65
30	1.85	3.8	2.45

Briefly outline how the proportionality constant between A_s and Γ^1 might be determined experimentally.

11. The following data* give the cmc at various temperatures for some commercial alkyl trimethylammonium bromides ("commercial" means C_n is a major alkyl homolog):

T (K)	cmc (mmol liter^{-1})		
	C_{12}	C_{14}	C_{16}
298.2	5.27	3.32	0.824
303.2	5.33	3.41	0.870
313.2	6.40	3.60	0.949
323.2	8.28	4.10	1.050
328.2	9.03	4.30	—
333.2	10.20	4.49	1.170

From a plot of ln(cmc) versus T^{-1} estimate ΔH for the micellization process for each of these compounds. Is the variation of ΔH with the number of carbons in the alkyl groups consistent with expectations? Is the variation of cmc values with chain length at 25°C consistent with expectations?

12. A quantity called the HLB (for hydrophile–lipophile balance) number has proved to be a useful way to match a surfactant to a particular application. For example, surfactants with HLB numbers in the range 4 to 6 produce water-in-oil emulsions; those in the range 7 to 9 are useful as wetting agents; those ranging between 8 and 18 produce oil-in-water emulsions; and those with values in the range 13 to 15 make good detergents. Use these considerations plus the following specific examples to formulate a generalization about the dependence of the HLB number on the molecular structure of the surfactant.

* B. W. Barry and G. F. J. Russell, *J. Colloid Interface Sci.* **40**:174 (1972).

Surfactant	HLB number
Sodium dodecyl sulfate	40
Sodium oleate	18
Tween 80 ($n = 20$, R = oleate)	15
Sorbitan monolaurate	8.6
Span 60	4.7
Sorbitan tristearate	2.1

(Data from Osipow [7].)

13. The adsorption of straight-chain fatty acids from n-heptane on Fe_2O_3 has been studied.* In all cases studied, the adsorption isotherms conform with the Langmuir equation. The following are values of the amount adsorbed at saturation:

Fatty acid	$\left(\frac{n_2^s}{w}\right)_{sat} \times 10^5$ (moles g^{-1})
Acetic	3.00
Propionic	2.36
n-Butyric	2.11
n-Hexanoic	
n-Heptanoic	
n-Octanoic	
Lauric	1.04
Myristic	0.97
Palmitic	0.91
Magaric	0.82
Stearic	0.81

Calculate the area per molecule (σ) of each on the saturated surface if the Fe_2O_3 is known to have a specific surface of 3.45 $m^2\ g^{-1}$

14. The adsorption of various aliphatic alcohols from benzene solutions onto silicic acid surfaces has been studied.† The experimental isotherms have an appearance consistent with the Langmuir isotherm. Both the initial slopes of a n/w versus c plot and the saturation value of n/w increase in the order methanol < ethanol < propanol < butanol. Discuss this order in terms of the molecular structure of the alcohols and the physical significance of the initial slope and the saturation intercept. Which of these two quantities would you expect to be most sensitive to the structure of the adsorbed alcohol molecules? Explain.

15. The Michaelis–Menton equation is an important biochemical rate law. It relates the rate of the reaction v to a substrate concentration [S] in terms of two

* T. Allen and R. M. Patel, *J. Colloid Interface Sci.* **35**:647 (1971).
† R. L. Hoffman, D. G. McConnell, G. R. List, and C. D. Evans, *Science* **157**:550 (1967).

constants v_{max} and K_m:

$$v = \frac{v_{max}[S]}{K_m + [S]}$$

It will be noted that this equation follows the same functional form as the Langmuir equation, specifically $v \to v_{max}$ as [S] increases. The biochemical literature contains three different graphical procedures to evaluate the constants v_{max} and K_m from kinetic data:

Name of method	Plotted on ordinate	Plotted on abscissa
Lineweaver–Burk	$1/v$	$1/[S]$
Hanes	$[S]/v$	$[S]$
Eadie	v	$v/[S]$

Describe how the three equivalent variations of the Langmuir equation would be plotted. Give the interpretation of the slope and intercept in each case.

16. In laboratory tests of flotation, fluorite (CaF_2) particles (range of particle diameters: 0.074–0.147 mm) with oleic acid as collector were aerated under three different conditions. These conditions and the percent CaF_2 recovery after 10 min are tabulated below (from Klassen and Mokrousov [5]):

Aeration conditions	% recovery after 10 min of aeration
Bubbles precipitated from solution	5
Bubbles produced by mechanical means	20
Combination of both means of bubble production	70

Suggest an interpretation for this variation in the efficiency of fluorite recovery.

Physical Adsorption at the Gas–Solid Interface

When I cut through your plane as I am now doing, I make your plane a section which you, very rightly, call a Circle. For even a sphere—which is my proper name in my own country—if he manifest himself at all to an inhabitant of Flatland—must needs manifest himself as a Circle. [from Abbott's *Flatland*]

8.1 INTRODUCTION

Adsorption at the solid–gas interface is traditionally subdivided into two broad classes: chemisorption and physical adsorption. As the name implies, the former comes very close to the formation of chemical bonds between the adsorbent and the adsorbate. Two consequences of this are that the associated heat effects are comparable to those which accompany ordinary chemical reactions and that the process is not always reversible. It is possible, for example, to adsorb (chemisorb) oxygen on carbon and desorb CO or CO_2. In physical adsorption, on the other hand, the energy effects are comparable to those which accompany physical changes such as liquefaction and are completely reversible for nonporous solids. Physical adsorption is the easier of the two types of adsorption, and provides much background needed for an understanding of chemisorption. We shall limit our discussion of the solid–gas surface to cases in which only physical adsorption occurs.

There are several different ways in which the topics pertaining to physical adsorption can be subdivided. We shall be primarily concerned with nonporous solids, briefly discussing porous materials only in Sec. 8.11. It is convenient to divide the extent of adsorption into three categories: submonolayer, monolayer, and multilayer, and we shall discuss them in this order. The thermodynamics of adsorption may be developed around experimental isotherms or around calorimetric data. In this presentation, we shall follow the former approach. Finally, isotherms may be derived from a consideration of two-dimensional equations of state, from partition functions by statistical thermodynamics, or from kinetic arguments. Even though these methods are not fundamentally different, they differ in ease of visualization. We shall consider examples of each method. Finally, much of this chapter will lead up to and be developed around the determination of specific areas by gas adsorption. Low-temperature N_2 adsorption and the BET method of analysis are so widely used for this purpose that these topics will receive special attention, particularly with respect to experimental methods.

8.2 GAS ADSORPTION: PRACTICAL VERSUS THEORETICAL ISOTHERMS

It is difficult to present a concise introduction to the adsorption from the gas phase onto solid substrates. The topic is complicated by the following facts: (a) solid surfaces are notoriously heterogeneous, (b) the possibility of chemical adsorption and physical adsorption both occurring must be admitted, and (c) there is the question of exactly what constitutes a satisfactory description of such adsorption.

What is measured experimentally is the quantity of gas adsorbed per unit weight of adsorbent as a function of pressure at constant temperature. The experiment may, of course, be repeated at several temperatures. From a practical point of view, any insight which permits the amount of material adsorbed to be related to the specific surface area of the adsorbent and which correctly predicts how this adsorption varies with temperature may be regarded as a success. From a theoretical point of view, what is desired is to describe adsorption in terms of molecular properties, particularly in terms of an equation of state for the adsorbed material, where the latter is regarded as a two-dimensional state of matter.

As we shall see in the course of the chapter, these two approaches frequently clash. The adsorption isotherm of Brunauer, Emmett, and Teller (BET), which is discussed in Sec. 8.6, is an ecellent example of this. The model upon which the BET isotherm is based has been criticized by a great many theoreticians. At the same time, the isotherm itself has become virtually the standard equation for determining specific areas from gas adsorption data. Ross [8] summarizes the situation effectively by comparing it to a master chef who concocts a palatable dish out of an old shoe. For some, the end result is what matters: a palatable dish. For others, the starting material dominates their opinions: the old shoe. To present a relatively accurate picture of the current state of affairs in this area, it is necessary to present both of these viewpoints. We shall attempt not to take too one-sided a position but to give some indication of each. As far as the connection between surface and colloid chemistry is concerned, the principal interest in gas adsorption measurements is to determine the specific area of particles, from which an equivalent radius may be obtained by Eq. (1.2).

Before turning our attention to individual isotherms and the various approaches to them, it will be helpful to say a few words about the experimental conditions under which isotherms are determined. We shall examine the experimental aspects of gas adsorption in detail in Sec. 8.8. For now, all that need be mentioned is that most experiments of this kind are conducted on solids of high specific area with gases below their critical temperature. Since the adsorbent is solid, we cannot measure γ and use the Gibbs equation to evaluate the surface excess. Instead, the amount adsorbed must be determined directly as was the case with adsorption from solution onto solid surfaces. The latter experiments are more difficult to interpret since both solute and solvent may adsorb and strang-looking isotherms such as that shown in Fig. 7.17 result. With gases, this kind of competition does not concern us (we shall not consider adsorption from mixed gases). Despite this simplification, the experiment still depends on measuring changes in bulk properties—in this case, gas pressure—which occur as the system comes to equilibrium. High-area solids and low temperatures are necessary so that the extent of adsorption will be sufficient to be accurately measured by difference. The range of pressures over which adsorption studies may be conducted is—in principle—from zero to p_0, the saturation pressure or the normal vapor pressure of the material at the temperature of the experiment. At the low-pressure end of this range, adsorption will be slight so the determination of the isotherm involves measuring small differences in pressure at low pressures. This is not easy to do experimentally, although relatively modern low-pressure techniques have greatly extended this region. As the pressure approaches p_0, adsorption often increases rapidly as if anticipating phase separation by occurring as multilayer adsorption in which most of the adsorbed molecules behave as if they were in the bulk liquid state. As a matter of fact, if the solid is porous, the vapor may actually condense in the small pores at $p < p_0$. We shall consider the phenomena associated with capillary condensation in Sec. 8.11.

With this preamble, let us now turn to some specific examples of the relationship between two-dimensional equations of state and adsorption isotherms.

8.3 TWO-DIMENSIONAL EQUATIONS OF STATE AND ISOTHERMS

To see how the equation of state of two-dimensional matter and the adsorption isotherm are related, we return to the Gibbs equation [Eq. (7.40)]:

$$-d\gamma = \Gamma_2 \, d\mu_2 \tag{1}$$

Since we are concerned here with adsorption from the gas phase, the chemical potential may be related to the pressure of the gas by

$$\mu_2 = \mu_2^0 + RT \ln fp \tag{2}$$

where f is the activity coefficient [cf. Eqs. (7.41) and (7.42)]. In this discussion we assume that the gas behaves ideally, although in analyzing experimental results it may be necessary to include the correction required by the nonideality of the gas. Combining Eqs. (1) and (2) leads to

$$-d\gamma = RT\Gamma_2 \, d \ln p \tag{3}$$

In the present context, it is convenient to use Eq. (7.37) to eliminate Γ_2 from Eq. (3) and express the surface excess as the number of moles adsorbed per unit area:

$$-d\gamma = \frac{nRT}{A} \, d \ln p \tag{4}$$

Next, we substitute the product of sample weight times specific area for A in this equation to obtain

$$-d\gamma = \frac{RT}{A_{sp}}\frac{n}{w}\,d\ln p \tag{5}$$

For a given adsorbent and an isothermal experiment, T and A_{sp} are constants. The ratio n/w is the equilibrium amount adsorbed which will be a function of p. Therefore, Eq. (5) may be integrated as follows:

$$-\int d\gamma = \frac{RT}{A_{sp}}\int \frac{n}{w}\,d\ln p \tag{6}$$

The constant of integration may be evaluated by recognizing that n/w goes to zero as p approaches zero. Under these circumstances $\gamma \to \gamma_0$; therefore, Eq. (6) becomes

$$\gamma_0 - \gamma = \pi = \frac{RT}{A_{sp}}\int_0^p \frac{n}{w}\,d\ln p \tag{7}$$

If the experimental isotherm (n/w as a function of p) is known, then Eq. (7) may be integrated either analytically or graphically to give the two-dimensional pressure as a function of coverage. This relationship, therefore, establishes the connection between the two- and three-dimensional pressures which characterize the surface and bulk phases.

As an illustration of the kind of information obtainable from Eq. (7), suppose we consider the situation in which the equilibrium adsorption of a gas is described by the isotherm

$$\frac{n}{w} = mp \tag{8}$$

where m is a constant. Equations (7) and (8) may be combined to give

$$\pi = \frac{RT}{A_{sp}}\int_0^p mp\frac{dp}{p} = \frac{RT}{A_{sp}}mp = \frac{N_A kTn}{A_{sp}w} \tag{9}$$

Since the quantity $A_{sp}w/nN_A$ equals σ, Eq. (9) may be written

$$\pi\sigma = kT \tag{10}$$

the two-dimensional ideal gas law for the surface phase!

The adsorption isotherm—Eq. (8)—which is associated with this surface equation of state is called the Henry's law limit, in analogy with the equation which describes the vapor pressure of dilute solutions. The constant m, then, is the adsorption equivalent of the Henry's law constant. When adsorption is described by the Henry's law limit, the adsorbed state behaves like a two-dimensional ideal gas.

Equation (8) may also be written

$$\theta = m'p \tag{11}$$

if the specific area of the adsorbent and the cross-sectional area of the adsorbate are known [Eq. (7.76)]. We see, therefore, that compliance with Henry's law implies that a log–log plot of θ versus p yields a straight line of unit slope. Figure 8.1 shows

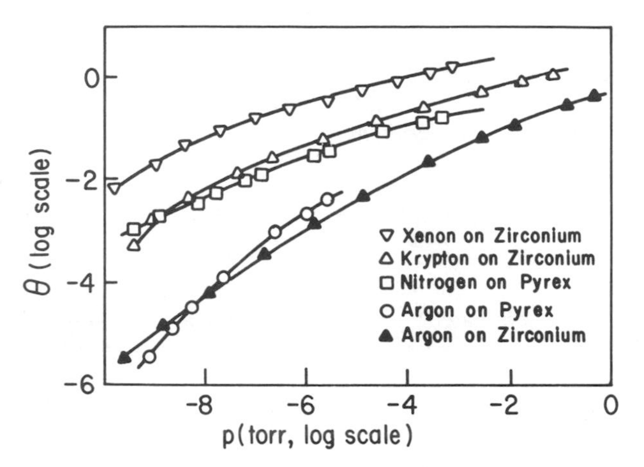

FIGURE 8.1 *A log–log plot of θ versus p for xenon, krypton, and argon on zirconium and nitrogen and argon on Pyrex. (Reprinted from Hobson* [6], *by courtesy of Marcel Dekker, Inc.*)

some experimental results for adsorption at 77.4 K plotted in this way for pressures down to 10^{-10} torr. Note that even at these low pressures the Henry's law limit is not yet reached (it would give a 45° line in Fig. 8.1), although argon on Pyrex appears to be approaching this limiting behavior. Note, further, that any errors introduced in evaluating θ would vertically shift the curves but would not change their slopes. We may conclude, therefore, that—just as with monolayers on aqueous substrates—the compliance of an adsorbed layer with the ideal gas law is a form of behavior which is extremely difficult to observe. The implication of this is that significant departures from two-dimensional ideality already set in at very low surface coverage.

In general, there are two different types of interactions in which adsorbed molecules may participate. There is the interaction between the adsorbed molecules and the adsorbent and the interaction between the adsorbed molecules themselves. In the Henry's law region, we have seen that the adsorbed layer behaves ideally. This is to be expected in view of their low surface concentration (θ very small). The adsorbed molecules definitely do interact with the adsorbent, however, and at very low coverage the interaction energy might be very sensitive to surface heterogeneity. Any "hot spots" on the surface would adsorb first, less energetic patches next, then the normal surface sites. In Sec. 8.5 and 8.10 we shall see how isotherms measured at several different temperatures may be interpreted to yield information on the energy of adsorption.

The second type of interaction possible for adsorbed molecules is direct adsorbate–adsorbate interaction. Interactions of this sort are expected to lead to deviations from ideality in the two-dimensional phase just as they lead to deviations from ideal behavior for bulk gases. In this case, surface equations of state which are analogous to those applied to nonideal bulk gases are suggested for the adsorbed molecules. The simplest of these allows for an excluded area correction [cf. Eq. (7.23)]:

$$\pi(\sigma - \sigma^0) = kT \tag{12}$$

Let us now consider the kind of adsorption isotherm associated with this two-dimensional equation of state.

Equation (12) may be used as a starting point in the evaluation of the isotherm from the equation of state. From this equation, $d\pi$ is given by

$$d\pi = -kT(\sigma - \sigma^0)^{-2}\, d\sigma \tag{13}$$

Next, we recall that, according to Eq. (7),

$$d\pi = -d\gamma \tag{14}$$

and that

$$\sigma = \frac{A_{sp} w}{n N_A} \tag{15}$$

Therefore, Eq. (5) may be written

$$\frac{\sigma\, d\pi}{kT} = d \ln p \tag{16}$$

Combining Eqs. (13) and (16) gives

$$-\frac{\sigma\, d\sigma}{(\sigma - \sigma^0)^2} = d \ln p \tag{17}$$

Suppose we divide numerator and denominator of the left-hand side of this equation by σ^0 and let $\sigma/\sigma^0 = x$ for the time being. Then Eq. (17) becomes

$$-\frac{x\, dx}{(x-1)^2} = d \ln p \tag{18}$$

in which the left-hand side is a standard integral. Evaluating the integral leads to the result

$$-\ln(x-1) + \frac{1}{x-1} = \ln p + C \tag{19}$$

The quantity we have called x may also be recognized as $1/\theta$:

$$x = \frac{\sigma}{\sigma^0} = \frac{1}{\theta} \tag{20}$$

Therefore, Eq. (19) may be written

$$\ln \frac{\theta}{1-\theta} + \frac{\theta}{1-\theta} = \ln p + C \tag{21}$$

Next, we evaluate the integration constant C. We know that $\theta \to 0$ as $p \to 0$. Therefore, at first glance, we are tempted to equate C to zero. It must be remembered, however, that Henry's law must apply as $p \to 0$. This condition is met if we let $C = \ln m'$ in which m' is defined by Eq. (11):

$$\ln \frac{\theta}{1-\theta} + \frac{\theta}{1-\theta} = \ln p + \ln m' \tag{22}$$

This may be readily verified by examining the limit of Eq. (22) as $\theta \to 0$.

Equation (22) may also be written

$$m'p = \frac{\theta}{1-\theta} \exp\left(\frac{\theta}{1-\theta}\right) \tag{23}$$

by taking the antilog of both sides of the equation. It is interesting to look at this form of the isotherm in the limit of small values of θ but still above Henry's limit. In this case, the exponential term approaches unity and Eq. (23) becomes

$$m'p = \frac{\theta}{1-\theta} \tag{24}$$

or

$$\theta = \frac{m'p}{1+m'p} \tag{25}$$

which is identical in form to the Langmuir equation [cf. Eq. (7.72)]. Equation (23) also reveals that for $m'p \gg 1$, θ approaches unity as an upper limit. Thus at both the upper and lower limits, Eq. (23) gives the same results as the Langmuir equation. At intermediate values the two functions differ slightly, but it would probably be difficult to distinguish between them in fitting experimental data.

Several points might be noted in summarizing the results of this section:

1. In principle, it is possible to correlate an adsorption isotherm and a two-dimensional equation of state by working from either direction. That is, we may start with an experimental isotherm and develop the associated equation of state [as in going from Eq. (8) to (10)] or we may proceed from the equation of state to the isotherm [as in the derivation between (12) and (23)].
2. Relatively small increases in complexity for the equation of state result in considerably more complex equations for the adsorption isotherms. The gross features of the more complex isotherms are also given by simpler isotherms. This means that it is very difficult to choose among various isotherms in terms of the goodness of fit to experimental data. Therefore, it is difficult to conclude from an experimental isotherm what the two-dimensional surface phases are like.
3. Two-dimensional equations of state are a useful source of isotherms, however, even though the test of the isotherm must be made in terms of some criterion other than an ability to describe adsorption. For example, the ability of an isotherm to predict the temperature dependence of adsorption or the specific area of an

adsorbent are more sensitive tests of an isotherm than merely describing the way n/w increases with p.

In the next section, we shall briefly describe some additional isotherms which have been generated by consideration of two-dimensional equations of state.

8.4 OTHER ISOTHERMS FROM SURFACE EQUATIONS OF STATE

Still greater nonideality in the two-dimensional equation of state might be represented by the van der Waals analog:

$$\left(\pi + \frac{a}{\sigma^2}\right)(\sigma - b) = kT \tag{26}$$

in which the b factor and σ^0 from Eq. (12) are identical in principle. What makes this expression especially interesting is the fact that there exists a temperature above which there is only one real root to Eq. (26) and below which some values of π correspond to three values of θ. The situation is shown schematically in Fig. 8.2a. With bulk gases and also insoluble monolayers on water, the three-root region is identified with a region of two-phase equilibrium. Is there any evidence for this type of phase equilibria in the two-dimensional layer of adsorbed gas on a solid substrate? Figure 8.2b shows several data for the adsorption of krypton on specially treated graphite over a range of temperatures from about 77 to 91 K. These plots are adsorption isotherms not π–σ diagrams, but nevertheless there is quite clear evidence of a two-phase region with a critical temperature at about 86 K. The bulk critical temperature for Kr is 210 K, so the two-dimensional value is about 41% of the three-dimensional one.

It is possible to develop a mathematical isotherm corresponding to the van der Waals equation for the two-dimensional state, following the procedure used in developing Eq. (23). The calculation is tedious, however, and yields no new insights into the nature of the adsorbed state. Therefore, we shall not go through the derivation in detail. The final result is given by the equation

$$m'p = \frac{\theta}{1-\theta} \exp\left(\frac{\theta}{1-\theta} - \frac{2a\theta}{bkT}\right) \tag{27}$$

Note than when a equals zero, Eq. (27) becomes identical to (23). A few additional assumptions lead to the prediction that the two-dimensional critical temperature should be one-half the value of the three-dimensional critical temperature according to this model. As just noted the two T_c values for Kr are close to this prediction, and the corresponding values for argon are 65 and 151 K.

There is another feature in Fig. 8.2b that deserves additional comment. It is the existence of a second set of vertical segments in the isotherms at values of θ in the range of 0.65 to 0.78. This suggests a second phase equilibrium in the two-dimensional matter, perhaps something analogous to the transition between the L_1 and L_2 states of monolayers on water or to a liquid–solid transition. Indeed, beyond

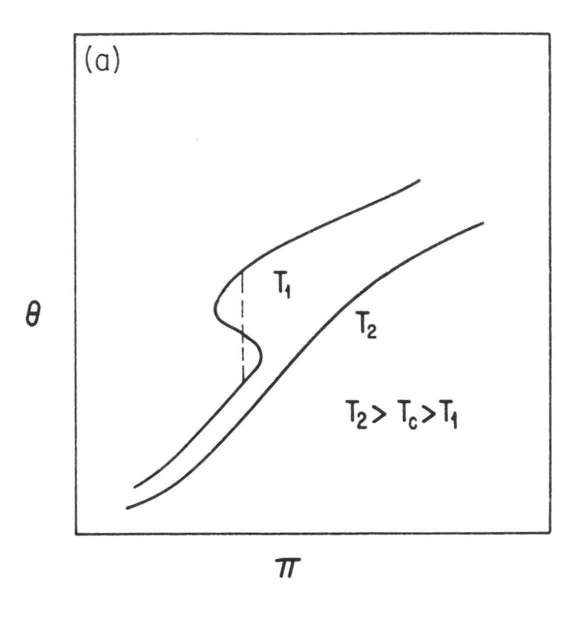

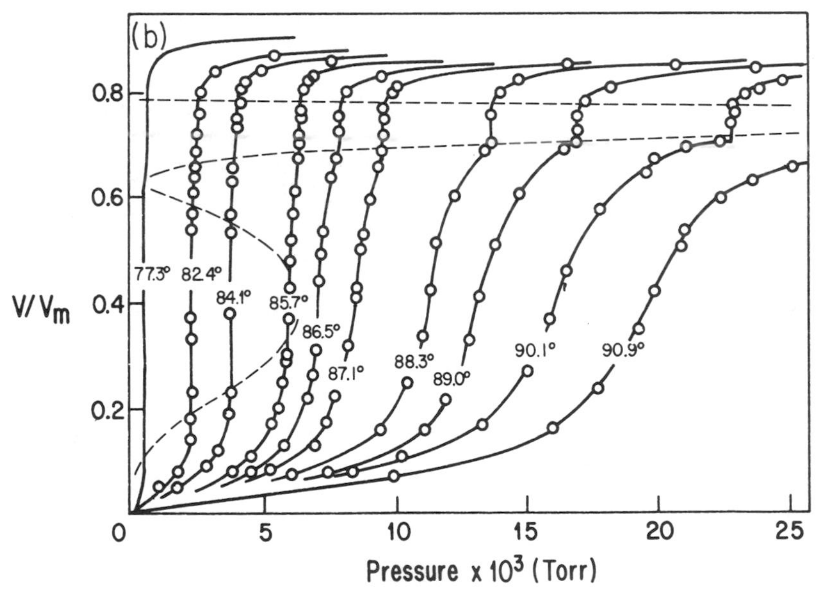

FIGURE 8.2 *(a) Schematic illustration of $\pi\sigma$ isotherms in the vicinity of a two-dimensional critical temperature. (b) Experimental data for the adsorption of krypton on exfoliated graphite showing similar features.* [*Source of data: A. Thomy and X. Duval,* J. Chim. Phys. **67**:1101 (1970)].

this second two-phase region the slope of the isotherm changes sharply in a way which corresponds to a much less compressible surface state.

Before exploring the consequences of this feature, it is first necessary to make certain that the result is not just an artifact arising from surface heterogeneities. We noted that the graphite substrate for which the data of Fig. 8.2b were collected was "specially treated." A few details of this treatment should be mentioned. Graphite and anhydrous $FeCl_3$ are introduced into opposite ends of a Pyrex tube; the evacuated tube is then sealed and brought to about 300°C. A slight temperature gradient is maintained between opposite ends of the tube so the $FeCl_3$ distils to the graphite where it reacts to form an interlamellar compound. Rapidly heating the latter to 900°C or higher results in the expulsion of the $FeCl_3$ and the attendant delamination of the graphite. The specific surface area of the final product—known as exfoliated graphite—is roughly two orders of magnitude greater than that of the initial graphite sample. The newly exposed graphite planes constitute a remarkably uniform solid adsorbant if all traces of $FeCl_3$ are removed. There is some evidence that the same vertical segments are observed in isotherms measured on the parent graphite as are observed in the exfoliated graphite. This argues that these features are not artifacts of the delamination process.

With these ideas in mind, another equation of state for the two-dimensional matter may be considered. The relatively incompressible surface state (e.g., above $\theta \simeq 0.70$ in Fig. 8.2b) may be approximately described by an equation of the type

$$\pi = -m\sigma + b \tag{28}$$

by analogy with the (approximately) linear π–σ isotherms for insoluble monolayers. The parameters m and b are merely empirical constants. From Eq. (28) it follows that

$$d\pi = -m\, d\sigma \tag{29}$$

which may be substituted into (16) to yield

$$-\frac{m\sigma\, d\sigma}{kT} = d \ln p \tag{30}$$

This expression is readily integrated to give

$$-\frac{m\sigma^2}{2kT} = \ln p + C \tag{31}$$

We shall not bother to evaluate the constant of integration in this equation. Instead we note that $\sigma = A_{sp}(w/n)$; therefore, Eq. (31) may be written

$$\ln p = -\frac{m}{2kT} A_{sp}^2 \left(\frac{n}{w}\right)^{-2} + C \tag{32}$$

This is one form of an adsorption isotherm known as the Harkins–Jura equation. It suggests that a plot of $\ln p$ versus $(n/w)^{-2}$ should be a straight line whose slope is proportional to the square of the specific area. In experiments with solids of known area, the linearity predicted by Eq. (32) has been observed. Furthermore, the proportionality constant relating the observed slopes to A_{sp}^2 is independent of the nature of the adsorbent to a first approximation. Thus, solids for which A_{sp} is known may be used to "calibrate" this method for a particular adsorbed species; then the specific area of an unknown may be determined using the same adsorbate.

The Harkins–Jura isotherm, therefore, introduces the possibility of determining specific areas by gas adsorption studies. This is probably the most important practical information to be derived from any study of the physical adsorption of gases. The Harkins–Jura equation is only one of many isotherms that permit the evaluation of A_{sp}. Although it may give satisfactory values for A_{sp}, Eq. (32) leaves something to be desired at the molecular level.

We have seen that the condensed state which characterizes the surface in the Harkins–Jura model appears only as $\theta \rightarrow 1$. In most instances of physical adsorption, however, no saturation limit of adsorption appears. As $p \rightarrow p_0$ the amount of material adsorbed increases asymptotically. Multilayer adsorption is the only reasonable model for this observation. Except for special cases, the neighborhood of $\theta = 1$ is obscured by the onset of multilayer adsorption. In summary, there is a mismatch between the situation described by the Harkins–Jura model and that suggested by macroscopic observations. We noted earlier that many isotherms are insensitive to the assumptions of their derivation. In line with that observation is the fact that the Harkins–Jura equation does fit a fairly wide range of experimental data and gives reasonable values of specific surfaces despite these objections. This is one example of an "old shoe-master chef" situation: We shall see presently that it is only one of several such cases.

Our approach until now has been to discuss adsorption isotherms on the basis of the equation of state of the corresponding two-dimensional matter. This procedure is easy to visualize and establishes a parallel with adsorption on liquid surfaces. However, it is not the only way to proceed. In the following section, we consider the use of statistical thermodynamics in the derivation of adsorption isotherms, and shall examine some other approaches in subsequent sections.

8.5 PARTITION FUNCTIONS AND ISOTHERMS

The partition function is the central feature of statistical thermodynamics. From the partition function, the various thermodynamic variables such as entropy, enthalpy, and free energy may be evaluated. It is also possible, in principle, to deduce the equation of state for a system from the partition function.

It should be apparent—since an adsorption isotherm can be derived from a two-dimensional equation of state—that an isotherm can also be derived from the partition function, since the equation of state is implicitly contained in the latter. The use of partition functions is very general, but it is also rather abstract, and the mathematical difficulties are often formidible (note the cautious "in principle" in the preceding paragraph). We shall not attempt any comprehensive discussion of the adsorption isotherms which have been derived by the methods of statistical thermodynamics. Instead, we shall derive only the Langmuir equation for adsorption from the gas phase by this method. The interested reader will find other examples of this approach discussed by Broeckhoff and van Dongen [2]. Statistical mechanics does not rely on pictorial models, but proceeds instead from mathematical statements about the energy states of the system. It turns out, however, that this freedom from more mechanical models helps reveal the underlying physical phenomena much more clearly.

A brief review of the statistical thermodynamics of ideal (bulk) gases will help us get started. In addition to reviewing some relevant physical chemistry, it will supply us with some expressions which may be useful since the two-dimensional ideal gas law applies to adsorbed molecules as a limiting case.

The partition function is defined by the equation

$$Q = \sum_i g_i \exp\left(-\frac{\varepsilon_i}{kT}\right) \tag{33}$$

in which g_i and ε_i represent the degeneracy and the energy, respectively, of the ith state. An important property of a partition function is its factorability into contributions arising from translation and internal degrees of freedom:

$$Q = Q_{\text{trans}} Q_{\text{int}} \tag{34}$$

The translational portion of the partition function is relatively easy to evaluate for ideal gases. Substituting its value into Eq. (34) gives

$$Q = \frac{1}{N!}(Q_{\text{trans}} Q_{\text{int}})^N = \left[V\left(\frac{2\pi mkT}{h^2}\right)^{3/2}\right]^N \left(\frac{e}{N} Q_{\text{int}}\right)^N \tag{35}$$

for N molecules of mass m in a volume V at temperature T.

The easiest quantity to evaluate from this expression is the Helmholtz free energy A:

$$A = -kT \ln Q \tag{36}$$

A variety of other thermodynamic functions may be evaluated from this. For example, the chemical potential—the quantity equalized in equilibrium calculations—is

$$\mu_i = \left(\frac{\partial A}{\partial N_i}\right)_T = -kT \frac{\partial \ln Q}{\partial N_i} \tag{37}$$

Also, to calculate the equation of state, we recall

$$p = -\left(\frac{\partial A}{\partial V}\right)_T \tag{38}$$

If we apply Eqs. (36) and (38) to (35), for example, we get the ideal gas law:

$$p = kT\left[\left(\frac{\partial \ln Q(V)}{\partial V}\right)_T + \left(\frac{\partial(\text{const})}{\partial V}\right)_T\right] = kT\frac{\partial \ln V^N}{\partial V} = \frac{NkT}{V} \tag{39}$$

The approach of statistical thermodynamics to the derivation of an adsorption isotherm goes as follows. First, suitable partition functions describing the bulk and surface phases are devised. The former is usually assumed to be that of an ideal gas. From the latter, the equation of state of the two-dimensional matter may be determined if desired, although this quantity ceases to be essential. The relationships just given are used to evaluate the chemical potential of the adsorbate in both the bulk and the surface. Equating the surface and bulk chemical potentials provides the equilibrium isotherm.

We shall apply this method to the derivation of the Langmuir isotherm both to illustrate the method and to see the assumed nature of the surface energy states upon which it is based.

The Langmuir isotherm is based on the assumption of localized adsorption. This means that an adsorbed molecule has such a high statistical preference for a certain surface site as to possess a negligible translational entropy in the adsorbed state. Localized adsorption is thus seen to be very plausible for chemisorption in which the adsorbed molecules and the adsorbent interact quite specifically. For nonspecific physical adsorption, a nonlocalized or mobile layer seems to be a more plausible picture. We have already discussed the Henry's law-type of isotherm and the ideal gas equation of state that are associated with the simplest type of mobile adsorption.

The adsorption sites on the surface are assumed to be uniform and to bind the adsorbate with an energy ε per molecule or E per mole. That is, the potential energy of a molecule in the gaseous state is zero and in the adsorbed state is $-\varepsilon$. Note that this adsorption energy is a characteristic of the interaction between the adsorbed molecules and the adsorbent. As such, it is the same not only for all parts of the surface but also for all degrees of surface coverage. This is equivalent to saying that the adsorbed molecules do not interact with each other.

The surface is assumed to consist of S adsorption sites. Suppose we consider the case in which N of the sites are occupied, that is, when N molecules are adsorbed. To write the partition function for the surface molecules, Q, we must ask how these molecules differ from those in the gas phase (g). Some of the internal degrees of freedom may be modified by the adsorption (Q^s_{int}), but the most notable difference will be in the translational degrees of freedom. From three equivalent translational degrees of freedom, the adsorbed molecule goes to two highly restrained translational degrees of freedom (remember the adsorption is localized) and one vibrational degree of freedom normal to the surface:

$$Q^g_{\text{trans,3d}} \rightarrow Q^s_{\text{trans,2d}} Q^s_{\text{vib}} \tag{40}$$

With these ideas in mind, we may assemble the partition function of the surface molecules as follows:

$$Q^s = g_N \{ Q^s_{\text{trans,2d}} Q^s_{\text{vib}} Q^s_{\text{int}} \exp[-(-\varepsilon)] \}^N \tag{41}$$

In this expression, the degeneracy factor g_N represents the number of ways the N molecules may be placed on S sites. The latter is given by the combinatorial formula (see Sec. 3.8):

$$g_N = \frac{S!}{N!(S-N)!} \tag{42}$$

Combining Eqs. (41) and (42) gives the following expression for the partition function of the adsorbed molecules:

$$Q^s = \frac{S!}{N!(S-N)!}(Q^s_{\text{trans,2d}} Q^s_{\text{vib}} Q^s_{\text{int}})^N \exp\left(\frac{N\varepsilon}{kT}\right) \tag{43}$$

Application of Eq. (36) to (43) gives the Helmholtz free energy of the adsorbed molecules:

$$A^s = -kT\left[\ln S! - \ln N! - \ln(S-N)! + \frac{N\varepsilon}{kT} + N \ln Q^s_{\text{trans,2d}} Q^s_{\text{vib}} Q^s_{\text{int}}\right] \tag{44}$$

Since N and S are large, the factorials may be expanded by Sterling's approximation ($\ln x \simeq x \ln x - x$) to give

$$A^s = -kT\left[S \ln S - N \ln N - (S-N)\ln(S-N) + \frac{N\varepsilon}{kT} + N \ln Q^s_{\text{trans,2d}} Q^s_{\text{vib}} Q^s_{\text{int}}\right] \tag{45}$$

The chemical potential of the adsorbed molecules is given, according to Eq. (37), by differentiating this quantity with respect to N:

$$\mu^s = kT\left(\ln \frac{N}{S-N} - \frac{\varepsilon}{kT} - \ln Q^s_{\text{trans,2d}} Q^s_{\text{vib}} Q^s_{\text{int}}\right) \tag{46}$$

The condition of equilibrium between the adsorbed molecules and molecules in the gas state requires that the chemical potential for the adsorbed species be the same in both the gas phase and the adsorbed state

$$\mu^s = \mu^g \tag{47}$$

Applying Eq. (36) and (37) to (35) shows the chemical potential for an ideal gas to be

$$\mu^g = -kT \ln \frac{kT}{p}\left(\frac{2\pi mkT}{h^2}\right)^{3/2} Q^g_{\text{int}} \tag{48}$$

since $V = RT/p$.

Equating (46) and (48) gives

$$\frac{N}{S-N} = \frac{p}{kT}\left(\frac{h^2}{2\pi mkT}\right)^{3/2} \exp\left(\frac{\varepsilon}{kT}\right) \frac{Q^s_{\text{trans,2d}} Q^s_{\text{vib}} Q^s_{\text{int}}}{Q^g_{\text{int}}} \tag{49}$$

We may group the following terms together to define a new quantity K:

$$K = \frac{1}{kT}\left(\frac{h^2}{2\pi mkT}\right)^{3/2} \exp\left(\frac{\varepsilon}{kT}\right) \frac{Q^s_{\text{trans,2d}} Q^s_{\text{vib}} Q^s_{\text{int}}}{Q^g_{\text{int}}} \tag{50}$$

in terms of which Eq. (49) becomes

$$\frac{N}{S-N} = Kp \tag{51}$$

Now if we divide both the numerator and denominator of the left-hand side by S and recognize that $N/S = \theta$, we obtain

$$\frac{\theta}{1-\theta} = Kp \tag{52}$$

or

$$\theta = \frac{Kp}{1+Kp} \tag{53}$$

two forms of the Langmuir adsorption isotherm [cf. Eq. (7.72)].

The quantity K defined by Eq. (50) may easily be expanded somewhat further. The two-dimensional partition function may be written by analogy with its three-dimensional counterpart, Eq. (35). To do this, V is replaced by the area accessible to the adsorbed molecule σ and the exponent $\frac{2}{2}$ ($=1$) rather than $\frac{3}{2}$ is used since two rather than three degrees of freedom are involved. Therefore, we write

$$Q^s_{\text{trans,2d}} = \sigma \frac{2\pi mkT}{h^2} \tag{54}$$

If the energy separating the vibrational quantum states is small, the partition function for vibration is approximately given by

$$Q^s_{\text{vib}} = \frac{kT}{\varepsilon_{\text{vib}}} \tag{55}$$

For physical adsorption, the approximation involved here is expected to be valid. Finally, the energy of vibration may be replaced by $h\nu$, where ν is the frequency with which the adsorbed molecules vibrate against the adsorbent:

$$Q^s_{\text{vib}} = \frac{kT}{h\nu} \tag{56}$$

Combining Eqs. (54) and (56) with (50) gives

$$K = \frac{\sigma}{\nu}(2\pi mkT)^{-1/2} \exp\left(\frac{\varepsilon}{kT}\right) \frac{Q^s_{\text{int}}}{Q^g_{\text{int}}} \tag{57}$$

It should be noted that this quantity has the cgs units of square centimeters per dyne, reciprocal pressure units, as required by Eq. (53).

We saw by Eq. (7.80) how to rearrange the Langmuir equation into a form which permits graphical evaluation of the parameters. For the adsorption of gases, this becomes

$$\frac{p}{n/w} = mp + b \tag{58}$$

and predicts a straight line when $p/(n/w)$ is plotted versus p, with

$$\text{Slope} = m = \frac{N_A \sigma}{A_{\text{sp}}} \tag{59}$$

and

$$\frac{\text{Slope}}{\text{Intercept}} = \frac{m}{b} = K \tag{60}$$

Figure 8.3 shows some data for the adsorption of ethyl chloride on charcoal. Since these data were collected at different temperatures, the ratio p/p_0 is used as the independent variable in fitting the data to Eq. (58). That is, to compare the adsorption at different temperatures, the pressure is expressed as a fraction of the equilibrium vapor pressure at that temperature. The data for each of the three temperatures in Fig. 8.3 give quite good straight lines when plotted according to the linear form of the Langmuir equation [Eq. (58)]. The slopes of the three plots agree to within 2.5%, indicating an approach to a common saturation adsorption of about 0.511 g ethyl chloride (g charcoal)$^{-1}$. The three slope–intercept ratios yield K values for each of the experimental temperatures. These K values may be fitted [by plotting ln $(K\sqrt{T})$ versus T^{-1}] by the following expression:

$$K = \frac{1.28}{\sqrt{T}} \exp\left(\frac{3330 \text{ cal mole}^{-1}}{RT}\right) \tag{61}$$

The range of temperatures is too narrow and the number of points too few for the numerical constants in this expression to be very accurate. Nevertheless, Eq. (57) describes the temperature variation of the isotherm fairly well. As noted previously, the ability to predict correctly the temperature dependence of adsorption is a more stringent test of an isotherm than the mere correlation of adsorption data. Nevertheless, an independent evaluation of the energy of adsorption is highly desirable. Although 3300 cal mole^{-1} is a plausible value for this quantity, it could still be an artifact, arising from several compensating deviations between the experimental system and the Langmuir model.

FIGURE 8.3 *Plot showing how the amount of ethyl chloride adsorbed on charcoal (in g g^{-1}) varies with pressure at* −15.3, 0, *and* 20°C. [*Data from F. Goldman and M. Polanyi,* Z. Phys. Chem. **132**:321 (1928)].

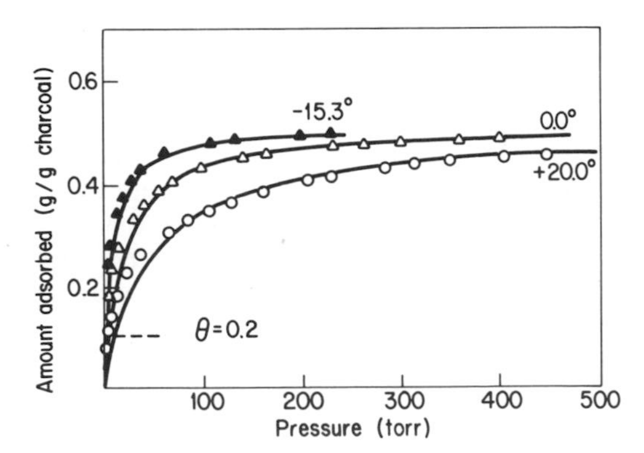

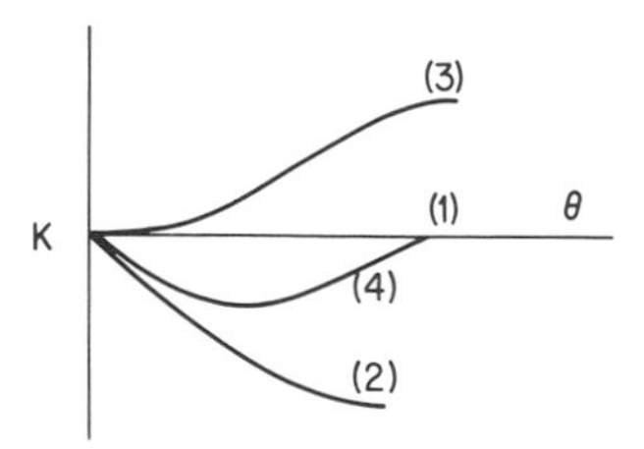

FIGURE 8.4 *Schematic illustration of the compensation between the effects on K of surface heterogeneity (curve 2) and lateral interactions among absorbed molecules (curve 3). In combination (curve 4), these effects may approach the assumption of the model (curve 1).*

An indication of how this may come about can be obtained from Fig. 8.4. The figure shows several different ways that K in Eq. (51) might vary with θ for a real system. Curve 1 shows no variation with θ and thus conforms to the Langmuir assumptions. Curve 2 shows K decreasing with increasing surface coverage. This suggests that "hot spots" for which ε is large are covered first, with the function flattening out once the most active sites are covered. Curve 3 describes the situation in which ε increases with increasing coverage. This may arise from lateral interactions between adsorbed molecules, an effect which increases with increased adsorption. Curve 4 may be regarded as a composite of the behavior displayed separately by curves 2 and 3. In each instance, there are ranges of coverage for which K is essentially constant. If this range is relatively wide and corresponds to a region in which many experimental points lie, then the Langmuir equation may fit the data quite satisfactorily, but the significance of the energy in (61) becomes quite obscure.

Until now, we have focused our attention on those adsorption isotherms which show a saturation limit, an effect usually associated with monolayer coverage. We have seen two ways of arriving at equations which describe such adsorption: from the two-dimensional equation of state via the Gibbs equation or from the partitition function via statistical thermodynamics. Next, we shall turn our attention to multilayer adsorption. At the same time, we shall introduce a third method for the derivation of isotherms, a kinetic approach.

8.6 MULTILAYER ADSORPTION: THE BET EQUATION

As noted previously, the range of pressures over which gas adsorption studies are conducted extends from zero to the normal vapor pressure of the adsorbed species, p_0. An adsorbed layer on a small particle may readily be seen as a potential nucleation center for phase separation at p_0. Thus at the upper limit of the pressure range, adsorption and liquefaction appear to converge. At very low pressures, it is plausible to restrict the adsorbed molecules to a monolayer. At the upper limit, however, the imminence of liquefaction suggests that the adsorbed molecules may be more than one layer thick. There is a good deal of evidence supporting the idea that multilayer adsorption is a very common form of physical adsorption on nonporous solids. In this section, we shall be primarily concerned with an adsorption isotherm derived by S. Brunauer, P. H. Emmett, and E. Teller in 1938; the theory and final equation are invariably known by the initials of the authors: BET.

The BET isotherm has subsequently been derived by a variety of methods, but we shall follow the approach of the original derivation: a kinetic description of the equilibrium state. Like the Langmuir isotherm, the BET theory begins with the assumption of localized adsorption. There is no limitation as to the number of layers of molecules which may be adsorbed, however; hence there is no saturation of the surface with increasing pressure. In general, the derivation assumes that the rates of adsorption and desorption from each layer are equal at equilibrium and that adsorption or desorption can occur from a particular layer only if that layer is exposed, that is, provided no additional adsorbed layers are stacked on top of it.

Suppose we consider a surface consisting of a total of S adsorption sites, S_1 of which are filled with adsorbed molecules to a depth of one molecule. Thus the number of bare sites S_0, is $S-S_1$. These two surface situations are defined to be in equilibrium when the rate at which molecules attach to bare spots is the same as the rate at which they escape from monolayer regions. The rate of the adsorption process is proportional to both the pressure and the number of available sites:

$$R_a = k_a p S_0 \tag{62}$$

The rate of desorption is proportional to the number of sites occupied:

$$R_{d,1} = k_{d,1} S_1 \tag{63}$$

The subscript 1 on R_d and k_d in Eq. (63) reminds us that these refer to layer 1 only. At equilibrium, the rates given by Eq. (62) and (63) are equal:

$$k_a p S_0 = k_{d,1} S_1 \tag{64}$$

It is profitable to interrupt the general BET development at this point to consider Eq. (64) in the special case in which the adsorption is restricted to a monolayer. In this case,

$$S_0 + S_1 = S \tag{65}$$

a result which may be substituted into Eq. (64) to give the Langmuir equation:

$$\frac{k_a}{k_{d,1}} p = \frac{S_1}{S - S_0} = \frac{\theta}{1-\theta} \tag{66}$$

Before returning to the case of multilayer adsorption, it is worthwhile to examine the expected form of the two rate constants k_a and $k_{d,1}$. The rate constant for desorption consists of a frequency factor and a Boltzmann factor. The former may be assumed to be proportional to a frequency ν, and the energy in the Boltzmann factor may be identified with the energy of interaction between the adsorbate and the adsorbent:

$$k_{d,1} \propto \nu \exp\left(-\frac{\varepsilon}{kT}\right) \tag{67}$$

From kinetic molecular theory, the number of collisions per unit area per unit time, Z, between gas molecules and a wall equals

$$Z = (2\pi mkT)^{-1/2} p \tag{68}$$

If this is multiplied by a surface area σ, the result is the rate of surface collisions. The coefficient of p may then be taken as proportional to k_a:

$$k_a \propto (2\pi mkT)^{-1/2} \sigma \tag{69}$$

The ratio $k_a/k_{d,1}$ may be set equal to the Langmuir K. Therefore,

$$K = \frac{k_a}{k_{d,1}} \propto (2\pi mkT)^{-1/2} \frac{\sigma}{\nu} \exp\left(\frac{\varepsilon}{kT}\right) \tag{70}$$

a result that is identical to Eq. (57) as far as translational factors are concerned.

This digression back to the Langmuir result provides no new information about the latter, but it does remind us of the common starting assumptions of both the Langmuir and BET derivations.

If we allow the possibility that a surface site may be covered to a depth of more than one molecule, then the following modifications are required. First, we define S_i to be the number of sites covered to a depth of i molecules. Second, Eq. (65) is modified to

$$S = S_0 + S_1 + S_2 + \cdots = \sum_{i=0}^{n} S_i \tag{71}$$

where the summation includes all thicknesses of coverage from zero to n, the maximum.

Now let us consider the composition of the second layer ($i = 2$). By analogy with Eq. (64), the rate at which molecules adsorb to form layer 2 is proportional to p and S_1. For simplicity, the proportionality constant is assumed to be the same as that for adsorption on the bare surface. Therefore, we write

$$R_a = k_a p S_1 \tag{72}$$

In a similar fashion, we assume the rate of desorption from the second layer to be proportional to S_2:

$$R_{d,2} = k_{d,2} S_2 \tag{73}$$

The constant $k_{d,2}$ is assumed to be of the same form as that given by Eq. (67) for the first layer *with one important modification.* While desorption from the first layer involves detaching a molecule from the adsorbent as a substrate, desorption from the second layer involves detachment from another adsorbed molecule of the same kind. The adsorption energy ε was used in the Boltzmann factor in the first case; the energy of vaporization ε_v is a more suitable value to use for the analogous quantity in the second case. We shall assume the frequency factor to be unchanged and write

$$k_{d,2} = \nu \exp\left(-\frac{\varepsilon_v}{kT}\right) \tag{74}$$

At equilibrium, the rate of adsorption and the rate of desorption from the second layer are also equal; therefore,

$$k_a S_1 p = k_{d,2} S_2 \tag{75}$$

As a matter of fact, the same expression also applies to the third, fourth, . . . , ith levels. As a first approximation, the "activation energy" for desorption is the same for all layers after the first. This leads to the generalization

$$k_a S_{i-1} p = k_{d,i} S_i \tag{76}$$

for $2 \leqslant i < n$.

Equation (76) enables us to relate S_i to S_{i-1} just as (64) relates S_1 to S_0. Therefore, we may write

$$S_i = \left(\frac{k_a}{k_{d,i}} p\right)^{i-1} S_{i-(i-1)} = \left(\frac{k_a}{k_{d,i}} p\right)^{i-1} S_1 = \left(\frac{k_a}{k_{d,i}}\right)^{i-1} \frac{k_a}{k_{d,1}} p^i S_0 \tag{77}$$

Since k_a is assumed to be the same for each layer and $k_{d,i}$ differs from k_d only in the value of the exponential energy, this may be written

$$S_i = \frac{k_a^i p^i S_0}{\nu^i [\exp(-\varepsilon_v/kT)]^{i-1} \exp(-\varepsilon/kT)} \tag{78}$$

Multiplying the numerator and denominator of the right-hand side of this equation by $\exp(-\varepsilon_v/kT)$ yields

$$S_i = \frac{k_a^i p^i S_0}{[\nu \exp(-\varepsilon_v/kT)]^i} \cdot \frac{\exp(-\varepsilon_v/kT)}{\exp(-\varepsilon/kT)} \tag{79}$$

which may be written

$$S_i = x^i c S_0 \tag{80}$$

with c and x defined as follows:

$$x = \frac{k_a p}{\nu \exp(-\varepsilon_v/kT)} = \frac{k_a}{k_{d,i \geqslant 2}} p \tag{81}$$

and

$$c = \exp\left(\frac{\varepsilon - \varepsilon_v}{kT}\right) \tag{82}$$

Equation (80) may now be substituted into (71) to give

$$S = S_0 + S_1 + S_2 + \cdots = S_0 + \sum_{i=1}^{n} x_i c S_0 \tag{83}$$

Under some circumstances, there may be a reason to restrict adsorption to a finite number of layers, that is, assign some specific value to n. In general, however, $n \to \infty$ as $p \to p_0$ is usually taken as the upper limit for this summation.

At this point we may define two other quantities in terms of the variables involved in Eq. (79): the total volume of gas adsorbed and the volume adsorbed at monolayer coverage, V and V_m, respectively. The total volume is obviously the sum of the volume held in each type of site V_i where the latter is proportional to iS_i:

$$V = \sum_{i=1}^{n} V_i \propto \sum_{i=1}^{n} iS_i \tag{84}$$

The volume adsorbed at monolayer coverage is simply proportional to the total number of sites irrespective of the depth to which they are covered:

$$V_m \propto S \propto S_0 + \sum_{i=1}^{n} S_i \tag{85}$$

Note that in writing Eq. (85), it is not assumed that the monolayer is completely filled before other layers are formed. On the contrary, the picture allows for the coexistence of all types of patches, with adsorbed molecules stacked to varying depths on each. Therefore, V_m equals the volume of gas which would be adsorbed *if* a monolayer were formed. There is no implication that a filled monolayer is a step along the way toward multilayer formation.

Taking the ratio of Eq. (84) to (85) eliminates the unspecified proportionality constant and gives

$$\frac{V}{V_m} = \frac{\Sigma_i iS_i}{S_0 + \Sigma_i S_i} = \frac{c\ \Sigma_i ix^i}{1 + c\ \Sigma_i x^i} \tag{86}$$

The ratio V/V_m may be identified with θ which may have values greater than unity in the case of multilayer adsorption. All that remains to be done to complete the BET derivation is to evaluate the summations in Eq. (86).

To assist in the evaluation of the summations, suppose we consider the quantity $x(1-x)^{-1}$. The factor in parentheses may be expanded as a power series (see Appendix A) to give

$$x(1-x)^{-1} = x(1 + x + x^2 + \cdots) = \sum_{i=1}^{n} x^i \tag{87}$$

Next, we consider the quantity $x(d/dx)(\Sigma_i x^i)$. Carrying out the indicated differentiation gives

$$x\frac{d}{dx}\sum_i x^i = x\sum_i ix^{i-1} = \sum_i ix^i \tag{88}$$

Equation (87) may now be used as a substitution for Σx^i in (88) to yield

$$\sum_i ix^i = x\frac{d}{dx}[x(1-x)^{-1}] = \frac{x}{(1-x)^2} \tag{89}$$

Substituting Eq. (87) and (89) into (86) gives

$$\frac{V}{V_m} = \frac{cx(1-x)^{-2}}{1 + cx(1-x)^{-1}} \tag{90}$$

This last result may be simplified further to become

$$\frac{V}{V_m} = \frac{cx}{(1-x)[1+(c-1)x]} \tag{91}$$

Equation (91) is the result generally defined as the BET equation.

The condition that $V \to \infty$ is seen to correspond to $x = 1$ by Eq. (91). Recalling the definition of x given by Eq. (81) and that $V \to \infty$ as $p \to p_0$ permits us to write

$$1 = \frac{k_a}{k_{d,i \geqslant 2}} p_0 \tag{92}$$

Therefore, Eq. (81) becomes

$$x = \frac{p}{p_0} \tag{93}$$

Thus the independent variable in the BET theory is the pressure relative to the saturation pressure. Therefore, the BET equation describes the volume of gas adsorbed at different values of p/p_0 in terms of two parameters V_m and c. Furthermore, the model supplies a physical interpretation to these two parameters.

8.7 TESTING THE BET EQUATION

The easiest way to evaluate the BET constants is to rearrange Eq. (91) into the following linear form:

$$\frac{1}{V}\frac{x}{1-x} = \frac{c-1}{cV_m}x + \frac{1}{cV_m} \tag{94}$$

This form suggests that a plot with $(1/V)[x/(1-x)]$ as the ordinate and x as the abscissa should yield a straight line, the slope and intercept of which have the following significance:

$$\text{Slope} = m = \frac{c-1}{cV_m} \tag{95}$$

and

$$\text{Intercept} = b = \frac{1}{cV_m} \tag{96}$$

Equations (95) and (96) may be solved to supply values of V_m and c from the experimental results:

$$V_m = \frac{1}{m+b} \tag{97}$$

and

$$c = \frac{m}{b} + 1 \tag{98}$$

In the following few paragraphs, we shall first examine some of the general features of gas adsorption as predicted by the BET theory. Next, we shall consider how well the theory actually fits experimental data. The use of experimental V_m values in the evaluation of A_{sp} is taken up in the following section.

Figure 8.5 enables us to observe some of the general features of the BET isotherm. The most apparent aspect concerning all the curves in this figure is their rapid increase as $p/p_0 \to 1$. It is also apparent, however, that the shape of the curve is sensitive to the value of c, especially for low values of c. Note particularly that for values of c equal to 2 or less, the curves show no inflection point, whereas the inflection becomes increasingly pronounced as c increases above 2. In view of the wide diversity of curve shapes that are consistent with the BET equation and the relative insensitivity of adsorption data to the model underlying a particular equation, we might expect that the BET equation will fit experimental data rather successfully.

Figure 8.6a is a plot of some actual experimental data showing the volume of N_2—expressed as cubic centimeters at STP per gram—adsorbed by a sample of nonporous silica at 77 K. In Sec. 8.9, we shall discuss the experimental techniques by which such results are obtained. Figure 8.6b shows these same results plotted according to the linear form of the BET equation, given by (94). The BET equation is seen to fit the adsorption data in the range $0.05 \lessapprox p/p_0 \lessapprox 0.30$. From the values of the slope (0.0257 g cm^{-3} at STP) and intercept (2.85×10^{-4} g cm^{-3} at STP), Eqs. (97) and (98) may be used to evaluate V_m and c for this system:

$$V_m = \frac{1}{(257 + 2.85) \times 10^{-4}} = 38.5 \text{ cm}^3 \text{ g}^{-1} \text{ at STP} \tag{99}$$

FIGURE 8.5 *Plots of V/V_m versus p/p_0 for several values of the parameter c, calculated according to the BET theory by Eq.* (91).

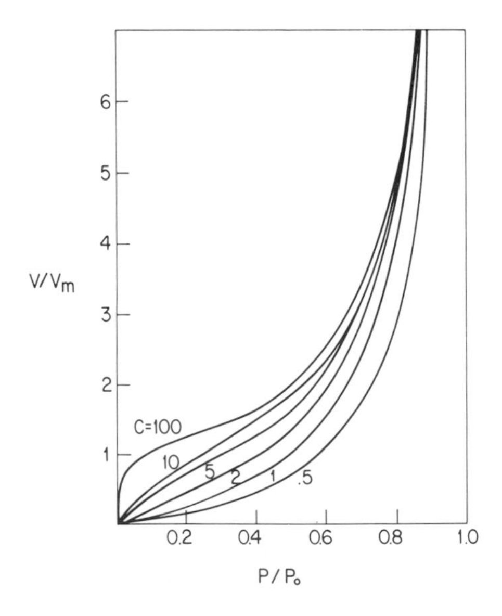

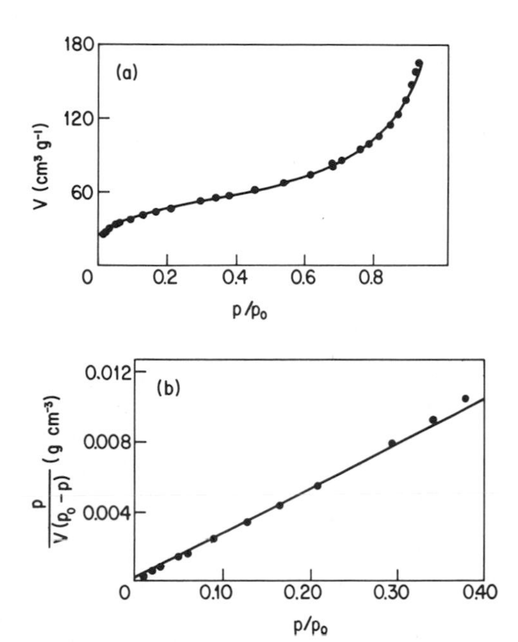

FIGURE 8.6 *Nitrogen adsorption on nonporous silica at* 77 K. *In (a), volume per gram (in* cm^3 *at* STP) *versus* p/p_0. *In (b), according to the linear form of the BET equation* [*Eq.* (94)]. [*Data from D. H. Everett, G. D. Parfitt, K. S. W. Sing, and R. Wilson*, J. Appl. Chem. Biotechnol. **24**:199 (1974).]

$$c = \frac{257 \times 10^{-4}}{2.85 \times 10^{-4}} + 1 = 91.2 \tag{100}$$

The range of pressures over which the linear form of the BET equation fits the experimental data in Fig. 8.6 is fairly typical for a variety of gases and adsorbents. At relative pressures below the range of fit, the BET equation underestimates the actual adsorption, whereas above $p/p_0 = 0.35$, the equation overestimates it. This range of p/p_0 values encompasses the region in which $V = V_m$ for $2 \lessapprox c \lessapprox 500$. In applying the BET equation to surface area determination, this is an important region of fit.

One way to appreciate the significance of V_m is to consider Eq. (91) under the following limiting conditions. If c is large and x is small, then the factors $(c-1)$ and $(1-x)$ in Eq. (91) become c and 1, respectively. In that case, the BET equation becomes

$$\frac{V}{V_m} = \theta = \frac{cx}{1+cx} \tag{101}$$

which is the same form predicted by the Langmuir equation. We have already seen how the saturation value of the Langmuir equation may be related to the specific area of the adsorbent [see Eq. (59)]. Equation (101) shows the relationship between $\theta = 1$ and $V = V_m$.

Inspection of the curves in Fig. 8.5 reveals a sharpening of the "knee" in the curve near $V/V_m = 1$ as c increases. One gets the impression in such a case that the volume adsorbed at monolayer coverage is almost recognizable, being just slightly obscured by the onset of multilayer adsorption.

At smaller values of c, the "knee" of the curve—or point B as it is often called—is much more definitely buried in the multilayer region. Another way to see this is to note that the fraction of surface covered when $V = V_m$ is 0.24, 0.50, 0.91, and 0.97 for values of c equal to 0.1, 1.0, 100, and 1000, respectively. That is, V indeed approaches monolayer coverage as c increases. Some critics of the BET theory are reluctant to accept Eq. (97) literally as a measure of the saturation capacity of the surface. They prefer instead to regard $(m + b)^{-1}$ as a precise way of locating point B. According to this interpretation, point B represents a degree of coverage at which the affinity of the adsorbed molecules for the adsorbent is changing most rapidly. Conceptually, this may be identified with the completion of the monolayer, without the quantitative restrictions assumed by the BET model.

Basically there are three criteria against which the success of the BET theory may be evaluated: its ability to "fit" adsorption data, correct prediction of the temperature dependence of adsorption, and correct evaluation of specific area. We have already commented on the data-fitting capabilities of the equation. In the following section, we shall discuss specific area determination exclusively. Therefore, we shall conclude this section with a few comments on the ability of the BET theory to predict the temperature dependence of adsorption.

The parameter c describes the temperature dependence of adsorption. In the derivation already presented, certain simplifying assumptions were made as to the constancy of k_a and ν_i in each of the layers. Various modifications of this assumption might be made, but they would involve minor temperature effects at best. To a first approximation, then, one should be able to evaluate $\varepsilon - \varepsilon_v$ from a knowledge of experimental c values, or vice versa. Proceeding in the first manner leads to values of $\varepsilon - \varepsilon_v$ which are too low, perhaps half their expected value. When we consider the region of fit on which the evaluation of c is based it is not difficult to see why values of c are too low. The data are not fitted to the earliest stages of adsorption which are associated with larger values of ε; hence, c is underestimated.

The final criterion for judging the success of a theoretical isotherm is its ability to measure specific surface areas. Since the BET equation has become a standard in this regard, we shall devote an entire section to the discussion of this topic.

8.8 SPECIFIC SURFACE AREA: THE BET METHOD

With monolayer adsorption, we saw how the saturation limit could be related to the specific surface area of the adsorbent. The BET equation permits us to extract from multilayer adsorption data [by means of Eq. (97)] that volume of adsorbed gas which would saturate the surface if the adsorption were limited to a monolayer. Therefore, V_m may be interpreted in the same manner that the limiting value of the ordinate is handled in the case of monolayer adsorption. Since it is traditional to express both V and V_m in cubic centimeters at STP per gram, we write [see Eq. (7.77)]

$$V_m = \left(\frac{n}{w}\right)_{\text{sat}} (22{,}414 \text{ cm}^3 \text{ mole}^{-1}) = \frac{A_{\text{sp}}\,(22{,}414)}{N_A \sigma^0} \tag{102}$$

Note that it is assumed that V_m has been expressed on a "per gram" basis in writing Eq. (102). If this is not the case, the area of the *actual sample* rather than its specific area is given by Eq. (102). If the area occupied per molecule on the surface is known, the specific surface may be evaluated:

$$A_{\text{sp}} = \frac{V_m N_A \sigma^0}{22{,}414} \tag{103}$$

This last quantity is something which may be determined by independent methods: therefore, the BET theory may be tested at this point. Two ways of proceeding come to mind. First, it is clear that the same value of A_{sp} must be obtained for a particular adsorbent, regardless of the nature of the adsorbate used. Studies of this sort lead to quite consistent values of A_{sp} provided the adsorbed species all have access to the same surface. On a porous surface, for example, large adsorbate molecules may not be able to enter small cavities which are accessible to smaller molecules.

If the particles are approximately spherical, the specific surface is given by Eq. (1.2). BET surface areas have been found to agree quite reasonably with values calculated from electron micrographs. Of course, proper statistical consideration of the poydispersity of the sample must be included. Also, if the surface is too rough, the gas adsorption area will be higher than a value based on the assumption of spherical particles.

In the preceding discussion, it has been assumed that a value of σ^0 is known unambiguously. This quantity is obviously the "yardstick" by which moles of adsorbed gas are converted to areas. Any error in this quantity will invalidate the determination of A_{sp}. What is generally done is to assume the adsorbed material has the same density on the surface that it has in the bulk liquid at the same temperature and to assume the molecules are close packed on the surface. In view of the earlier discussion of surface phases, this is seen to be a somewhat risky procedure. A safer way to proceed would be to evaluate σ^0 for a particular adsorbate from independent measurements of V_m and A_{sp}.

In view of the difficulty in translating measured gas adsorption into absolute specific surface areas, it is not surprising that self-consistency is often normative in this matter. In this sense, at least, an area of 16.2 Å^2 for nitrogen has become something of a standard. Values for other common adsorbed species may be found in the works of Adamson [1], Broeckhoff and van Dongen [2], and Kantro, et al. [7]. It is probably not surprising that polar molecules such as water display values of σ^0 which are sensitive to the nature of the substrate. Using nitrogen adsorption to evaluate A_{sp}, then using the latter to evaluate σ^0 for water has led to values of 12.5, 10.4, and 11.4 Å^2 for amorphous silica, calcium hydroxide, and calcium silicate hydrate, respectively.

In spite of a variety of objections to the BET theory, V_m values from N_2 adsorption studies have become a very common means for determining specific surface areas. As a matter of fact, a IUPAC commission was organized in 1969 to study N_2 adsorption with the objective of preparing reference standards for surface area determinations. In this project, four silicas and four carbon blacks of different particle size were investigated independently in 13 different laboratories. Nitrogen adsorption data were analyzed by the BET method over the best fit region, and values of V_m were converted to A_{sp} using 16.2 Å^2 as the value of σ^0. Table 8.1. shows the values of A_{sp} and c obtained in the 10 laboratories that studied the silica of Fig. 8.6. The average value of the specific surface from these determinations is 163.4 ($\pm 6\%$) $m^2\,g^{-1}$. This value agrees with our analysis of Fig. 8.6 obtained by combining Eqs. (99) and (103):

$$A_{sp} = \frac{(38.5)(6.02 \times 10^{23})(16.2 \times 10^{-20})}{(22{,}400)} = 168\ m^2\,g^{-1} \tag{104}$$

Samples of this material as well as others investigated are now available for calibration purposes as gas adsorption standards. Details may be found in the reference cited for the data of Fig. 8.6.

At first glance, it may seem surprising that the BET method is as successful as it is in the evaluation of A_{sp}. The model upon which the isotherm is derived includes some improbable assumptions, and the values of c are not particularly close to theoretical predictions. Some understanding of its success can be found in the following observation. With multilayer adsorption, the particular differences arising from different adsorbents get quickly masked after the first layer is fairly full, that

TABLE 8.1 *Values of* A_{sp} and c *as determined in* 10 *different laboratories by the BET method for the same silica sample shown in Fig.* 8.6[a]

Laboratory	A_{sp} ($m^2\,g^{-1}$)	c
A	166.4	92
B	162.8	101
C	174.0	100
D	148.5	166
E	173.5	70
F	166.0	98
G	167.9	91
H	143.6	113
I	169.5	62
J	161.7	122
Average	163.4	102
Standard deviation	10.0 (6%)	29 (28%)

[a] From D. H. Everett, G. D. Parfitt, K. S. W. Sing, and R. Wilson, *J. Appl. Chem. Biotechnol.* **24**:199 (1974).

is, above $p/p_0 \simeq \frac{1}{3}$ or so. With this idea in mind, it is not difficult to understand that a log–log plot of V/V_m versus p/p_0 gives a common curve when nitrogen adsorption is studied on a variety of nonporous solids. Another way to say this is that log–log plots of V versus p/p_0 would describe a family of curves of the same shape but displaced vertically from one another by an amount equal to log V_m. The only significant contributor to the differences in adsorption for $p/p_0 \gtrsim \frac{1}{3}$ is the difference in V_m for the solids involved. The BET method provides a relatively easy way to evaluate this quantity.

Until now, we have limited our discussion of gas adsorption to theories and results with no mention of the experimental techniques used. In the following section, we shall turn our attention to experimental methods.

8.9 BET SURFACE AREAS: EXPERIMENTAL METHODS

In this section, we shall discuss the experimental aspects of surface area determination by gas adsorption. The discussion is oriented toward the BET theory since it is most commonly used. Actually, all that is done is to measure the volume of gas adsorbed as a function of pressure, so any method of analyzing the data may be applied subsequently.

A gas adsorption apparatus is designed to enable us to measure the pressure, temperature, and volume of gas samples before and after contact with the adsorbent. From these data, the number of moles of gas initially exposed to the sample and the number at equilibrium can be calculated. The actual practice is somewhat more complicated that this summary indicates. Furthermore, many students are unfamiliar with high vacuum techniques. Therefore, we shall discuss a typical procedure in some detail. In doing this, however, we restrict ourselves to a consideration of those manipulations which are involved in the actual collection of data. The purification of the gases, evacuation of the system to less than 10^{-3} torr, filling the gas reservoirs, installation of the sample, and a variety of checks of the high vacuum system are assumed to be already completed.

Figure 8.7 is a partial sketch of an apparatus which could be used for gas adsorption measurements. The high vacuum pumps and low-pressure measuring gauges (McLeod) are not shown. Further, the apparatus is not drawn to scale. The gas storage bulbs, for example, are of the order of several liters and the calibrated gas buret has a capacity of about 0.1 liter. In the figure, the region within the dotted lines which includes the sample space has been especially enlarged.

In Table 8.2 a typical adsorption experiment is outlined in terms of the apparatus shown in Fig. 8.7. The experiment is described by considering the objective of various subroutines, the actual manipulations of the apparatus, the nature of the data collected, and the quantities calculated at each stage. The following notation is used in outlining this procedure:

1. p, V, and T correspond to pressure, volume, and temperature.
2. The subscripts b and d refer to the buret and "dead space," respectively; the dead space is the volume of the sample space beyond stopcock 4.

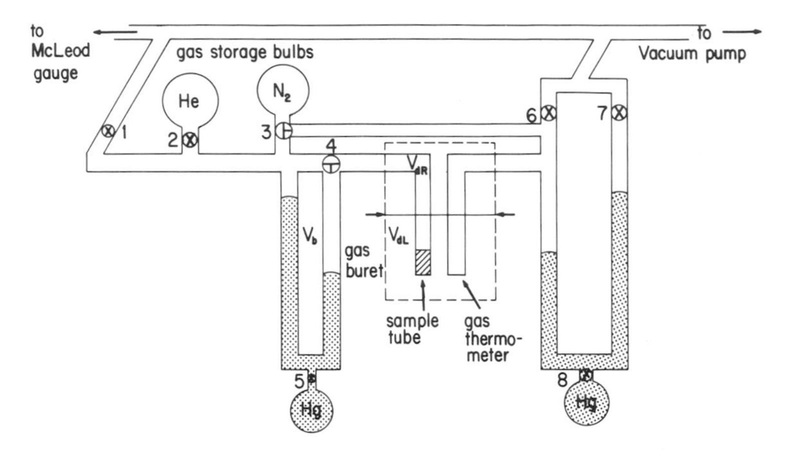

FIGURE 8.7 *A schematic illustration of a high vacuum apparatus which could be used for low-temperature gas adsorption studies, V_b, V_{dR}, and V_{dL} are defined in the text.*

3. The subscripts R and L refer to room and liquid nitrogen (low) temperatures, respectively.
4. The numerical subscripts refer to a set of readings which apply under the same conditions; they are numbered consecutively in the order in which they are determined.
5. Primed measurements refer to helium (or some other gas which is not adsorbed at low temperatures) and the unprimed quantities refer to nitrogen (or some other adsorbed gas).

The procedure merely consists of successive applications of the ideal gas law. With helium as the working gas, the volume of the dead space and its distribution between room temperature and liquid nitrogen temperature are determined. Then the helium is replaced by nitrogen. Any difference between the pVT behavior of the nitrogen and the helium is attributed to the specific adsorption of the nitrogen. Successive increments of nitrogen are added to give the desired number of data points for the BET (or some other) analysis.

In each case, enough time must be allowed for equilibrium to be established. A way of checking whether equilibrium conditions have been achieved is to compare adsorption and desorption measurements. Desorption is measured by the same procedure outlined in Table 8.2 except that gas is removed from the buret rather than added between subsequent determinations.

Throughout this discussion, we have assumed that the temperature of the liquid nitrogen bath was known. This is measured during the experiment by means of the gas thermometer which is also submerged in the liquid nitrogen bath next to the sample tube. The gas thermometer works in the following way. After initial evacuation, the second manometer is filled with N_2 (close 6 and 7, open three-way stopcock 3 to connect N_2 reservoir with manometer). The mercury level in this manometer may be adjusted (at 8) so that $p > p_0$ to assure that liquid–vapor equilibrium is established in the submerged side-arm. The vapor pressure of the N_2 is then determined from the difference of column heights in the manometer, and the temperature corresponding to this nitrogen vapor pressure is determined from a suitable calibration curve.

Although we shall not discuss their use quantitatively, several additional factors might be mentioned:

1. A correction for the nonideality of the nitrogen at low temperatures might be required unless the volume of the low-temperature dead space is minimized.
2. Gas burets should be thermostated unless the laboratory has very good temperature control.
3. Sufficient time must be allowed for thermal equilibration before readings are made if the mercury in the manometer and that in the reservoir are at different temperatures.
4. The gases used for adsorption must be of high purity.

The prior history of the solid sample may have an effect on its specific surface. After the solid is introduced into the sample tube and the latter is attached to the vacuum line, it is generally pretreated by some sort of degassing procedure. This is a combination of heating and pumping to ensure the removal of physically adsorbed contaminants. Consideration must be given to the possibility of changing the surface area of the solid when the temperature of degassing is selected. A heat treatment which is too vigorous may result in changes in any chemisorbed layer, which—from our point of view, at least—amounts to a change in the adsorbent itself. If extensive enough, such changes may alter the surface area of a solid. Even less drastic changes are sufficient to alter the adsorption energy of a surface.

Although we have often mentioned the adsorption energy in this chapter, we have not yet discussed any procedure by which this can be measured quantitatively, except as some sort of an average quantity. Since most solid surfaces are heterogeneous, it is desirable to be able to examine adsorption energy in a more discriminating way, for example, as a function of coverage or pretreatment. In the following section, we shall see how this can be done.

8.10 HEATS OF ADSORPTION

As we have seen, an adsorption isotherm is one way of describing the thermodynamics of gas adsorption. However, it is by no means the only way. Calorimetric measurements can be made for the process of adsorption and thermodynamic parameters may be evaluated from the results. To discuss all of these in detail would require another chapter. Rather than develop all the theoretical and experimental aspects of this subject, therefore, it seems preferable to continue focusing on adsorption isotherms, extracting as much thermodynamic insight from this topic as possible. Within this context, results from adsorption calorimetry may be cited for comparison, without a full development of this latter topic.

TABLE 8.2 *Summary of an experiment in which low-temperature gas adsorption is measured. See text for definition of symbols*

Objective	General procedure	Stopcock manipulations in terms of Fig. 8.7	Quantities meaured	Equation relating measured variables
Determine dead space volume (V_d)	Admit He to buret	1 and 3 closed, 2 open, 4 open to connect manifold and buret only	p'_{b1}, V'_{b1}, T_R	—
	Admit He to sample	4 open to connect buret and sample only	p'_{b2}, V'_{b2}, T_R	$\frac{p'_{b1}V'_{b1}}{T_R}=\frac{p'_{b2}(V'_{b2}+V_d)}{T_R}$ (solve for V_d)
Determine distribution of dead space between room temperature (V_{dR}) and low temperature (V_{dL})	Position liquid nitrogen bath	Liquid N_2 maintained at constant (but arbitrary) level of arrows	$p'_{b3}, V'_{b3}, T_R, T_L$	$\frac{p'_{b1}V'_{b1}}{T_R}=\frac{p'_{b3}V'_{b3}}{T_R}+\frac{p'_{b3}V_{dR}}{T_R}+\frac{p'_{b3}V_{dL}}{TL}$ $V_{dR}+V_{dL}=V_d$ (solve for V_{dL} and V_{dR})

Replace gases	Remove He	4 open to both buret and sample, 1 open	—	—
	Introduce N_2	1 and 2 closed, 3 and 4 open to connect storage bulb to buret	p_{b4}, V_{b4}, T_R	—
Determine moles N_2 adsorbed at equilibrium pressure p_{b5}		4 open to connect buret and sample	p_{b5}, V_{b5}, T_R, T_L	$\frac{p_{b4}V_{b4}}{T_R}=\frac{p_{b5}V_{b5}}{T_R}+\frac{p_{b5}V_{dR}}{T_R}+\frac{p_{b5}V_{dL}}{T_L}+Rn_{ads}$ (R = gas constant; solve for n_{ads})
Determine moles N_2 adsorbed at equilibrium pressure p_{b7}		Close 4, open 5, apply pressure to Hg reservoir	p_{b6}, V_{b6}, T_R	—
		Open 4 to connect buret and sample	p_{b7}, V_{b7}, T_R, T_L	$\frac{p_{b6}V_{b6}}{T_R}=\frac{p_{b7}V_{b7}}{T_R}+\frac{p_{b7}V_{dR}}{T_R}+\frac{p_{b7}V_{dL}}{T_L}+Rn_{ads}$ (solve for n_{ads})

The approach we shall follow is essentially that used to derive the Calpeyron equation. Suppose we consider an infinitesimal temperature change for a system in which adsorbed gas and unadsorbed gas are in equilibrium. The criterion for equilibrium is that the free energy of both the adsorbed (s) and unadsorbed (g) gas change in the same way:

$$dG_s = dG_g \tag{105}$$

The following equations may be written for these two quantities if the temperature change is assumed to cause no change in the amount of adsorbed material:

$$dG_g = -S_g\,dT + V_g\,dp \tag{106}$$

and

$$dG_s = -S_s\,dT + V_s\,dp \tag{107}$$

Substituting Eqs. (106) and (107) into (105) and rearranging gives

$$\left(\frac{\partial p}{\partial T}\right)_{n_s} = \frac{S_g - S_s}{V_g - V_s} \tag{108}$$

In writing this last result, it has been explicitly noted that the number of moles of adsorbed gas n_s is constant. If the process under consideration is carried out reversibly, $S_g - S_s$ may be replaced by q_{st}/T where q_{st} is known as the isosteric (the same coverage) heat of adsorption:

$$S_g - S_s = \frac{q_{st}}{T} \tag{109}$$

Combining Eqs. (108) and (109) leads to the result

$$\left(\frac{\partial p}{\partial T}\right)_{n_s} = \frac{q_{st}}{T(V_g - V_s)} \tag{110}$$

Equation (110) may be integrated if the following assumptions are made: (a) that $V_g \gg V_s$, so that V_s may be neglected; (b) that the gas behaves ideally so that the substitution $V_g = RT/p$ may be used; (c) that q_{st} is independent of T. With these assumptions, Eq. (110) integrates to

$$\ln\frac{p_1}{p_2} = -\frac{q_{st}}{R}\left(\frac{1}{T_1} - \frac{1}{T_2}\right) \tag{111}$$

Equation (111) shows that the isosteric heat of adsorption is evaluated by comparing the equilibrium pressure at different temperatures for samples showing the same amount of surface coverage. The data of Fig. 8.3 may be used as an example to see how this relationship is applied.

For an arbitrarily chosen extent of adsorption, a horizontal line such as the dashed line in Fig. 8.3 may be drawn which cuts the various isotherms at different pressures. The pressure coordinates of these intersections can be read off the plot. According to Eq. (111), a graph of $\ln p$ versus $1/T$ should be linear with a slope of

TABLE 8.3 *Values of the isosteric heat of adsorption at different values of θ for the data shown in Fig.* 8.3 *as evaluated by Eq.* (111)

θ	q_{st} (kcal mole^{-1})	Differential energy of adsorption $E_d = q_{st} - RT$ (kcal mole^{-1})	$E_d - 5.4$ for comparison with Langmuir E_{ads} (kcal mole^{-1})
0.06	13.6	13.1	7.7
0.08	11.3	10.8	5.4
0.10	11.1	10.6	5.2
0.20	10.6	10.1	4.7
0.30	9.9	9.4	4.0
0.40	9.6	9.1	3.7
0.50	9.6	9.1	3.7
0.60	9.8	9.3	3.9
0.70	9.1	8.6	3.2
0.80	8.9	8.4	3.0

$-q_{st}/R$. From Fig. 8.3, for example, when the adsorption is 0.102 g ethyl chloride (g charcoal)$^{-1}$ (which corresponds to $\theta = 0.2$) the equilibrium pressures are 0.20, 0.63, and 2.40 torr at -15.3, 0, and 20°C, respectively. When plotted in the manner just described, these data yield a line of slope -5330 K. Multiplication by R gives $q_{st} = 10.6$ kcal mole^{-1} as the isosteric heat of adsorption for this system at $\theta = 0.2$. Table 8.3 lists values of q_{st} for different θ values as calculated from the data in Fig. 8.3.

Figure 8.3 contains the same data that were analyzed according to the Langmuir equation in Sec. 8.5. Equation (61) shows that the data are consistent with an adsorption energy of about 3.3 kcal mole^{-1} according to the Langmuir interpretation. In presenting this result initially, the remark was made that an independent determination of the adsorption energy would be desirable. Although the present reinterpretation of the same data is not exactly an "independent" determination, it does extract an energy quantity from the experimental results which is free of any assumed model for the mode of adsorption. Accordingly, it is informative to compare the two interpretations. At first, a large discrepancy seems to exist: $q_{st} \simeq 10$ kcal mole^{-1} whereas $E_{ads} \simeq 3.3$ kcal mole^{-1}. Several considerations must be included, however, before a valid comparison can be made.

1. Equation (111) yields an enthalpy, not an energy. Comparison with the analogous expression for the temperature variation of vapor pressure makes this clear. Subtracting RT from the enthalpy converts it to an energy at the same level of approximation that is involved in going from Eq. (110) to (111). This amounts to subtracting about 0.5 kcal mole^{-1} from each value of q_{st} to give a quantity known as the differential energy of adsorption, E_d. These values are also listed in Table 8.3.
2. The interpretation of the experimental data according to the Langmuir equation in Sec. 8.5 was done with p/p_0 as the pressure variable rather than p itself.

Accordingly, the constant which is evaluated is c not K. This is seen by examination of Eq. (101). Equation (82) shows that the exponential energy in c is the difference between the adsorption energy and the energy of vaporization. For ethyl chloride near 0°C the energy of vaporization is about 5.4 kcal mole^{-1}. This should be subtracted from the differential energies of adsorption to give a quantity which is comparable to the energy in Eq. (61). Values resulting from this subtraction are also listed in Table 8.3.

With these considerations included, the Langmuir adsorption energy and the isosteric heats of adsorption can be compared. The Langmuir model assumes that a single energy applies to all adsorption sites. Therefore, any data which are analyzed according to this model cannot yield more than one energy. The isosteric heat of adsorption, on the other hand, is evaluated at different degrees of surface coverage. The data in Table 8.3 show that this quantity definitely varies with coverage, tending to level off as $\theta \rightarrow 1$. This is consistent with the picture of the more active "hot spots" being covered first. The average energy which the Langmuir analysis yields is approximately the same as the energy toward which q_{st} converges, after the suitable corrections are considered.

There are several additional thermal quantities besides q_{st} and E_d which may be generically called "heats of adsorption." One of these is the integral heat of adsorption Q_n which is related to the isosteric heat of adsorption as follows:

$$Q_n = \int_0^n q_{st}\, dn \tag{112}$$

It applies to the process in which n moles of adsorbate are transferred from the bulk gas to the surface, starting from a bare surface. The integral heat of adsorption is determined from data such as those contained in Table 8.3 by graphical integration.

In addition, there are a number of different calorimetric methods to determine heats of adsorption. For example, we may distinguish between isothermal and adiabatic heats, depending on the type of colorimeter involved. Of course, thermodynamic relationships exist between these various quantities. We shall not pursue these topics, but the reader should be aware of the differences and seek precise definitions if the need arises.

The data shown in Fig. 8.8 indicate both the kind of data that may be obtained by direct calorimetric study of gas adsorption and some evidence of the effect of preheating on the properties of surfaces. The figure shows the colorimetric heat of adsorption of argon on carbon black. The broken line indicates the behavior of the untreated black, and the solid line is the "same" adsorbent after heating at 2000°C in an inert atmosphere, a process known as graphitization. The horizontal line indicates the heat of vaporization of argon.

There are several interesting aspects of this figure which are quite generally observed.

1. The untreated carbon black shows the effect of surface heterogeneity, an effect which becomes smeared out as the coverage increases. The surface of the untreated black contains a certain amount of oxygen in a variety of functional

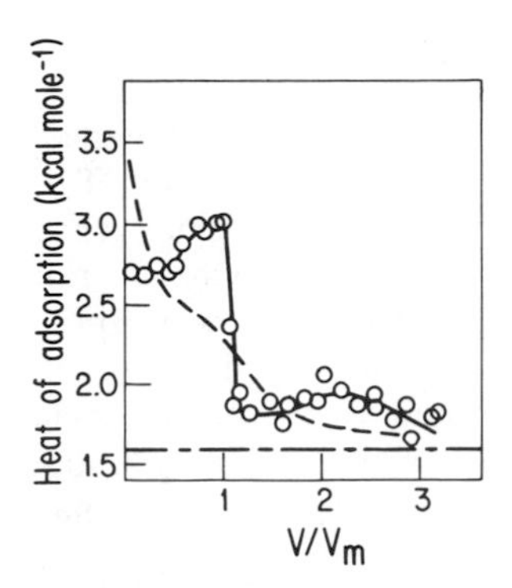

FIGURE 8.8 *Calorimetric heats of adsorption as a function of coverage for argon on carbon black at* 78 K. *Dashed line represents untreated black, solid line is after graphitization at* 2000°C. *Horizontal line is heat of vaporization of argon.* [*Reprinted from R. A. Beebe and D. M. Young,* J. Phys. Chem. **58**:93 (1954), *copyright by the American Chemical Society.*]

groups (e.g., ether, carbonyl, hydroxyl, and carboxyl). Graphitization results in both the reduction of these oxygen-containing groups and the sharpening of both the basal and prismatic crystallographic planes. Electron micrographs of carbon black before and after graphitization are shown in Fig. 1.5.

2. After graphitization, the most notable feature is the sharp discontinuity at monolayer coverage. Beyond the monolayer, the heat of adsorption is close to the heat of vaporization (as required by the BET theory) but does show some influence of the surface as well. At coverage below $V/V_m = 1$, the heat of adsorption increases with increasing coverage, probably due to lateral interactions between the adsorbed molecules.

Figure 8.8 is an extreme example of the effect of heating on the properties of an adsorbent. Degassing prior to measuring an isotherm is done at far less severe conditions; nevertheless, these conditions should always be reported when adsorption studies are conducted because of the possibility of surface modification upon heat treatment.

Another calorimetric technique for measuring the heat of adsorption consists of comparing the heat of immersion (see Sec. 6.10) of bare solid with that of a solid which has been pre-equilibrated with vapor to some level of coverage. Table 8.4 summarizes some results of this sort. The experiment consisted of measuring the heats of immersion of anatase (TiO_2) in benzene with the indicated amount of water vapor preadsorbed on the solid. Small quantities of adsorbed water increase the heat of immersion more than threefold so that it approaches the value for water itself. Most laboratory samples will be contaminated with adsorbed water. Unless the material has been carefully pretreated, the actual nature of the surface may be quite different from what would be expected nominally.

8.11 CAPILLARY CONDENSATION AND ADSORPTION HYSTERESIS

High specific area solids of the type that are studied by gas adsorption consist of small particles for which the radius of an equivalent sphere is given by Eq. (1.2). This figure may be a fairly reasonable measure of a characteristic linear dimension even of irregularly shaped particles, provided the gas adsorption is restricted to the exterior surface of the particles. Any cracks or pores in the solid particles will expose additional adsorbing surfaces and increase the total specific area of the material. Because of this complication, we have specified nonporous solids at several places in this chapter. The particles of high specific area solids are small to begin with, so it follows immediately that any pores in these solids will necessarily have very small dimensions.

Porous solids have been the object of considerable research. In this section, however, we shall restrict our attention to only one aspect of their adsorption behavior: the hysteresis they display in their adsorption isotherms. A schematic illustration of the phenomenon is shown in Fig. 8.9. Although the region enclosed by the hysteresis loop may have a variety of shapes, in all cases there are two quantities of adsorbed material for each equilibrium pressure in the hysteresis range. That branch of the loop which corresponds to adsorption (increasing pressure) inevitably displays less adsorption at any given pressure than the desorption branch (decreasing pressure). In many cases, hysteresis loops such as this are reproducible, although they are not reversible in the thermodynamic sense. Since irreversible processes are involved, the substitution of q_{st}/T for ΔS in the derivation of Eq. (111) is not valid. One can go through the motions of evaluating $\partial p/\partial T$ for a system which displays hysteresis, but the "apparent q_{st} values" so obtained (one for each branch) are not easily related to calorimetric adsorption energies.

Adsorption hysteresis is definitely associated with porous solids, so we must examine porosity for an understanding of the origin of this effect. As a first

TABLE 8.4 *Effect of traces of adsorbed water on the heat of immersion of TiO_2 in benzene*[a]

Amount of water adsorbed ($mmol\ kg^{-1}$)	Heat of immersion ($ergs\ cm^{-2}$)
0.0	150
2.0	250
4.0	320
10.0	450
17.0	506
Pure H_2O	520

[a] Reprinted with permission from G. E. Boyd and W. D. Harkins, *J. Am. Chem. Soc.* **64**:1195 (1942), copyright by the American Chemical Society.

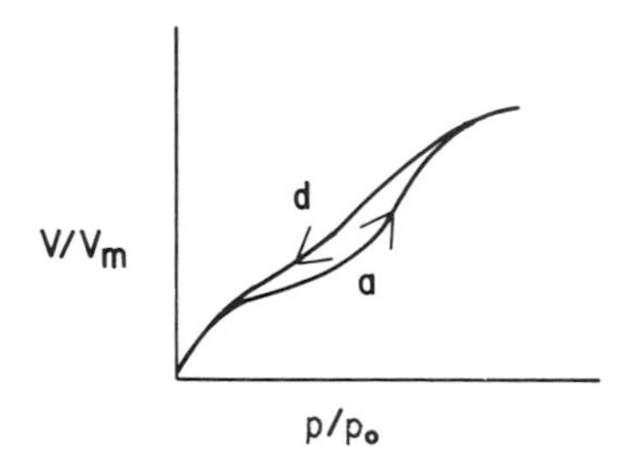

FIGURE 8.9 *A schematic illustration of hysteresis between the adsorption (a) and desorption (d) branches of an experimental isotherm.*

approximation, we may imagine a pore to be a cylindrical capillary of radius r. As just noted, r will be very small. The surface of any liquid condensed in this capillary will be described by a radius of curvature which is related to r. According to the Laplace equation [Eq. (6.30)], the pressure difference across a curved interface increases as the radius of curvature decreases. This means that vapor will condense in small capillaries at pressures less than the normal vapor pressure p_0 which is defined for flat surfaces. The condensation of vapors in small capillaries is an equilibrium phenomenon, however, so capillary condensation in itself does not account for hysteresis. It does point out the fact that a liquid–vapor surface is also involved in the adsorption on porous solids for $p < p_0$.

For simplicity, let us assume that the liquid condensed in a pore has a surface which is part of a sphere of radius R with $R > r$. For a spherical surface, we may use the Kelvin equation [Eq. (6.3)] to calculate p/p_0:

$$N_A kT \ln \frac{p}{p_0} = -\frac{2M\gamma}{\rho R} \tag{113}$$

A minus sign has been introduced in the Kelvin equation since the radius is measured outside the liquid in this application whereas it was inside the liquid in the derivation of Chap. 6. In hysteresis, adsorption occurs at relative pressures that are higher than those for desorption. According to Eq. (113), this is as if adsorption–condensation took place in larger pores than desorption–evaporation. Since the pore dimensions are presumably constant, we must seek some mechanism consistent with this observation to explain hysteresis.

Figure 8.10 contains sketches for several different models of pores which will be useful in our discussion of capillary condensation. Figure 8.10a is the simplest, attributing the entire effect just described to variations in pore radius with the depth of the pore. That is, when liquid first begins to condense in the pore, the larger radius r_a determines the pressure at which the adsorption–condensation occurs. Once the pore has been filled and the desorption–evaporation branch is being studied, the smaller radius r_d determines the equilibrium pressure. Although bottle-necked pores of this sort may exist in some cases, this model seems far too specialized to account for the widespread occurrence of hysteresis.

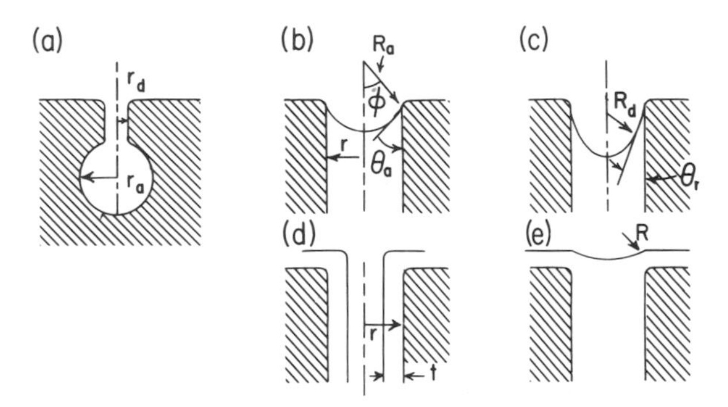

FIGURE 8.10 *Five models for capillary condensation. Radius of pore equals r, radius of curvature of spherical meniscus is* ***R****, and t is thickness of adsorbed layer. Subscripts a and d refer to adsorption and desorption.*

Figures 8.10b and 8.10c represent another model, based on contact angle hysteresis. These sketches represent the surface of liquid in a capillary during the adsorption and desorption stages of hysteresis, respectively. In Fig. 8.10b, the capillary is filling; in 8.10c, it is emptying. Accordingly, the advancing and receding values of the contact angle apply to adsorption and desorption, respectively. The radius of the spherical surface R and the radius of the capillary r are related through the contact angle θ and its complementary angle ϕ (see Fig. 8.10):

$$r = R \sin \phi = R \cos \theta \tag{114}$$

Substituting Eq. (114) into (113) gives

$$N_A kT \ln \frac{p}{p_0} = -\frac{2M\gamma \cos \theta}{\rho r} \tag{115}$$

For a pore of constant radius, the equilibrium pressure decreases as $\cos \theta$ increases or as θ decreases. It will be recalled from Chap. 6 that advancing contact angles are larger than receding ones. Therefore, this model is consistent with the observation that desorption–evaporation occurs at lower relative pressures than adsorption–condensation. The only objection to this explanation of adsorption hysteresis is that it makes no reference whatsoever to adsorption!

Figures 8.10d and 8.10e illustrate another model for adsorption hysteresis which considers multilayer adsorption explicitly. During adsorption, the capillary is viewed as a cylinder of radius $r - t$, where t is the thickness of the adsorbed layer at that pressure. This is represented by Fig. 8.10d. For such a surface, the Kelvin equation becomes

$$N_A kT \ln \frac{p}{p_0} = -\frac{M\gamma}{\rho(r-t)} \tag{116}$$

since one of the radii of curvature is infinite. As multilayer adsorption proceeds, however, the thickness of the adsorbed layer equals r. Once the pore is thus filled, its surface becomes a meniscus which may be treated by Eq. (115) as a portion of a sphere, as shown in Fig. 8.10e. In the event that $\theta = 0$, Eq. (113) applies with $R = r$. Comparison of adsorption and desorption in this situation is particularly easy if $t \ll r$ in Eq. (116). In that case, the right-hand side of Eq. (116) is proportional to $1/r$, and comparison with Eq. (113) reveals that

$$\left(\frac{p}{p_0}\right)_d^{1/2} = \left(\frac{p}{p_0}\right)_a \tag{117}$$

Since both of these ratios are less than unity, $p_a > p_d$. This model is qualitatively consistent with the observed hysteresis, but is difficult to apply quantitatively because it neglects differences between the adsorbed material on the surface and that in the bulk liquid.

This third mechanism for adsorption hysteresis clearly imposes some sort of upper limit to the formation of a multilayer: the condition in which $t = r$. Suppose we assume that this limiting thickness occurs after n layers have been deposited. In that case, the summations involved in the BET derivation may not be carried out for $1 < i < \infty$, but must be truncated at $i = n$. This leads to a modification of the BET equation to the following form:

$$\frac{V}{V_m} = \frac{cx[1-(n+1)x^n + nx^{n+1}]}{(1-x)[1+(c-1)x - cx^{n+1}]} \tag{118}$$

where $x = p/p_0$ and V_m and c have the same meaning as before. In this form, the BET theory results in a three-parameter equation; the maximum number of layers n must be specified. Equation (118) reduces to (91) as $n \to \infty$ and to (101) as $n \to 1$. The three-parameter equation is virtually indistinguishable from the two-parameter equation for $n > 3$ in the range of applicability of the BET theory. For values of n less than this, however, it extends the range of relative pressures over which the BET theory applies. An even better fit of the data with a more logical justification is obtained by truncating only the numerator of Eq. (86) after n terms and allowing the numerator to take on values to $i = \infty$. Only in the latter case is the total surface of the adsorbent represented. This leads to

$$\frac{V}{V_m} = \frac{cx(1-x^n)}{(1-x)[1+(c-1)x]} \tag{119}$$

an even simpler result than Eq. (118).

Gas adsorption data may be analyzed for the distribution of pore sizes. What is generally done is to interpret one branch of the isotherm and use an appropriate equation to calculate the effective pore radius at a given pressure. The amount of material adsorbed or desorbed for each increment or decrement in pressure measures the volume of pores with that effective radius.

8.12 WRAP-UP: CHAPTERS 6 THROUGH 8

In Chaps. 6 to 8 attention has been focused on the properties of the surfaces which separate phases. This contrasts with the first five chapters which emphasized finely subdivided particles as kinetic units. The obvious connection is simply this: The properties of surfaces play an increasingly important role as the size of the dispersed particles decreases.

Many of the topics we have discussed in Chaps. 6 through 8 appeared in more than one place. Some of these recurring themes might be mentioned again by way of summary:

1. We saw that there is a general thermodynamic equivalency among all types of interfaces even though there are very real differences in the way they are handled experimentally. The idea, for example, of treating γ as a force is useful when discussing the shapes assumed by liquid surfaces; treating it as a free energy is more satisfying for solids. Likewise, it is easier to measure γ than Γ at liquid surfaces, whereas for rigid surfaces the reverse is true.
2. We saw that surface excess concentrations lower the free energy of an interface. This not only provides us with an insight as to the driving force responsible for adsorption but also shows us how to study adsorption by means of surface tension or vice versa.
3. We also encountered the notion of two-dimensional states of matter. This point of view permits us to visualize many phenomena in terms of the more familiar properties of bulk matter.
4. We saw that solid surfaces are likely to be rough and chemically heterogeneous, both of which factors make them energetically heterogeneous as well. Thus, measurements on solids as well as liquid surfaces are especially susceptible to errors arising from contamination.

A number of important topics have been omitted from this survey of surface chemistry which probably would have been included had the topic of the book been surface chemistry alone. Subjects such as chemisorption and catalysis or the connection between adsorption and chromatography are marginal to the *combined* discussion of colloid and surface chemistry and have been omitted for this reason. Other subjects which would nicely supplement the material presented—such as surface viscosity, surface spectroscopy, and adsorption calorimetry—were not developed for lack of space. Finally, there are topics that just marginally approach our area of interest which are also omitted because of space limitations. Liquid crystals, for example, can possess order in two dimensions, which is reminiscent of insoluble monolayers, and molecular sieves may be regarded as extreme examples of porous solids.

In these chapters, two topics have been mentioned—one has been partly developed, the other totally postponed—which will play major roles in the remaining chapters. We saw in Chap. 6 how one component of surface tension, the dispersion component, could be evaluated. We also saw that this component is influential at distances somewhat removed from the actual surface. In Chap.

10, we shall return to this concept when we discuss the flocculation of dispersions. At several points in Chap. 7, we mentioned the possibility that ions might be adsorbed, especially those of amphipathic structure. The adsorption of ions will obviously impart an electrical charge to the surface. These electrical effects are the concern of Chaps. 9 and 11, and—along with van der Waals forces—Chap. 10 as well.

In the first five chapters of this book, we were concerned with dispersed particles almost exclusively. Chapters 6 to 8 involved surfaces exclusively, without regard to the state of subdivision of the phases per se. In the remaining three chapters, properties originating in the surface region which affect the behavior of dispersions are discussed. This will complete the merger of surface and colloidal considerations.

REFERENCES

1. A. W. Adamson, *Physical Chemistry of Surfaces* (2nd ed.), Wiley-Interscience, New York, 1967.
2. J. C. P. Broeckhoff and R. H. van Dongen, "Mobility and Adsorption on Homogeneous Surfaces," in *Physical and Chemical Aspects of Adsorbents and Catalysts* (B. G. Linsen, ed.), Academic Press, New York, 1970.
3. S. Brunauer, L. E. Copeland, and D. L. Kantro, "The Langmuir and BET Theories," in *The Solid–Gas Interface*, Vol. 1 (E. A. Flood, ed.), Marcel Dekker, New York, 1967.
4. J. H. de Boer, *The Dynamical Character of Adsorption* (2nd ed.), Oxford University Press, London and New York, 1968.
5. S. J. Gregg, *The Surface Chemistry of Solids* (2nd ed.), Chapman & Hall, London, 1961.
6. J. P. Hobson, "Physical Adsorption at Extremely Low Pressures," in *The Solid–Gas Interface*, Vol. 1 (E. A. Flood, ed.), Marcel Dekker, New York, 1967.
7. D. L. Kantro, S. Brunauer, and L. E. Copeland, "BET Surface Areas—Methods and Interpretations," in *The Solid–Gas Interface*, Vol. 1 (E. A. Flood, ed.), Marcel Dekker, New York, 1967.
8. S. Ross, "Monolayer Adsorption on Crystalline Surfaces," in *Progress in Surface and Membrane Science*, Vol. 4 (J. F. Danelli, M. D. Rosenberg, and D. A. Cadenhead, eds.), Academic Press, New York, 1971.

PROBLEMS

1. An isotherm which is not too difficult to derive by the methods of statistical mechanics assumes an adsorbed layer which obeys the two-dimensional analog of the van der Waals equation. The result of such a derivation is the equation [cf. Eq. (27)]

$$\ln p = \ln \frac{\theta}{1-\theta} + \frac{\theta}{1-\theta} - \frac{2a}{bkT}\theta + \ln K$$

where a, b, and K are constants, the first two being the two-dimensional van der Waals constants. Like its three-dimensional counterpart, this equation predicts that at the critical point for the two-dimensional matter,

$$\left(\frac{\partial \ln p}{\partial \theta}\right)_{T_c} = 0 \qquad \text{and} \qquad \left(\frac{\partial^2 \ln p}{\partial \theta^2}\right)_{T_c} = 0$$

From the second of these derivatives, evaluate θ_c as predicted by this model. Use this value of θ_c and the first of these derivatives to evaluate the relationship between T_c and the two-dimensional a and b constants. How does this result compare with the three-dimensional case? The van der Waals b constant is four times the volume of a hard sphere molecule. What is the relationship between the two dimensional b value and the area of a hard disk molecule?

2. Use the linear form of the Langmuir equation to evaluate $(n/w)_{sat}$ and K for the adsorption of pentane on carbon black from the following data. Use the ratio p/p_0 rather than p only to normalize pressures relative to the equilibrium vapor pressure of pentane at different temperatures.*

T(°C)	−63.7		0		5.24		20.5	
p_0(mm)	3.48		187.5		235.6		445.1	
	p(mm)	g C_5 g^{-1}	p(mm)	g C_5 g^{-1}	p(mm)	g C_5 g^{-1}	p(mm)	g C_5 g^{-1}
	0.024	0.2827	0.0533	0.1062	3.8	0.2299	0.284	0.1061
	0.028	0.2904	0.1405	0.1322	7.2	0.2535	0.675	0.1317
	0.067	0.3162	0.203	0.1427	19.6	0.2939	0.950	0.1421
	0.103	0.3276	0.890	0.1908	36.7	0.3147	3.62	0.1884
	0.233	0.3459	2.5	0.2305	53.6	0.3231	9.7	0.2262
	0.288	0.3507	5.0	0.2506	88.5	0.3313	15.9	0.2469
	0.671	0.3581	14.6	0.2964	199.9	0.3418	62.9	0.3002
	1.46	0.3647	29.2	0.3184			90.4	0.3107
			45.4	0.3272			155.2	0.3204
			80.7	0.3353			328.7	0.3321
			161.0	0.3433			428.7	0.3372

3. Examine the temperature variation of K values from the preceding problem by means of the form given by Eq. (61). (To decrease computational effort, various members of the class may be assigned different temperatures to analyze in Problem 2. The K values may then be pooled for this problem.)

4. The accompanying data give the volume of N_2 at STP adsorbed on colloidal silica at the temperature of liquid nitrogen as a function of the ratio p/p_0.† Plot these results according to the linear form of the BET equation. Evaluate c, V_m, and A_{sp} from these results, using 16.2 Å^2 as the value for σ^0.

* M. Polanyi and F. Goldmann, *Z. Phys. Chem.* **132**:321 (1928).

† D. H. Everett, G. D. Parfitt, K. S. W. Sing, and R. Wilson, *J. Appl. Chem. Biotechnol.* **24**:199 (1974).

V @ STP (cm^3 g^{-1})	p/p_0	V @ STP (cm^3 g^{-1})	p/p_0
44	0.008	117	0.558
52	0.025	122	0.592
57	0.034	130	0.633
61	0.067	148	0.692
64	0.075	165	0.733
65	0.083	194	0.775
70	0.142	204	0.792
77	0.183	248	0.825
78	0.208	296	0.850
85	0.275		
90	0.333		
96	0.375		
100	0.425		
109	0.505		

5. The following data give the volume at STP of nitrogen and argon adsorbed on the samc nonporous silica at $-196°C$*:

	V @ STP (cm^3 g^{-1})	
p/p_0	Nitrogen	Argon
0.05	34	23
0.10	38	29
0.15	43	32
0.20	46	38
0.25	48	41
0.30	51	43
0.35	54	45
0.40	58	50
0.45	58	54
0.50	61	55
0.60	68	62
0.70	77	69
0.80	89	79
0.90	118	93

Using 16.2 Å^2 as the N_2 cross section, calculate A_{sp} for the silica by the BET method. What value of σ^0 is required to give the same BET area for the argon data?

6. Use the accompanying data† to criticize or defend the following proposition: "Self-consistent A_{sp} values for nonporous solids are obtained at 77 and 90 K by using values of σ^0 for N_2 equaling 16.2 and 17.0 Å^2, respectively. These are consistent with the density of liquid N_2 at these two temperatures. For the same

* D. A. Payne, K. S. W. Sing, and D. H. Turk, *J. Colloid Interface Sci.* **43**:287 (1973).

† K. M. Hanna, I. Odler, S. Brunauer, J. Hagymassy, and E. E. Bodor, *J. Colloid Interface Sci.* **45**:27 (1973); also Osipow [7], in Chap. 7.

self-consistency in A_{sp} using O_2 as the adsorbate, σ^0 values of 14.3 and 15.4 Å^2 must be used at these two temperatures. This suggests that O_2 is somewhat more loosely packed on the surface than in the liquid state at 90 K compared to 77 K."

	N_2		O_2	
T (K)	ρ (g cm^{-3})	V_{sp} (cm^3 g^{-1})	ρ (g cm^{-3})	V_{sp} (cm^3 g^{-1})
77	0.808	1.238	1.204	0.831
90	0.751	1.332	1.14	0.877

7. Use the data from Problem 2 to estimate the isosteric heat of adsorption of pentane on carbon black at $\theta \simeq 0.3, 0.6$, and 0.9. Under what conditions would greater variation of q_{st} be expected? What prevents these conditions from being examined in this problem? How does q_{st} compare with the energy of adsorption for this system as determined in Problem 3? How does the difference between q_{st} and ε compare with ΔH_v for pentane?
8. Colloidal carbon was formed by the slow pyrolysis of polyvinylidene chloride. Surface oxidation occurs during subsequent storage in air. By heating at elevated temperatures, the following gases are evolved.* Heats of immersion were also measured after the different thermal pretreatments.

	Cumulative amount desorbed (mmol g^{-1})			
T (°C)	CO	CO_2	H_2	Heat of immersion (cal g^{-1})
100	0.03	0.02	—	11.0
200	0.05	0.14	—	9.8
300	0.15	0.33	—	9.2
400	0.35	0.44	—	8.7
500	0.63	0.51	—	8.2
600	0.95	0.55	—	7.7
700	1.30	0.55	—	7.0
800	1.58	0.55	—	6.2
900	1.80	0.55	0.03	6.0
1000	1.823	0.556	0.269	5.9

Using 7.9 and 9.1 Å^2 as the areas of oxide desorbing as CO and CO_2, respectively, estimate the area (in m^2 g^{-1}) occupied by each of these oxide types. If the specific area of the carbon is 1100 m^2 g^{-1}, what percentage of the surface is covered with oxide? Discuss the variation of the heat of immersion with the removal of surface oxides.
9. Samples of rutile TiO_2 were outgassed at elevated temperatures: In some instances this was followed by subsequent exposure to O_2 at 150°C. The

* S. S. Barton, M. J. B. Evans, and B. H. Harrison, *J. Colloid Interface Sci.* **45**:542 (1973).

following results were obtained by the BET analysis of N_2 adsorption after various preliminary treatments*:

Pretreatment	A_{sp} ($m^2\ g^{-1}$)	c
150°C+O_2	11.0	200
210°C+O_2	11.7	100
200°C, no O_2	11.9	520
240°C, no O_2	11.3	370
470°C+O_2	11.7	450
560°C+O_2	11.9	1210

Criticize or defend the following proposition: "After the initial removal of physically adsorbed water, partial dehydroxylation of the surface occurs with increasing temperature. In the absence of O_2, Ti^{3+} cations are present which interact with N_2 pretty much the same way as the isolated hydroxyl groups on the surface. At still higher temperatures, surface diffusion permits the hydroxyl groups to migrate into patches, which show stronger interactions with N_2. The specific surface is not essentially changed throughout treatment."

* G. D. Parfitt, D. Urwin, and T. J. Wiseman, *J. Colloid Interface Sci.* **36**:217 (1971).

The Electrical Double Layer

. . . Suppose that I had the power of passing through . . . things, so that I could penetrate my subjects, one after another, even to the number of a billion, verifying the size and distance of each by the sense of feeling. [From Abbott's *Flatland*]

9.1 INTRODUCTION

When ions are present in a system that contains an interface, there will be a variation in the ion density near that interface which is described by a profile like that shown in Fig. 7.11. The boundary we identify as *the* surface defines the surface excess charge. Suppose that it were possible to separate the two bulk phases at this boundary in the manner shown in Fig. 6.13. Then each of the separated phases would carry an equal and opposite charge. These two charged portions of the interfacial region are called the electrical double layer.

The electrical double layer is the subject of this chapter. We shall begin by a phenomenological discussion of two specific charged interfaces: those occurring in the reversible AgI electrode and those occurring in the polarizable mercury electrode. These electrode systems will introduce us to some of the variables which are useful in a discussion of the double layer.

After these phenomenological considerations, we shall turn our attention to various models for the distribution of charge near the surface. Although we shall examine several different models under several limiting conditions, most of the theoretical developments of this chapter will involve the following assumptions: (a) planar surfaces, (b) isolated surfaces, and (c) constant potential surfaces, and will be examined specifically for (d) the variation with distance from the surface of the potential and (e) the effect of added electrolyte on the potential.

In discussing these points, we shall see the following:

1. How to apply results derived for planar surfaces to curved surfaces.
2. How the results for isolated surfaces are modified when a pair of surfaces are brought together.
3. How the potential can be used to evaluate the charge density at the interface.

This chapter is primarily concerned with the double layer per se. The application of these ideas to the stability of lyophobic colloids and to electrokinetic phenomena are the topics of Chaps. 10 and 11, respectively.

9.2 REVERSIBLE ELECTRODES: THE SILVER IODIDE ELECTRODE

To arrive at an understanding of the distribution of charge and potential near an interface, it is helpful to consider some specific electrodes. First of all, it is convenient to distinguish between two types of electrodes: reversible and polarizable. A reversible electrode is one in which each of the phases contains a common ion which is free to cross the interface. The system Ag–AgI–aqueous solution is an example of a reversible electrode. A polarizable electrode, on the other hand, is impermeable to charge carriers, although charge may be brought to the surface by the application of an external potential. The system metallic Hg–aqueous solution is an example of a polarizable electrode, if the aqueous solution is free of ions which are easily oxidized or reduced. Of course, there are many intermediate situations between completely reversible and completely polarizable electrodes, but we shall restrict our attention to these two extremes for simplicity.

It is also convenient to divide ions into two categories: potential determining and indifferent ions. The terminology here is self-explanatory. For example, we can say that Ag^+ is potential determining for the Ag–Ag^+ electrode and that $NaNO_3$ is an indifferent electrolyte as far as this potential is concerned. This obviously neglects any effects of $NaNO_3$ on the activity of the Ag^+. Such an approximation increases in accuracy as the concentration of electrolyte decreases. We shall consistently neglect activity corrections in this chapter.

Now let us turn our attention to each of these electrodes separately to describe how they behave with respect to charge and potential. After considering these two actual systems, we shall develop some models to describe charged interfaces.

The solubility product constant for AgI is about 7.5×10^{-17} at 25°C. This means that the equilibrium concentration of Ag^+ and I^- in a saturated solution of AgI in pure water equals about 8.7×10^{-9} mole liter^{-1}. Electrokinetic experiments (Chap. 11) on AgI particles under these conditions reveal that the particles carry a negative charge in this case. Common ion sources such as $AgNO_3$ or KI may be added to the solution to vary the proportions of the Ag^+ and I^- ions in solution, subject to the condition that the ion product equals K_{sp}. When this is done, it is found that the AgI particles reverse charge at a Ag^+ concentration of about 3.0×10^{-6} mole liter^{-1}. When the concentration of Ag^+ is greater than this, the particles are positively charged. For Ag^+ concentrations less than 3.0×10^{-6} *M*, they are negatively charged.

One way of understanding these results is to consider the Ag^+ and I^- ions competing for adsorption sites on the surface. The tendency of both kinds of ions to adsorb at the AgI interface is not hard to understand. After all, the solid crystals would continue to grow if more ions were present. At the point of zero charge, the two kinds of ions are adsorbed equally (in stoichiometric proportion). Negatively charged particles imply the adsorption of excess I^- ions, whereas positively charged particles imply excess Ag^+ adsorption. Since the zero point of charge and the saturation concentration in pure water do not coincide, we infer that the I^- ions have a greater affinity for the surface. In terms of surface excesses, we may define the charge density of the surface σ (not to be confused with the area per molecule for which the same symbol was used in earlier chapters)

$$\sigma = e(\Gamma_{Ag} - \Gamma_I) \tag{1}$$

where e is the unit (positive) charge.

The Nernst equation provides us with a relationship which permits an electrical potential difference to be associated with a concentration difference. We shall adopt the convention that the potential at the AgI–solution interface is zero at the zero point of charge (zp), a point at which the ion molarities will be symbolized c_{zp}. Our interest is to express the potential at the interface ψ_0 in terms of the concentration of ions in solution for conditions other than the zero point of charge. The Nernst equation gives

$$\psi_0 = \frac{kT}{e} \ln \frac{c}{c_{zp}} = \frac{2.303RT}{F} \log \frac{c}{c_{zp}} \tag{2}$$

where F is the Faraday constant. We are accustomed to using the second form of Eq. (2) in physical and analytical chemistry. The quantity $2.303RT/F$ has the familiar numerical value 0.05917 V at 25°C. Multiplying this by 10^3 and dividing by 2.303 gives 25.7 mV as the value of kT/e. We shall verify this result shortly, but this is a convenient way to relate a familiar numerical constant to the units which are most often used in surface and colloid chemistry.

Suppose we apply Eq. (2) to AgI in "pure water" (i.e., no common ion source present). We use $c_{Ag} = 8.7 \times 10^{-9}$ and $c_{Ag,zp} = 3.0 \times 10^{-6}$ to calculate:

$$\psi_0 = 25.7 \ln \frac{8.7 \times 10^{-9}}{3.0 \times 10^{-6}} = -150 \text{ mV} \tag{3}$$

Identical results would be obtained if the calculation had been based on I^- concentrations rather than Ag^+. As noted, the surface is negatively charged at this concentration since I^- is preferentially adsorbed.

In principle, part of the potential of any cell may be attributed to each interface. That is, if ψ_i is the potential drop associated with the ith interface, we can write for the total potential difference ψ_T

$$\psi_T = \psi_1 + \psi_2 + \cdots + \psi_i \tag{4}$$

Any electrochemical cell containing a Ag–AgI electrode automatically includes the AgI–solution interface and the potential associated therewith. Generally speaking, we are not able to assign absolute numerical values to the various contributions in Eq. (4). We can, however, design cells such that only one of the interfaces is sensitive to a particular ion. Clearly, Ag^+ and I^- are the potential-determining ions at the AgI–solution interface. If none of the other interfaces in the cell are appreciably affected by changes in the concentrations of these ions, then variations in the experimental cell potential ψ_T measure changes in ψ_0. This may be expressed

$$d\psi_T = d\psi_0 = \frac{kT}{e}\frac{dc}{c} \tag{5}$$

where c is the concentration of the potential-determining ion. This result is important because it shows how *changes* in ψ_0 can be measured even if ψ_0 itself is unknown. That is, to integrate Eq. (5) back to an absolute value of ψ_0, an integration constant must be evaluated. According to our convention, this involves knowing the zero point of charge ($\psi_0 = 0$ at $c = c_{zp}$).

9.3 POLARIZABLE ELECTRODES: THE MERCURY ELECTRODE

Next, let us turn our attention to the polarizable Hg–solution interface. In this case, no specific forces are recognizable, which could result in the preferential adsorption of one kind of ion at the interface. However, the metallic mercury can be used as an inert electrode along with some reference electrode, and an external voltage source can be added to examine the effects of variations in the electrolyte solution. The advantage of using mercury over other noble metals is that the surface tension of mercury may be readily measured as a function of the applied voltage and the ion content of the solution.

Figure 9.1 illustrates an apparatus which is convenient for this purpose. Since mercury has a contact angle (measured in the mercury) which is greater than 90°, the mercury–solution interface is depressed (again in reference to the mercury). If an etched mark is placed on the capillary, it is possible to bring the mercury–solution meniscus to that mark by varying the height of the mercury reservoir. It turns out that this is quite sensitive to the potential between the electrodes, E. The height of the capillary depression (a negative capillary "rise") can be readily converted to the interfacial tension through Eq. (6.49). We see, therefore, that γ for the Hg–solution interface depends on the electrical potential across the system. In addition, the detailed shape of the so-called electrocapillary curve—a plot of γ versus E—depends on the concentration and nature of the electrolyte present. Figures 9.2a and 9.2b show examples of typical electrocapillary curves.

Several generalizations are evident from an inspection of Fig. 9.2:

1. The general shape of the curves is roughly parabolic. The coordinates of the maximum in the electrocapillary curve depend on the electrolyte content of the

FIGURE 9.1 *Schematic illustration of an apparatus to measure the electrocapillary effect.*

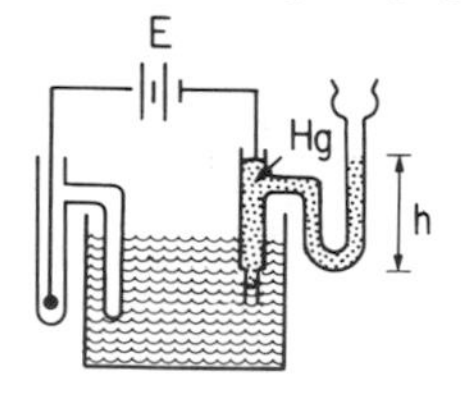

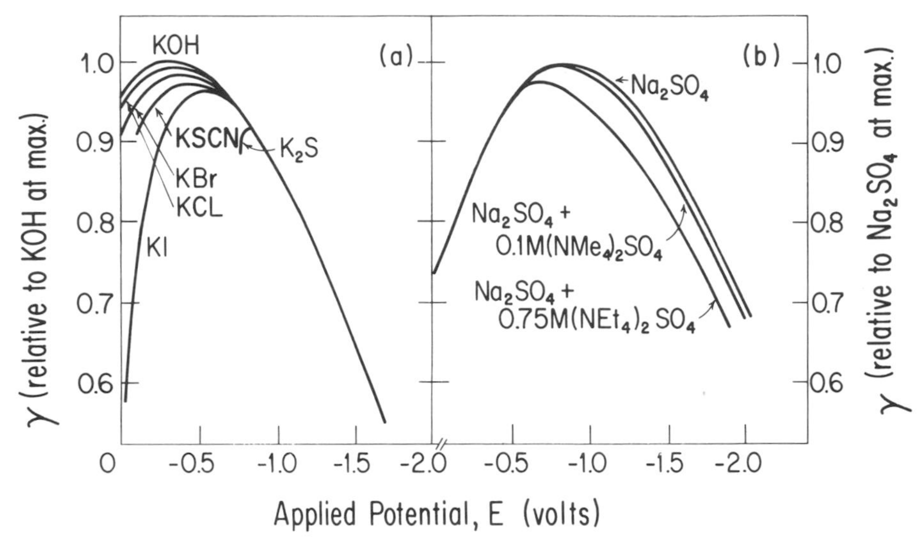

FIGURE 9.2 *Typical electrocapillary curves. (a) Anions are adsorbed; (b) cations. (From Adam* [1], *used with permission.)*

system. Since γ decreases on both sides of the capillary maximum and since reductions in γ are associated with adsorption, we conclude that adsorption increases as we move in either direction from the maximum. That is, the electrocapillary maximum seems to be a point of minimum adsorption.

2. The left-hand branch of the electrocapillary curves (also called the rising or ascending branch) is sensitive to the chemical nature of the anion present. In Fig. 9.2a, for example, different potassium salts are used as the electrolytes. Although the ascending branches of these curves differ, the descending branches (to the right of the maximum) lie on a common curve. Figure 9.2b shows that the right-hand branch of the curves is sensitive to the nature (and concentration) of the cation. These observations, coupled with the preceding remarks about adsorption, suggest that the anion is preferentially adsorbed on the ascending branch of the curve, the cation is preferentially adsorbed on the descending branch, and that neither ion is preferentially adsorbed at the maximum, although both may be adsorbed in equivalent amounts.
3. The charge carriers in the system—ions in the solution and electrons in the mercury—come to an equilibrium distribution which is consistent with the applied potential. To the left of the maximum, the mercury surface is a positively charged anode (this branch is also called the anodic branch) toward which anions are attracted. Electrons in the mercury are repelled from the surface by the anions in the water, so the potential on the mercury must become progressively more positive (relative to the maximum) for increased anion adsorption to occur. To the right of the maximum, the mercury is negative (the cathodic branch), attracts cations, and must become progressively more negative (relative to the maximum) for increased cation adsorption.

4. The potential at which the maximum occurs is different for different ions. In light of item (3), the differences in the voltage coordinates of the maxima must reflect differences in the chemical (as opposed to purely electrostatic) affinities of the ions for the interface. Nonionic solutes have also been investigated extensively, but we shall not go into this aspect of the subject.

Electrocapillary phenomena have been studied for a long time; the apparatus shown in Fig. 9.1 is essentially that used by G. Lippmann in 1875 in his comprehensive studies of electrocapillarity. We shall not examine either the experimental or the theoretical aspects of this system in great detail. However, an interpretation of the results that is more quantitative than that just outlined qualitatively is possible with relatively little additional effort.

As a starting point, it is convenient to return to the kind of argument that leads to the Gibbs–Duhem equation [Eq. (7.33)] for one-phase systems and to the Gibbs adsorption equation [Eq. (7.38)] for systems containing an interface. In the present context, we are interested not only in the interface but also in possible charge effects at that interface. Accordingly, it is convenient to distinguish between charged and uncharged components in the system; we shall use the subscript i to identify the former and j for the latter. In this notation, Eq. (7.38) may be written

$$-d\gamma = \sum_i \Gamma_i \, d\bar{\mu}_i + \sum_j \Gamma_j \, d\mu_j \tag{6}$$

In this expression μ_j is the usual chemical potential for the uncharged species. The quantity $\bar{\mu}_i$ is called the *electro*chemical potential and is related to μ_i as follows:

$$\bar{\mu}_i = \mu_i + z_i e\psi \tag{7}$$

where ψ is the potential of the charged species in the phase in question, and z_i is the valence number of the ith charged species.

In this discussion, the mercury–solution interface is assumed to be completely polarizable; that is, there is no transfer of charge across the interface. Therefore, each of the charged species (including electrons) occurs in only one of the phases. The requirement of complete polarizability means that the surface excess may be divided between that lying in the aqueous phase (W) and that in the mercury (Hg). In view of Eqs. (6) and (7), we write

$$-d\gamma = \sum_i (\Gamma_i \, d\mu_i)_{\mathrm{W}} + \sum_i (z_i \Gamma_i e\psi)_{\mathrm{W}} + \sum_i (\Gamma_i \, d\mu_i)_{\mathrm{Hg}}$$

$$+ \sum_i (z_i \Gamma_i e\psi)_{\mathrm{Hg}} + \sum_j \Gamma_j \, d\mu_j \tag{8}$$

Since the system as a whole is electrically neutral, the charge density on each of the phases must be equal and opposite:

$$\sigma_{\mathrm{W}} = -\sigma_{\mathrm{Hg}} \tag{9}$$

Adopting the present notation in Eq. (1) and combining it with (9), we obtain

$$\sum_i (z_i e \Gamma_i)_{\mathrm{W}} = -\sum_i (z_i e \Gamma_i)_{\mathrm{Hg}} \tag{10}$$

Substituting Eq. (10) into (8) yields

$$-d\gamma = \sum_i (\Gamma_i \, d\mu_i)_W + \sum_i (\Gamma_i \, d\mu_i)_{Hg} + \sum_j \Gamma_j \, d\mu_j + \sum_i (z_i \Gamma_i e)_W (\psi_W - \psi_{Hg}) \tag{11}$$

The difference in potential between the two phases varies as dE, the externally applied potential difference. Therefore, Eq. (11) may also be written

$$-d\gamma = \sum_i (\Gamma_i \, d\mu_i)_W = \sum_i (\Gamma_i \, d\mu_i)_{Hg} + \sum_j \Gamma_j \, d\mu_j + \sum_i (z_i \Gamma_i e)_W \, dE \tag{12}$$

For a completely polarizable electrode, the concentration of each of the components in both solutions is constant and, therefore, so is the chemical potential for each. Therefore, we obtain

$$-\left(\frac{\partial \gamma}{\partial E}\right)_\mu = \sum_i (ze\Gamma_i)_W = \sigma_W \tag{13}$$

This result is known as the Lippmann equation. The charge density on the right-hand side of this equation refers exclusively to the solution phase; therefore, the subscript is no longer retained.

For monovalent ions, Eq. (13) is simply

$$\left(\frac{\partial \gamma}{\partial E}\right)_\mu = -e(\Gamma_+ - \Gamma_-) \tag{14}$$

When the surface excess of anions exceeds that of cations, $d\gamma/dE$ is positive, as is observed in Fig. 9.2a. Negative values of $d\gamma/dE$ correspond to larger surface excesses of cations, as shown by Fig. 9.2b. Finally, the condition $d\gamma/dE = 0$ corresponds to equal amounts of positive and negative adsorbed charge, that is, surface neutrality. Note that this is not the same as saying no ions are adsorbed. The slope of the electrocapillary curve measures the *difference* between the cation and anion surface excesses.

Another interpretation of the electrocapillary curve is easily obtained from Eq. (11). We wish to investigate the effect of changes in the concentration of the aqueous phase on the interfacial tension at constant applied potential. Several assumptions are made at this point to simplify the desired result. More comprehensive treatments of this subject may be consulted for additional details, for example the chapter by Overbeek [6]. We assume (a) that the aqueous phase contains only 1:1 electrolyte, (b) that the solution is sufficiently dilute to neglect activity coefficients, (c) that the composition of the metallic phase (therefore μ_i^{Hg}) is constant, (d) that only the potential drop at the mercury-solution interface is affected by the composition of the solution, and (e) that the Gibbs dividing surface can be located in such a way as to make the surface excess equal to zero for all uncharged components ($\Gamma_i = 0$). With these assumptions, Eq. (11) becomes

$$-d\gamma = RT(\Gamma_+ + \Gamma_-)_W \, d \ln c + e(\Gamma_+ - \Gamma_-)_W \, dE \tag{15}$$

where c is the concentration of the electrolyte. If we specify constant applied potential, Eq. (15) becomes

$$-\left(\frac{\partial \gamma}{\partial \ln c}\right)_E = RT(\Gamma_+ + \Gamma_-)_W \tag{16}$$

This result shows that the vertical displacements (at fixed potential) of the electrocapillary curve with changes in electrolyte concentration measure the *sum* of the surface excesses at the solution surface. Curves such as those in Fig. 9.2b may be interpreted by this result. We have already seen that $\Gamma_+ = \Gamma_-$ at the electrocapillary maximum (where $E = E_{max}$); therefore,

$$-\left(\frac{\partial \gamma}{\partial \ln c}\right)_{E_{max}} = 2RT\Gamma_{+,w} \tag{17}$$

A final result that can be extracted from the Lippmann equation is readily obtained by differentiating Eq. (13) with respect to E at constant μ:

$$-\left(\frac{\partial^2 \gamma}{\partial E^2}\right)_\mu = \left(\frac{\partial \sigma}{\partial E}\right)_\mu = C \tag{18}$$

The quantity $d\sigma/dE$ is called the differential capacitance C of an interface. Although experimental values of surface capacitance provide valuable information about the distribution of charge at an interface, we shall not consider the experimental aspects of this topic. Our interest in Eq. (18) is primarily in the suggestion it offers that a double layer may be viewed as a capacitor. In the following section, we shall use the parallel plate capacitor as a model for the electrical double layer.

In an abbreviated fashion, we have described both the charge density and the potential drop at two different kinds of interfaces: reversible and polarizable. We have seen that both the potential drop and the charge density at an interface depend on the concentration of ions, their valence, and their specific tendencies to adsorb. The abundant relationships of thermodynamics make it possible to derive phenomenological equations relating these various quantities. One thing that thermodynamics does not do, however, is to provide information about the details of these phenomena at the molecular level. Although we have seen that there is a potential drop across certain interfaces and a charge density at the face of adjoining phases, we have no information as to how the potential or the density of charge carriers varies as we move through a small distance (the order of molecular dimensions) perpendicular to the surface. Again we are reminded of Fig. 7.11 in which the profile of the variation of some general property in the immediate vicinity

FIGURE 9.3 *The variation of electrochemical potential in the vicinity of the interface between two phases, α and β. (a) According to a schematic profile, (b) according to the parallel plate capacitor model.*

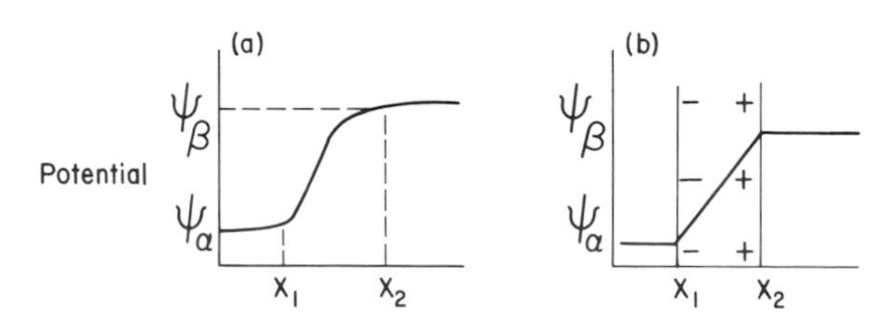

of a surface is shown. Figure 9.3a is essentially the same drawing with the property under discussion specified to be the potential.

The question we shall consider in the next few sections is this: How does the potential vary with distance across an interface? This question cannot be answered by thermodynamics alone, but it can be examined in terms of various models. We shall consider a succession of models for a planar surface between two phases. The models will become progressively more complex—and therefore realistic—as we proceed. As far as this presentation is concerned, models will be proposed and modified in an intuitive way, rather than by critique of each in terms of experimental results.

9.4 THE CAPACITOR MODEL OF THE DOUBLE LAYER

Figure 9.3a shows schematically the situation in which the potential equals ψ_α at a position x_1 a small distance into the α phase, and equals ψ_β at x_2, a small distance into the β phase. One of our major goals in this chapter is to study the details of the potential variation between x_1 and x_2.

A vastly oversimplified model of how this potential variation might occur on a molecular scale (remember that the distance between x_1 and x_2 is on the order of molecular dimensions) is shown in Fig. 9.3b. In this representation, two smeared-out planes of charge are situated at x_1 and x_2. Note that the model shown in Fig. 9.3b resembles a parallel plate capacitor in which two charged conducting surfaces separated by a dielectric occur with a potential difference $\Delta\psi$ between them.

Our interest in this chapter and in Chaps. 10 and 11 is centered primarily on that part of the double layer which extends into the aqueous solution, which is the continuous phase in many important systems. There may be some interfaces between water and a second phase in which the charge on the nonaqueous phase is essentially concentrated on the surface plane. The rigid alignment of a second layer of counterions in the aqueous solution is implausible, however, because of thermal agitation, which tends to diffuse the ions throughout the solution. For now, the parallel plate capacitor model will get us started by reviewing some basic relationships and units from elementary physics. The diffuse model of the double layer will be taken up in later sections.

Coulomb's law is the basic point of departure. It may be written

$$F = \frac{kqq'}{\varepsilon r^2} \tag{19}$$

to describe the force operating between two charges q and q' separated by a distance r. The factor ε is the dielectric constant of the medium and the proportionality constant k depends on the units chosen for the calculation. When cgs units are used for F and r, Coulomb's law itself defines the unit of charge, known as the electrostatic unit or the statcoulomb. In this case, $k = 1$ in Eq. (19). If mks units are used, charge is measured in coulombs, which is an independently defined unit. Therefore, to accommodate the coulomb along with other mks units in Coulomb's law, the constant k has a value of about 9×10^9 N m^2 C^{-2}.

Potential is also measured in different units in the two systems: volts in the mks system and statvolts in the cgs–esu system. Their magnitudes are such that about 300 (ordinary) volts equals one statvolt. Table 9.1 summarizes some of these differences between the two systems. Some additional notes on the two systems of units can be found in Appendix C.

Volts and coulombs are practical electrical units: A point in favor of the mks system. The mks system also has the advantage of being favored by international commissions. On the other hand, the proportionality factor k in Coulomb's law is unity for the cgs system: This is an advantage of the latter system. Since much of the existing literature in colloid and surface chemistry is based on cgs–electrostatic units, we shall continue to use these units. There is no way to escape the confusion which results from multiple systems of units. The only thing to do is to be aware of the differences and use the required conversions when necessary.

A final point of confusion is that the factor 4π appears in various ways in the two systems because of the two different relationships between the unit of charge and the unit of force. Although we certainly cannot afford to be too causal about the factor 4π, the reader should be prepared to encounter formulas in other places which differ in this regard. Beginning derivations from different definitions of charge comprise the fundamental source of this complication.

Next, let us consider the definition of the strength of an electric field $\bar{E}$ (care must be taken not to confuse this quantity with applied potential E). The field $\bar{E}$ describes the force per unit charge in an electrically influenced environment

$$\bar{E} = \frac{F}{q} \tag{20}$$

Now suppose we bring two identical $+q$ charges toward one another to a distance of separation r. Combining Eqs. (19) and (20) enables us to calculate the field at that

TABLE 9.1 *The relationship between cgs–electrostatic units and mks–SI units for some common electrical quantities*

	Electrostatic–cgs	MKS–SI
Charge, q	esu or statcoulombs	coulombs
	(2.998×10^9 statcoul = 1 coulomb)	
Electron	4.8×10^{-10} esu	1.6×10^{-19} coulomb
Potential, V	statvolt	volt
	(1 statvolt = 299.8 volts)	
Energy, E (qV)	erg (statvolt × statcoulomb)	joule (volt × coulomb)
Coulomb's law proportionality constant ($F = kqq'/r^2$)	1 dyne cm^2 statcoulomb^{-2}	8.987×10^9 newtons meter^2 coulomb^{-2}

separation:

$$\bar{E} = \frac{q}{\varepsilon r^2} \tag{21}$$

This is precisely the same as the force that a unit positive charge would experience at the same location. Since force is the gradient of the potential, Eq. (21) also supplies a second definition of field:

$$\bar{E} = \frac{d\psi}{dx} \tag{22}$$

where ψ is the potential and x is the separation of the plates.

A fiction which helps us understand electric fields is the notion of lines of force. Suppose we imagine 4π lines of force as emanating from each unit of positive charge. The factor 4π is associated with spherical symmetry and we may readily imagine an isolated charge to produce a spherically symmetrical field. If the charge has a magnitude of $+q$, then $4\pi q$ would be the total number of lines of force produced by this particular charge.

A radial distance r from this central charge, the lines of force cut across a spherical surface of area $4\pi r^2$. If we divide the number of lines of force by the cross-sectional area, we obtain

$$\frac{\text{Number of lines}}{\text{Area}} = \frac{4\pi q/\varepsilon}{4\pi r^2} = \frac{q}{\varepsilon r^2} \tag{23}$$

This result is identical to the field as given by Eq. (21). Hence, the number of lines of force per unit area is a convenient way of describing an electric field. A unit field, in this representation, corresponds to one line per square centimeter. Now suppose we apply this idea to a parallel plate capacitor.

In the case of a capacitor—taken here as a prototype of a double layer—the charges are not isolated. Instead, the lines of force emanating from one charged surface terminate at an opposite charge on the other plate of the capacitor. Figure 9.4a represents such a situation when the plates are separated by a vacuum. Suppose a plate of area A carries q charges; then we define the charge density σ as

$$\sigma = \frac{q}{A} \tag{24}$$

Since 4π lines of force are associated with each unit of charge, there are $4\pi q/A$ lines of force crossing the evacuated gap between the two plates of the capacitor. As we have already seen, the number of lines of force measures the field; therefore, we write for a capacitor which contains a vacuum

$$\bar{E}_0 = \frac{4\pi q}{A} = 4\pi\sigma \tag{25}$$

If a substance with a dielectric constant ε is placed between the plates, the field will be less by a factor equal to the dielectric constant. This is because of the partially compensating field which is induced within the dielectric by dipole orientation, as

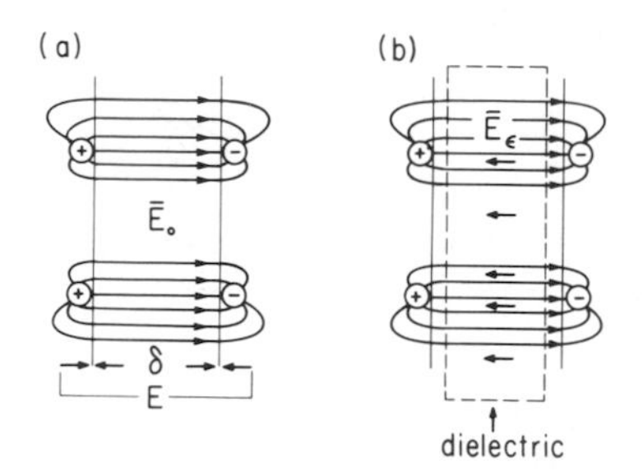

FIGURE 9.4 *The electric field in a parallel plate capacitor. (a) The dielectric is a vacuum; (b) a material of dielectric constant ε is present.*

suggested by Fig. 9.4b. Therefore, in the presence of the dielectric, the field is given by

$$\bar{E}_\varepsilon = \frac{4\pi q}{\varepsilon A} = \frac{4\pi}{\varepsilon}\sigma \tag{26}$$

Next, suppose we equate (22) and (26) to obtain

$$\frac{d\psi}{dx} = \frac{\Delta\psi}{\delta} = \frac{4\pi}{\varepsilon}\sigma \tag{27}$$

where $\Delta\psi$ is the potential drop between plates separated by a distance δ. This equation relates the charge density, voltage difference, and distance of separation of the capacitor. Since this is the model we are using for the double layer, it is of interest to check whether Eq. (27) agrees—at least qualitatively—with what we know about the double layer.

We saw in Chap. 7 that charged monolayers are likely to obey the two-dimensional ideal gas law, and we also saw that areas per molecule of 10^3 Å^2 or so were also required for this ideal law to apply. Hence we may estimate σ for a monovalent ion to be

$$\sigma = \frac{\text{ion}}{10^3\ \text{Å}^2} \times \frac{10^{16}\ \text{Å}^2}{1\ \text{cm}^2} \times \frac{4.8 \times 10^{-10}\ \text{esu}}{\text{ion}} = 4800\ \text{esu cm}^{-2} \tag{28}$$

Taking the dielectric constant of water to be about 80, its bulk value, Eq. (26) permits the field strength to be estimated:

$$\bar{E} = \frac{4\pi(4800)}{80}\frac{\text{statvolts}}{\text{cm}} \times \frac{300\ \text{V}}{\text{statvolt}} = 2.26 \times 10^5\ \text{V cm}^{-1} \tag{29}$$

Even allowing for an order of magnitude error in this estimate, we see that there is an exceptionally strong field in the vicinity of a charged interface. In Sec. 11.8, we shall examine this in greater detail and consider whether we are justified in using bulk values for such parameters as ε and η within the double layer.

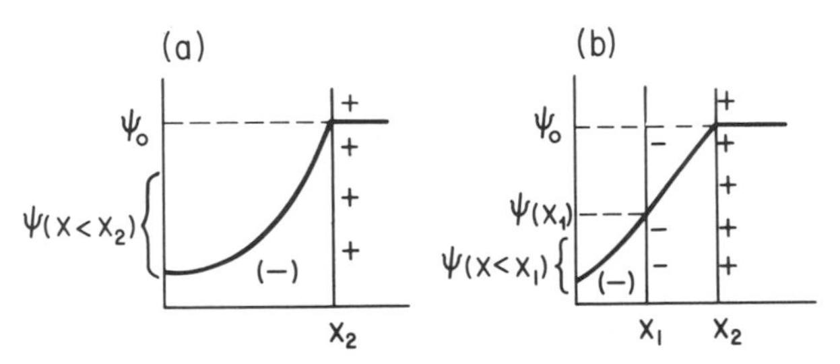

FIGURE 9.5 *Two models for the double layer. (a) A diffuse double layer; (b) charge neutralization due partly to a parallel plate charge distribution and partly to a diffuse layer.*

If we estimate the potential drop between the two phases, we may determine the distance over which the potential drop occurs from the value of $\bar{E}$ given by Eq. (29). Taking the potential difference to be 0.10 V—an arbitrary but reasonable value—Eq. (27) shows that the plate separation of an equivalent capacitor is

$$\delta = \frac{\Delta\psi}{\bar{E}} = \frac{0.10}{2.26 \times 10^5} = 4.4 \times 10^{-7} \text{ cm} \tag{30}$$

Considering the simplicity of the model and the arbitrariness of the numerical estimates made in this calculation, 45 Å seems like a reasonable estimate of the distance over which surface charge neutralization is accomplished.

Throughout this section, we have examined the distribution of charge at an interface as if the charge were constrained to two planes. When one of the phases is an aqueous electrolyte solution, the inadequacy of this model is apparent. An immediate improvement of the model is anticipated if we allow for a diffuse double layer, that is, a situation in which the charge density varies with distance from the interface, as shown in Fig. 9.5a. Alternatively, we might combine features from both the parallel plate distribution and the diffuse distribution to give a still more elaborate picture of the double layer, as shown in Fig. 9.5b. We shall consider this latter situation in Sec. 9.8. According to this picture, each part of the double layer is analyzed independently, and the effects combined according to the rule for adding capacitors in series:

$$\frac{1}{C_T} = \frac{1}{C_1} + \frac{1}{C_2} + \cdots + \frac{1}{C_i} \tag{31}$$

where C_i is the capacitance of the ith element in series and C_T is the total capacitance.

9.5 THE DIFFUSE DOUBLE LAYER: THE DEBYE–HÜCKEL APPROXIMATION

In the preceding section, we discussed the problem of the variation of potential with distance from an interface from the highly artificial perspective of a parallel plate capacitor. The variation of potential with distance from a charged surface of arbitrary shape is a classical electrostatic problem. The general problem is described by the Poisson equation

$$\frac{\partial^2\psi}{\partial x^2}+\frac{\partial^2\psi}{\partial y^2}+\frac{\partial^2\psi}{\partial z^2}=-\frac{4\pi}{\varepsilon}\rho \tag{32}$$

or in terms of the Laplacian operator ∇^2

$$\nabla^2\psi=-\frac{4\pi}{\varepsilon}\rho \tag{33}$$

In these expressions, ρ is the charge density in the system, a quantity which itself is a function of x, y, and z. The solution to this differential equation, therefore, is an expression for the potential ψ which satisfies Eq. (32) and also the boundary conditions of the specific problem. We shall adopt the convention of measuring all distances outward from the interface where the potential has the value ψ_0. As the distance from an isolated surface increases to infinity, the value of ψ approaches zero. The stipulation of an "isolated" surface means that we are concerned with only one interface at this time. In Sec. 9.9, we shall consider the case in which electrical double layers overlap.

The Poisson equation is a fundamental relationship of classical electrostatics and really need not be proved here. However, since we are using it as a starting point, it seems desirable to explore the meaning of this important equation to some extent.

Equation (21) describes the field a distance r from a charge $+q$. A basic law of electrostatics is that this field describes any distribution of charge which results in q units of positive charge being enclosed by a sphere of radius r. It is not critical that a single $+q$ charge be situated at the center for this expression to apply. Suppose, therefore, we consider a portion of solution in which the charge is distributed with a uniform density ρ. In this case,

$$q=\tfrac{4}{3}\pi r^3\rho \tag{34}$$

and

$$\bar{E}=\frac{4\pi r\rho}{3\varepsilon} \tag{35}$$

Next, we multiply both sides of Eq. (35) by r^2 and then differentiate with respect to r:

$$\frac{d}{dr}(r^2\bar{E})=\frac{4\pi r^2\rho}{\varepsilon} \tag{36}$$

In the present notation, Eq. (22) becomes

$$\bar{E}=-\frac{d\psi}{dr} \tag{37}$$

where the minus sign is included since ψ decreases as r increases. Substituting this result into Eq. (36) gives

$$\frac{1}{r^2}\frac{\partial}{\partial r}\left(r^2\frac{\partial\psi}{\partial r}\right)=-\frac{4\pi}{\varepsilon}\rho \tag{38}$$

Remember that the operator $\nabla^2\psi$ in Eq. (33) transforms into the following form in spherical coordinates:

$$\nabla^2\psi = \frac{1}{r^2}\frac{\partial}{\partial r}\left(r^2\frac{\partial\psi}{\partial r}\right) + \frac{1}{r^2 \sin\theta}\frac{\partial}{\partial\theta}\left(\sin\theta\frac{\partial\psi}{\partial\theta}\right) + \frac{1}{r^2\sin^2\theta}\left(\frac{\partial^2\psi}{\partial\phi^2}\right) \tag{39}$$

Thus the left-hand side of Eq. (38) is seen to be identical to $\nabla^2\psi$ for the case (spherical symmetry) in which ψ is independent of θ and ϕ. Although this presentation does not constitute the most general proof of the Poisson equation, it does give it some plausibility.

There are many situations in which the spherically symmetrical case is specifically invoked, as in the Debye–Hückel theory of electrolyte nonideality for example. We shall consider situations for which this is the case in Chap. 11. For now, however, we consider the potential distribution adjacent to a planar wall which carries a positive charge. We shall define the direction perpendicular to the wall as the x direction, and shall consider the wall as extending to infinity in the positive and negative y and z directions. In this case, the operation $\nabla^2\psi$ becomes $d^2\psi/dx^2$, and Eq. (32) is written

$$\frac{d^2\psi}{dx^2} = -\frac{4\pi}{\varepsilon}\rho \tag{40}$$

The next problem is to express the charge density as a function of the potential so the differential equation (40) can be solved for ψ. The procedure is to describe the ion concentrations in terms of the potential by means of a Boltzmann factor in which the work required to bring an ion to a position where the potential is ψ is given by $z_i e\psi$. The probability of finding an ion at this position is given by the Boltzmann factor, with this work appearing as the exponential energy:

$$\frac{n_i}{n_{i0}} = \exp\left(\frac{-z_i e\psi}{kT}\right) \tag{41}$$

In this expression, n_i is the number of ions of type i per cubic centimeter near the surface, and n_{i0} is the concentration far from the surface, that is, the bulk concentration. The valence number z_i is either a positive or negative integer.

The charge density is related to the ion concentrations as follows:

$$\rho = \sum_i z_i e n_i = \sum_i z_i e n_{i0} \exp\left(\frac{-z_i e\psi}{kT}\right) \tag{42}$$

Combining Eqs. (40) and (42) gives a result known as the Poisson–Boltzmann equation:

$$\frac{d^2\psi}{dx^2} = -\frac{4\pi e}{\varepsilon}\sum_i z_i n_{i0} \exp\left(\frac{-z_i e\psi}{kT}\right) \tag{43}$$

This same relationship is the starting point of the Debye–Hückel theory of electrolyte nonideality, except that the latter uses the value of $\nabla^2\psi$ required for spherical symmetry. It is interesting to note that G. Gouy (in 1910) and D. L. Chapman (in 1913) applied this relationship to the diffuse double layer a decade before the Debye–Hückel theory appeared.

The derivation of the Poisson equation implies that the potentials associated with various charges combine in an additive manner. The Boltzmann equation, on the other hand, involves an exponential relationship between the charges and the potential. In this way, a fundamental inconsistency is introduced when Eqs. (40) and (42) are combined. Equation (43) does not have an explicit general solution anyhow, and must be solved for certain limiting cases. These involve approximations which—at the same time—overcome the objection just stated.

We introduce the first of these approximations by considering only those situations in which $z_i e\psi < kT$. In this case, the exponentials in Eq. (42) may be expanded (see Appendix A) as a power series. If only first-order terms in $z_i e\psi/kT$ are retained, Eq. (42) becomes

$$\rho = \sum_i z_i e n_{i0}\left(1 - \frac{z_i e\psi}{kT}\right) \tag{44}$$

Because of electroneutrality, two of the terms in Eq. (44) cancel:

$$\sum_{+\text{and}-} z_i e n_{i0} = 0 \tag{45}$$

so that (44) becomes

$$\rho = -\sum_i \frac{z_i^2 n_{i0} e^2 \psi}{kT} \tag{46}$$

In this approximation, the ion potentials are additive, so Eq. (46) may be consistently substituted into (40) to give

$$\frac{d^2\psi}{dx^2} = \frac{4\pi e^2 \psi}{\varepsilon kT} \sum_i z_i^2 n_{i0} \tag{47}$$

The assumption of low potentials made in reaching this result is also made in the Debye–Hückel theory and prompts us to call this model the Debye–Hückel approximation. Equation (47) has an explicit solution. Since potential is the quantity of special interest in Eq. (47), let us evaluate the potential at 25°C for a monovalent ion which satisfies the condition $ze\psi = kT$:

$$\psi = \frac{kT}{e} = \frac{(1.38\times 10^{-16})(298)}{(4.80\times 10^{-10})} \text{statvolts} \times \frac{300 \text{ V}}{\text{statvolt}}$$

$$= 0.0257 \text{ V} = 25.7 \text{ mV} \tag{48}$$

Thus at 25°C, potentials may be regarded as low or high, depending on whether they are less or more than about 25 mV. The factor $e\psi/kT$ appears often in double layer calculations, so this conversion factor is worth remembering. The relationship of kT/e to RT/F was noted in Sec. 9.2.

It is convenient to identify the cluster of constants in Eq. (47) by the symbol κ^2 which is defined as follows:

$$\kappa^2 = \frac{4\pi e^2 \sum_i z_i^2 n_{i0}}{\varepsilon kT} \tag{49}$$

With this change in notation, Eq. (47) becomes simply

$$\frac{d^2\psi}{dx^2} = \kappa^2\psi \tag{50}$$

Equation (50) has the solution

$$\psi = \psi_0 \exp(-\kappa x) \tag{51}$$

Note that Eq. (51) satisfies the required boundary conditions inasmuch as $\psi \to \psi_0$ as $x \to 0$ and $\psi \to 0$ as $x \to \infty$. In the following section we shall examine the implications of Eq. (51) in detail.

9.6 THE DEBYE–HÜCKEL APPROXIMATION: RESULTS

The Debye–Hückel approximation is strictly applicable only in the case of low potentials. Nevertheless, there are several reasons why the significance of Eq. (51) should be fully appreciated:

1. It is simpler to understand than any of the modifications we shall consider subsequently.
2. It is a limiting result to which all equations that are more general must reduce in the limit of low potentials.
3. The effects of electrolyte concentration and valence in this approximation are qualitatively consistent with the results of more elaborate calculations.

One of the most important quantities to emerge from the Debye–Hückel approximation is the parameter κ. This quantity appears throughout double layer discussions and not merely at this level of approximation. Since the exponent κx in Eq. (51) is dimensionless, κ must have units of reciprocal length. This means that κ^{-1} has units of length. This latter quantity is often (imprecisely) called the "thickness" of the double layer. All distances within the double layer are judged large or small relative to this length. Note that the exponent κx may be written x/κ^{-1}, a form which emphasizes the notion that distances are measured relative to κ^{-1} in the double layer.

Since κ and κ^{-1} are such important quantities, we shall examine them in greater detail, first verifying their dimensions, then considering their numerical magnitude. Especially important is the dependence of κ and κ^{-1} on the concentration and valence of the electrolyte in solution.

It is an easy matter to verify that κ^2 as defined by Eq. (49) does indeed have units of length^{-2} or cm^{-2} in the cgs system. This is done by recalling that q^2/r has units of energy, hence e^2 has units of ergs · cm in the cgs system. Likewise, n has units of cm^{-3} and for κ they are cm^{-1}.

The parameter κ is concentration dependent. Accordingly, it is desirable to express κ in practical concentration units. We begin by noting that n_i is related to the molar concentration of the ions M_i by the expression

$$n_i = \frac{M_i N_A}{1000} \tag{52}$$

Therefore, Eq. (49) yields

$$\kappa = \left(\frac{4\pi e^2 N_A}{1000\varepsilon kT}\sum_i z_i^2 M_i\right)^{1/2} = \left(\frac{8\pi e^2 N_A I}{1000\varepsilon kT}\right)^{1/2} \tag{53}$$

where I is the ionic strength of the solution:

$$I = \tfrac{1}{2}\sum_i z_i^2 M_i \tag{54}$$

We are primarily concerned with aqueous solutions at 25°C for which ε equals 78.54. Table 9.2 lists numerical values for the constants in Eq. (53) to facilitate evaluation of κ and κ^{-1}. The table also lists numerical values for κ and κ^{-1} for several different electrolytes and electrolyte concentrations.

Several things should be noted about Table 9.2:

1. The tabulated values of κ^{-1} multiplied by 10^8 give the double layer "thicknesses" in angstrom units. For example, in a 0.01 M solution of a 1:1 electrolyte, κ^{-1} equals 30.4 Å.
2. This "thickness" is the same magnitude as the prediction based on the capacitor model [Eq. (30)]. The diffuse model is clearly superior, however, since it shows how the double layer "thickness" depends on the concentration and valence of the ions in the solution.
3. The "thickness" of the double layer varies inversely with z and inversely with $M^{1/2}$ for a symmetrical $z:z$ electrolyte. Therefore, κ^{-1} equals 15 Å for a 0.01 M solution of a 2:2 electrolyte and is about 96 Å for a 0.001 M solution of a 1:1 electrolyte.

Figure 9.6 shows how the potential drops with distance from the surface according to Eq. (51). The curves in this figure are drawn for two different variations in κ: in (a), a 1:1 electrolyte at 0.1, 0.01, and 0.001 M concentrations; in (b), a 0.001 M solution of 1:1, 2:2, and 3:3 electrolytes. Again, it is important to recognize that the curves drop off more rapidly for either higher concentrations or higher valences of the electrolyte.

The curves in Fig. 9.6 are marked at the x value that corresponds to κ^{-1}. Note that the potential has dropped to the value ψ_0/e at this point. Calling κ^{-1} the double layer "thickness" is clearly a misnomer. We shall see presently, however, that there is some logic underlying this terminology.

Although the potential is fundamentally a more important quantity than charge density, examining the latter will enable us to compare the capacitor and diffuse models for the double layer.

The condition of electroneutrality at a charged interface requires that the density of charge at the two faces be equal. Note that this does not require the charges to be physically situated at the interface. When one of the phases contains a diffuse layer, the total charge contained in a volume element of solution of unit cross section and extending from the wall to infinity must contain the same amount of charge—although of opposite sign—as a unit area of wall contains. Stated in formula, this

TABLE 9.2 *Values of κ and κ^{-1} for several different electrolyte concentrations and valences. Numerical formulas for these quantities also given for aqueous solutions at 25°C*

		Symmetrical electrolyte			Unsymmetrical electrolyte	
General formulas		$\kappa\ (\text{cm}^{-1}) = 3.29\times10^{7}\lvert z\rvert M^{1/2}$	$\kappa^{-1}\ (\text{cm}) = 3.04\times10^{-8}\lvert z\rvert^{-1}M^{-1/2}$		$\kappa\ (\text{cm}^{-1}) = 2.32\times10^{7}(\sum_i z_i^2 M_i)^{1/2}$	$\kappa^{-1}\ (\text{cm}) = 4.30\times10^{-8}(\sum_i z_i^2 M_i)^{-1/2}$
Molarity	$z_+ : z_-$			$z_+ : z_-$		
0.001	1:1	1.04×10^{6}	9.61×10^{-7}	1:2, 2:1	1.80×10^{6}	5.56×10^{-7}
	2:2	2.08×10^{6}	4.81×10^{-7}	3:1, 1:3	2.54×10^{6}	3.93×10^{-7}
	3:3	3.12×10^{6}	3.20×10^{-7}	2:3, 3:2	4.02×10^{6}	2.49×10^{-7}
0.01	1:1	3.29×10^{6}	3.04×10^{-7}	1:2, 2:1	5.68×10^{6}	1.76×10^{-7}
	2:2	6.58×10^{6}	1.52×10^{-7}	1:3, 3:1	8.04×10^{6}	1.24×10^{-7}
	3:3	9.87×10^{6}	1.01×10^{-7}	2:3, 3:2	1.27×10^{7}	7.87×10^{-8}
0.1	1:1	1.04×10^{7}	9.61×10^{-8}	1:2, 2:1	1.80×10^{7}	5.56×10^{-8}
	2:2	2.08×10^{7}	4.81×10^{-8}	1:3, 3:1	2.54×10^{7}	3.93×10^{-8}
	3:3	3.12×10^{7}	3.20×10^{-8}	2:3, 3:2	4.02×10^{7}	2.49×10^{-8}

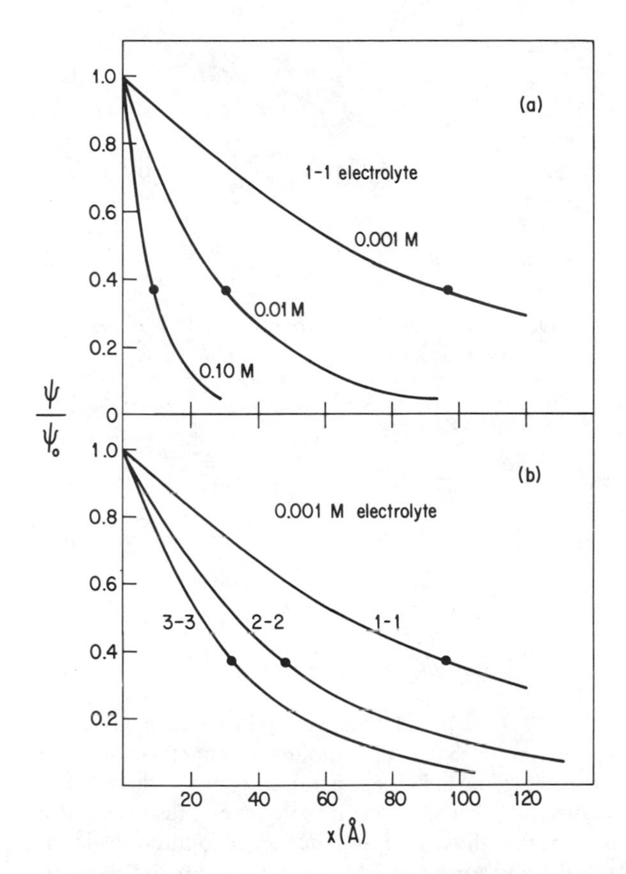

FIGURE 9.6 *Fraction of double layer potential versus distance from a surface according to the Debye–Hückel approximation, Eq.* (51). *(a) Curves drawn for* 1 : 1 *electrolyte at three concentrations; (b) curves drawn for* 0.001 *M symmetrical electrolytes of three different valence types.*

becomes

$$\sigma = -\int_0^\infty \rho \, dx \tag{55}$$

We shall now examine the implications of Eq. (55) for the situation in which one of the adjoining phases contains the diffuse half of a double layer. Combining Eqs. (40) and (55) gives

$$\sigma = \frac{\varepsilon}{4\pi} \int_0^\infty \frac{d^2\psi}{dx^2} \, dx \tag{56}$$

a result which is easily integrated to yield

$$\sigma = \frac{\varepsilon}{4\pi} \frac{d\psi}{dx}\bigg|_0^\infty \tag{57}$$

The quantity $d\psi/dx$ is zero at infinity, and we shall define its value at the wall by $(d\psi/dx)_0$; therefore, Eq. (57) becomes

$$\sigma = -\frac{\varepsilon}{4\pi}\left(\frac{d\psi}{dx}\right)_0 \tag{58}$$

Next, we turn to Eq. (51)—the Debye–Hückel approximation for ψ—to evaluate $(d\psi/dx)_0$. Differentiation leads to the value

$$\left(\frac{d\psi}{dx}\right)_0 = \lim_{x\to 0} -\kappa\psi_0 \exp(-\kappa x) = -\kappa\psi_0 \tag{59}$$

Substituting Eq. (59) into (58) gives

$$\sigma = \frac{\varepsilon}{4\pi}\kappa\psi_0 \tag{60}$$

Rewriting Eq. (60) in terms of κ^{-1}, the double layer "thickness," yields

$$\sigma = \frac{\varepsilon}{4\pi}\frac{\psi_0}{\kappa^{-1}} \tag{61}$$

Equation (61) is identical to Eq. (27) for a parallel plate capacitor with ψ_0 replacing $\Delta\psi$ and κ^{-1} replacing d. This result show that a diffuse double layer at low potentials behaves like a parallel plate capacitor in which the separation between the plates is given by κ^{-1}. This explains why κ^{-1} is called the double layer thickness. It is important to remember, however, that the *actual* distribution of counterions in the vicinity of a charged wall is diffuse, and approaches the unperturbed bulk value only at large distances from the surface.

Even allowing for the fact that the Debye–Hückel approximation applies only for low potentials, this analysis reveals some features of the electrical double layer that are general and are of great importance as far as stability with respect to flocculation of dispersions and electrokinetic phenomena are concerned. In summary three specific items might be noted:

1. The distance away from the wall that an electrostatic potential persists may be comparable to the dimensions of colloidal particles themselves.
2. The distance over which significant potentials exist decreases with increasing electrolyte concentration.
3. The range of electrostatic potentials decreases as the valence of the ions in solution increases.

We shall encounter these same trends in discussing overlapping double layers. These ideas will have considerable value in explaining the effect of electrolytes on the stability of dispersions, which we shall consider in detail in Chap. 10.

The Debye–Hückel approximation to the diffuse double layer problem produces a number of relatively simple equations which introduce a variety of double layer topics as well as a number of qualitative generalizations. In order to extend the range of the quantitative relationships, however, it is necessary to return to the Poisson–Boltzmann equation and the unrestricted Gouy–Chapman theory.

9.7 THE ELECTRIC DOUBLE LAYER: GOUY–CHAPMAN THEORY

The theoretical inconsistencies inherent in the Poisson–Boltzmann equation were shown in Sec. 9.5 to vanish in the limit of very small potentials. It may also be shown that errors arising from this inconsistency will not be too serious under the conditions that prevail in many colloidal dispersions, even though the potential itself may no longer be small. Accordingly, we return to the Poisson–Boltzmann equation as it applies to a planar interface, Eq. (43), to develop the Gouy–Chapman result without the limitations of the Debye–Hückel approximation.

If both sides of Eq. (43) are multiplied by $2\,d\psi/dx$, we obtain

$$2\frac{d\psi}{dx}\frac{d^2\psi}{dx^2} = -\frac{8\pi e}{\varepsilon}\sum_i z_i n_{i0} \exp\left(\frac{-z_i e\psi}{kT}\right)\frac{d\psi}{dx} \tag{62}$$

The left-hand side of this equation is the derivative of $(d\psi/dx)^2$; therefore,

$$\left(\frac{d\psi}{dx}\right)^2 = \frac{8\pi kT}{\varepsilon}\sum_i z_i n_{i0} \exp\left(\frac{-z_i e\psi}{kT}\right) + \text{const} \tag{63}$$

The integration constant in this expression is easily evaluated if we define the potential in the solution at $x = \infty$ to be zero. At the same limit, $d\psi/dx$ also equals zero. In view of these conventions, Eq. (63) becomes

$$\left(\frac{d\psi}{dx}\right)^2 = \frac{8\pi kT}{\varepsilon}\sum_i n_{i0}\left[\exp\left(\frac{-z_i e\psi}{kT}\right) - 1\right] \tag{64}$$

This result may be integrated further if we restrict the electrolyte in solution to the symmetrical $z:z$ type. In that case, Eq. (64) can be written

$$\left(\frac{d\psi}{dx}\right)^2 = \frac{8\pi kT n_0}{\varepsilon}\left[\exp\left(\frac{-ze\psi}{kT}\right) + \exp\left(\frac{ze\psi}{kT}\right) - 2\right] \tag{65}$$

in which z is the absolute value of the valence number, the sign having been incorporated into the algebraic form. The bracketed term is readily seen to equal $[\exp(-ze\psi/2kT) - \exp(ze\psi/2kT)]^2$; therefore, Eq. (65) may be written

$$\left(\frac{d\psi}{dx}\right)^2 = \frac{8\pi kT n_0}{\varepsilon}\left[\exp\left(\frac{-ze\psi}{2kT}\right) - \exp\left(\frac{ze\psi}{2kT}\right)\right]^2 \tag{66}$$

Identifying $ze\psi/kT$ as y permits the simplification of notation to

$$\frac{dy}{dx} = \left(\frac{8\pi e^2 z^2 n_0}{\varepsilon kT}\right)^{1/2}(e^{-y/2} - e^{y/2}) = \kappa(e^{-y/2} - e^{y/2}) \tag{67}$$

This last result may be written in an integrable form by defining u as $e^{y/2}$, in which case $dy = 2e^{-y/2}\,du$, and the following relationships hold:

$$\frac{dy}{e^{-y/2} - e^{y/2}} = \frac{2\,du}{e^{y/2}(e^{-y/2} - e^{y/2})} = \frac{2\,du}{1 - e^{y}}$$

$$= \frac{2\,du}{1 - u^2} = \frac{du}{1+u} + \frac{du}{1-u} \tag{68}$$

Combining Eqs. (67) and (68) gives

$$\frac{du}{1+u} + \frac{du}{1-u} = \kappa\,dx \tag{69}$$

which is easily integrated to yield

$$\ln\frac{1+u}{1-u} = \kappa x + \text{const} \tag{70}$$

The integration constant is evaluated from the fact that $\psi = \psi_0$, $y = y_0$, and $u = u_0$ at $x = 0$; therefore,

$$\ln\left(\frac{1+u}{1-u}\right)\left(\frac{1-u_0}{1+u_0}\right) = \kappa x \tag{71}$$

In terms of the physical variables, Eq. (71) may be written

$$\ln\frac{[\exp(ze\psi/2kT)+1][\exp(ze\psi_0/2kT)-1]}{[\exp(ze\psi/2kT)-1][\exp(ze\psi_0/2kT)+1]} = \kappa x \tag{72}$$

Equation (72) describes the variation in potential with distance from the surface for a diffuse double layer without the simplifying assumption of low potentials. It is obviously far less easy to gain a "feeling" for this relationship than for the low potential case. Anticipation of this fact is why so much attention was devoted to the Debye–Hückel approximation in the first place.

Note that Eq. (72) may be written

$$\frac{\exp(ze\psi/2kT)-1}{\exp(ze\psi/2kT)+1} = \frac{\exp(ze\psi_0/2kT)-1}{\exp(ze\psi_0/2kT)+1}\exp(-\kappa x) \tag{73}$$

Equation (73) is the Gouy–Chapman expression for the variation of potential within the double layer. For simplicity, Eq. (73) may be written

$$\gamma = \gamma_0 \exp(-\kappa x) \tag{74}$$

where γ is defined by the relationship

$$\gamma = \frac{\exp(ze\psi/2kT)-1}{\exp(ze\psi/2kT)+1} \tag{75}$$

Equation (74) shows that it is the complex ratio γ that varies exponentially with x in the Gouy–Chapman theory rather than ψ, as is the case in the Debye–Hückel

TABLE 9.3 *Variation of the parameter γ_0 with ψ_0 at 25°C. γ is defined by Eq.* (75)

ψ_0 (mV)	γ_0
260	0.9874
240	0.9814
220	0.9727
200	0.9600
180	0.9415
160	0.9149
140	0.8765
120	0.8230
100	0.7500
80	0.6528
60	0.5249
40	0.3711
20	0.1968

approximation. Some values of γ calculated for a variety of ψ_0 values are listed in Table 9.3.

As a check on the consistency of our mathematics, it is profitable to verify that Eq. (73) reduces to Eq. (51) in the limit of low potentials. Expanding the exponentials in γ, and truncating the series so that only one term survives in both the numerator and denominator, results in the Debye–Hückel expression, Eq. (51).

Another situation of interest in which Eq. (73) simplifies considerably is the case of large values of x at which ψ has fallen to a small value regardless of its initial value. Under these conditions, the exponentials of the left-hand side are expanded to give

$$\frac{ze\psi}{4kT} = \gamma_0 \exp(-\kappa x) \tag{76}$$

or

$$\psi = \frac{4kT\gamma_0}{ze} \exp(-\kappa x) \tag{77}$$

For very large values of ψ_0, $\gamma_0 \rightarrow 1$. In this case, Eq. (77) becomes

$$\psi = \frac{4kT}{ze} \exp(-\kappa x) \tag{78}$$

which shows that the potential in the outer (i.e., well removed from the wall) portion of the diffuse double layer is independent of the potential at the wall for larger potentials. In colloidal dispersions, $ze\psi_0/kT$ is generally greater than unity, but not too much greater. This means that approximations (51) and (77) will generally bracket the true potential versus distance curve given by (73). The situation is shown graphically in Fig. 9.7. In drawing these curves, a value of ψ_0 equal to 77.1 mV was chosen and the electrolyte was arbitrarily selected to be a 0.01 M solution of a 1:1 electrolyte for which κ^{-1} is 30.4 Å. Equation (78) is a poor approximation in this case because ψ_0 is not large enough. Values of the abscissa are readily converted into dimensionless variables which apply to any solution by dividing the x coordinate in angstroms by 30.4 Å.

We shall complete this section by considering the expression for charge density, Eq. (55), as it applies in the Gouy–Chapman model.

As we saw in the preceding section, the charge density expression integrates to Eq. (58), with no assumptions as to the nature of the potential function. Accordingly, we may combine Eqs. (58) and (66) to obtain

$$\sigma = \frac{\varepsilon}{4\pi}\left(\frac{8\pi kTn_0}{\varepsilon}\right)^{1/2}\left[\exp\left(\frac{ze\psi_0}{2kT}\right) - \exp\left(\frac{-ze\psi_0}{2kT}\right)\right] \tag{79}$$

Equation (79) describes the variation of charge density with potential at the surface

FIGURE 9.7 *Variation of the double layer potential versus distance from the surface according to four expressions from this chapter for* $\psi_0 = 77.1$ mV *and* $\kappa = 3.29 \times 10^6$ cm^{-1} *(or* 0.01 M *solution of* 1:1 *electrolyte). Curves drawn according to Eqs.* (51), (73), (77), *and* (78).

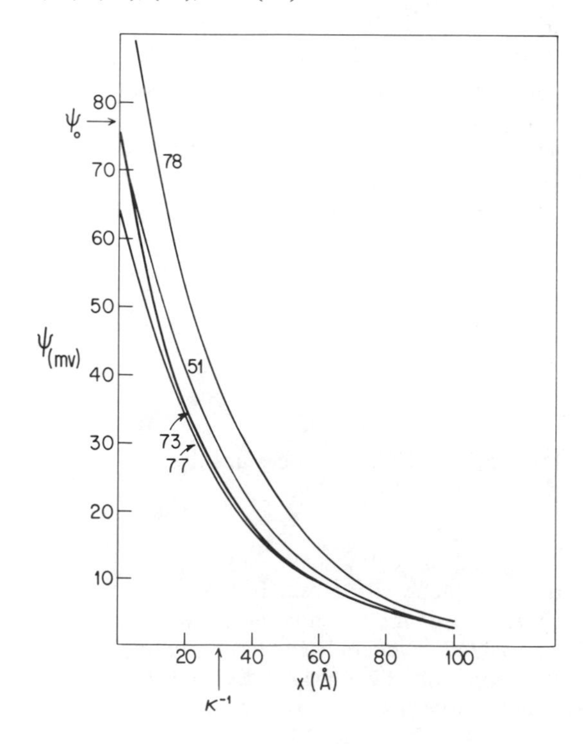

J/K . K mol/m³

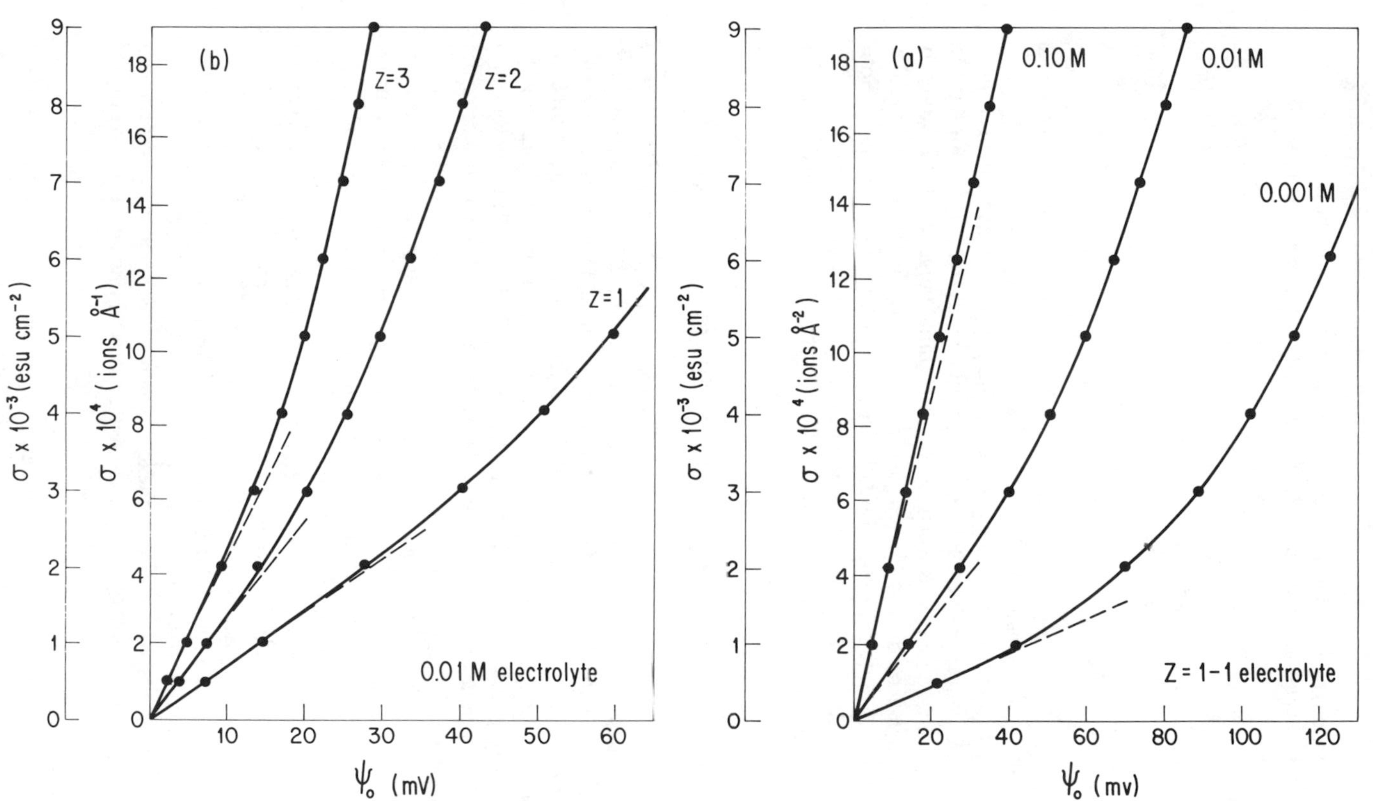

FIGURE 9.8 *Plot of σ versus ψ_0 according to Eq.* (81). (*a*) 0.1, 0.01, *and* 0.001 *M solutions of* 1 : 1 *electrolyte are represented*; (*b*) 0.01 *M solutions of* 1 : 1, 2 : 2, *and* 3 : 3 *electrolytes are shown. Initial tangents correspond to Eq.* (61).

with no limitations as to the value of the potential. Plots of this relationship are given in Fig. 9.8. The construction of such plots is facilitated by recognizing that

$$2 \sinh x = e^{x} - e^{-x} \tag{80}$$

Therefore, Eq. (79) may be written

$$\sigma = \frac{\varepsilon kT\kappa}{2\pi ze} \sinh\left(\frac{ze\psi_0}{2kT}\right) \tag{81}$$

and tables of the hyperbolic sine function may be consulted to evaluate σ for a particular value of ψ_0 or vice versa.

Figure 9.8a shows the relationship between σ and ψ_0 for a monovalent electrolyte at three different concentrations; Fig. 9.8b shows the equivalent curves for 0.01 M solutions of 1 : 1, 2 : 2, and 3 : 3 electrolytes. The initial tangents of the curves in Fig. 9.8 are drawn according to Eq. (61) as a reminder that Eq. (81) must have the same low potential limit as predicted by the Debye–Hückel approximation. Note that the Gouy–Chapman results allow for much higher charge densities at high potentials than would be permitted by the Debye–Hückel approximation, if the latter applied at these potentials.

9.8 CORRECTIONS FOR SPECIFIC ADSORPTION: THE STERN THEORY

The values for σ which are plotted in Fig. 9.8 according to the Gouy–Chapman theory are all reasonable numbers. After all, even when $\sigma = 20 \times 10^{-4}$ ions Å^{-2} the area per molecule is still 500 Å^2. The hyperbolic sine function increases rapidly with increasing values of ψ_0, so reasonable results from this model cannot go on indefinitely. As a matter of fact, for a 0.1 M solution of a 1 : 1 electrolyte, Eq. (81) predicts a value of σ corresponding to 23 ions Å^{-2} at ψ_0 equaling 250 mV. It is obvious that this kind of surface density is physically impossible because of the finite size of the ions. Theory leads to such results only because it treats ions as point sources of charge. The assumption that the ions have no volume is accepted at low concentrations, but clearly must break down as ion crowding occurs. Since the conditions under which an obvious breakdown occurs are within a physically accessible range, we must seek a way to correct this situation.

One way of handling this—due to O. Stern—is to divide the aqueous part of the double layer by a hypothetical boundary known as the Stern surface. The Stern surface is situated a distance δ from the actual surface. Figure 9.9 schematically illustrates the way this surface intersects the double layer potential and how it divides the charge density of the double layer.

The Stern surface is drawn through the ions which are assumed to be adsorbed on the charged wall. There are several consequences of assuming an adsorbed layer of ions at the surface:

1. An adsorption isotherm may be written for these ions which allows for surface saturation and thus introduces the idea of finite ionic size. The Langmuir

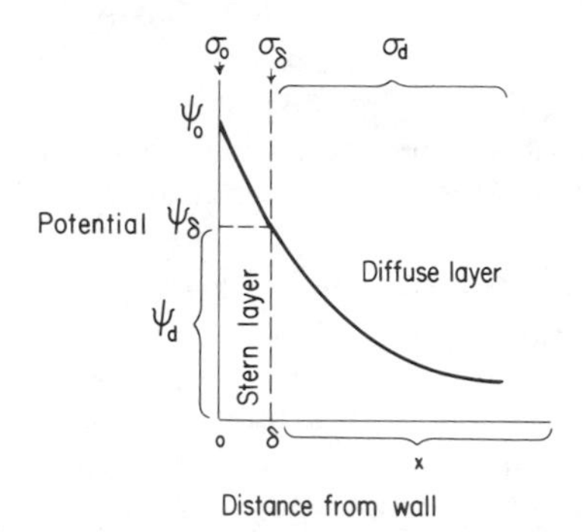

FIGURE 9.9 *Schematic illustration of the variation of potential with distance from a charged wall in the presence of a Stern layer. Significance of subscripts*: 0 *at wall*, δ *at Stern surface*, *d in diffuse layer.*

isotherm [Eq. (7.72)] is one expression that can be used for this purpose:

$$\theta = \frac{Kn_0}{1+Kn_0} \tag{82}$$

In this expression, θ is the fraction of surface adsorption sites occupied, n_0 is the concentration of the adsorbed ions in the solution, and K is a constant.

2. The constant in Eq. (82) is easily shown to be proportional to a Boltzmann factor in which the exponential energy consists of two contributions: $ze\psi_\delta$, the electrical energy associated with the ion in the Stern layer, and ϕ, the specific chemical energy associated with the adsorption:

$$K \simeq \exp\left(\frac{ze\psi_\delta + \phi}{kT}\right) \tag{83}$$

3. The Stern layer resembles the parallel plate capacitor model for the double layer. Therefore, Eq. (27) may be applied to this region:

$$\frac{\psi_0 - \psi_\delta}{\delta} = \frac{4\pi}{\varepsilon_s}\sigma_s \tag{84}$$

4. The fraction of surface sites occupied equals the ratio σ_s/σ_{s0} where σ_{s0} is the charge density at surface saturation. Therefore, Eqs. (82) and (84) may be combined to give

$$\frac{\psi_0 - \psi_\delta}{\delta} = \frac{4\pi}{\varepsilon_s}\frac{\sigma_{s0}Kn_0}{1+Kn_0} \tag{85}$$

This equation shows that the potential drop in the Stern layer increases with the concentration of the adsorbed ion, and ultimately approaches a constant value when the surface is saturated.

Outside of the Stern surface, the double layer continues to be described by Eq. (73) or one of its approximations. The only modifications of the analysis of the diffuse double layer required by the introduction of the Stern surface are that x is measured from δ rather than from the wall and that ψ_δ is used instead of ψ_0 as the potential at the inner boundary of the diffuse layer.

The Stern theory is difficult to apply quantitatively because several of the parameters it introduces into the picture of the double layer cannot be evaluated experimentally. For example, the dielectric constant of the water is probably considerably less in the Stern layer than it would be in bulk because the electric field is exceptionally high in this region. This effect is called dielectric saturation and has been measured for macroscopic systems, but it is difficult to know what value of ε_s applies in the Stern layer. The constant K is also difficult to estimate quantitatively, principally because of the specific chemical interaction energy ϕ. Some calculations have been carried out, however, in which the various parameters in Eq. (85) were systematically varied to examine the effect of these variations on the double layer. The following generalizations are based on these calculations:

1. As ϕ increases K increases and the equilibrium amount adsorbed for any n_0 value short of saturation also increases.
2. As the electrolyte concentration increases, increasing amounts of the potential drop occur in the Stern layer. This is true even if $\phi = 0$, which shows that specific chemical effects are not necessary for this result.
3. Values of ψ_δ which are much less than ψ_0 are possible in dilute solutions only if ϕ is relatively large.
4. ψ_δ varies only slightly with ϕ—although it is highly sensitive to n_0—until ϕ is relatively large.

Values of the parameter ϕ may be experimentally evaluated for the mercury–water surface from electrocapillary studies. The displacement of the coordinates of the electrocapillary maxima in Fig. 9.2 reflects differences in the intrinsic adsorbability of various ions. Electrocapillary studies reveal that the strength of specific adsorption at the mercury–water interface for some monovalent anions follows the order

$$I^- > SCN^- > Br^- > Cl^- > OH^- > F^-$$

whereas for some monovalent cations the order is

$$N(C_2H_5)_4{}^+ > N(CH_3)_4{}^+ > Tl^+ > Cs^+ > Na^+$$

In general, the specific adsorption of an ion is enhanced by larger size—and therefore larger polarizability—and by lower hydration, which itself is a function of ion size. For example, among the ions just listed, the large I^- ion is the most strongly adsorbed and the small but highly hydrated Na^+ ion is adsorbed least.

By allowing for surface saturation, the Stern theory overcomes the objection to the Gouy–Chapman theory of excessive surface concentrations. In so doing, however, it trades off one set of difficulties for another. In the Gouy–Chapman theory, the functional dependence of ψ on x involves only the parameters κ and ψ_0.

The former is known and the latter may be evaluated—at least for some surfaces—by Eq. (2) when the point of zero charge is known. The Stern modification of the double layer picture introduces parameters which are not only difficult to estimate—such as δ and K (or ϕ)—but which are also specific characteristics of different ions. The generality of the Gouy–Chapman model is thus lost when the specific adsorption effects of the Stern theory are considered.

It is the outer portion of the double layer that interests us most as far as colloidal stability is concerned. The existence of a Stern layer does not invalidate the expressions for the diffuse part of the double layer. As a matter of fact, by lowering the potential at the inner boundary of the diffuse double layer, the validity of low potential approximations is enhanced. The only problem is that specific adsorption effects make it difficult to decide what value to use for ψ_δ.

In subsequent discussions, it will be the potential in the diffuse double layer that concerns us. It can be described relative to its value at the inner limit of the diffuse double layer which may be either the actual surface or the Stern surface. We shall continue to use the symbol ψ_0 for the potential at this inner limit. It should be remembered, however, that specific adsorption may make this quantity lower than the concentration of potential-determining ions in the solution would indicate. We shall see in Chap. 11 how the potential at some (unknown) location close to this inner limit can be measured. It is called the zeta potential.

Until now, we have been exclusively concerned with the double layer in the aqueous phase adjacent to an isolated planar interface. In the next few sections, we shall consider how these results must be modified by the presence of a second surface. How the influence of the second surface varies with the distance of separation between the two surfaces is our principal concern. We shall continue to concentrate on the case of planar surfaces.

9.9 OVERLAPPING DOUBLE LAYERS AND INTERPARTICLE REPULSION

From the viewpoint of explaining the stability of lyophobic colloids, this and the following section are of central importance. In this section, we shall examine the force per unit area—that is, the pressure—that operates on two charged surfaces as a result of the overlapping of their double layers. It is more convenient to compare attraction and repulsion between particles in terms of potential energy rather than force. In Sec. 9.10, we shall see how to express double layer repulsion in terms of potential energy. As was the case with a single double layer, it is easier to treat overlapping double layers if the potential is low. Accordingly, the detailed derivation we consider will assume this condition. We shall generalize to the case of higher potentials later, but without presenting all the mathematical details of that situation.

Figure 9.10 is a schematic representation of the situation with which we are concerned. It shows two plates of unspecified thickness; the planar faces of these plates are separated by a distance $2d$. The plates are immersed in an infinite reservoir of electrolyte, the bulk concentration of which is n_0. The potential at the surface of the plates is defined to be ψ_0. It will be convenient to distinguish between the inner and outer regions of the solution. By the former, we refer to the region between the plates; by the latter, the region influenced by only one of the faces.

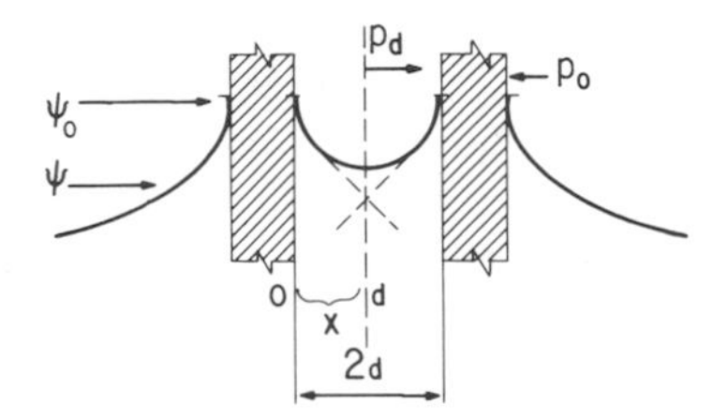

FIGURE 9.10 *Schematic representation of the overlap of two double layers when a pair of plates are brought to a surface separation 2d.*

When the distance between the plates is large, the potential on both the inner and outer faces will drop with distance from the surface according to Eq. (73) or one of its approximations. The profile of the potential drop in this case is shown by the solid curve in the outer region of the figure and by the dashed line in the inner region. As the distance between the plates decreases, the potential from each of the inner faces begins to overlap in the inner region. Therefore, the net potential in the inner region varies as the solid line in the figure for this region. The potential in the outer region is unaffected by the separation of the plates. Since we already know how to handle the potential in the outer region, our interest now focuses on the potential in the inner region where double layer overlap occurs.

At equilibrium, all forces on any volume element of a solution must balance. Suppose we apply this equilibrium criterion to a volume element of solution which lies in the plane parallel to the face of the plates in Fig. 9.10 and lies a distance x from the face of one of these plates. Two kinds of force must operate on this volume element: pressure and electrostatic forces. According to Eq. (2.13) the x component of the pressure force acting on the volume element (i.e., per unit volume) is given by

$$F_x = \frac{\partial p}{\partial x} \tag{86}$$

The electrical force per unit volume is given by the product of the charge density times the field strength according to Eq. (20):

$$F_{\text{el}} = \rho \frac{d\psi}{dx} \tag{87}$$

Combining these results with the criterion for equilibrium leads to the expression

$$\frac{dp}{dx} + \rho \frac{d\psi}{dx} = 0 \tag{88}$$

When Eq. (40) is substituted for ρ, this becomes

$$\frac{dp}{dx} - \frac{\varepsilon}{4\pi} \frac{d^2\psi}{dx^2} \frac{d\psi}{dx} = 0 \tag{89}$$

Since

$$\frac{d^2\psi}{dx^2}\frac{d\psi}{dx}=\frac{1}{2}\frac{d}{dx}\left(\frac{d\psi}{dx}\right)^2$$

Eq. (89) can be written

$$\frac{d}{dx}\left[p-\frac{\varepsilon}{8\pi}\left(\frac{d\psi}{dx}\right)^2\right]=0 \tag{90}$$

This result shows that the condition for equilibrium is equivalent to requiring that

$$p-\frac{\varepsilon}{8\pi}\left(\frac{d\psi}{dx}\right)^2=\text{const} \tag{91}$$

at all locations in the solution. Equation (91) shows that two influences—the pressure and the electric field—operate within the solution, but that the difference between them is a constant that still remains to be evaluated. We are specifically interested in evaluating this constant in the inner region.

At this point, the symmetry of the situation shown in Fig. 9.10 becomes helpful. We realize that ψ goes through a minimum at the midpoint position; that is, $d\psi/dx=0$ at $x=d$. Thus, the constant in Eq. (91) equals the pressure at the midpoint p_d. The difference between the pressure and the field effect is equal to this quantity at all locations between the plates. Because of this constancy, the entire inner region is characterized by the parameters that apply at the midpoint. Therefore, ψ_d is the potential that governs the repulsion between the surfaces. Next, we write Eq. (88) as

$$dp=-\rho\, d\psi \tag{92}$$

Now Eq. (42) may be substituted for ρ for a $z:z$ electrolyte, giving

$$dp=-zen_0\left[\exp\left(-\frac{ze\psi}{kT}\right)-\exp\left(\frac{ze\psi}{kT}\right)\right]d\psi \tag{93}$$

Since $e^x-e^{-x}=2\sinh x$, Eq. (93) may also be written

$$dp=2kTn_0\sinh\left(\frac{ze\psi}{kT}\right)d\psi \tag{94}$$

Equation (94) is easily integrated between the following limits: $p=p_0$ (the outer reference pressure) at $\psi=0$ and $p=p_d$ at $\psi=\psi_d$. Integration of Eq. (94) gives

$$p_d-p_0=2kTn_0\left[\cosh\left(\frac{ze\psi}{kT}\right)-1\right]=F_R \tag{95}$$

Equation (95) gives the excess pressure at $x=d$ and therefore the force per unit area with which the plates are pushed apart, F_R.

This analysis is one of several treatments of double layer repulsion presented in Verwey and Overbeek's classical book [8]. The reader will find the topics of this chapter developed in great detail in that source.

Although Eq. (95) is correct, it is not particularly helpful. The problem is that F_R is expressed in terms of the potential at the midpoint, which is itself an unknown quantity. For the special case in which d is relatively large, the approximation given by Eq. (77) may be applied to the potential from each of the two approaching surfaces. The potential at the midpoint then becomes

$$\psi_d = \psi_1 + \psi_2 \simeq 2\left(\frac{\psi kT\gamma_0}{ze}\right)\exp(-\kappa d) \tag{96}$$

according to this approximation. Since this result applies well away from the surface, the potential is low when Eq. (96) holds. Therefore, cosh $ze\psi_d/kT$ may be expanded as a power series (see Appendix A) with only the leading term retained. This leads to the result

$$F_R \simeq n_0 kT\left(\frac{ze\psi_d}{kT}\right)^2 = n_0 kT[8\gamma_0 \exp(-\kappa d)]^2 \tag{97}$$

or

$$F_R \simeq 64 n_0 kT\gamma_0{}^2 \exp\left(\frac{-2d}{\kappa^{-1}}\right) \tag{98}$$

An issue of considerable practical importance is how the force of repulsion described by Eq. (95) and its approximation (98) varies with the electrolyte content of a solution. Since κ varies with $n_0^{1/2}$, Eq. (98) is of the form

$$F_R = (\text{const } 1)n_0 \exp[-(\text{const } 2)n_0^{1/2}] \tag{99}$$

The exponential factor is clearly the more sensitive involvement of n_0 in Eq. (99). Therefore, this expression shows that the force of repulsion decreases with increasing electrolyte concentration between two surfaces compared at the same separation at least at relatively large separations. We shall see in the following chapter that the addition of indifferent electrolyte to an aqueous dispersion of a lyophobic colloid may, indeed, increase the flocculation of that colloid. Equation (99) is therefore an important step toward understanding this behavior.

Figure 9.11 shows how the force of repulsion varies with separation according to Eq. (98) for different concentrations of 1:1 electrolyte. Figure 9.11 is drawn with $\gamma_0 = 1$ and, therefore, assumes that ψ_0 is large. These forces are decreased by the factor $\gamma_0{}^2$ with decreasing values of ψ_0. Table 9.3 shows the relationship between γ_0 and ψ_0.

9.10 POTENTIAL ENERGY OF INTERPARTICLE REPULSION

In order to evaluate fully the effect of electric charge on the stability of a colloidal dispersion, it is necessary to compare the magnitude of the repulsion between particles with the magnitude of the attraction between them. The latter is discussed in Chap. 10. The attraction is most readily described in terms of potential energy; therefore, the repulsion should be expressed in this form as well. For the approximation we have just discussed, this is not particularly difficult to evaluate. Since

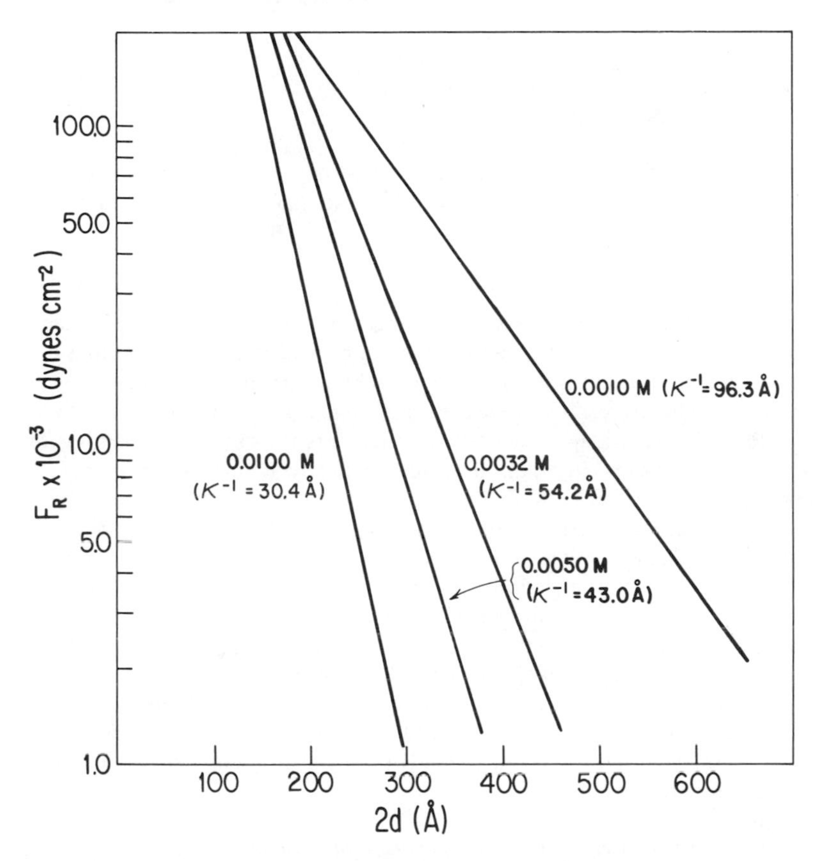

FIGURE 9.11 *The force of repulsion between two plates as a function of the separation of their surfaces. Curves drawn for several concentrations of* 1 : 1 *electrolyte according to Eq.* (98).

potential energy is given by the force times the distance through which it operates, we may write

$$d\Phi_R = -F_R \, d(2d) \tag{100}$$

In this expression, $d\Phi_R$ is the increment in potential energy arising from a change in the separation. The minus sign arises from the fact that the potential energy decreases with increasing separation.

Substituting the approximation given by Eq. (98) for F_R gives

$$d\Phi_R = -64 n_0 k T \gamma_0{}^2 \exp\left(\frac{-2d}{\kappa^{-1}}\right) d(2d) \tag{101}$$

a result that is easily integrated by recalling that $\Phi_R = 0$ when $2d = \infty$. Integration yields

$$\Phi_R = \frac{64 n_0 k T \gamma_0^2}{\kappa} \exp\left(\frac{-2d}{\kappa^{-1}}\right) \tag{102}$$

This particular form is limited in applicability to situations in which the separation of the surfaces is large compared to κ^{-1} and ψ_0 is large so that $\gamma_0 \simeq 1$. As we did with the force of repulsion, we may write Φ_R as

$$\Phi_R = (\text{const } 1) n_0^{1/2} \exp[-(\text{const } 2) n_0^{1/2}] \tag{103}$$

since κ varies with $n_0^{1/2}$. Again we see that at large separations the potential energy of repulsion decreases with increasing electrolyte concentration. The continued emphasis on large separations is formally imposed by the use of approximation (77) in the development of this result. There are practical reasons for interest in this limit also. As colloidal particles approach one another, it is the outermost portions of their double layers that first interact. The outcome of such an encounter, then, is influenced by the interaction between the particles at large separations.

The derivation of Eq. (102) is possible only because relatively simple approximations are available which permit ψ_d to be solved explicitly and generate an integrable expression for Φ_R. This is not generally the case, however, so the approach used to reach Eq. (102) is not applicable to most situations. E. J. W. Verwey and J. Th. G. Overbeek have found another method for evaluating Φ_R, but the mathematics are tedious by this approach as well, involving numerical integrations. The method consists of calculating the free energy difference between particles separated by a distance $2d$ and infinitely separated. The interested reader will find details of this method discussed by Verwey and Overbeek [8]. We shall merely summarize the conclusions obtained from this study.

FIGURE 9.12 *Potential energy of repulsion versus d for various values of surface potential ψ_0. Solid lines drawn according to full theory; dashed lines according to Eq.* (102). *See Table* 9.4 *for scale factors of coordinates. (From Verwey and Overbeek* [8].)

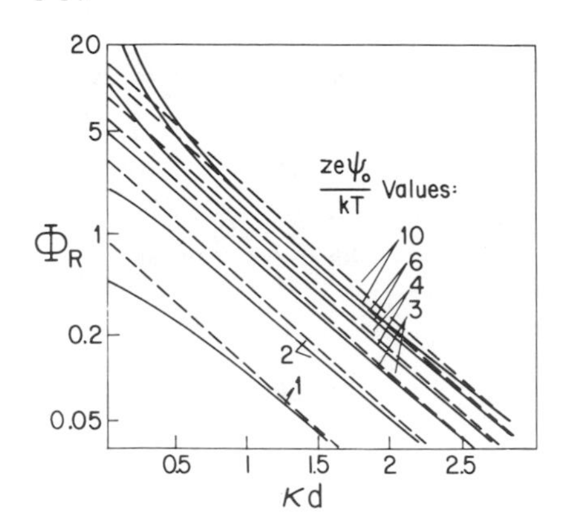

TABLE 9.4 *Scale factors for the interpretation of Fig. 9.12. Also included are the concentrations of various z : z electrolytes which give κ values which are integral powers of* 10

κ (cm^{-1})		10^7	10^6	10^5
Scale factor for Φ_R (ergs cm^{-2}) in Fig. 9.12:		$\times 1$	$\times 10^{-1}$	$\times 10^{-2}$
Scale factor for d (Å) in Fig. 9.12:		$\times 10$	$\times 10^2$	$\times 10^3$
Molarity of aqueous electrolyte at 25°C				
	1:1	9.24×10^{-2}	9.24×10^{-4}	9.24×10^{-6}
	2:2	2.31×10^{-2}	2.31×10^{-4}	2.31×10^{-6}
	3:3	1.03×10^{-2}	1.03×10^{-4}	1.03×10^{-6}

Figure 9.12 shows the variation of Φ_R with κd for different values of the surface potential ψ_0. The solid lines are the values given by the full theory, and the dashed lines are drawn according to Eq. (102). Several features are apparent from inspection of this figure:

1. The dashed and solid lines indeed converge for large values of κd.
2. As the parameter $ze\psi_0/kT$ increases, the difference decreases between the curves describing Φ_R for various values of ψ_0. For large values of ψ_0 ($\gamma_0 \rightarrow 1$), the potential energy is independent of ψ_0.
3. As κd decreases, the full theory and the approximation deviate in functional form as well as numerical magnitude. The potential varies more or less than exponentially with κd, depending on whether $ze\psi/kT$ is more or less than 3.

Figure 9.12 is drawn with "floating" coordinates which permit pairs of Φ_R, d values to be read off the graph for different κ values. Table 9.4 lists the scale factors that are used to obtain numerical values for Φ_R and d from the curves in Fig. 9.12. The coordinates from the figure are multiplied by the scale factors from Table 9.4 to give the potential energy in ergs per square centimeter and the distance d in angstroms for the value of κ under consideration. Using the $ze\psi_0/kT = 4$ curve as an example:

1. If $\kappa = 10^7$ cm^{-1}, $\Phi_R = 1.0$ erg cm^{-2} at $d = 10$ Å.
2. If $\kappa = 10^6$ cm^{-1}, $\Phi_R = 0.1$ erg cm^{-2} at $d = 100$ Å.
3. If $\kappa = 10^5$ cm^{-1}, $\Phi_R = 0.01$ erg cm^{-2} at $d = 1000$ Å.

In summary, these results show the following:

1. A potential energy of repulsion may extend appreciable distances from surfaces, but its range is compressed by increasing the electrolyte content of the system.

2. The conditions under which approaching particles first influence one another are at large distances of separation for which the approximate relationship given by Eq. (102) holds.
3. The sensitivity of aqueous lyophobic colloids to electrolyte content is due to the dependence of interparticle repulsion on this concentration.

Next, a few words are in order about the double layer phenomena associated with nonplanar surfaces, specifically with spherical particles. Particularly when these concepts are applied to dispersed particles, it is preferable to use relationships for spheres rather than for (infinite) planes. At the very least, we must inquire whether applying these results to curved surfaces introduces any qualitative differences not evident from studying planar surfaces.

9.11 DOUBLE LAYER INTERACTION FOR SPHERICAL PARTICLES

To examine the interaction of double layers around spherical particles, the following strategy, due to B. Derjaguin, is very helpful. As shown in Fig. 9.13, the facing surfaces of two spheres may be regarded as a set of circular rings of radius h and centered at the line of centers between the spheres. As the value of h increases, the curvature of the surface assures that the distance of separation between the corresponding segments on the two surfaces also increases. The potential energy functions which we discussed in the preceding section may be applied to the planar faces of each of the rings. The total of all these interactions—for rings of infinitesimal size—will give the net interaction.

To implement this strategy, it is necessary to relate the area of a ring to the radius of the spheres R and the distance of closest approach of their surface, s_0. Figure 9.13 shows that the separation of the ith ring s_i is related to the distance of closest approach by the formula

$$\frac{s_i - s_0}{2} = R - \sqrt{R^2 - h^2} \tag{104}$$

The factor 2 enters since part of the effect occurs at each surface. Since R is a

FIGURE 9.13 *Schematic illustration which shows how the repulsion between spheres may be calculated from the interaction between flat plates.*

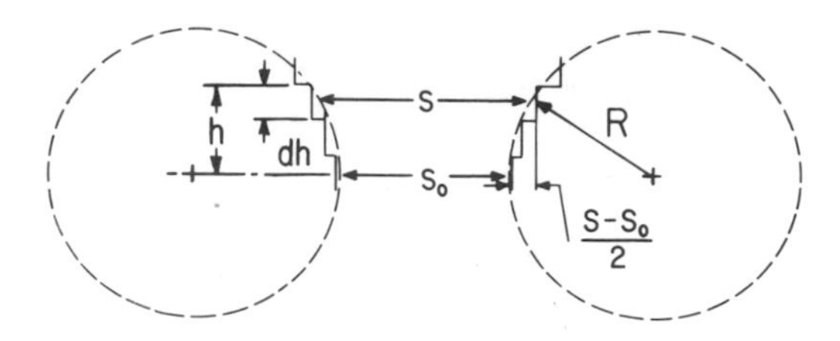

constant, Eq. (104) may be differentiated and rearranged to give

$$R\left(1-\frac{h^2}{R^2}\right)^{1/2} ds = 2h\,dh \tag{105}$$

The area of the face of the ith ring equals $2h\pi\,dh$; therefore, the increment of the interaction due to the ith ring is

$$d\Phi_R = \Phi_i\,dA_i = 2\pi h\Phi(s_i)\,dh = \pi R\left(1-\frac{h^2}{R^2}\right)^{1/2}\Phi(s)\,ds \tag{106}$$

where the functional notation has been written for $\Phi(s)$ as a reminder that it will be different for each ring since the distance of separation increases with h. The total potential of repulsion may be obtained by substituting a suitable expression for $\Phi(s)$ into Eq. (106) and integrating over all values of s.

Once again, we encounter a situation in which further progress depends on finding a substitution for $\Phi(s)$ which leads to an integrable function. If Eq. (102) is used for this purpose, we obtain

$$d\Phi_R = \pi R\left(1-\frac{h^2}{R^2}\right)^{1/2}\frac{64n_0kT\gamma_0^2}{\kappa}\exp\left(-\frac{s}{\kappa^{-1}}\right)ds \tag{107}$$

If we further assume that $h^2/R^2 \ll 1$, Eq. (107) becomes

$$d\Phi_R = \frac{64\pi Rn_0kT\gamma_0^2}{\kappa}\exp(-\kappa s)\,ds \tag{108}$$

The total potential energy of repulsion will be given by integrating the right-hand side of Eq. (108) over all values of s. This is most readily done by assuming that s varies between s_0 and infinity. This upper limit is justified because the potential function drops off exponentially with distance. Therefore, large separations make a negligible contribution to the total. Integrating between these limits, we obtain

$$\Phi_R = \frac{64\pi Rn_0kT\gamma_0^2}{\kappa^2}\exp(-\kappa s_0) \tag{109}$$

This result is further simplified by assuming low potentials and expanding the exponentials in γ_0. If only the leading terms of these expansions are retained

$$\gamma_0 \simeq \frac{ze\psi_0}{4kT} \tag{110}$$

and Eq. (109) becomes

$$\Phi_R \simeq \tfrac{1}{2}\varepsilon R\psi_0^2\exp(-\kappa s_0) \tag{111}$$

Although the assumptions of this derivation limit Eq. (111) to uniform spheres of large radius and low potentials, the result shows that the repulsion increases with R and decreases exponentially with s_0. This result must be obtained in the appropriate limits regardless of the particular expression used to evaluate $\Phi(s)$ for a particular ring. The corresponding results for more general situations have been evaluated numerically by computer and are given in tabular form by Loeb et al. [5].

We have now completed our discussion of the double layer per se. The presentation itself has been abstract and will benefit from some applications. These are presented in Chaps. 10 and 11.

In Chap. 10, we shall discuss the stability of aqueous lyophobic colloids with respect to flocculation. Double layer considerations are not the whole story of flocculation. Since overlapping double layers result in a repulsion between particles, these dispersions would be stabilized against flocculation unless attractions of comparable magnitude also exist. The van der Waals forces which are responsible for the dispersion component of surface tension are such attractions. Therefore, to examine flocculation quantitatively, these two opposing interactions must be compared. As already noted, it is the potential energies of attraction and repulsion which are easiest to compare.

In this sense, we have come only halfway toward understanding flocculation, having discussed only the repulsion part of the interaction. Nevertheless, we can make some generalizations about stability, at least from the perspective of the electrostatic repulsion. All other things being equal, stability is enhanced by values of Φ_R between approaching particles which are as large as possible at large separations. This may be accomplished by a high potential at the inner limit of the diffuse part of the double layer, ψ_0 or ψ_δ, as well as by large values of the reference distance in the double layer κ^{-1} which, in turn, is increased by low concentrations of low-valence electrolytes. These conclusions are explicitly present in Eq. (111) and continue to apply qualitatively even under conditions in which this approximation breaks down. Conversely, flocculation is promoted by conditions opposite to those enumerated above: low ψ_0, high concentrations, and high valences.

In Chap. 11, we shall see how the migration of charged colloids in an electric field, electrophoresis, provides a direct measure of the potential within the double layer. However, there is a catch: We have no way of knowing the exact location within the double layer at which this potential applies. Nevertheless, electrokinetic measurements are important in colloid and surface chemistry. The results developed in this chapter about the structure of the double layer will find ample application in Chap. 11.

REFERENCES

1. N. K. Adam, *The Physics and Chemistry of Surfaces*, Dover, New York, 1968.
2. C. A. Barlow, Jr., "The Electrical Double Layer," in *Physical Chemistry*, Vol. IXA (H. Eyring, D. Henderson, and W. Jost, eds.), Academic Press, New York, 1970.
3. P. Delahay, *Double Layer and Electrode Kinetics*, Wiley-Interscience, New York, 1965.
4. D. A. Haydon, "Electrical Double Layers and Electrokinetics," in *Recent Progress in Surface Science*, Vol. 1 (J. F. Danielli, K. A. G. Parkhurst, and A. C. Riddiford, eds.), Academic Press, New York, 1964.

5. A. L. Loeb, J. Th. G. Overbeek, and P. H. Wiersema, *The Electrical Double Layer around a Spherical Colloid Particle*, MIT Press, Cambridge, Massachusetts, 1960.
6. J. Th. G. Overbeek, in *Colloid Science*, Vol. 1 (H. R. Kruyt, ed.), Elsevier, Amsterdam, 1952.
7. H. van Olphen, *An Introduction to Clay Colloid Chemistry*, Wiley-Interscience, New York, 1963.
8. E. J. W. Verwey and J. Th. G. Overbeek, *Theory of the Stability of Lyophobic Colloids*, Elsevier, Amsterdam, 1948.

PROBLEMS

1. D. C. Grahame* gives the following data for γ_{max} versus $\log c$ (corrected for activity) for the interface between water and aqueous KI at 18°C:

γ_{max} (dynes cm^{-1})	390	398	407	414	419	422
$\log c$ (c in moles liter^{-1})	0.5	0	−0.5	−1.0	−1.5	−2.0

Use these results to estimate Γ, the surface excess of KI at the electrocapillary maximum, for 1.0 and 0.1 M KI. Express your results as moles KI adsorbed per square centimeter and as total electrostatic units of charge per square centimeter.

2. Show by the double integration of Eq. (18), that is $-(\partial^2\gamma/\partial E^2)_\mu = C$, that a parabolic relationship between $\gamma_{max} - \gamma$ and $E - E_{max}$ is expected if the capacitance of the double layer is constant. Use the equation you derive to test the assumed constancy of C for the interface between mercury and 1 M NaCl from the following data*:

γ (dynes cm^{-1})	340	376	396	410	418	423[a]	421
E (V)	0.02	−0.08	−0.18	−0.28	−0.38	−0.56[a]	−0.68
γ (dynes cm^{-1})	415	405	396	384	373	358	340
E (V)	−0.78	−0.88	−0.98	−1.08	−1.18	−1.28	−1.38

[a] Maximum.

Double layer capacitance values are usually expressed as microfarads per square centimeter; remember that practical electrical units, including the farad, are consistent with mks units. Comment on these results in terms of anion and cation adsorption.

3. It has been observed that AgI sols at pH 3.5 are flocculated by 4.5×10^{-5} M Al $(NO_3)_3$ in the absence of a second salt and by 1.7×10^{-4} M $Al(NO_3)_3$ in the

* D. C. Grahame, *Chem. Rev.* **41**:441 (1947).

presence of 0.009 M K_2SO_4.* Calculate the value of κ for each of these solutions. Is the threshold of instability consistent with your expectations in terms of the values of κ for these two systems? The authors of this research suggest that $K = 370$ for the equilibrium $Al^{3+} + SO_4^{2-} \rightleftarrows AlSO_4^{+}$. Calculate the concentration of Al^{3+} in the system containing K_2SO_4. Does the behavior of the AgI appear to correlate with the concentration of "free" Al^{3+}? Is the specific adsorption of Al^{3+} expected on AgI in the presence of 4×10^{-4} M excess KI? Discuss in terms of ψ_0 and ϕ.

4. The viscosity of negatively charged colloidal agar (0.14% at 50°C) was studied with a variety of different electrolytes added.† The ratio of the specific viscosity ($\eta_{sp} = \eta/\eta_0 - 1$) in the presence of salt to η_{sp} without salt is given below for some of these salts at several concentrations:

	$(\eta_{sp})_{salt}/(\eta_{sp})_{H_2O}$							
c (meq liter^{-1})	KCl	K_2SO_4	$K_4Fe(CN)_6$	$BaCl_2$	$SrCl_2$	$MgSO_4$	$La(NO_3)_3$	$Pt(en)_3(NO_3)_4^a$
0.25	—	—	—	—	—	—	—	79.2
0.50	90.8	90.7	90.5	—	—	—	76.6	69.7
1.00	—	—	—	78.2	77.9	78.6	—	64
2.00	81.2	81.4	81.1	—	—	—	68.1	60.6
4.00	77.3	77.5	76.7	70.6	70.9	71.6	67.2	60

[a] en = ethylenediamine.

Discuss this electroviscous effect in terms of the concepts of this chapter and Chap. 2.

5. The deficiency of positive ions ($\Gamma_+ < 0$) adjacent to a positively charged planar surface may be evaluated as follows ($y = ze\psi/kT$):

$$\Gamma_+ \overset{(1)}{=} \int_0^\infty (c_0 - c_x)\,dx \overset{(2)}{=} c_0 \int_0^\infty (1 - e^{-y})\,dx \overset{(3)}{=} c_0 \int_0^\infty (1 - e^{-y}) \frac{dx}{dy}\,dy$$

$$\overset{(4)}{=} c_0 \int_{\psi_0}^0 \frac{(1 - e^{-y})\,dy}{dy/dx} \overset{(5)}{=} \frac{c_0}{\kappa} \int_{\psi_0}^0 \frac{1 - e^{-y}}{e^{-y/2} - e^{y/2}}\,dy$$

$$\overset{(6)}{=} -\frac{c_0}{\kappa} \int_{\psi_0}^0 e^{-y/2}\,dy \overset{(7)}{=} \frac{2c_0}{\kappa}\left[1 - \exp\left(\frac{ze\psi_0}{2kT}\right)\right]$$

Present the physical and/or mathematical justification for equalities (1) through (7) in this sequence. Show that the final result is equivalent, at high surface potentials, to emptying a region of thickness $2\kappa^{-1}$ of ions possessing the same charge as the wall.

* L. J. Stryker and E. Matijević, *J. Phys. Chem.* **73**:1484 (1969).
† H. R. Kruyt and H. G. deJong, *Kolloid Z.* **100**:250 (1922).

6. Show that for a 1:1 electrolyte in water at 25°C, Eq. (81) can be rearranged to give

$$\psi_0 = 51.4 \sinh^{-1}\left(\frac{137}{\sigma^0\sqrt{c}}\right)$$

in which ψ_0 is expressed in millivolts, c is in moles per liter, and σ^0 is the area (in $Å^2$) per charge at the surface. J. T. Davies* measured the potential across an air–aqueous NaCl interface which carried a monolayer of $C_{18}H_{37}N(CH_3)_3^+$. When the quaternary octadecylamine was at a pressure corresponding to $\sigma^0 = 85$ $Å^2$, the following potentials were measured at different concentrations of NaCl:

E (mV)	240	280	325	340	380
c (moles NaCl liter^{-1})	2.0	0.5	0.1	0.033	0.01

About 200 mV of these potential differences arises from dipole effects at the interface and should be subtracted from each value to give the double layer contribution to the measured potentials. Compare these corrected values with the values of ψ_0 calculated by the equation just given.

7. The interfacial tension at the electrocapillary maximum for several electrolytes in dimethylformamide (DMF) solutions has been measured as a function of the electrolyte concentration†:

	γ_{max} (dynes cm^{-1})		
log c	KI	LiCl	KSCN
0	354	366	370
−0.3	356	367	371
−1.0	361	369	373
−1.3	364	370	374

Use these results to estimate the relative adsorbabilities of the I^-, Cl^-, and SCN^- ions from DMF. How does the sequence of anion absorbabilities compare with that from aqueous solution as given in Sec. 9.8? More comprehensive electrocapillary data suggest that SCN^- is more solvated in DMF than in water. Is this consistent with the adsorbability series just compared?

8. A negatively charged AgI dispersion was caused to flocculate by the addition of various electrolytes. The concentrations of several divalent metal nitrates

* J. T. Davies, *Proc. Roy. Soc.* **208A**:224 (1951).
† V. D. Bezuglyi and L. A. Korshikov, *Electrokhimiya* **3**:390 (1967).

needed to produce flocculation are as follows*:

Salt	$Mg(NO_3)_2$	$Ca(NO_3)_2$	$Sr(NO_3)_2$
$c \times 10^3$ (mole liter^{-1})	2.60	2.40	2.38
Salt	$Ba(NO_3)_2$	$Zn(NO_3)_2$	$Pb(NO_3)_2$
$c \times 10^3$ (mole liter^{-1})	2.26	2.50	2.43

That these different compounds produce the same effect at so nearly the same concentration argues that the principal cause of the effect is electrostatic. Use the average of these concentrations to calculate: (a) the value of κ at which this system flocculates; (b) the approximate charge density at the interface using Eq. (60); (c) the force of repulsion [Eq. (98)]; and (d) the potential energy of repulsion [Eq. (102)] when two planar surfaces are separated by a distance of 100 Å. For the purpose of calculation in parts (b), (c), and (d), ψ_0 may be taken as 100 mV. Comment on the applicability of these equations to the physical system under consideration.

9. The slight differences in the concentrations of divalent cations required to flocculate the AgI sol of the preceding problem may be attributed to differences in the adsorbability of these cations at the AgI–solution interface. Use the data of Problem 8 to rank the cations with respect to their tendency to adsorb. Is there a correlation between adsorbability and ion size and/or hydration? List any references consulted for data concerning the latter two quantities.
10. Once the significance of the midpoint between two parallel plates for the force between those plates is established, there are several ways of arriving at Eq. (95). One argument is that the plates shown in Fig. 9.10 function as a semipermeable membrane, sustaining a concentration difference between $x = d$ and the outer region of the solution. Use Eq. (4.37) and Eq. (41) of this chapter to show that the osmotic pressure across this "semipermeable membrane" is given by Eq. (95).
11. J. Lyklema and K. J. Mysels† measured the equilibrium thickness of a soap bubble to be 731 Å when the bubble was stabilized by 8.7×10^{-4} M sodium dodecyl sulfate and the hydrostatic pressure on the surface of the film was 660 dynes cm^{-2}. In this experiment, the thickness satisfies the condition of equilibrium between the hydrostatic pressure and the force of repulsion between double layers on the adjacent faces of the film, as given by Eq. (98). Assuming that the adsorption of dodecyl sulfate ions at the air–solution interface gives a very high value of ψ_0, calculate the equilibrium bubble thickness predicted by this model. Criticize or defend the following propositions: "If an appreciable concentration of indifferent electrolyte is added to the

* H. R. Kruyt and M. A. Klompe, *Kolloid Beihefte* **54**:484 (1942).
† J. Lyklema and K. J. Mysels, *J. Am. Chem. Soc.* **87**:2539 (1965).

soap solution, the calculation just given is no longer feasible because (a) Eq. (98) ceases to be valid for F_R, (b) van der Waals attraction between the two air masses also promotes film thinning, and (c) steric repulsion between adsorbed surfactant molecules enhances film thickness."

van der Waals Attraction and Flocculation

Already the difficulties of avoiding a collision in a crowd are enough to tax the sagacity of even a well-educated Square; but if no one could calculate the Regularity of a single figure . . . all would be chaos and confusion, and the slightest panic would cause serious injuries. [From Abbott's *Flatland*]

10.1 INTRODUCTION

This chapter is concerned with two major topics: the van der Waals attraction between colloidal particles and one of its principal manifestations, flocculation.

Chapter 9 indicated that the stability of a dispersion with respect to flocculation depends on the relative magnitude of the potential energy of attraction and that of repulsion of the particles involved. Until now, we have taken a very one-sided viewpoint and focused attention exclusively on the electrostatic repulsion component of this interaction. In this chapter, the picture is rounded out, and a quantitative comparison between attraction and repulsion becomes possible.

The division of this discussion between two chapters is convenient pedagogically, but somewhat artificial physically, at least as far as the total phenomena of flocculation are concerned. This is because flocculation is governed by the balance between the forces of attraction and repulsion and really cannot be understood in terms of either one alone. The full theory of stability—hence the contents of Chaps. 9 and 10—is known as the Derjaguin, Landau, Verwey, Overbeek (DLVO) theory. In its totality, the DLVO theory is a remarkable accomplishment of physical chemistry: It successfully disentangles many different phenomena and is able to explain (at least in broad concepts) many complex observations.

The first half of this chapter is devoted to developing an expression for the potential energy of attraction between two colloidal particles. As in the preceding chapter for repulsion, this topic is developed in stages and for several approximations. We shall begin by considering the interaction between a pair of isolated atoms or molecules, then "scale up" this effect to the interaction of bodies in the colloidal size range.

After developing expressions for the potential energy of attraction, these are combined with their counterparts from Chap. 9 to give a net interaction potential. After we have generated a function which describes the net interaction between colloidal particles as a function of their separation, we shall apply these results to the phenomenon of flocculation. This development occupies the second half of the chapter.

Several important aspects of this topic have either been given abbreviated treatments or have been omitted altogether. The relationship between flocculation and rheology in dispersions of interacting particles is an example of the former, and the use of soap films to study colloidal interactions is an example of the latter. The interested reader will find these and other topics more fully developed in the references at the end of this chapter.

10.2 MOLECULAR INTERACTIONS AND POWER LAWS

To understand the origin of the attraction between colloidal particles, it is necessary to back off a bit and consider the interactions between individual molecules. Macroscopic interactions—as we shall call the interactions between colloids, since these particles are large compared to atomic dimensions—are the summation of the pairwise interactions of the constituent molecules in the individual particles. Therefore, we begin by examining the interactions between a pair of isolated molecules.

To discuss the interaction between a pair of isolated molecules, it is convenient to deal with the potential energy rather than the force exerted between the molecules. This energy is clearly a function of the distance of separation between the molecules, decreasing as the distance increases. In keeping with the usual conventions, we shall associate attractions with negative potential energies and repulsions with positive potential energies. This convention permits us to speak of the "height" of energy "barriers" and the "depth" of energy minima.

Our primary interest in this section is to discuss the functional form which relates potential energy to the distance of separation x for various types of interactions. For many interactions, an inverse power dependence on the separation describes the potential energy. Several examples of this are shown in Table 10.1. The main point to be observed for now is that the value of the exponent in the inverse power dependence on the separation differs widely for the various types of interactions. An immediate consequence of this is that the range of the interactions is quite different also.

It is those functions with an inverse sixth power dependence on the separation that are our main concern in this chapter. Those power laws with exponents greater or less than 6 are included in Table 10.1 mainly to emphasize the point that many types of interactions exist and that these are governed by different relationships. The interactions listed are by no means complete: Interactions of quadrupoles, octapoles, and so on, might also be included as well as those due to magnetic moments. However, all of these are less important than the interactions listed. Let us now examine Table 10.1 in greater detail.

TABLE 10.1 *Partial list of interactions between pairs of isolated ions and/or molecules. Functions which describe the potential energy versus separation are listed along with the appropriate proportionality constants*

Description	Φ	Definitions and restrictions	Attributed to	Value of n in $\Phi \propto x^{-n}$
Ion 1–ion 2	$\frac{(ze)_1(ze)_2}{x}$	z = valence, e = electron charge under vacuum (otherwise ε in denominator) (sign depends on the z value)	Coulomb	1
Ion 1–permanent dipole 2	$\frac{(ze)_1\mu_2 \cos\theta}{x^2}$	μ = dipole moment, θ = angle between line of centers and axis of dipole. Length of dipole small compared to x (sign depends on z and orientation)	(Coulomb)	2
Permanent dipole 1–permanent dipole 2	$(\text{const})\frac{\mu_1\mu_2}{x^3}$	Numerical constant (including sign) depends on orientation. Const = $\sqrt{2}$ for average over all orientations. Const = +2 for parallel and −2 for antiparallel alignment	(Coulomb)	3

Permanent dipole 1–induced dipole 2	$-\frac{(\alpha_{0,1}\mu_2^2+\alpha_{0,2}\mu_1^2)}{x^6}$	α_0 = polarizability (always negative)	Debye	6
Permanent dipole 1–permanent dipole 2	$-\frac{2}{3}\frac{\mu_1^2\mu_2^2}{kTx^6}$	Free rotation of dipoles (always negative)	Keesom	6
Induced dipole 1–induced dipole 2	$-\frac{3}{2}\frac{h}{x^6}\frac{\nu_1\nu_2}{\nu_1+\nu_2}\alpha_{0,1}\alpha_{0,2}$	ν = characteristic vibrational frequency of electrons (always negative)	London	6
Induced dipole 1–induced dipole 2 (retarded)	$-\frac{23}{8\pi^2}hc\frac{\alpha_{0,1}\alpha_{0,2}}{x^7}$	h = Planck's constant, c = speed of light. Applies if $x > c/\nu$ (always negative)	Casimmir and Polder	7
Repulsion	$+\frac{\xi}{x^{12}}$	Exponent in range of 9–15. 12 mathematically convenient (always positive)		12

The first three entries in Table 10.1 include Coulomb's law and two results which follow directly from it by treating stationary dipoles as a pair of charges and adding all pairwise interactions. What is important to note about these results is that the sign may be positive or negative—corresponding to repulsion or attraction—depending on the charge of ions or the orientation of the dipoles, or both.

By contrast, those results which involve an inverse sixth power law are always negative. That is, attraction always results from interactions of the following types:

1. Permanent dipole–induced dipole interaction (Debye equation).
2. Permanent dipole–permanent dipole interaction (Keesom equation).
3. Induced dipole–induced dipole interaction (London equation).

The inverse seventh power law is a special case of the induced dipole–induced dipole interaction which applies to the case of large separations. This set of attractive interactions is collectively known as van der Waals attraction. In the following section we shall discuss in greater detail the significance and the origin of the van der Waals attractions listed in Table 10.1. In this section, it is only the exponent in the power law that we are considering.

The last entry in Table 10.1 is the least well defined of those listed. This is of little importance to us, however, since our interest is in attraction, and the final entry always corresponds to repulsion. The reader may recall that so-called hard sphere models for molecules involve a potential energy of repulsion which sets in and rises vertically when the distance of closest approach of the centers equals the diameter of the spheres. A more realistic potential energy function would have a finite, though steep, slope. An inverse power law with an exponent in the range of 9 to 15 meets this requirement. For reasons of mathematical convenience, an inverse 12th power dependence on the separation is frequently postulated for the repulsion between molecules.

In general, the combined effects of van der Waals attraction and interparticle repulsion may be represented by the equation

$$\Phi = \xi x^{-12} - \beta x^{-6} \tag{1}$$

in which the constant β has been used to represent the various constants in the Debye, Keesom, and London equations. Since the two terms on the right-hand side of Eq. (1) correspond to opposing tendencies, the total potential energy function will display a minimum, the coordinates of which will describe an equilibrium situation. By differentiating Eq. (1) with respect to x and setting the result equal to zero, the coordinates of the minimum can be evaluated. These are readily found to be

$$x_m = \left(\frac{2\xi}{\beta}\right)^{1/6} \tag{2}$$

and

$$\Phi_m = -\frac{\beta}{2}x_m^{-6} = -\xi x_m^{-12} \tag{3}$$

where the subscript reminds us that these are values at the minimum (equilibrium).

Equation (3) can be used to eliminate β and ξ from (1) to obtain

$$\Phi = -\Phi_m\left[\left(\frac{x}{x_m}\right)^{-12} - 2\left(\frac{x}{x_m}\right)^{-6}\right] \tag{4}$$

Equations (1) and (4) or other variations of the 6–12 power law are often called the Lennard–Jones potential. The numerical values of the constants in the Lennard-Jones potential may be obtained from studies of the compressibility of condensed phases, the virial coefficients of gases, and by other methods. A summary of these methods and other expressions for the molecular interaction energy can be found in the book by Moelwyn-Hughes [4].

Figure 10.1 is a plot of the Lennard-Jones function for methane for which $\xi = 6.2 \times 10^{-103}$ erg cm^{12} and $\beta = 2.3 \times 10^{-58}$ erg cm^{6} ($x_m = 42$ Å and

FIGURE 10.1 *Potential energy versus distance of separation for two methane molecules. Curve* 1, *repulsion according to an inverse* 12*th power law*; *curve* 2, *attraction according to an inverse sixth power law*; *curve* 3, *resultant of curves* 1 *and* 2.

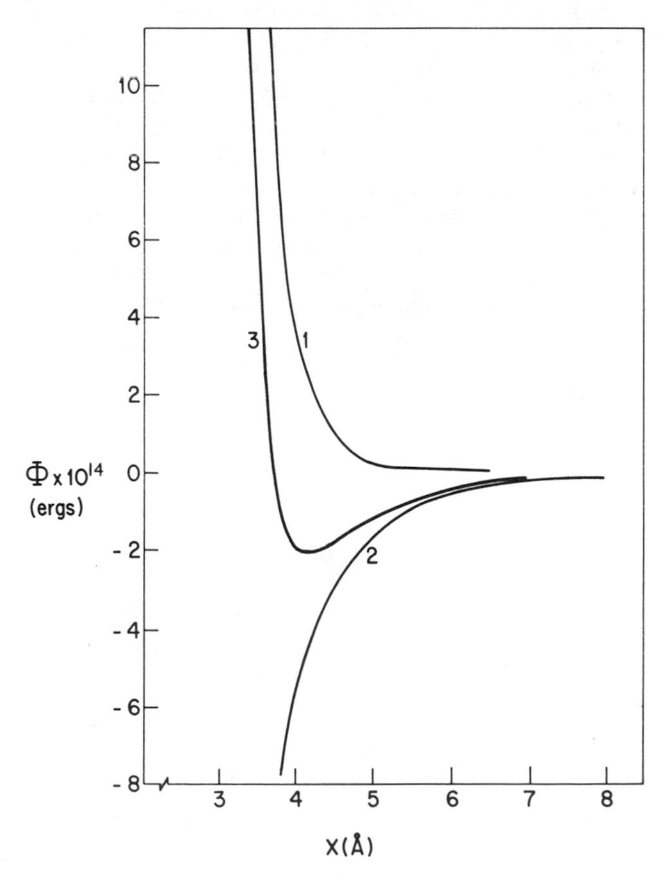

$\Phi_m = 2.1 \times 10^{-14}$ erg). The curves in this figure may be compared with those of Fig. 6.14 which were qualitative representations of the same effect.

Curves of this sort occur in many places in physical chemistry and it is important to realize that they are the resultant of two contributions: a very short range repulsion and a relatively long range attraction. It is the latter with which we are primarily concerned, so we shall turn next to an examination of the origins of these inverse sixth power attractions.

10.3 MOLECULAR ORIGINS OF VAN DER WAALS ATTRACTIONS

In Table 10.1 we saw that all random dipole–dipole interactions follow the inverse sixth power law, except the so-called retarded van der Waals attraction which varies with the inverse seventh power of the separation. We shall discuss retarded van der Waals attraction in Sec. 10.7. In this section, we shall examine briefly the physical basis of the three different inverse sixth power laws which describe intermolecular attractions. Space limitations prevent us from deriving the Debye, Keesom, and London expressions in detail. More complete derivations may be found in many physical chemistry textbooks, such as that by Moelwyn-Hughes [4]. The abbreviated discussion we present should be sufficient, however, to indicate the connection between these attractions and molecular parameters.

Interactions between dipoles, whether permanent or induced, are the result of the electric field produced by one dipole (subscript 1) acting on the second dipole (subscript 2). Therefore, the first factor to consider in discussing such interactions is the field $\bar{E}$ produced by a dipole and measured a distance x from the dipole, where x is large compared to the length of the dipole. The field is a vector quantity and may be resolved into components as shown in Fig. 10.2. We define θ to be the angle between the axis of the dipole and the line which connects the point under consideration with the center of the dipole and along which x is measured. The field may be resolved into the following components:

1. Parallel to the line of centers:

$$\bar{E}_{\parallel} = -2\mu_1 x^{-3} \cos\theta \tag{5}$$

FIGURE 10.2 *The electric field a distance x from a dipole. Field is resolved into components parallel and perpendicular to the line of centers along which x is measured.*

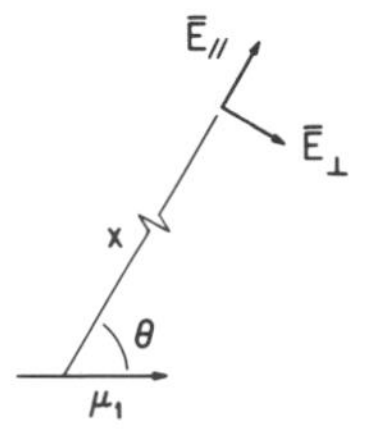

2. Perpendicular to the line of centers:

$$\bar{E}_{\perp} = -\mu_1 x^{-3} \sin\theta \tag{6}$$

The total field is the square root of the sum of the squares of Eqs. (5) and (6), or

$$\bar{E} = \mu_1 x^{-3}(\sin^2\theta + 4\cos^2\theta)^{1/2} = \mu_1 x^{-3}(1 + 3\cos^2\theta)^{1/2} \tag{7}$$

Now let us consider the effect of such a field on a particle with no permanent dipole of its own. The field will induce a dipole in the second molecule; the magnitude of the induced dipole moment μ_2 is proportional to the field with the polarizability of the second molecule, α_2, the proportionality constant:

$$\mu_2 = \alpha_2 \bar{E} \tag{8}$$

The potential energy of the second dipole due to this field is given by $-\mu_2\bar{E}$ or $-\alpha_2\bar{E}^2$. To this must be added the energy necessary to induce the dipole, $\frac{1}{2}\alpha_2\bar{E}^2$, since the second is not a permanent dipole. Therefore, the total potential energy Φ of the second dipole is

$$\Phi_2 = -\tfrac{1}{2}\alpha_2\bar{E}^2 \tag{9}$$

Substituting Eq. (7) into this expression and averaging over all orientations yields

$$\Phi_2 = -\alpha_2\mu_1^2 x^{-6} \tag{10}$$

The second dipole acts on the original dipole in a similar fashion, giving a second contribution to the interaction energy which is identical to Eq. (10) except that the subscripts are interchanged. The total potential energy of attraction is the sum of these two contributions:

$$\Phi_D = -(\alpha_2\mu_1^2 + \alpha_1\mu_2^2)x^{-6} \tag{11}$$

This is the Debye equation in Table 10.1 which describes the attraction between a permanent dipole and an induced dipole.

If this argument is applied to two permanent dipoles, the polarizability may be regarded as the sum of two contributions: one which is independent of the presence of the permanent dipole, α_0, and a second which is the average effect of the rotation of the molecules in the electric field. The molar polarization P of a substance is given by

$$P = \tfrac{4}{3}\pi N_A\left(\alpha_0 + \frac{1}{3}\frac{\mu^2}{kT}\right) \tag{12}$$

The second term inside the parentheses thus gives the average value of the orientation contribution to the polarizability. It is the latter that is used as a substitution for α in Eq. (11) for the interaction of two permanent dipoles:

$$\Phi_K = -\frac{2}{3}\frac{\mu_1^2\mu_2^2}{kT}x^{-6} \tag{13}$$

This is the Keesom equation from Table 10.1; it applies to the interaction of two permanent dipoles.

Finally, we turn our attention to the third contribution to van der Waals attraction, London or dispersion forces between a pair of induced dipoles. It will be noted that (at least one) permanent dipole is needed for the preceding sources of attraction to be operative. No such restriction is present for the dispersion component. Therefore, this latter quantity is present between molecules of all substances whether or not they have a permanent dipole. These are the same forces that we considered in Chap. 6 when we discussed the dispersion component of surface tension.

The London equation for the attraction between a pair of induced dipoles is a quantum mechanical result which represents one of the contributions to the "bond" between a pair of particles. Like other quantum mechanical results, the interaction energy emerges as part of the solution to the Schrödinger equation. We shall dispense with a rigorous examination of the situation and consider only the physical model and the final results.

Figure 10.3 represents the situation we wish to consider. It represents two dipoles, the length of which, l_i, is negligible compared to the distance between their centers. The dipoles are formed by the symmetrical vibration of electrons in the two particles. According to Table 10.1, the potential energy of two dipoles in this arrangement is $\pm 2\mu^2 x^{-3}$ or $\pm(el_1)(el_2)x^{-3}$ since $\mu_i = el_i$. In addition, each of the vibrating dipoles may be regarded as a harmonic oscillator for which the potential energy is given by

$$\Phi = \tfrac{1}{2} k l_i^2 \tag{14}$$

in which

$$k = \frac{e^2}{\alpha_0} \tag{15}$$

Combining these various energy contributions gives the following expression to be used as the potential energy of this system:

$$\Phi_T = \tfrac{1}{2} k_1 \pm 2(el_1)(el_2)x^{-3} + \tfrac{1}{2} k_2 l_2^2 \tag{16}$$

When this net potential energy function is substituted into the one-dimensional Schrödinger equation and the suitable mathematical operations are carried out, the allowed energies E are found to be

$$E = (n_i + \tfrac{1}{2})h\nu_i + (n_j + \tfrac{1}{2})h\nu_j \tag{17}$$

FIGURE 10.3 *A linear arrangement of two dipoles used to define the potential energy in the Schrödinger equation for the London interaction energy.*

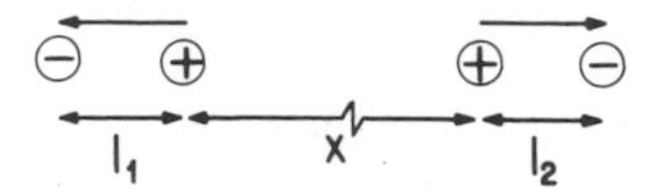

in which

$$\nu_i = \nu\left(1 - \frac{2\alpha_0}{x^3}\right)^{1/2} \tag{18}$$

and

$$\nu_j = \nu\left(1 + \frac{2\alpha_0}{x^3}\right)^{1/2} \tag{19}$$

and the n terms are integers. Note that the terms in Eq. (17) are formally identical to the energy of quantized harmonic oscillators of frequency ν_i and ν_j. In addition, we observe that ν_i and ν_j approach ν as $x \to \infty$. Thus ν is identified as the frequency of vibration for the system in the case where the electrons vibrate independently.

The lowest energy for the two coupled oscillators is the situation in which $n_i = n_j = 0$, in which case Eq. (17) becomes

$$E = \tfrac{1}{2}h(\nu_i + \nu_j) \tag{20}$$

On the other hand, the energy of two independent oscillators in their ground state is given by

$$E = 2(\tfrac{1}{2}h\nu) \tag{21}$$

The difference between Eqs. (20) and (21) gives the contribution of the interaction to the potential energy:

$$\Phi_L = \tfrac{1}{2}h[(\nu_i + \nu_j) - 2\nu] \tag{22}$$

This is one way of writing the potential energy of attraction between two induced dipoles.

If the expressions for ν_i and ν_j given by Eqs. (18) and (19) are substituted into (22), the following result is obtained:

$$\Phi_L = \tfrac{1}{2}h\nu\left[\left(1 - \frac{2\alpha_0}{x^3}\right)^{1/2} + \left(1 + \frac{2\alpha_0}{x^3}\right)^{1/2} - 2\right] \tag{23}$$

Expanding the square roots by the binomial expansion (Appendix A) and retaining no terms of higher than second order yields

$$\Phi_L = -\tfrac{1}{2}h\nu\alpha_0^2 x^{-6} \tag{24}$$

Several modifications of Eq. (24) are also important:

1. When the molecules are capable of vibration in all three dimensions, the numerical constant in Eq. (24) becomes $\frac{3}{4}$ rather than $\frac{1}{2}$:

$$\Phi_L = -\tfrac{3}{4}h\nu\alpha_0^2 x^{-6} \tag{25}$$

2. When unlike molecules are involved, their individual frequencies and polarizabilities are involved, and the expression equivalent to Eq. (24) is

$$\Phi_L = -\tfrac{3}{2}h\left(\frac{\nu_1\nu_2}{\nu_1 + \nu_2}\right)\alpha_{0,1}\alpha_{0,2}x^{-6} \tag{26}$$

The latter is the result shown in Table 10.1. Note that this result becomes identical to the three-dimensional version of Eq. (24) when the atoms are identical.

3. The quantity $h\nu$ in Eq. (25) may be regarded as some energy which characterizes the system and is sometimes approximated by the ionization energy I:

$$\Phi_L \simeq -\tfrac{3}{2}\left(\frac{I_1 I_2}{I_1+I_2}\right)\alpha_{0,1}\alpha_{0,2}x^{-6} \tag{27}$$

4. The frequency of a harmonic oscillator, the model for the two dipoles, equals $(1/2\pi)\sqrt{k/m_e}$ where m_e is the mass of the electron. Substituting the value of k given by Eq. (15) yields

$$\Phi_L = -\frac{3}{8}\frac{he}{\pi}\frac{\alpha_0^{3/2}}{m_e^{1/2}}x^{-6} \tag{28}$$

for two identical molecules, since

$$\nu = \frac{1}{2\pi}\sqrt{\frac{e^2}{\alpha_0 m_e}} \tag{29}$$

Equations (25) through (28) are widely encountered expressions for the London attraction between two molecules.

In examining the Debye, Keesom, and London equations we see (a) that they share as a common feature an inverse sixth power dependence on the separation, and (b) that the molecular parameters which describe the polarization of a molecule, polarizability and dipole moment, serve as proportionality factors in these expressions. For a full discussion of the experimental determination of α_0 and μ, a textbook of physical chemistry should be consulted. For our purposes, it is sufficient to note that the Clausius–Mosotti equation relates the molar polarization of a substance to its dielectric constant ε:

$$P = \frac{M}{\rho}\frac{\varepsilon-1}{\varepsilon+2} \tag{30}$$

where M and ρ are the molecular weight and density.

Combining Eqs. (12) and (30) gives the general result:

$$\frac{M}{\rho}\frac{\varepsilon-1}{\varepsilon+2} = \tfrac{4}{3}\pi N_A\left(\alpha_0 + \frac{1}{3}\frac{\mu^2}{kT}\right) \tag{31}$$

Thus studies of ε as a function of T may be analyzed to yield values of both α_0 and μ. For substances with no permanent dipole moment, $\mu = 0$ and $\varepsilon = n^2$ where n is the refractive index at long wavelengths. For such a system, Eq. (31) becomes

$$\frac{M}{\rho}\frac{n^2-1}{n^2+2} = \tfrac{4}{3}\pi N_A \alpha_0 \tag{32}$$

To use Eq. (32) it is necessary to extrapolate to infinite wavelength (or zero frequency) to obtain the unperturbed polarizability, since the electric field of the light also alters the molecule. Failure to carry out such an extrapolation introduces far less error, however, than is introduced by an approximation such as Eq. (27).

In general, we may think of any molecule as possessing a dipole moment and a polarizability. This means that each of the three types of interaction may operate between any pair of molecules. Of course, in nonpolar molecules where $\mu = 0$, two of the three sources of attraction make no contribution.

As we have already noted, all molecules display the dispersion component of attraction since all are polarizable and that is the only requirement for the London interaction. Not only is the dispersion component the most ubiquitous of the attractions but it is also the most important in almost all cases. Only in the case of highly polar molecules such as water is the dipole–dipole interaction greater than the dispersion component. Likewise, the mixed interaction described by the Debye equation is generally the smallest of the three.

For a pair of identical molecules, Eqs. (11), (13), and (25) may be combined to give the *net* van der Waals attraction Φ_A:

$$\Phi_A = -\left(2\alpha_{0,1}\mu_1^2 + \frac{2}{3}\frac{\mu_1^4}{kT} + \tfrac{3}{4}h\nu_1\alpha_{0,1}^2\right)x^{-6} = -\beta_{11}x^{-6} \tag{33}$$

The interaction parameter β_{11} is defined

$$\beta_{11} = \frac{2}{3}\frac{\mu_1^4}{kT} + 2\alpha_{0,1}\mu_1^2 + \tfrac{3}{4}h\nu_1\alpha_{0,1}^2 \tag{34}$$

where the subscript 11 has been added to β as a reminder that this result applies to a pair of identical molecules.

Dividing Eq. (34) through by β_{11} gives the fractional contribution made to the total attraction by the Debye (D), Keesom (K), and London (L) components of potential energy:

$$f_D + f_K + f_L = 1 \tag{35}$$

Table 10.2 shows these fractions calculated for a variety of molecules. In virtually all cases except the highly polar water molecule, the London or dispersion component is the largest of the contributions to attraction. In the case of water, hydrogen bonding is also possible and contributes an additional strong interaction, so the role of dispersion is even less than shown in the table.

In this section, we have examined the three major contributions to what is generally called the van der Waals attraction between molecules. All three show the same functional dependence on the molecular separation and are, therefore, conveniently considered together. The following statements summarize the status of our development of these ideas up to this point:

1. The composite van der Waals attraction is characterized by a proportionality factor β which is given by Eq. (34) or some similar expression.
2. Equation (34) is itself an approximation, and the molecular parameters which it contains are often unavailable for substances of interest. Therefore, this relationship is more useful for the insights it supplies to the origin of van der Waals attraction than for actual computational purposes.

TABLE 10.2 *Percentage contribution of the Debye, Keesom, and London contributions to the van der Waals attraction between various molecules.*[a]

Compound	$\mu \times 10^{18}$ (esu cm)	$\alpha \times 10^{24}$ (cm^3)	$\beta \times 10^{58}$ (ergs cm^6)	Percentage contribution of		
				Keesom (permanent–permanent)	Debye (permanent–induced)	London (induced–induced)
CCl_4	0	10.7	4.41	0	0	100
Ethanol	1.73	5.49	3.40	42.6	9.7	47.6
Thiophene	0.51	9.76	3.90	0.3	1.3	98.5
t-Butanol	1.67	9.46	5.46	23.1	9.7	67.2
Ethyl ether	1.30	9.57	4.51	10.2	7.1	82.7
Benzene	0	10.5	4.29	0	0	100
Chlorobenzene	1.58	13	7.57	13.3	8.6	78.1
Fluorobenzene	1.35	10.3	5.09	10.6	7.5	81.9
Phenol	1.55	11.6	6.48	14.5	8.6	76.9
Aniline	1.56	12.4	7.06	13.6	8.5	77.9
Toluene	0.43	11.8	5.16	0.1	0.9	99.0
Anisole	1.25	13.7	7.22	5.5	6.0	88.5
Diphenylamine	1.08	22.6	14.25	1.5	3.7	94.7
Water	1.82	1.44	2.10	84.8	4.5	10.5

[a] Dipole moments and polarizabilities from A. L. McClellan, *Tables of Experimental Dipole Moments*, Freeman, San Francisco, California, 1963.

3. For this reason, we shall (temporarily, at least) continue to speak in terms of β rather than any development of this quantity in terms of polarization parameters.
4. Likewise, we shall seek an alternative way to relate β to measurable quantities. In Sec. 10.6, we shall see how β is related to the dispersion component of surface tension.
5. In order to complete this development and to establish the general usefulness of these ideas to colloid chemistry, we must consider the application of these ideas to macroscopic particles. This is the topic of the following section.

10.4 ATTRACTION BETWEEN MACROSCOPIC BODIES

The interaction between individual molecules obviously plays an important role in determining, for example, the nonideality of a gas. It is less clear how to apply this insight to dispersed particles in the colloidal size range. If atomic interactions are assumed to be additive, however, then the extension to macroscopic particles is not particularly difficult.

We begin by considering two spherical particles of the same composition and the same radius R. As Fig. 10.4 shows, we shall consider two situations. In case a, we shall consider spheres of radius R_a; and in case b, spheres of radius R_b. The two radii are related as follows:

$$R_b = fR_a \tag{36}$$

where f is a factor greater than unity.

Assume that every atom in sphere 1a attracts every atom in sphere 2a with an energy given by Eq. (33). If $\rho N_A/M$ is the number of atoms per cubic centimeter in the material, then there are $\rho N_A/M\, dV_{1a}$ atoms in a volume element of sphere 1a and $\rho N_A/M\, dV_{2a}$ atoms in a volume element of sphere 2a. The number of pairwise interactions between the two volume elements is $\frac{1}{2}(\rho N_A/M)^2\, dV_{1a}\, dV_{2a}$. The factor $\frac{1}{2}$ enters since each pair is counted twice. This number times the interaction per pair gives the increment in potential energy for the two interacting volume elements in

Figure 10.4 *Two spheres of equal radius R separated by a distance d. All linear dimensions in (b) (i.e., R and d) are larger than those in (a) by a constant factor f.*

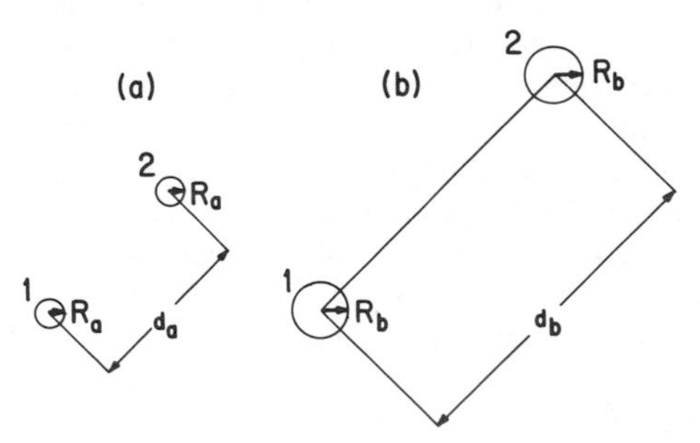

case a:

$$d\Phi_a = -\frac{1}{2}\left(\frac{\rho N_A}{M}\right)^2 \beta x_a^{-6}\, dV_{1a}\, dV_{2a} \tag{37}$$

By including additional geometrical considerations, the volume elements and their separation may be expressed in a common set of variables and Eq. (37) can be integrated over the volume of both spheres. Rather than complicate the issue by specific mathematical substitutions at this time, we indicate this procedure as follows:

$$\Phi_A = -\frac{1}{2}\left(\frac{\rho N_A}{M}\right)^2 \beta \iint_{1\text{ and }2} \frac{dV_{1a}\, dV_{2a}}{x_a{}^6} \tag{38}$$

Now let us turn our attention to Fig. 10.4b. In this case, the spheres are larger by a factor f as just noted. In addition to this, the distance between their centers is also assumed to be larger than that in case a by the factor f. If we let the separation of centers be represented by d, then

$$d_b = f d_a \tag{39}$$

Following the same procedure as used for case a, we can evaluate the total interaction potential in case b. It is given by

$$\Phi_b = -\frac{1}{2}\left(\frac{\rho N_A}{M}\right)^2 \beta \iint_{1\text{ and }2} \frac{dV_{2a}\, dV_{2b}}{x_b{}^6} \tag{40}$$

Since the linear scale in case b is larger by the factor f than that in case a, Eq. (40) can also be written

$$\Phi_b = -\frac{1}{2}\left(\frac{\rho N_A}{M}\right)^2 \beta \iint_{1\text{ and }2} \frac{(f^3\, dV_{1a})(f^3\, dV_{2a})}{(f x_a)^6} \tag{41}$$

Note that the factor f cancels out of this expression entirely, so that

$$\Phi_a = \Phi_b \tag{42}$$

That is, the potential energy of attraction is identical in the two cases. This is an important result as far as the extension of molecular interactions to macroscopic spherical bodies is concerned. What it says is that two molecules, say 3 Å in diameter and 10 Å apart, interact with exactly the same energy as two spheres of the same material which are 300 Å in diameter and 1000 Å apart. Furthermore, an inspection of Eq. (41) reveals that this is a direct consequence of the inverse sixth power dependence of the energy on the separation. Therefore, the conclusion applies equally to all three contributions to the van der Waals attraction. Precisely the same forces that are responsible for the association of individual gas molecules to form a condensed phase operate—over a suitably enlarged range—between colloidal particles and are responsible for the flocculation of the latter.

In the preceding chapter, we saw that an electrostatic potential energy of repulsion could have considerable magnitude some distance from the surface of a particle. Now we discover that van der Waals attraction can also be significant at appreciable distances from particles. The quantitative comparison of the two will determine the behavior of the dispersion when two particles encounter one another as a result of random movements. If the repulsion outweighs the attraction, the particles will rebound from the collision and go their separate ways. On the other hand, if the attraction is greater than the repulsion, then the two will adhere and behave as a single kinetic unit.

This synopsis summarizes the direction of the remainder of this chapter. Before we can quantitatively apply these ideas to flocculation in real systems, we must explicitly carry out the kinds of integration indicated formally in Eqs. (38) and (41). We shall illustrate the method only for the case of two blocks of material with facing planar surfaces such as we examined in the preceding chapter for repulsion. Results from other geometrical situations will not be derived but merely summarized after we derive the case of the blocks.

Figure 10.5a represents a molecule at O located a normal distance z from the surface of a bulk sample of the same material. The bulk portion is assumed to have a planar face but otherwise be of infinite extension. The molecule is located a distance x from all the molecules in the ring-shaped volume element shown in the figure. The volume of this ring is given by $dV = 2\pi y\, dy\, d\zeta$. Therefore, the increment of interaction between the molecule and the block due to the molecules a distance x from the point O is given by

$$d\Phi = -\frac{\rho N_A}{M}\beta\frac{2\pi y\, dy\, d\zeta}{x^6} \tag{43}$$

We assume the ring is located a distance ζ inside the surface of the block; then

$$x^2 = (z+\zeta)^2 + y^2 \tag{44}$$

FIGURE 10.5 *Schematic illustration showing the interaction* (*a*) *between a molecule and a block of material and* (*b*) *between two blocks of material.* [*Reprinted with permission from P. C. Hiemenz,* J. Chem. Educ., **49**:164 (1972), *copyright by the American Chemical Society.*]

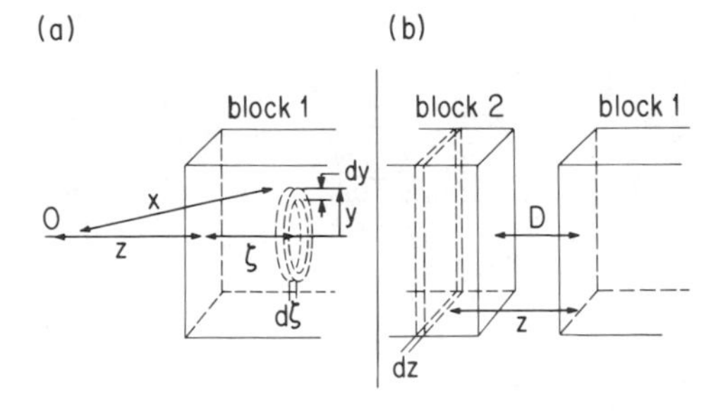

Combining Eqs. (43) and (44) gives

$$d\Phi = -\frac{\rho N_A}{M}\beta 2\pi \frac{y\,dy\,d\zeta}{[(z+\zeta)^2+y^2]^3} \tag{45}$$

Equation (45) is now integrated over the volume of the block, that is, for $0<y<\infty$ and $0<\zeta<\infty$.

Integration over y yields

$$\int_0^\infty \frac{y\,dy}{[(z+\zeta)^2+y^2]^3} = \frac{1}{2}\int_0^\infty \frac{du}{[(z+\zeta)^2+u]^3} = -\frac{1}{2}\left\{\frac{1}{2[(z+\zeta)^2+u]^2}\right\}\Bigg|_0^\infty$$

$$= \frac{1}{4}\frac{1}{(z+\zeta)^4} \tag{46}$$

and over ζ yields

$$\frac{1}{4}\int_0^\infty \frac{d\zeta}{(z+\zeta)^4} = -\frac{1}{4}\left[\frac{1}{3(z+\zeta)^3}\right]\Bigg|_0^\infty = \frac{1}{12z^3} \tag{47}$$

Therefore, the integration of Eq. (45) over the block gives

$$\Phi = -\frac{\rho N_A}{M}\frac{\beta\pi}{6z^3} \tag{48}$$

Now suppose point O is located inside a second block of material, as shown in Fig. 10.5b. We recognize that all atoms in a slice of the second block a distance z from the first block will be attracted toward the latter with an energy given by Eq. (48). If we position a volume element of thickness dz at this location in the second block, we realize that it contains $\rho N_A/M$ molecules per cubic centimeter. Per unit area this is $(\rho N_A/M)\,dz$ molecules; therefore, the increment of attraction per unit area due to this slice is

$$d\Phi = -\left(\frac{\rho N_A}{M}\right)^2 \frac{\beta\pi}{6} z^{-3}\,dz \tag{49}$$

Equation (49) may now be integrated over values of z between the distance of closest approach D and infinity. The result of this integration gives the potential energy of attraction per unit area between two blocks of infinite extension:

$$\Phi_A = -\left(\frac{\rho N_A}{M}\right)^2 \frac{\beta\pi}{12} D^{-2} \tag{50}$$

where the subscript A has been appended to Φ to emphasize the fact that this energy is always attractive.

It is traditional to designate the cluster of constants $(\rho N_A/M)^2\pi^2\beta$ as the Hamaker constant A in honor of one of the first workers to investigate the application of these ideas to macroscopic bodies:

$$A = \left(\frac{\rho N_A \pi}{M}\right)^2 \beta \tag{51}$$

With this change of notation, Eq. (50) becomes

$$\Phi_A = -\frac{A}{12\pi}D^{-2} \tag{52}$$

The derivations of this section show that the Hamaker constant plays an equivalent role in describing the van der Waals attraction between macroscopic bodies that the constant β plays in the interaction of individual molecules. From now on, we shall emphasize A more heavily than β in describing interparticle attractions.

Before proceeding any further, it may be worthwhile to examine the quantitative magnitude of the interaction described by Eq. (52). A particularly convenient form for an order of magnitude estimation is obtained by making the following stipulations:

1. Assume that the dispersion component is the dominant contributor to the attraction; therefore, $\beta = \frac{3}{4}h\nu\alpha^2$.
2. Recognize that $\rho N_A/M$ is the reciprocal of the volume per molecule and that polarizabilities are typically about 10% the magnitude of atomic volumes; therefore, $A \simeq \frac{3}{4}\pi^2 h\nu(0.1)^2$.
3. The quantity $h\nu$ is the same order of magnitude as the ionization potential, typically about 10^{-11} erg.

Combining these results leads us to estimate the Hamaker constant to lie in the range 10^{-13} to 10^{-12} erg. Taking A to be 10^{-12} erg, Eq. (52) shows that Φ_A equals about 3 ergs cm^{-2} at a separation of 10 Å and about 0.03 erg cm^{-2} at 100 Å. For $A = 10^{-13}$ erg, $\Phi_A = 0.3$ and 0.003 erg cm^{-2} at 10 and 100 Å, respectively. These values are seen to be of roughly the same order of magnitude as the potential energies of repulsion calculated in Sec. 9.10 for a similar model. It should be noted that the derivations of Sec. 9.10 assumed the separation of the surfaces to be $2d$ whereas the present treatment defines the separation to be D. That is, when the results of these two sections are compared, we must remember that

$$D \text{ (Sec. 10.4)} = 2d \text{ (Sec. 9.10)} \tag{53}$$

We saw in Sec. 9.10 that the magnitude of the potential energy of repulsion is highly sensitive to the electrolyte content of the system. The latter plays a large role in determining whether attraction or repulsion will be the dominant interaction when dispersed particles approach each other. We shall return to a more detailed examination of the relative magnitude of these effects in Sec. 10.9.

The interaction of two infinite blocks separated at the surface by a distance D is one of the easiest possible situations to consider and, therefore, was chosen as the example to develop in detail. Bodies of different geometries have also been analyzed in much the same way that we have done for the blocks. The results of several such derivations are shown in Table 10.3. The expressions for Φ_A become more complicated for more complex geometrical situations, but considerable simplification results for spheres in which the separation is considerably less than the radius. This is the same situation to which the expression for repulsion given by Eq. (9.109) applies. Note also that the expression for interacting plates of thickness δ in

TABLE 10.3 *Potential energy of attraction between two particles with the indicated geometries*

Particles	Φ_A	Definitions/limitations
Two spheres	$-\frac{A}{6}\left[\frac{2R_1R_2}{s_0^2+2R_1s_0+2R_2s_0}+\frac{2R_1R_2}{s_0^2+2R_1s_0+2R_2s_0+4R_1R_2} + \ln\left(\frac{s_0^2+2R_1s_0+2R_2s_0}{s_0^2+2R_1s_0+2R_2s_0+4R_1R_2}\right)\right]$	R_1, R_2 = radii, s_0 = separation of surfaces
Two spheres of equal radius	$-\frac{A}{6}\left[\frac{2R^2}{s_0^2+4Rs_0}+\frac{2R^2}{s_0^2+4Rs_0+4R^2}+\ln\left(\frac{s_0^2+4Rs_0}{s_0^2+4Rs_0+4R^2}\right)\right]$	$R_1=R_2=R$
Two spheres of equal radius	$\frac{AR}{12s_0}$	$R \gg s_0$
Two spheres of unequal radius	$\frac{AR_1R_2}{6s_0(R_1+R_2)}$	R_1 and $R_2 \gg s_0$
Two plates of equal thickness	$\frac{A}{12\pi}\left[\frac{1}{D^2}+\frac{1}{(D+2\delta)^2}-\frac{2}{(D+\delta)^2}\right]$	δ = thickness of plates
Identical blocks	$\frac{A}{12\pi D^2}$	$\delta \to \infty$

Table 10.3 reduces to Eq. (52) as $\delta \to \infty$. The feature just described for interacting spheres is also evident in the results given in the table. It is most readily seen by examining the case of equal spheres separated by small distances. Note that both R and s_0 may be multiplied by any common factor without changing the magnitude of the interaction energy.

Before these results can be applied quantitatively to the phenomena of flocculation, there are several additional points which must be discussed:

1. The assumption—implicit in the derivation of Eq. (52) and the other results in Table 10.4—that matter is continuous.
2. The phase difference between the fields of interacting dipoles that are separated by large distances.
3. The effect of the intervening medium on the interaction between particles.

These topics are discussed in the following sections.

10.5 MACROSCOPIC INTERACTIONS AT SMALL SEPARATIONS

In the preceding section, we examined the additivity of intermolecular attractions between macroscopic bodies. Pairwise interaction energies were multiplied by the number of molecules at that distance of separation and the results were totaled by integration. In doing this, the assumption is made implicitly that the density of matter is the same at all locations within a phase, that is, that matter is continuous. As long as the distances involved are large, this assumption is justified. From a great distance, the molecules cannot "see" the detailed structure of the other particle at the molecular level. This is no longer true, however, when the separation between the particles is the same order of magnitude as the spacing between the molecules in an individual particle. In the latter case, the discreteness of individual molecules must be taken into account.

If we wish to consider the specific coordinates of the molecules in calculating the interaction between phases, we enter the realm of molecular spacing, packing, and the like. It is not possible to proceed along these lines in any completely general way; however, assumptions about structure may be introduced in the form of specific models which—although they may not be totally accurate—probably represent improved approximations at small separations.

One such approximation is based on the following model. Instead of picturing the matter of a block as being continuous, this model assumes that the atoms are "smeared out" in planes which run parallel to the interacting face of the block, as shown in Fig. 10.6. As with Fig. 10.5a, we consider the interaction of a molecule located at O with an infinite stack of these smeared-out sheets, each having a thickness δ and separated by a distance d. If the sheets are of infinite extension, the interaction between the ith sheet and a molecule at O is given by

$$d\Phi_i = \frac{\rho' N_A}{M}\beta\delta \int_0^\infty \frac{2\pi y\, dy}{\{[z+i(d+\delta)]^2 + y^2\}^3} = \frac{\rho' N_A}{2M}\beta\pi\delta[z+i(d+\delta)]^{-4} \tag{54}$$

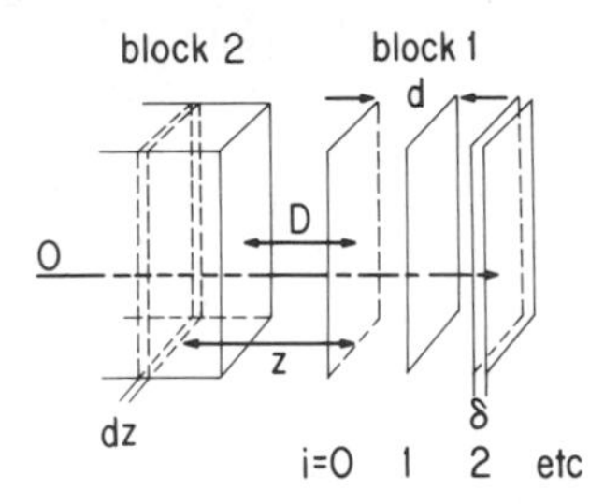

FIGURE 10.6 *Schematic illustration analogous to Fig.* 10.5 *except that molecules in the block are "smeared out" in planes of thickness δ and separated by a distance d.*

using Eqs. (45) and (46) and adopting the present notation. In this expression, ρ' is the density in the smeared-out sheets. Since the total amount of material must be the same whether a continuous or sheet model is considered,

$$\rho(d+\delta)=\rho'\delta \tag{55}$$

Incorporating this result into Eq. (54) and considering the limit of $\delta \to 0$ gives

$$d\Phi_i = \frac{\rho N_A}{2M}\beta\pi d(z+id)^{-4} \tag{56}$$

The total interaction of the molecule with an infinite stack of these sheets is obtained by adding increments given by Eq. (56) for all values of i:

$$\Phi_A = \frac{\rho N_A}{2M}\beta\pi d \sum_{i=1}^{\infty} (z+id)^{-4} \tag{57}$$

It will be recalled that our reason for considering this approach is an interest in values of z which are comparable to molecular dimensions. A particularly simple result is obtained if z is set equal to d, in which case Eq. (57) becomes

$$\Phi_A = \frac{\rho N_A \beta\pi}{2Md^3} \sum_{i=1}^{\infty} (1+i)^{-4} = \frac{1.082\rho N_A \beta\pi}{2Md^3} \tag{58}$$

This result is directly comparable to Eq. (48) with $z=d$ except that the atoms are assumed to be continuous in two-dimensional sheets rather than continuous in three dimensions.

Now suppose that rather than an isolated molecule interacting with such a block, we consider a second block (subscript j) structured in an analogous fashion but possessing its own characteristic spacing d_j. In this extension, we shall designate the spacing of the first block d_i and the spacing of the two surface sheets ($i=0$ and $j=0$) D. Again ignoring the thickness of the sheets compared to their spacing, the total energy of interactions may be written

$$\Phi_A = \left(\frac{\rho N_A}{M}\right)_i \left(\frac{\rho N_A}{M}\right)_j \frac{\beta_{ij}\pi}{2} d_i d_j \sum_i \sum_j (D+id_i+jd_j)^{-4} \tag{59}$$

by analogy with Eq. (57). This result is greatly simplified if it is assumed that $d_i = d_j = D$, in which case Eq. (59) becomes

$$\Phi_A = 1.20\left(\frac{\rho N_A}{M}\right)_i\left(\frac{\rho N_A}{M}\right)_j\frac{\beta_{ij}\pi}{2D^2} \tag{60}$$

This result is comparable to Eq. (50) except that D is identical to the intermolecular spacing in the present result and to much larger distances in Eq. (50).

Comparing Eqs. (58) and (60), which apply at small separations only, with their counterparts for large separations, (48) and (50), reveals that the two have the same functional forms. However, the numerical coefficients are larger for the smaller separations. The inverse sixth power relationship which underlies all these results means that the nearest pairs of molecules make the largest contributions. Roughly 80% of the total potential energy of attraction arises from the interaction between the first sheet in each block. Therefore, the increased density in the sheets compared to the continuous bulk model [Eq. (55)] accounts for the stronger interaction in this case.

As far as flocculation is concerned, it is the interaction at larger separations that is most important. However, the results of this section suggest an alternative method for numerical evaluation of the Hamaker constant. We shall examine this connection in the following section.

10.6 VAN DER WAALS FORCES AND SURFACE TENSION

Throughout our discussion of van der Waals attraction between particles up to this point, it has been implicitly assumed that all the molecular parameters are available for the asking. This implies that the attraction between macroscopic bodies—at least those of well-defined geometry—may be evaluated at will. This picture is overly optimistic for several reasons, some of which are the following:

1. The contribution of permanent dipoles to the total interaction is almost always assumed to be zero. To evaluate this contribution, some estimation of the extent of orientation has to be made, and such information is not readily available.
2. The characteristic frequencies and/or polarizabilities needed to calculate the London attraction are not available for many substances of interest.
3. Chemical heterogeneities at the surface—which are very common near an interface—introduce uncertainties as to the composition of the neighboring faces. This is especially troublesome at small separations where, according to Eq. (60), actual intermolecular distances enter the picture.

In light of these considerations, another approach seems desirable: namely, to work backward from some observed quantity to an average value of the interaction parameter for the system. If the concepts we have been discussing provide the basis for the evaluation of these constants, then the method at least has a certain amount of internal consistency even if it deviates from "absolute" predictions. As it turns out, interaction energies may be deduced from surface tension data which are quite reasonable compared to those calculated by the absolute approach.

To see how this works, it is convenient to consider the work of cohesion defined by Eq. (6.72) and illustrated schematically by Fig. 6.13a. The figure represents a column of liquid (designated L here) of unit cross section being separated from a single column into two segments. For our present purposes, it is desirable to picture the initial state of a "single column" as being two columns also—but two columns separated by the normal intermolecular distance for that material. With this picture in mind, the work of adhesion becomes

$$W_{LL} = 2\gamma_L = \Phi_{D=\infty} - \Phi_D \tag{61}$$

Now Eq. (60) may be used as a substitution for Φ_D and (52) as a substitution for Φ_∞. The latter vanishes anyhow since it is evaluated at $D = \infty$. Making these substitutions we obtain

$$2\gamma_L = 1.2\left(\frac{\rho N_A}{M}\right)_L^2 \frac{\beta_L \pi}{2} d_L^{-2} \tag{62}$$

Since this result applies to a homogeneous substance, the subscripts are redundant and will not be carried any further.

It is obvious from Eq. (61) that γ is being related to the *long-range* attraction between the phases. Recall that surface tension was subdivided into contributions in Chap. 6, for example, in Eq. (6.79).

Therefore, it seems that Eq. (62) would be improved if only the dispersion component of surface tension γ^d were used instead of γ itself. Incorporating this consideration and the definition of A provided by Eq. (51) permits us to write

$$A = \frac{4\pi}{1.2}\gamma^d d^2 \tag{63}$$

Remember that d is the intermolecular spacing of the bulk material in this expression.

Equation (63) provides an alternate means in addition to (34) and (51) for the evaluation of the Hamaker constant. Note that this method does not resolve any of the difficulties enumerated at the beginning of this section. Either the same objections apply or else one inaccessible parameter is substituted for another:

1. Dispersion forces are still assumed to predominate.
2. Reliable values of γ^d are not always available.
3. The details of molecular spacing are still uncertain in the interfacial region.

Thus the choice of a method for the evaluation of A depends very much on the kind of data that are available for the system under consideration.

The primary objective of the present discussion is to show the intrinsic connection between the dispersion component of surface tension and the van der Waals energy of attraction between macroscopic bodies. The connection not only provides computational options but also—and more importantly—unites two apparently separate phenomena and strengthens our confidence in the correctness of our understanding.

Table 10.4 shows a few numerical examples of how well this attempt at unification succeeds. Equations (28), (32), and (51) have been used to calculate values of the Hamaker constant from refractive index data at visible wavelengths. These values have then been used along with γ^d values from Chap. 6 to calculate d values according to Eq. (63). The resulting values of d are seen to be physically reasonable. That such plausible values for d are obtained is especially noteworthy in view of the crudeness of the model underlying Eq. (63) and the other approximations made in the calculations.

The applicability of this procedure receives a far more stringent test in the case of water and SiO_2 than for the hydrocarbons. Dispersion forces are assumed to be the only contributors to γ^d (i.e., $\gamma^d = \gamma$) for hydrocarbons. This is definitely not the case for quartz or water, so the d values obtained for these substances are quite satisfactory.

It is extremely difficult to measure the Hamaker constant directly, although this has been the object of considerable research effort. Direct evaluation, however, is complicated either by experimental difficulties or by uncertainties in the values of other variables which affect the observations.

The direct measurement of van der Waals forces has been undertaken by literally measuring the force between macroscopic bodies as a function of their separation. The distances, of course, must be very small so optical interference methods may be used to evaluate the separation. The force has been measured from the displacement of a sensitive spring and also from capacitance-type measurements. The two principal sources of difficulty in these methods are external vibrations and surface roughness. Nevertheless, it has been possible to verify directly the functional dependence on radii for the attraction between dissimilar spheres (see Table 10.3), the retardation of van der Waals forces (see the following section), and to evaluate the Hamaker constant for several solids, including quartz. Values in the range of 6 to 7×10^{-13} erg have been found for quartz by this method. This is remarkably close to the value listed in Table 10.4 for SiO_2.

A less direct but also useful source of information on the value of Hamaker constants is the study of thin liquid films. If the film is arranged so that excess liquid can drain away, the film will thin until an equilibrium is established between (a) the van der Waals attraction between the phases on opposite sides of the film and (b) the repulsion between opposite faces of the film. The latter may be double layer forces such as those discussed in the preceding chapter or they may be purely steric or some combination of the two. The mechanism of steric repulsion arises as follows: Surfaces with oriented adsorbed molecules can approach no closer than twice the length of the adsorbed molecule. The thinning of hydrocarbon films which are surrounded by water with lipids adsorbed at the surfaces has been analyzed in this way. The effective Hamaker constant in this case (for two masses of water separated by hydrocarbon) is about 5.6×10^{-14} erg. We shall discuss the effect of the intervening medium on these measurements in Sec. 10.8. For the present, it is sufficient to note that this measurement is consistent with a Hamaker constant for water of about 5.1 to 6.3×10^{-13} erg. In view of all the difficulties involved, this too must be regarded as acceptable agreement with the value of A for water in Table 10.4.

TABLE 10.4 *Calculations intended to show that Eq.* (63) *predicts values of the Hamaker constant which are compatible with those evaluated from Eqs.* (28), (32), *and* (51)[a]

Compound	M (g mole^{-1})	ρ (g cm^{-3})	n	A (ergs)	γ^d (ergs cm^{-2})	d (Å)
Heptane	100.2	0.684	1.39	1.05×10^{-13}	20.3	2.2
Dodecane	170.3	0.749	1.42	9.49×10^{-14}	25.4	1.8
Eicosane	282.5	0.789	1.44	2.07×10^{-13}	29.0	2.6
SiO_2 (quartz)	60	2.65	1.54	4.14×10^{-13}	78.0	2.2
Polystyrene	(104)[b]	1.05	1.59	2.2×10^{-13}	41.0	2.3
Water	18	1.00	1.33	2.43×10^{-13}	21.3	3.3

[a] Equations (28), (32), and (51) are used to evaluate A, then A and γ^d values from Chap. 6 are used to evaluate d.
[b] Monomer.

Aqueous films stabilized by ionic surfactants have also been studied extensively and have provided estimates of A for water. We shall not discuss these here, however, since such a discussion would anticipate some developments which we have not yet considered. The interested reader will find this method as well as the others cited described more fully in the works by Israelachvili and Tabor [2] and Sonntag and Strenge [6].

In summary, the following generalizations can be made:

1. Direct experimentation confirms that the London attraction between macroscopic bodies does indeed follow the predictions presented in Sec. 10.4.
2. The experimental difficulties associated with the direct determination of the Hamaker constant make this prohibitive as a routine procedure at this time. Either the molecular parameters α_0 and ν or the combination of empirical and molecular parameters γ^d and d can be used to obtain reasonable estimates of A.
3. The dispersion component of surface tension is quite reasonably consistent with the models we have discussed for the interaction of macroscopic bodies. In fact, γ^d can be calculated by this approach if reliable values of a_0, ν, and d are available.

10.7 INTERACTIONS AT LARGE SEPARATIONS: RETARDED VAN DER WAALS FORCES

The electric field which is responsible for the London attraction between molecules propagates itself with the speed of light between the particles. Thus, if a pair of molecules are widely separated, a time lag or a phase difference develops between vibrations at the two locations. The situation is analogous in many ways to the scattering of light by particles whose dimensions are large compared to the wavelength of the light (see Sec. 5.9). In the present situation, we find that the importance of this time lag or retardation increases as the separation becomes comparable to the wavelength of the propagating field.

Equation (29) provides us with an expression for the frequency of the interaction. Therefore, its wavelength is given by

$$n\lambda = \frac{c}{\nu} = 2\pi c\sqrt{\frac{\alpha_0 m_e}{e^2}} \tag{64}$$

where n is the refractive index. For typical values of n and α_0, λ is about 2000 Å. As was the case with light scattering, separations less than about one-twentieth of this distance may be considered "small compared to the wavelength." This means that at separations of about 100 Å or so the effects of retardation begin to enter the picture. Our reason for interest in this is the fact that the comparison of attraction and repulsion between colloidal particles is made at this distance in some cases.

We shall not present the detailed analysis of this complication. In essence, it involves the time-dependent Schrödinger equation rather than the time-independent equation which resulted in (22). H. B. G. Casimir and D. Polder have investigated this situation. They found that for values of $x \gg \lambda$, the potential energy of attraction according to the modified London treatment is given by

$$\Phi_A = -\frac{23}{8\pi^2}\frac{hc\alpha_0^2}{x^7} \tag{65}$$

This is the one entry in Table 10.1 that has not yet been discussed. The direct measurement of the force of attraction between macroscopic bodies that we discussed in the preceding section reveals a crossover from an inverse sixth to an inverse seventh power law at separations in the range 10^2 to 10^3 Å.

For separations at which the retardation cannot be neglected, we must multiply the London potential energy of attraction by a correction factor $K(\alpha)$ which is a function of the dimensionless variable $\alpha = 2\pi x/\lambda$. This factor may be approximated by the following expressions in the indicated ranges:

1. For $0<\alpha<3$:

$$K(\alpha)= 1.01-0.14\alpha \tag{66}$$

2. For $\alpha>3$:

$$K(\alpha)=\frac{2.45}{\alpha}-\frac{2.04}{\alpha^2} \tag{67}$$

Note that Eq. (67) applied to (25) reduces to (65) for large values of α.

Equation (41) is particularly convenient to show the effects of retardation on the attraction between spherical particles at large separations. Correcting the potential function for retardation according to Eq. (65), (41) becomes

$$\Phi_b \simeq \iint_{1\text{ and }2} \frac{(f^3\,dV_{1a})(f^3\,dV_{2a})}{(fx_a)^7} = \frac{1}{f}\Phi_a \tag{68}$$

at large separations. This result shows that the scale factor f does not drop out of the expression in the case of the retarded van der Waals forces. Therefore, the potential energy of the attraction decreases as the scale increases (i.e., as f increases).

In subsequent discussions, we shall not consider the effect of retardation any further. Additional details are given by Israelachvili and Tabor [2] and Sonntag and Strenge [6].

10.8 EFFECT OF THE MEDIUM ON THE VAN DER WAALS ATTRACTION

Until now, we have considered the interaction between isolated molecules or macroscopic bodies when the particles are separated by a vacuum. The former may be reasonably applied to molecules in the gas phase. However, for dispersions of one phase in another, the effect of the medium must be taken into account. The easiest way to do this is to consider the pseudochemical reaction illustrated in Fig. 10.7. The particles numbered 2 (solid lines) represent the dispersed phase and those numbered 1 (dashed lines) are the solvent. In the initial condition, each dispersed

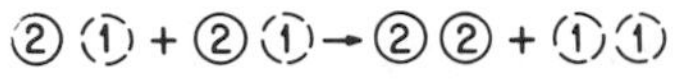

② ⓵ + ② ⓵ → ② ② + ⓵ ⓵

FIGURE 10.7 *The flocculation process as a pseudochemical reaction. Solid lines indicate particles of the dispersed phase; dashed lines, satellite particles of the solvent.*

particle and its satellite solvent particle comprise an independent kinetic unit. Figure 10.7 represents the process in which the two dispersed particles come together to form a doublet and the two solvent particles form a kinetically independent doublet.

The change in the potential energy which accompanies this process is given by

$$\Delta\Phi = \Phi_{11} + \Phi_{22} - 2\Phi_{12} \tag{69}$$

where the subscripts apply to the two types of particles. Each of the terms for Φ on the right-hand side of Eq. (69) depends in the same way on the size and distance parameters and differs only in molecular parameters which are fully contained in the Hamaker constant. Therefore, $\Delta\Phi$ follows the appropriate function for the interaction from Table 10.3 with the following value of the Hamaker constant:

$$A_{212} = A_{11} + A_{22} - 2A_{12} \tag{70}$$

The subscript 212 indicates two particles of type 2 separated by the medium of type 1.

Equation (6.80) suggests an approximation that results in a useful simplification:

$$A_{12} \simeq \sqrt{A_{11}A_{22}} \tag{71}$$

Equation (71) says that the interaction between dissimilar bodies is given by the geometrical mean of the homogeneous interactions for the two species considered separately. This geometrical mixing rule is widely used in solution theory to calculate heterogeneous interactions. We invoked this type of averaging procedure in Sec. 6.12 in discussing the attraction between two phases and we shall use it again in this chapter for the same purpose.

Combining Eqs. (70) and (71) leads to

$$A_{212} = (A_{11}^{1/2} - A_{22}^{1/2})^2 \tag{72}$$

This is the effective value of the Hamaker constant to be used in evaluating the attraction between (like) particles embedded in a medium. Equation (72) leads to three important generalizations about the value of A_{212}:

1. The effective Hamaker constant A_{212} is always positive regardless of the relative magnitudes of A_{11} and A_{22}. Thus, dispersed particles exert a net attraction on one another due to van der Waals forces in a medium as well as under vacuum.
2. Embedding particles in a medium generally diminishes the van der Waals attraction between them. Table 10.4 shows that the Hamaker constants for homogeneous interactions, A_{ii}, are generally of the same order of magnitude for different substances. Therefore, the effective Hamaker constant—which depends on their difference according to Eq. (72)—will be smaller than A_{ii} for either of the homogeneous interactions. For example, if A_{11} and A_{22} equal

8.1×10^{-13} and 6.4×10^{-13} erg, then A_{212} is 10^{-14} erg, according to Eq. (72). We saw in Sec. 10.6 that A_{212} is about 5.6×10^{-14} erg when 2 is water and 1 is a hydrocarbon. For the hydrocarbon layer A_{11} is estimated to be in the range 2.3 to 3.1×10^{-13} erg; therefore, A_{22} is in the range 5.1 to 6.3×10^{-13} erg. This is the experimental value of A for water that we compared with the theoretical value for that quantity in Table 10.4.

3. For $A_{11} = A_{22}$, $A_{212} = 0$ and $\Phi_A = 0$. From the viewpoint of van der Waals forces, this condition corresponds to no net interaction between particles.

Since van der Waals forces are responsible for the flocculation of lyophobic colloids, we see that the continuous phase itself imparts a certain measure of stability to the dispersed phase by diminishing the energy of attraction between the particles. The quantitative magnitude of this effect for dispersions in which water is the continuous phase is illustrated by Fig. 10.8. This figure shows how precipitously the

FIGURE 10.8 *A_{212} versus A_{22} for $A_{11} = 5.8 \times 10^{-13}$ erg, the value for water. γ^d values equivalent to these A_{22} values if $d = 5$ Å [calculated by Eq. (63)] are also included.*

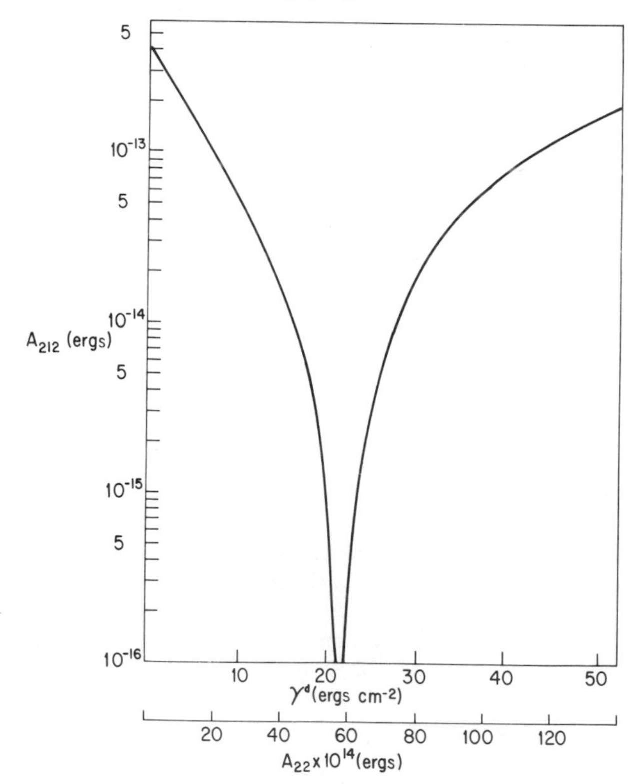

effective Hamaker constant A_{212} drops for particles having a value of A_{22} near that of water. Using Eq. (63) and arbitrarily assuming $d = 5$ Å, the corresponding values of γ^d can be estimated for the dispersed particles having these values of A_{212}. Any substance for which γ^d exceeds about 50 ergs cm^{-2} will almost certainly have an effective Hamaker constant greater than 10^{-13} erg in water.

In Chap. 9 and in this chapter, we have discussed—separately—the two principal sources of interaction between colloidal particles. Since the two work in opposition, it is clear that the actual behavior of a colloidal system will depend on the resultant of the attraction and repulsion. In the following few sections, we shall assemble the various components to arrive at a quantitative theory for the stability of lyophobic colloids. The resulting theory is generally called the DLVO theory after B. Derjaguin, L. D. Landau, E. J. W. Verwey, and J. Th. G. Overbeek, who—working independently in Russia and the Netherlands—brought the diverse elements together into a unified picture.

10.9 POTENTIAL ENERGY CURVES AND THE DLVO THEORY

The Derjaguin, Landau, Verwey, and Overbeek theory of colloid stability is best approached by considering curves which show the potential energy of a pair of colloid particles as a function of their separation. We have thus come full circle: We started this chapter with a discussion of the potential energy versus separation of two atoms. Now we are concerned with the same sort of information about a pair of colloidal particles. As we have seen, the attraction has the same origin in both cases, although the functional dependence on separation is different as a result of the integration over the macroscopic particles. The repulsion in the case of a pair of atoms originates from the overlap of their respective electron clouds. With colloidal particles, it is the overlap of their electrical double layers that is responsible for the repulsion.

Any quantitative comparison of attractive and repulsive potential energies requires that some geometry be specified for the particles. We shall consider mainly the case of interacting blocks, because the appropriate expressions are somewhat simpler for this geometry than for the more realistic case of interacting spheres. The more complicated spherical geometry contains no fundamental insights beyond those which can be obtained from the model of interacting blocks. Only in considering the effect of varying the electrolyte concentration will we turn to the model of interacting spheres. Some artifacts with which we need not be concerned enter the picture when this effect is examined for interacting blocks.

There is one difference between the interaction of blocks versus the interaction of spheres which should be noted. In the former, the expressions for potential energy give the energy per unit area whereas the latter give total potential energy. Thus, in order to compare the two in the same units, a cross-sectional area in the gap between the blocks must be specified.

For the case of two blocks with flat faces, Eq. (9.102) shows that the energy of repulsion varies exponentially with the separation ($2d$). For the same geometry, Eq.

(52) of this chapter shows that the attraction varies inversely with the square of the separation (D). Two observations about these functions will prove helpful:

1. As $2d = D \rightarrow 0$: $\Phi_R \rightarrow$ const and $\Phi_A \rightarrow -\infty$.
2. As $2d = D \rightarrow \infty$: Φ_R and $\Phi_A \rightarrow 0$ with Φ_R decreasing more rapidly.

These show that attraction will predominate at very small and very large separations. At intermediate separations the details of the two contributions must be considered. Figure 10.9 is a qualitative sketch of a potential energy curve which is the resultant of a repulsive and an attractive component:

$$\Phi_{\text{net}} = \Phi_R - \Phi_A \tag{73}$$

Such a curve may show a maximum and two minima as sketched, although some of these features may be masked if one contribution greatly exceeds the other. The height of the maximum above $\Phi = 0$ is what we shall call the height of the energy barrier. The deeper minimum is called the primary minimum; and the more shallow one, the secondary minimum. Their depth is measured from $\Phi = 0$ also. Although attraction predominates at large distances—that is, the secondary minimum is generally present—it may be quite shallow, especially in view of the effects of retardation and the medium on attraction.

Now let us qualitatively consider the implications of the potential energy curve shown in Fig. 10.9 for two particles approaching one another as a result of diffusion:

1. If no barrier is present or the height of the barrier is negligible compared to thermal energy, then the net force of attraction will pull the particles together into the primary minimum, after which the two behave as a single kinetic unit. Flocculation has occurred.
2. If the height of the potential energy barrier is appreciable compared to thermal energy, the particles are prevented from flocculating in the primary minimum. If the depth of the secondary minimum is small compared to thermal energy, then

FIGURE 10.9 *Qualitative sketch of potential energy versus separation between blocks. Curve* 1, *double layer repulsion*; *curve* 2, *van der Waals attraction*; *curve* 3, *resultant of curves* 1 *and* 2.

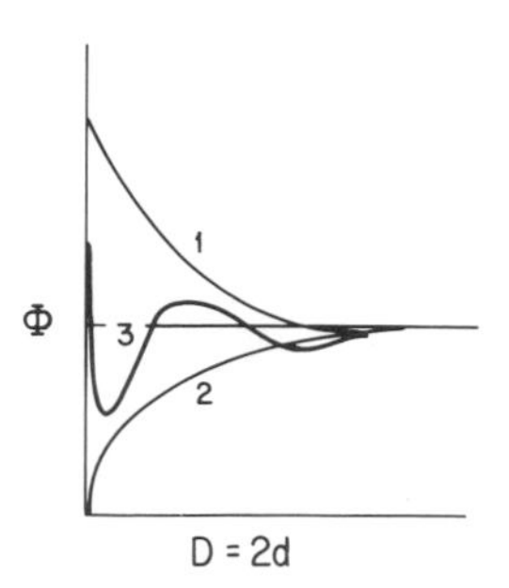

the particles will simply diffuse apart. This system is sta
flocculation.

3. There may be situations in which flocculation occurs in the se
but the flocs in this case will clearly be much more easily di
which form by flocculation in the primary minimum.
4. Generally speaking, it is the height of the barrier that determines whether a colloid will be stable or undergo flocculation.

With this qualitative background in mind, we may now turn to more quantitative considerations. Equation (9.102) and Eq. (52) may be substituted into (73) to give the net potential energy between two blocks as a function of their separation D:

$$\Phi = \frac{64 n_0 k T \gamma_0^2}{\kappa} \exp(-\kappa D) - \frac{A}{12\pi} D^{-2} \tag{74}$$

In the next few paragraphs, the effect on the net potential energy curves of the Hamaker constant, the surface potential, and the electrolyte content—considered separately and in this order—will be examined.

1. *The effect of A* It is understood that A in Eq. (74) is the effective Hamaker constant A_{212} for the system. Of the variable parameters in this equation, it is the one over which we have least control; its value is determined by the chemical nature of the dispersed and continuous phases. The presence of small amounts of solute in the continuous phase leads to a negligible alteration of the value of A for the solvent.

   The effect of variations in the value of A_{212} on the net potential energy is shown in Fig. 10.10. Each of the curves in the figure is drawn for a different value of A, but at identical values of κ (10^7 cm^{-1} or 0.093 M for 1 : 1 electrolyte) and ψ_0 (103 mV). As might be expected, the height of the potential energy barrier decreases and the depth of the secondary minimum increases with increasing values of A. If the cross-sectional area of interaction is 400 Å^2, each unit on the ordinate scale corresponds to kT at 25°C. This is the unit of thermal energy against which all interactions are judged to be large or small. Thus for the curves shown in Fig. 10.10, the depth of the secondary minimum is indeed slight and only for the smallest A value is the barrier height significant for particles with this interaction cross section.
2. *The effect of ψ_0* The potential at the inner limit of the diffuse part of the double layer—taken to be ψ_0—enters Eq. (74) through γ_0, defined by Eq. (9.75). For large values of ψ_0, $\gamma_0 \simeq 1$, so sensitivity to the value of ψ_0 decreases as ψ_0 increases. Figure 10.11 shows the effect of variations in the value of ψ_0 on the total interaction potential energy with κ (10^7 cm^{-1} or 0.093 M for a 1 : 1 electrolyte) and A (2×10^{-12} erg) constant. The height of the potential energy barrier is seen to increase with increasing values of ψ_0 as would be expected in view of the increase of repulsion with this quantity. For some systems, ψ_0 is adjustable by varying the concentration of potential-determining ions, as in a AgI dispersion for example. Complications due to specific adsorption in the Stern layer often make

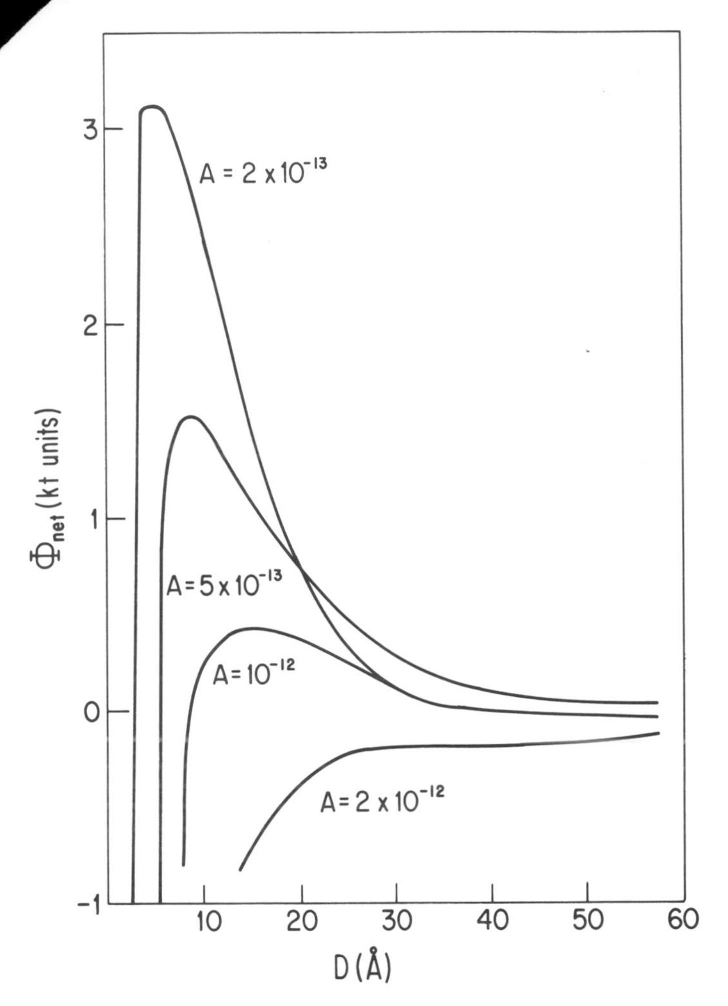

FIGURE 10.10 Φ_{net} *versus D according to Eq.* (74) *for flat blocks. Curves drawn for different values of* A_{212} *with constant values of* κ (10^7 cm^{-1}) *and* ψ_0 (103 mV). *Units of ordinate: multiples of kT at* 25°C *for an interaction area of* 400 Å^2.

the proper numerical value for this parameter uncertain. As we shall see in the following chapter, electrokinetic experiments measure a potential within the double layer—the ζ potential—but it is not entirely clear at which location within the double layer this potential applies. However, the experimental ζ potential does establish a lower limit for ψ_0.

3. *The effect of electrolyte concentration* Of the various quantities which affect the shape of the net interaction potential curve, none is as accessible to empirical

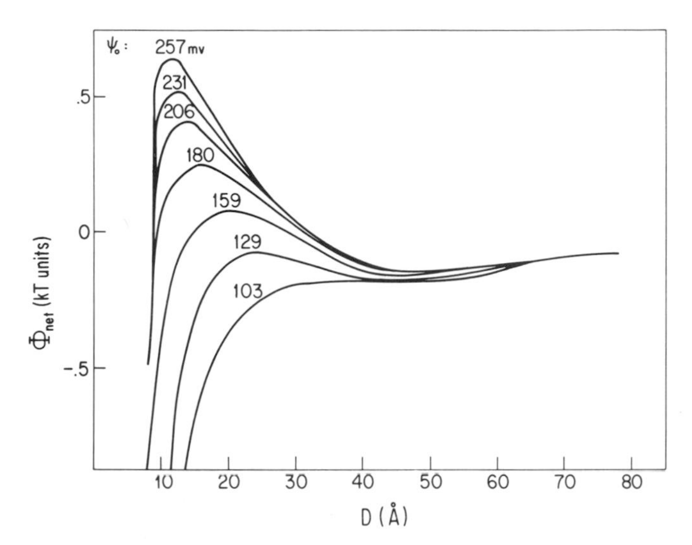

FIGURE 10.11 Φ_{net} *versus D according to Eq.* (74) *for flat blocks. Curves drawn for different values of* ψ_0 *with constant values of* κ (10^7 cm^{-1}) *and A* (2×10^{-12} erg). *Units of ordinate: multiples of kT at* 25°C *for an interaction area of* 400 Å^2.

adjustment as κ. This quantity depends on both the concentration and valance of the indifferent electrolyte as shown by Eq. (9.53). For the present, we examine only the consequences of concentration changes on the total potential energy curve. We shall consider the valence of electrolytes in the following section. As noted earlier, to consider the effect of electrolyte concentration on the potential energy of interaction, it is best to use the more elaborate expressions for interacting spheres. Figure 10.12 is a plot of Φ_{net} for this situation as a function of separation of surfaces with κ as the parameter which varies from one curve to another.

Figure 10.12 has been drawn for spheres with $R = 1000$ Å, $A = 10^{-12}$ erg, and $\psi_0 = 25.7$ mV. The ordinate in this figure has been labeled both in ergs and in multiples of kT at 25°C. For the system described by these curves, a significant energy barrier is present at all concentrations of a 1:1 electrolyte less than about 10^{-3} *M*. For concentrations between 10^{-3} and 10^{-2} *M*, however, the barrier vanishes. This particular colloid is thus expected to undergo a transition from a stable dispersion to a flocculated one with additions of an indifferent 1:1 electrolyte to a concentration in this range.

It has long been known that the addition of an indifferent electrolyte can cause a lyophobic colloid to undergo flocculation. The DLVO theory provides a quantitative explanation for this fact. Furthermore, it is known that for a particular salt a fairly sharply defined concentration is needed to induce flocculation. This concentration may be called the critical flocculation concentration. The DLVO theory in

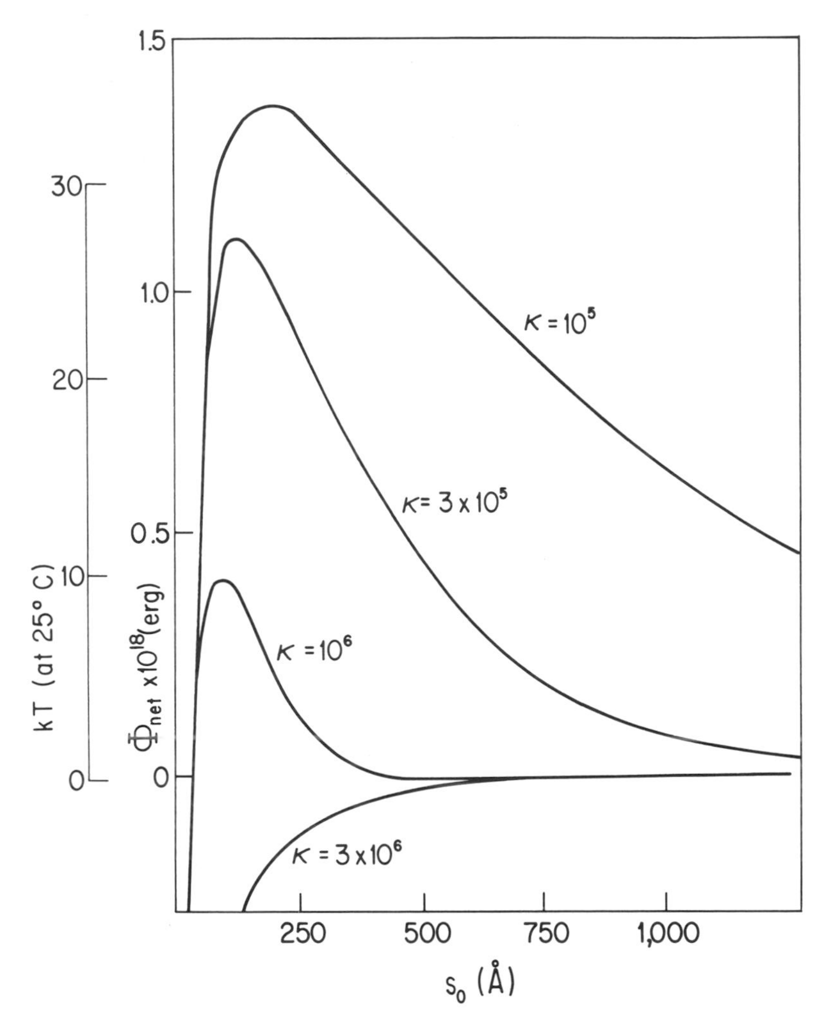

FIGURE 10.12 Φ_{net} *versus* s_0, *the separation of surfaces, for two spheres of equal radius* (1000 Å). *Curves drawn for different values of* κ *with constant values of* A (10^{-12} erg) *and* ψ_0 (25.7 mV). (*From Verwey and Overbeek* [7], *used with permission.*)

general and Figs. 10.10, 10.11, and 10.12 in particular can be summarized by the following statements:

1. The higher the potential at the surface of a particle—and therefore throughout the double layer—the larger the repulsion between the particles will be.
2. The lower the concentration of indifferent electrolyte, the longer is the distance from the surface before the repulsion drops significantly.

3. The larger the Hamaker constant, the larger is the attraction between molecules and, by extension, between macroscopic bodies.

Although these generalization present a satisfying picture of the variables affecting the stability of a colloid, they are a skimpy dividend from two chapters of preliminary discussion. What is needed is a more definitive test of the theory and a method to extract numerical values of the various parameters from experimental observations. In the following section, we shall discuss the critical flocculation concentration as a simple quantitative test of the theory. In subsequent sections, we shall see how studies of the rate of flocculation provide still more stringent tests of the theory and means for evaluating parameters of interest.

10.10 THE CRITICAL FLOCCULATION CONCENTRATION AND THE SCHULZE–HARDY RULE

One of the easiest tests that can be performed on an aqueous colloid is to determine the critical concentration of electrolyte required to flocculate the colloid. We shall use the notation CFC (for critical flocculation concentration) to indicate this quantity. This experiment is conducted by introducing the dispersion into a series of test tubes and adding to each various proportions of water and electrolyte solution. In this way, the total dilution of the dispersed particles is held constant while different amounts of salt are added to each. After mixing and waiting an arbitrary but consistent length of time, the tubes are visually inspected for evidence of the effect of the added salt. There will generally be clear evidence of flocculation (e.g., the settling out of the dispersed phase) in some of the tubes whereas others appear unchanged. Thus the highest concentration of salt which leaves the colloid unchanged and the lowest concentration which causes flocculation bracket the CFC. A second series of experiments may be conducted within this range to narrow the range of the CFC still further.

The actual concentration of electrolyte at the CFC depends on

1. the time allowed to elapse before the evaluation is made,
2. the uniformity or, more likely, the polydispersity of the sample,
3. the potential at the surface,
4. the value of A, and
5. the valence of the ions.

In a series of tests on any particular system, items 1 through 4 remain constant, so the CFC is a quantitative measure of the effect of the valence of the added ions. Table 10.5 summarizes some experimental results of this sort.

The results in Table 10.5 have been collected for colloids bearing both positive and negative surface charges. One of the earliest (1900) generalizations about the effect of added electrolyte is a result known as the Schulze–Hardy rule. This rule states that it is the valence of the ion of opposite charge to the colloid that has the principal effect on the stability of the colloid. The CFC value for a particular electrolyte is essentially determined by the valence of the counterion regardless of the nature of the ion with the same charge as the surface. The numbers listed in

TABLE 10.5 *Numbers in parentheses give* CFC *values* (*in moles liter*$^{-1}$) *for mono-, di-, tri-, and tetravalent ions acting on both positive and negative colloids. Numbers outside parenthesis are* CFC *values relative to the value for monovalent electrolytes on the same system. Theoretical values given by* Eq. (83).[a]

	Negative colloids			Positive colloids		
Valence of counterion	As_2S_3	Au	AgI	Fe_2O_3	Al_2O_3	Theory
1	(5.5×10^{-2})	(2.4×10^{-2})	(1.42×10^{-1})	(1.18×10^{-2})	(5.2×10^{-2})	
	1	1	1	1	1	1
2	(6.9×10^{-4})	(3.8×10^{-4})	(2.43×10^{-3})	(2.1×10^{-4})	(6.3×10^{-4})	
	1.3×10^{-2}	1.6×10^{-2}	1.7×10^{-2}	1.8×10^{-2}	1.2×10^{-2}	1.56×10^{-2}
3	(9.1×10^{-5})	(6.0×10^{-6})	(6.8×10^{-5})	—	(8×10^{-5})	
	1.7×10^{-3}	0.3×10^{-3}	0.5×10^{-3}	—	1.5×10^{-3}	1.37×10^{-3}
4	(9.0×10^{-5})	(9.0×10^{-7})	(1.3×10^{-5})	—	(5.3×10^{-5})	
	17×10^{-4}	0.4×10^{-4}	1×10^{-4}	—	10×10^{-4}	2.44×10^{-4}
Potential-determining ion	S^{2-}	Cl^-	I^-	H^+	H^+	

[a] Data from Overbeek [5].

parentheses in Table 10.5 are the CFC values in moles per liter for counterions of the indicated valence. That is, about $7\times10^{-4}\,M$ of divalent cation is needed to flocculate the negative As_2S_3 sols whereas about $6\times10^{-4}\,M$ of divalent anion is required to flocculate positive Al_2O_3 sols.

The actual values of these concentrations depend on a whole array of unknown parameters, but their relative values depend only on the valence of the counterions. The entries without parentheses in Table 10.5 are the values of the CFC relative to the value for the monovalent electrolyte in the same set of experiments. These are seen to be remarkably consistent for the divalent ions and acceptably close together for trivalent and tetravalent counterions.

Now let us see how this result is to be understood in terms of the DLVO theory. At first glance, it seems remarkable that any consistency at all can be found in tests as arbitrary as the CFC determination. It is not difficult, however, to show that these results are quite close to the values predicted in terms of the DLVO model for interacting blocks with flat faces. From an inspection of Fig. 10.12, we concluded that the system at $\kappa = 10^6\ \text{cm}^{-1}$ would be stable with respect to flocculation whereas the one at $\kappa = 3\times10^6\ \text{cm}^{-1}$ would flocculate. Furthermore, we examined the energy barrier to draw these conclusions. Next, we must ask how the qualitative criteria we used in discussing the curves can be translated into an analytical expression.

One way of doing this is to assume that the demarcation between stable and unstable colloids occurs at the value of κ for which the height of the "barrier" is zero. Physically, this is a somewhat arbitrary choice: Thermal energy is sufficient to allow particles to overcome a barrier of low but nonzero height. Mathematically, however, the assumption that the maximum in the potential energy curve occurs at zero permits us to write

$$\Phi_{\text{net}} = 0 \tag{75}$$

and

$$\frac{d\Phi_{\text{net}}}{dD} = 0 \tag{76}$$

as the conditions for stability.

Applying Eqs. (75) and (76) to (74) gives

$$\frac{64 n_0 k T \gamma_0^2}{\kappa} \exp(-\kappa D_m) = \frac{A}{12\pi} D_m^{-2} \tag{77}$$

and

$$64 n_0 k T \gamma_0^2 \exp(-\kappa D_m) = \frac{A}{6\pi} D_m^{-3} \tag{78}$$

where the subscript m reminds us that this describes the maximum. From these equations, it is readily apparent that

$$\kappa D_m = 2 \tag{79}$$

is the criterion for stability according to this model. This may also be written

$$D_m = 2\kappa_m^{-1} \tag{80}$$

in terms of the "thickness" of the double layer. Again we see an important distance measured in terms of κ^{-1}.

It is the dependence of the CFC values on the valence of the electrolyte that we seek to obtain rather than the absolute value of the CFC. Therefore, it is sufficient to proceed from this point by merely retaining those factors which involve either the concentration ($n_0 \propto M$) or the valence (z). Substituting Eq. (79) back into (78) yields

$$n_0 \propto \kappa^3 \tag{81}$$

The dependence of γ_0 on z has been neglected in writing this result, a procedure that is entirely justified for the level of approximation involved here (recall $\gamma_0 \simeq 1$). From the definition of κ [Eq. (9.53)], we obtain

$$n_0 \propto z^3 n_0^{3/2} \tag{82}$$

or

$$M \propto z^{-6} \tag{83}$$

which is the desired result. According to Eq. (83), the CFC value varies inversely with the sixth power of the valence of the ions in solution. The column of numbers in Table 10.5 labeled "theory" follows the progression z^{-6}: $1, 2^{-6}, 3^{-6}, 4^{-6}$. The actual CFC values are seen to be in quite reasonable accord with these predictions.

Although this test of the DLVO theory itself introduces some additional approximations, it is a workable unification of experimental and theoretical points of view. The threshold of stability in terms of the concentration and valence of indifferent electrolyte is easily measured. Theoretical models describe the interaction between a pair of particles in terms of potential energy diagrams. The reconciliation of these two approaches constitutes an important step toward obtaining still more quantitative information from the study of flocculation. This is taken up in the following sections concerned with the kinetics of flocculation. First, however, a few remaining comments about the critical flocculation concentrations must be made.

In Chap. 9 when we discussed the structure of the double layer, we implicitly anticipated that the ion opposite in charge from the wall plays the predominant role in the double layer, the central observation of the Schulze–Hardy rule. This enters the mathematical formalism of the Gouy–Chapman theory in Eq. (9.62) in which a Boltzmann factor is used to describe the relative concentration of the ions in the double layer compared to the bulk solution. For those ions which have the same charge as the surface (positive), the exponent in the Boltzmann factor is negative. This reflects the Coulombic repulsion of these ions from the wall. Ions with the same charge as the surface are thus present at lower concentration in the double layer than in the bulk solution. The signs are reversed for oppositely charged ions; therefore, the concentration of the latter is increased in the double layer. It may be shown—at least for high ψ_0 values—that the result of these considerations is essentially equivalent to emptying a region $2\kappa^{-1}$ thick of ions having the same charge as the wall.

Thus in terms of the model for flocculation just presented, it is essentially only the counterions that contribute to the diffuse double layer at the critical separation for flocculation.

The CFC values reported in Table 10.5 in many cases are average values for several compounds of similar valence. The use of averages to compare CFC values is justified since the valance primarily determines the CFC for an electrolyte.

However, a closer inspection of the data reveals that there are second-order differences between different ions. For example, 0.058 and 0.051 *M* are the Li^+ and Na^+ concentrations required to flocculate As_2S_3 sols, and 0.165 and 0.140 *M* are the concentrations required to flocculate AgI sols. Although both sets of values are acceptably close to the mean (which includes a number of other compounds), it is also clear that Li^+ is consistently slightly less effective than Na^+ in inducing flocculation. A more complete sequence of these variations in effectiveness is as follows:

1. For monovalent cations:

$$Cs^+ > Rb^+ > NH_4^+ > K^+ > Na^+ > Li^+$$

2. For monovalent anions:

$$F^- > Cl^- > Br^- > NO_3^- > I^- > SCN^-$$

These sequences may be compared with the adsorbability sequences presented in Sec. 9.8. For the most part, there is a correspondence between the two series suggesting that the second-order effect described here is a consequence of specific adsorption in the Stern layer and not a purely electrostatic effect arising from the diffuse part of the double layer. As might be expected, amphipathic ions show CFC values of an entirely different magnitude than truly indifferent electrolytes. The adsorption characteristics of the amphipathic ions outweigh their contribution as electrolytes.

10.11 THE KINETICS OF FLOCCULATION

The rates of ordinary chemical reactions are often described in terms of an energy profile along the so-called reaction coordinate. In such a representation, the initial and final states are separated by an energy barrier which is associated with a reaction intermediate or a transition state. The activation energy of such a reaction is then identified as the difference between the energy of the maximum and the energy of the initial state. The potential energy curves of the type discussed in Sec. 10.9 are the equivalent profiles for the flocculation process. The initial or dispersed state corresponds to large values of D for which $\Phi = 0$; therefore, this analogy suggests that the height of the potential barrier may be regarded as the equivalent to an activation energy for flocculation.

With ordinary reactions, the activation energy is fixed by the nature of the reactants and is measured by studying the effect of temperature variations on the rate of the reaction. In earlier discussions, we implied that a colloidal particle's ability to overcome a potential energy barrier depended on its thermal energy, so at first glance it appears that the analogy is valid at this point as well. A bit of additional reflection, however, suggests that the "activation energy" for flocculation is more efficiently investigated by varying the height of the barrier through the addition of electrolyte. Since the operational definition of activation energy is linked to temperature variation, we have thus departed from traditional kinetics. Nevertheless, the realization that studies of the rate of flocculation may provide a method for measuring the potential barrier is a valuable insight. In this section, we examine the principles underlying the rate of flocculation and in the following section we apply these ideas to the DLVO theory.

We begin by considering an array of uniform spherical particles whose motion is totally governed by Brownian movement. We assume the spheres to be totally noninteracting except on contact, in which case they adhere, forming a doublet. Although this is a highly oversimplified picture, it provides a model from which more realistic models can be developed in subsequent stages of the presentation. Figure 10.13 shows a schematic illustration of the formation of a doublet. Since the primary particles adhere on contact, the rate of doublet formation equals the rate at which molecules diffuse across the dashed surface. The latter corresponds to a spherical surface of radius $2R$ inscribed around one of the spheres which, for the present, is assumed to be stationary. After flocculation, the number of independent kinetic units is locally decreased in the neighborhood of this flocculation site. Therefore, we may imagine a concentration gradient around the fixed particle as responsible for the diffusion toward it.

In 1917, M. Smoluchowski applied the theory of diffusion to this situation to evaluate the rate of doublet formation. According to Fick's first law [Eq. (3.27)] the number of particles crossing a unit area toward the reference particle per unit of time, J, is given by

$$J = -D\frac{dN}{dr} \tag{84}$$

where D is the diffusion coefficient of the spheres and N is their total number. The total number crossing a spherical surface of radius r and area A around the reference particle per unit time, JA, is given by

$$\frac{\text{Total number}}{\text{Time}} = JA = -(4\pi r^2)D\frac{dN}{dr} \tag{85}$$

FIGURE 10.13 *Flocculation of uniform spherical particles to form a doublet.*

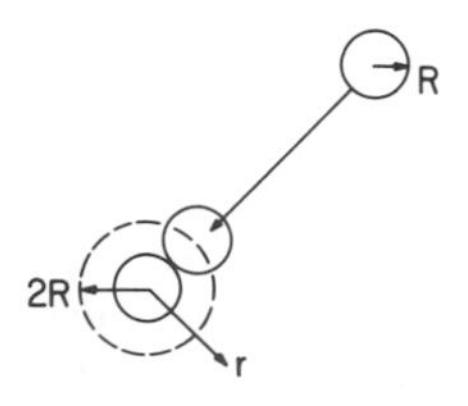

Under steady state conditions, the left-hand side of this equation is a constant, so the expression is readily integrated once the appropriate boundary conditions are considered. We assume that $N = N_0$, the initial concentration of the spheres, at $r = \infty$ and $N = 0$ at $r = 2R$, since the particles lose their kinetic identity on contact. With these conditions in mind, Eq. (85) is integrated to give

$$JA = -8\pi RDN_0 \tag{86}$$

This equation gives the number of particles which move toward the fixed particle per unit time. If we remove the restriction that the "target" particle is stationary, the effect is to replace the diffusion coefficient of the single moving particle with the relative diffusion coefficient of the two, which is simply $2D$ when the particles are the same size. Thus the frequency with which uniform spheres collide by diffusion with a moving reference sphere is given by

$$(JA)' = -16\pi RDN_0 \tag{87}$$

Instead of looking at a single sphere, we may regard any one of the N_0 spheres initially present as the "reference" particle. Therefore, the total rate at which doublets are formed is given by

$$\text{Rate} = -(16\pi RDN_0)N_0 = -k_r N_0^2 \tag{88}$$

where k_r is effectively a second-order rate constant. It is important to realize that Eq. (88) applies strictly only at the outset of flocculation when the only particles present are N_0 uniform spheres. In applying these ideas to actual kinetic experiments, therefore, we are limited to the initial rates of flocculation.

We saw in Chap. 3 [Eq. (3.37)] that, for uniform spheres,

$$D = \frac{kT}{f} = \frac{(kT)}{6\pi\eta R} \tag{89}$$

Substituting this result into Eq. (88) yields

$$k_r = \frac{8(kT)}{3\eta} \tag{90}$$

It is unfortunate that we have two k terms in the same equation, but this seems preferable to introducing new symbols for quantities which are already familiar in these symbols. To minimize confusion, however, we attach the subscript r (for rapid) to the rate constant and enclose the Boltzmann constant along with T in parentheses.

The rate law represented by Eqs. (88) and (90) applies to what are known as rapid flocculation conditions. It is rapid in the sense that no barriers are present to interfere with the flocculation process. Next, we turn our attention to the situation in which barriers are present: the case of slow flocculation.

To handle this situation, we retrace our steps through the preceding derivation, this time including a term for the resistance to the flocculation process. Thus Eq. (85) becomes

$$JA = -(4\pi r^2)D\frac{dN}{dr} + \text{resistance term} \tag{91}$$

for the case of a sphere diffusing toward a stationary particle. The problem now is to define the resistance term.

The following considerations will permit us to arrive at the proper form for the flux away from the stationary particle due to the potential energy barrier:

1. The derivative of the potential gives the force which opposes the approach of the particles: $d\Phi/dr$.
2. This force divided by the friction factor gives the velocity of "rebound" off the potential barrier: $(1/f)\, d\Phi/dr$.
3. This velocity times the concentration of the particles at r gives the flux of particles away from the stationary particle: $(N/f)\, d\Phi/dr$.
4. This flux times the area of the spherical shell at r gives the resistance term: $(4\pi r^2 N/f)\, d\Phi/dr$.
5. The sign of this term must be opposite to the sign of the first term. Since r is measured from the stationary particle, dN/dr is positive and $d\Phi/dr$ is negative beyond the maximum, which is the region of interest since it is the domain of the second particle. Therefore, the required difference in sign is covered by the signs of the gradients of N and Φ.

Incorporating considerations (1) through (5) into Eq. (91) yields

$$JA = -4\pi r^2\left(D\frac{dN}{dr} + \frac{N}{f}\frac{d\Phi}{dr}\right) \tag{92}$$

Although it was derived for the case of repulsion, there is nothing to prevent this expression from being applied to situations in which the energy is attractive, in which case the sign of $d\Phi/dr$ would be reversed and the flux toward the stationary particle would be augmented. Since the particles are spheres, we again substitute Eq. (3.37) for f to obtain

$$JA = -4\pi r^2\left(D\frac{dN}{dr} + \frac{ND}{kT}\frac{d\Phi}{dr}\right) \tag{93}$$

Removing the restriction that the reference particle is stationary again requires us to replace D by $2D$ to obtain

$$(JA)' = -8\pi r^2 D\left(\frac{dN}{dr} + \frac{N}{kT}\frac{d\Phi}{dr}\right) \tag{94}$$

or

$$\frac{dN}{dr} + \frac{N}{kT}\frac{d\Phi}{dr} = -\frac{(JA)'}{8\pi r^2 D} \tag{95}$$

This is the differential equation that must be solved to give N as a function of r for steady state diffusion in an energy gradient.

The solution to this equation is easily verified to be

$$N\exp\left(\frac{\Phi}{kT}\right) = -\int \exp\left(\frac{\Phi}{kT}\right)\frac{(JA)'}{8\pi D}r^{-2}\,dr + \text{const} \tag{96}$$

We eliminate the constant of integration as follows: At $r=\infty$, $N=N_0$ and $\Phi=0$; therefore, Eq. (96) becomes

$$N_0=-\left[\int \exp\left(\frac{\Phi}{kT}\right)\frac{(JA)'}{8\pi D}r^{-2}\,dr\right]\bigg|_{r=\infty}+\text{const} \tag{97}$$

At $r=2R$, $N=0$; therefore,

$$0=-\left[\int \exp\left(\frac{\Phi}{kT}\right)\frac{(JA)'}{8\pi D}r^{-2}\,dr\right]\bigg|_{r=2R}+\text{const} \tag{98}$$

In writing this last result it is implicitly recognized that the primary minimum has a finite depth and does not drop to $\Phi=-\infty$ as implied by Eq. (74). Recall that at very small separations the interaction between molecules becomes one of repulsion; we have simply ignored this region until now. Subtracting Eq. (98) from (97) gives

$$N_0=-\frac{(JA)'}{8\pi D}\int_{2R}^{\infty} \exp\left(\frac{\Phi}{kT}\right)r^{-2}\,dr \tag{99}$$

We may verify that we have proceeded correctly by noting that Eq. (99) reduces to (87) in the event that $\Phi=0$.

As before, the product $(JA)'$ times N_0 gives the initial rate at which doublets are formed:

$$\text{Rate}=(JA)'N_0=\frac{-8\pi D N_0^{\,2}}{\int_{2R}^{\infty}\exp(\Phi/kT)r^{-2}\,dr}=-k_s N_0^{\,2} \tag{100}$$

where k_s is the rate constant for slow flocculation. Comparing eqs. (88) and (100) reveals the following relationship between the rate constants for rapid and slow flocculation:

$$k_s=\frac{k_r}{2R\int_{2R}^{\infty}\exp(\Phi/kT)r^{-2}\,dr}=\frac{k_r}{W} \tag{101}$$

Equation (101) shows that the effect of the energy barrier is to decrease the rate constant for flocculation by a factor called the stability ratio W:

$$W=2R\int_{2R}^{\infty} \exp\left(\frac{\Phi}{kT}\right)r^{-2}\,dr \tag{102}$$

Since Φ itself is a complicated function of r for spherical particles, the integration of Eq. (102) is generally carried out graphically on the basis of potential energy curves such as those discussed in Sec. 10.9, unless the approximation described later is invoked.

As we shall see in the following section, it is possible to measure the stability ratio directly. What is desirable, therefore, is a method for deducing from the value of W some information about the potential energy of interaction instead of the reverse procedure as provided by Eq. (102). Unfortunately, this is not easy to accomplish because of the complex functional relationship between Φ and W. By making additional assumptions and approximations, however, some progress along these lines has been made.

In order to simplify Eq. (102) the following observations will be helpful:

1. For slow flocculation, the potential energy curves always show a maximum (coordinates: Φ_m and r_m).
2. The function $\exp(\Phi/kT)$ has its maximum value at Φ_m and drops very rapidly for $\Phi < \Phi_m$. That is, the exponential term in Eq. (102) is primarily determined by Φ_m.
3. The entire barrier portion of the potential energy curve may be approximated as a normal distribution function, Eq. (1.25).

On the basis of these ideas, it has been shown that

$$W \simeq \frac{2\sqrt{\pi}R}{r_m^2 p} \exp\left(\frac{\Phi_m}{kT}\right) \tag{103}$$

with

$$p = \left[-\frac{\partial^2\Phi/\partial r^2}{2kT}\right]^{1/2} \tag{104}$$

Note that $\partial^2\Phi/\partial r^2$ is negative since the point of evaluation is a maximum. As a first approximation, at least, the pre-exponential factors in Eq. (103) may be treated as a constant, which shows that Eqs. (101) and (103) may be combined to give

$$k_s \propto k_r \exp\left(-\frac{\Phi_m}{kT}\right) \tag{105}$$

This confirms the intuition expressed at the start of this section that the height of the potential barrier might be regarded as the activation energy for the flocculation process.

Using the approximation given by Eq. (103) and the interaction energy for spherical particles in the limit of small separations, it has been possible to predict from the DLVO theory that W should vary with the electrolyte concentration of the solution as follows:

$$\log W = K_1 \log c + K_2 \tag{106}$$

where K_1 and K_2 are constants and c is the concentration of the ions in moles per liter. For water at 25°C, the value of K_1 has been calculated to be $-2.15 \times 10^7\, \gamma_0^2/z^2$, where γ_0 is given by Eq. (9.75) and z is the valence of the counterions.

The significance of this last result should not be underestimated:

1. If W is measured for a colloid with particles of known radius as a function of electrolyte concentration, then Eq. (106) can be used to evaluate γ_0.
2. From this value of γ_0 the potential at the inner surface of the diffuse double layer, ψ_0, can be evaluated from Eq. (9.75).
3. Once a suitable value for ψ_0 is known, potential energy curves for that value of ψ_0 and a measured concentration of electrolyte (i.e., κ) can be constructed for different values of A.

4. The plot of Φ versus D which is most consistent with the observed value of W may be used to identify the value of A_{212} for the system under consideration.

In the following section, we shall examine the topic of flocculation kinetics from an experimental point of view.

10.12 THE KINETICS OF FLOCCULATION: EXPERIMENTAL

In principle, there is no difference between an ordinary chemical reaction and a flocculation reaction as far as the determination of the rate constants for the processes is concerned. Equations (88) and (100) show that both rapid and slow flocculation are characterized by a bimolecular rate-determining step. The integration of such a rate law shows that

$$\frac{1}{N}-\frac{1}{N_0}=kt \tag{107}$$

The most reliable way to evaluate a rate constant for flocculation, therefore, is to measure N, the number of independent kinetic units per unit volume, as a function of time. The time into the experiment is measured from the addition of indifferent electrolyte to the colloid. Although this is an easy statement to make, it is not an easy thing to do experimentally. One technique for doing this is literally to count the particles microscopically. In addition to particle size limitations, this is an extraordinarily tedious procedure. Light scattering is particularly well suited to kinetic studies since, in principle, experimental turbidities can be interpreted in terms of the number and size of the scattering centers.

A variation of the light scattering technique is to measure the change of absorbance with time, taking $d(\text{Abs})/dt$ to be proportional to dN/dt. The numerical

FIGURE 10.14 *Log W versus log c for AgI sols of five different particle sizes flocculated with the electrolytes shown. The mean particle radii in the different sols are* ●, 520 Å; ○, 225 Å; □, 535 Å; ▲, 650 Å; *and* △, 1580 Å. [*From H. Reerink and J. Th. G. Overbeek*, Discuss. Faraday Soc. **18**:74 (1954).]

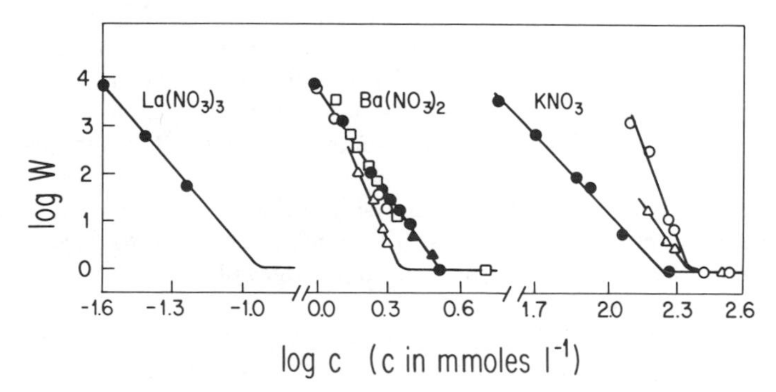

value of the proportionality constant—the total extinction coefficient [see Eq. (5.103)]—need not be known. It is the ratio between rapid and slow flocculation in which we are interested, and the proportionality factor cancels out of that quantity.

Once a suitable method for following the flocculation process has been determined, the procedure is to measure the rate constant (or the rate if the concentration of the particles is held constant) for a series of different salt concentrations. As the concentration of electrolyte is increased, the rate increases until the CFC is reached. The latter marks the transition from slow to rapid flocculation. Once rapid flocculation is established, the rate is insensitive to additional increases in concentration. The ratio of the rate constants for slow and rapid flocculation gives W, according to Eq. (101).

Figure 10.14 is a plot of log W versus log c for AgI sols of several different particle sizes. The experimental W values in this figure were determined by the absorbance technique described previously. According to the preceding section, data of this sort not only test the DLVO theory but also permit the evaluation of several important colloidal parameters. From the data in Fig. 10.14, the following conclusions can be drawn:

1. A plot of log W versus log c is linear as required by Eq. (106).
2. The concentrations at which $W = 1$ (where the breaks in the curves appear) measure the CFC values for the electrolyte involved. The CFC values for mono-, di-, and trivalent ions are about 0.199, 2.82×10^{-3}, and 1.3×10^{-4} mole liter^{-1}, respectively. These are in the ratio $1:1.42 \times 10^{-2}:0.7 \times 10^{-3}$. These figures compare very favorably with the other experimental data for AgI and the theoretical values presented in Table 10.5.
3. Slow flocculation is observed for log $W < 4$ or $W < 10^4$. For a typical potential energy curve, this corresponds to a value of Φ_m of about $15kT$. From this we may conclude that the height of an energy barrier must be *at least* $15kT$ if the colloid is to have any appreciable stability. Likewise, we may assume that unless the secondary minimum is approximately this deep, particles will be able to "escape" from it. In view of the general shape of the potential energy curves, the retardation effect, and this assessment of what constitutes a "high" barrier or a "deep" well, it seems likely that rigid aggregates are not formed in the secondary minimum.
4. Equation (106) can be used to analyze the slopes of the curves in Fig. 10.14 since the mean size of the AgI particles is known. In this way, Reerink and Overbeek found ψ_0 values in the range 12 to 53 mV and A values in the range 0.2 to 10×10^{-13} erg. Both of these are of the proper order of magnitude, no minor accomplishment in itself in light of the diverse assumptions required to get to this point.
5. The values of A and ψ_0 obtained from this analysis are slightly less satisfying in detail: The values of A show a lot of scatter and ψ_0 appears to be too low. Recall that the variation of the CFC with z^{-6} implies large values of ψ_0; for lower ψ_0 values a different dependence on z is expected.
6. Least satisfactory of all is the correlation with particle size. The results shown in Fig. 10.14 were determined for AgI sols covering a tenfold range of particle sizes.

TABLE 10.6 *Values of the effective Hamaker constant for various sols*[a]

Phase 2	$A_{212} \times 10^{13}$ (ergs)
AgI	0.2–10; 10–70; 3
Au	6; 0.5–1
Se	0.5–5; 2
SiO_2	0.2–0.5
Arachidic acid	0.03
Polystyrene	0.1–1; 0.5–2
Paraffin	1.5

[a] H. Sonntag and K. Strenge, *Coagulation and Stability of Disperse Systems*, copyright 1964, Halsted Press. Reprinted by permission of John Wiley & Sons, Inc.

It is evident from the figure that the slopes do not vary over a similar range, as required by Eq. (106). As a matter of fact, there are examples for which the steepest slope is associated with the coarsest particles (as required by theory) and others where it occurs with the smallest particles. The quantitative predictions fail on this particular point, but, as we shall see presently, there are some discrepancies between the theoretical model and the actual experimental system that may account for this apparent insensitivity to particle size.

The results presented in Fig. 10.14 represent a remarkable synthesis of physical chemical concepts. In terms of the organization of this book, the concepts of two chapters—which include statistical thermodynamics, electrochemistry, wave mechanics, and kinetics—are all drawn together into a unique attack on a complicated phenomenon. There can be no doubt as to the essential correctness of the approach. At the same time, one must be aware that a number of physical and mathematical approximations have been made along the way. Each of these makes it increasingly difficult to apply the full theory to any particular experimental system.

Table 10.6 shows some values of A_{212} which have been determined from flocculation studies along the lines of the presentation just made. Multiple values indicate the current lack of reproducibility among independent determinations. The corresponding A_{22} values may be calculated from Eq. (72) or read from Fig. 10.8 if 5.8×10^{-13} erg is accepted as a value of A_{11} for water. With additional assumptions about intermolecular distances near the surface, γ^d values for the dispersed particles may also be calculated from these data. All in all, a coherent picture emerges from these diverse sources of information, although it is obvious that considerable room for improvement still exists.

Before concluding this section, it seems desirable to comment a bit more fully on some possible sources of the discrepancy between the predictions of Eq. (106) and the data shown in Fig. 10.14. It is convenient to divide these remarks into those which involve the interaction potential explicitly and those which pertain to the kinetic part of the discussion.

As far as the interaction potential is concerned:

1. It may not be adequate to describe the interaction between AgI particles—especially at relatively close range—in terms of the radii of the dispersed units. In fact, the radii of surface protuberances rather than the dimensions of the particle as a whole may affect the short-range interaction.
2. Throughout this discussion, only nonspecific effects have been considered. That is, we have totally neglected to consider ion adsorption and the contribution of the Stern layer to the overall picture. This can be a serious source of complication, at least in some systems. This is evident from the fact that the arachidic acid sols described in Table 10.6 show a reversal of charge (from negative to positive) with the addition of La^{3+} and Th^{4+}, indicating the adsorption of these ions.

Several aspects of the kinetic part of this discussion also warrant additional comment:

3. For highly unsymmetrical particles, the probability of collision is greater than that predicted by Eqs. (87) or (94). This may be understood by thinking that the diffusion coefficient is most influenced by the smaller dimensions of the particles (therefore increased) and the "target radius" is most influenced by the longer dimension (also increased, relative to the case of symmetrical particles).
4. The frequency of collisions is also expected to be greater in a polydisperse system than in a monodisperse system by the same logic as presented in item 3.
5. The presence of velocity gradients in the system may also increase the rate of flocculation above the value given by Eqs. (87) or (94).

The ratio of the probability of a collision induced by a velocity gradient (mechanical agitation) to the collision probability under the influence of Brownian motion (thermal) has been shown to be

$$\text{Ratio} = \frac{\eta R^3 \, dv/dx}{2kT} \tag{108}$$

Since this increases with the cube of particle size, it may be the dominant mechanism for the flocculation of larger particles. Note that all the kinetic complications—items (3) through (5)—tend to cancel out of the evaluation of the stability ratio since the rates of both rapid and slow flocculation are determined on the same colloid.

Our emphasis in the latter part of this chapter has been on the flocculation *process*. A great deal of valuable information about interparticle forces is also obtained from studies of systems which are *already* flocculated. Studies of the viscous behavior of flocculated colloids are especially important in this regard. By varying the velocity gradient or rate of shear across a dispersion, the level of floc structure is altered and may thus be monitored experimentally. There is an extensive literature in this area, due—at least in part—to the numerous practical applications of this type of information. Space limitations require us to treat these studies only briefly. This is the topic of the following section.

10.13 VISCOUS BEHAVIOR AND FLOCCULATION

In Chap. 2, we were concerned exclusively with Newtonian flow, although examples of non-Newtonian behavior were presented in Fig. 2.4 for purposes of comparison. Our primary purpose in this section is to examine the consequences of instability with respect to flocculation on the flow properties of a dispersion. The viscous behavior of such systems is non-Newtonian for the most part. Viscosity and flocculation are each—taken separately—complex topics. Taken together, things get even worse! We shall not attempt any quantitative treatment of this subject, but shall merely describe qualitatively some of the phenomena involved. There are numerous practical aspects to this important topic, however, and an extensive literature exists.

Strictly speaking, *the* coefficient of viscosity is meaningful only for Newtonian fluids, in which case it is the slope of a plot of stress versus rate of shear, as shown in Fig. 2.4. For non-Newtonian fluids, such a plot is generally nonlinear, so the slope varies from point to point. In actual practice, the data are traditionally represented with the rate of shear (dv/dx) as the ordinate and the stress (F/A) as the abscissa, as shown in Fig. 10.15. In this case, the apparent viscosity at some particular point is given by the *co*tangent of the angle which defines the slope at that particular point.

It is convenient to distinguish between three different types of behavior: (a) The apparent viscosity increases as the rate of shear increases; (b) the apparent viscosity decreases as the rate of shear increases; and (c) the apparent viscosity is very high—effectively infinite—until the applied force reaches a critical value, then it decreases abruptly and may be either constant or continue to decrease beyond this. The critical stress in this third case is called the yield value since the system behaves as

FIGURE 10.15 *Rate of shear versus shearing stress (a) for* 7% *aqueous carbon black dispersions [data from A. I. Medalia and E. Hagopian,* Ind. Eng. Chem. Prod. Res. Develop. **3**:120 (1964)]; *(b) for* 11% *aqueous bentonite dispersion* (*pH* = 8.7). *Time of cycle* 70 *s.* [*Schematic based on data of W. F. Gabrysh, H. Eyring, P. Lin-Sen, and A. F. Gabrysh,* J. Am. Ceramic Soc. **46**:523 (1963).]

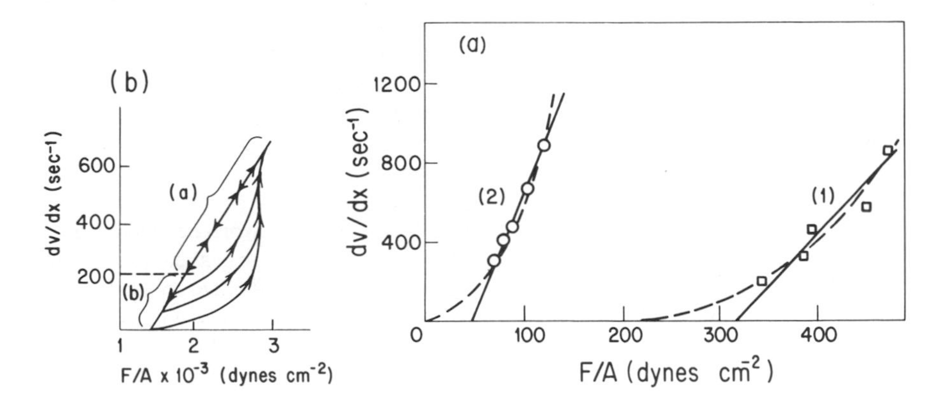

an elastic body for lesser stresses. The two regions of behavior in the third situation may merge more or less gradually in the vicinity of the yield value, giving rise to several different ways of defining this latter parameter.

These phenomena have different names, depending on whether they are observed under time-dependent or stationary state (time-independent) conditions. If the apparent viscosity increases with increasing rate of shear, the effect is called rheopexy if time dependent and dilatancy if time independent. If the apparent viscosity decreases with increasing rate of shear, the effect is called thixotropy if time dependent and plasticity (with a yield value) or pseudoplasticity (without a yield value) if time independent.

As we shall see presently, thixotropic behavior is associated with unsymmetrical particles which are unstable with respect to flocculation. In certain instances, quicksand apparently operates through a thixotropic mechanism. The struggles of the victim merely decrease the viscosity of the trap and worsen his plight. By contrast, wet beach sand is an example of a dilatant system in which the apparent viscosity increases with shear. Anyone who has wiggled his toes in the latter knows that under these conditions (low shear) the wet sand is very fluid. However, the same sand is hard to a firm footstep (high shear). In contrast to thixotropy, dilatancy is favored by colloidal stability and particle symmetry. It is almost always observed at concentrations in the neighborhood of 40% dispersed particles. At this concentration and in the absence of disruptable flocs, the only way such a system can flow is by the gradual rolling of particles past one another. If the rate of shear is too great, this deformation is impossible.

Next let us see how thixotropic behavior relates to the phenomena of flocculation. The data shown in Fig. 10.15a were obtained for a 7% slurry of carbon black in water. Curve 1 shows the results obtained immediately after the dispersion is prepared in a high-speed blender. After additional mild agitation, the results shown in curve 2 are obtained. The latter are independent of further agitation. With shorter periods of mild agitation, a family of curves lying between 1 and 2 would be obtained.

The data presented in Fig. 10.15a are consistent with the following mechanism. The dispersion which emerges from the blender is fundamentally unstable with respect to flocculation (no surfactant has been added to build a significant surface potential) and flocculates rapidly to form a volume-filling network throughout the continuous phase. Except for the size and structure of the "chains," the situation is comparable to a cross-linked polymer swollen by solvent. In both, the liquid is essentially immobilized by the network of chains, and the system behaves as an elastic solid under low stress. The term gel is used to describe such systems whether the dispersed particles are lyophilic or lyophobic.

As the force applied to the surface of the gel (see Fig. 2.3) is increased, however, a point is ultimately reached—the yield value—at which the network begins to break apart and the system begins to flow (curve 1). Increasing the rate of shear may result in further deflocculation, in which case the apparent viscosity would decrease further with increased shear. Highly unsymmetrical particles can form volume-filling networks at low concentrations and are thus especially well suited to display these phenomena.

As the system is subjected to on-going, low-level mechanical agitation, the network structure is rearranged to a dispersion of more compact flocs which display both a lower yield value and a lower apparent viscosity than the initial dispersion (curve 2). A certain amount of time is required for the dispersed units to acquire a size and structure which are compatible with the prevailing low level of agitation. This is why intermediate (not shown in Fig. 10.15a) cases are observed before the actual stationary state condition is obtained.

If the time for measurement of a stress/shear curve is short compared to the time required for rearrangement of the structure of the dispersed particles, then different results are obtained, depending on whether the rate of shear is increasing or decreasing. Figure 10.15b is an example of such hysteresis for 11% dispersions of bentonite (a montmorillonite clay with plate-shaped primary particles) in which the entire cycle is measured in 70 s. These data show the sensitivity of such experiments to the level of shear and to the time of observation. If the direction of the cycle is reversed along the descending branch of the curve, different results are obtained, depending on whether the reversal is done at a rate of shear above (region a in Fig. 10.15b) or below (region b) about 200 s^{-1}. This shows that the structure within the colloid builds up rapidly (compared to the cycle time of 70 s) at rates of shear below 200 s^{-1}, with negligible buildup at greater rates of shear.

These complicated observations are difficult to interpret in terms of fundamental interactions between particles; nevertheless, they have tremendous practical significance. For example, hundreds of millions of pounds of carbon black are mixed annually into rubber latex and dispersions of clay in oils are used as lubricating greases. Obviously, the viscosity of these substances under various conditions of shear is an important consideration. Printing inks, drilling muds (circulated around the shaft in well-drilling operations to cool the bit and flush away cuttings), paper coatings, paints, and innumerable industrial slurries may all be considered examples of areas in which these considerations are vitally important.

In a paint, for example, a controlled level of flocculation is important in both the actual application of the paint and its storage in the container during application. In the former, thixotropy (the word means "changing with touch") permits the paint to "thin" under the shearing influence of the paintbrush or spraygun. Once applied, it thickens, preventing the drip or sag of the paint on the surface. In addition, the time required for this yield value to develop should be sufficient to allow for the leveling of brushmarks. Thixotropy is an important property of paint in the bucket as well as on the wall. The buildup of a yield value interferes with the sedimentation of the pigment and eliminates the need to stir the paint continuously to assure uniformity. The fact that these requirements are well met by commercial paints indicates the success of paint chemists in regulating thixotropy.

REFERENCES

1. F. M. Fowkes, "Calculation of Work of Adhesion by Pair Potential Summation," in *Hydrophobic Surfaces* (F. M. Fowkes, ed.), Academic Press, New York 1969.

2. J. N. Israelachvili and D. Tabor, "Van der Waals Forces: Theory and Experiment," in *Progress in Surface and Membrane Science*, Vol. 7 (J. F. Danielli, M. D. Rosenberg, and D. A. Cadenhead, eds.), Academic Press, New York, 1973.
3. D. H. Kaelble, *Physical Chemistry of Adhesion*, Wiley–Interscience, New York, 1971.
4. E. A. Moelwyn-Hughes, *Physical Chemistry* (2nd ed.), MacMillan, New York, 1964.
5. J. Th. G. Overbeek, in *Colloid Science*, Vol. 1 (H. Kruyt, ed.) Elsevier, Amsterdam, 1952.
6. H. Sonntag and K. Strenge, *Coagulation and Stability of Disperse Systems*, Halsted Press, New York, 1964.
7. E. J. W. Verwey and J. Th. G. Overbeek, *Theory of the Stability of Lyophobic Colloids*, Elsevier, Amsterdam, 1948.

PROBLEMS

1. The parameter β_{12} for heterogeneous (12) interactions plays a similar role as β_{11} [Eq. (34)] does for homogeneous (11) interactions. Use entries from Table 10.1 to write an expression for β_{12}. If the Debye interaction makes a negligible contribution to β_{12} and $\nu_1\nu_2/(\nu_1+\nu_2) \simeq \frac{1}{2}(\nu_1\nu_2)^{1/2}$, show that

$$\beta_{12} \simeq (f_{11L}\beta_{11})^{1/2}(f_{22L}\beta_{22})^{1/2} + (f_{11K}\beta_{11})^{1/2}(f_{22K}\beta_{22})^{1/2}$$

where the f terms are defined by Eq. (35). If $f_{11L} = f_{22L}$ and $f_{11K} = f_{22K}$, show that this last result becomes

$$\beta_{12} \simeq (\beta_{11}\beta_{22})^{1/2}$$

Comment on the relevancy of this result to Eq. (71). Criticize or defend the following proposition: "The geometrical mixing rule does not require the absence of permanent dipoles, only that 11 and 22 interactions both consist of the same fraction of dispersion and permanent dipole contributions. Specific interactions, such as hydrogen bonding, must also be absent in the 11, 22, and 12 systems."

2. A gas adsorption isotherm may be derived by comparing the adsorbed layer around a solid particle to a planetary atmosphere, with an equilibrium pressure p_0 at the surface. The change in free energy for a molecule going from the bulk pressure p to the surface is $kT \ln(p_0/p)$. Equating this with Eq. (48), the potential energy of attraction which is responsible for the adsorption, gives

$$kT \ln \frac{p_0}{p} = \frac{\rho N_A}{M} \frac{\beta\pi}{6} z^{-3}$$

Since $z \propto V$, this may be written $\ln(p_0/p) = \text{const } V^{-3}$, where V is the volume of gas adsorbed. A more general form of this isotherm, called the Frenkel–Halsey–Hill (FHH) isotherm, treats the power dependence of V as an unknown n and writes

$$\left(\frac{V}{V_m}\right)^n = \frac{K}{\ln(p_0/p)}$$

Prepare a plot of the FHH isotherm using $n = 3$ and $K = 0.1$ and comment on the resemblance of this isotherm to actual gas adsorption isotherms as shown in Chap. 8.

3. If a soap film is sufficiently thin, its equilibrium thickness is the resultant of the double layer repulsion, given by Eq. (9.98), and van der Waals attraction, given by

$$F_A = \frac{\partial \Phi_A}{\partial D} = \frac{A}{6\pi} D^{-3}$$

according to Eq. (52). These conditions are satisfied by certain films studied by Lyklema and Mysels* who obtained the following results:

Concentration (moles liter^{-1})	0.103	0.066	0.0197
Thickness of aqueous layer (Å)	91	94	153
Thickness of entire film (Å)	123	126	185

Use the thickness of the aqueous layer in Eq. (9.98) to calculate F_R per unit area (assume $\gamma_0 = 1$). By equating this quantity to F_A per unit area (just given) and using the total film thickness, estimate A for each of these data. Explain why two different values are used for D in this calculation. How does the average value for A compare with the A value of water presented in this chapter?

4. Using the average value of A you determined in Problem 3, criticize or defend the proposition that van der Waals forces are negligible compared to hydrostatic forces when the latter equal 660 dynes cm^{-2} in a film for which the total thickness is 763 Å. Note that this is the assumption made in Problem 9.11. Qualitatively re-examine the latter question in light of the results of this problem.

5. Criticize or defend the following propositions: "The DLVO theory should apply to particles dispersed in nonaqueous media once ε and A for the solvent have been included in the relevant expressions. Since ion concentrations are low in media with low dielectric constant, κ^{-1} will be very large for such systems. For a concentrated colloid, the mean interparticle spacing may be less than κ^{-1}. In such a case, it is more plausible to picture a particle approaching a 'target' as traveling along a potential energy plateau rather than facing a potential energy barrier." Comment on the relevancy of these

* J. Lyklema and K. J. Mysels, *J. Am. Chem. Soc.* **87**:2539 (1965).

propositions to the observation* that a 15% (by volume) water-in-benzene emulsion stabilized by the calcium salt of didodecylsalicylic acid has a ψ_0 value of ~130 mV, yet breaks immediately after preparation.

6. Verify that combining Eqs. (78) and (79) with the definition of κ [Eq. (9.49)] leads to the following expression (purely numerical factors may be omitted):

$$\text{CFC} \propto n \propto \frac{\varepsilon^3 (kT)^5 \gamma_0^4}{e^6 A^2 z^6}$$

By the series expansion of γ_0 [definition of γ_0 in Eq. (9.75)], verify that $\gamma_0^4 \propto (ze\psi_0/kT)^4$ if ψ_0 is low. Use these two results to predict the dependence of the CFC on the ionic valence if ψ_0 is small. Compare this result with the same quantity in the limit of large ψ_0.

7. Use the accompanying data† and the conclusions of Problem 6 to criticize or defend the following propositions:
   (a) "Since the CFC for positively charged AgBr is less (regardless of counter-ion valence) than that for polyvinyl chloride latex (PVC), ψ_0 must be less for the AgBr."
   (b) "Since the CFC depends on a higher power of the counterion valence for AgBr than for PVC, ψ_0 must be greater for AgBr."

	CFC (moles liter^{-1}) of ions opposite in charge to ψ_0	
Colloid	$z = 1$	$z = 2$
PVC latex	2.3×10^{-1}	1.2×10^{-2}
AgBr	1.6×10^{-2}	2.3×10^{-4}

8. H. Müller studied by dark field microscopy the flocculation of colloidal gold upon the addition of NaCl to the aqueous sol. For a sample in which the gold particles have a 36.9 Å radius, the following particle counts were observed at different times after the colloid was made about 0.2 M with NaCl‡:

t (s)	120	195	270	390	450	570
$N \times 10^{-8}$ (cm^{-3})	11.2	7.3	5.4	4.5	3.7	2.7

Determine the second-order rate constant, k_{exp}, which describes this flocculation process. How does k_{exp} compare with k_r as given by Eq. (90)?

9. Polystyrene latex particles were flocculated by the addition of $Ba(NO_3)_2$. The number of dispersed particles deposited onto a planar polystyrene surface was determined 15 min after the addition of salt by optical microscopy. The light microscope does not permit the aggregation of the deposited particles to be determined; subsequent examination by the electron microscope gives this

* W. Albers and J. Th. G. Overbeek, *J. Colloid Sci.* **14**:501 (1959).
† Data cited by E. Matijević, *J. Colloid Interface Sci.* **43**:217 (1973).
‡ H. Müller, *Kolloid Z.* **38**:1 (1926).

information. G. E. Clint et al.* obtained the following results:

$Ba(NO_3)_2$ conc. $\times 10^3$ (mole liter^{-1})	Total deposition cm$^{-2} \times 10^{-5}$ (after 15 min)	% Deposit		
		Single	Double	Triple
9.1	8.04	94.7	3.3	0.3
15.4	14.25	95.0	4.3	0.4
22.7	14.43	82.9	11.3	3.5
57.0	11.25	75.0	15.8	5.6

Discuss these data in terms of the following points:

(a) For all salt concentrations, the order of particle abundance in the deposit is single > double > triple.

(b) The higher the $Ba(NO_3)_2$ concentration, the smaller is the difference indicated by these inequalities.

(c) The decrease in total deposition with increasing concentration is not offset by the higher aggregation state of the deposit, but arises from the slower diffusion of more highly aggregated kinetic units.

10. If flocculation involves two noninteracting spheres of different radii, R_i and R_j, Eq. (88) predicts

$$k_r = \frac{2}{3}\frac{kT}{\eta}(R_i + R_j)\left(\frac{1}{R_i} + \frac{1}{R_j}\right)$$

Retrace the development of Eq. (88) to verify this result. Show that this expression is identical to

$$k_r = \frac{2}{3}\frac{kT}{\eta}\left(4 + \left[\sqrt{\frac{R_i}{R_j}} - \sqrt{\frac{R_j}{R_i}}\right]^2\right)$$

Estimate the ratio R_i/R_j needed to account for a k_r value of 2.9×10^{-11} cm^3 s^{-1} as observed for arachidic acid sols.† Does this expression reduce to the proper limit when $R_i = R_j$?

11. Verify (a) that Eq. (96) is a solution to (95), (b) that Eq. (99) reduces to (87) if $\Phi = 0$, and (c) that Eq. (100) leads to a rate constant larger than k_r, if $\Phi = 0$ for $r > \Delta$ and $\Phi = -\infty$ for $r \leqslant \Delta$ where $\Delta > 2R$. Show that the experimental k_r value cited in Problem 10 is consistent with a value of Δ equaling $5.4R$ for an aqueous colloid at 20°C ($\eta = 0.01$ P). Discuss the relevancy of this last result to the rapid flocculation of particles between which van der Waals attraction exists.

12. Arachidic acid sols were studied with different concentrations of La^{3+} added. The stability ratio W and the direction of particle migration in an electric field

* G. E. Clint, J. H. Clint, J. M. Corkill, and T. Walker, *J. Colloid Interface Sci.* **44**:121 (1973).
† R. H. Ottewill and D. J. Wilkins, *Trans. Faraday Soc.* **58**:608 (1962).

(i.e., particle charge) were observed* and the following results obtained:

c (moles La^{3+} liter^{-1})	10^{-5}	3×10^{-5}	10^{-4}	3×10^{-3}	10^{-3}
W	7.9	4.5	~1	1.6	15.8
Particle charge	−	−	~0	+	+

Taking 10^{-4} M as the CFC value for La^{3+}, CFC values of 7.29×10^{-2} M and 1.1×10^{-3} M would be predicted for monovalent and divalent cations, respectively, according to Eq. (83). In view of the observed behavior of La^{3+}, would you expect these calculated CFC values to be correct or too low or too high? Explain briefly.

13. Colloidal gold stabilized by citrate ions and having a mean particle radius of 103 Å was flocculated by the addition of $NaClO_4$. The kinetics of flocculation were studied colorimetrically and the stability ratio W for different $NaClO_4$ concentrations was determined†:

$c\times10^3$ (moles liter^{-1})	2	3	5	8	10.5
W	48	31	17	8.9	0.84

Graph these data according to the form suggested by Eq. (106), and evaluate γ_0 from the slope of the linear portion. Verify that this value of γ_0 corresponds to a value of ψ_0 equal to about 25 mV. Estimate what the CFC value would be for this system if W continued to vary according to the same function of c both above and below 10^{-2} M. Suggest an explanation for the abrupt decrease in W near 10^{-2} M.

14. A. Kitahara and H. Ushiyama‡ flocculated a polystyrene latex of radius 665 Å with KCl. The stability ratio W was found to vary with the KCl concentration as follows:

$\log c$ (c in moles liter^{-1})	0	−0.13	−0.33	−0.44	−0.60
$\log W$	0	0	0.30, 0.46	0.73	1.20

From a plot of $\log W$ versus $\log c$ determine the CFC value and γ_0 [by means of Eq. (106)]. Use the approximation for γ_0 given in Problem 6 to estimate ψ_0 for this colloid. Use the values of the CFC and γ_0 determined in Eqs. (78) and (79) to estimate the effective Hamaker constant A_{212} for polystyrene dispersed in water. How does the value obtained compare with the value listed in Table 10.6? Describe how A might be estimated using a more realistic model than that used in the derivation of Eqs. (78) and (79).

* R. H. Ottewill and D. J. Wilkins, *Trans. Faraday Soc.* **58**:608 (1962).
† B. V. Enüstün and J. Turkevich, *J. Am. Chem. Soc.* **85**:3317 (1963).
‡ A. Kitahara and H. Ushiyama, *J. Colloid Interface Sci.* **43**:73 (1973).

Electrophoresis and Other Electrokinetic Phenomena

11

There is a constant attraction to the South . . . yet the hampering effect of the southward attraction is quite sufficient to serve as a compass in most parts of our earth.

[From Abbott's *Flatland*]

11.1 INTRODUCTION

The word "electrokinetic" implies the combined effects of motion and electrical phenomena. Specifically, our interest in this chapter centers on those processes in which a relative velocity exists between two parts of the electrical double layer. This may arise from the migration of a particle relative to the continuous phase which surrounds it. In this case, the resulting electrokinetic effect is called electrophoresis. Alternatively, it could be the solution phase which moves relative to stationary walls, in which case either electro-osmosis or streaming potential is the phenomenon observed.

These three electrokinetic processes are our concern in this chapter, with the emphasis on electrophoresis. In each case, the electrokinetic measurements can be interpreted to yield a quantity known as the zeta (ζ) potential. It is important to note that this is an *experimentally* determined potential measured in the double layer. Therefore, it is the empirical equivalent to the double layer potentials discussed theoretically in Chap. 9. We saw in Chap. 10 how the stability of a hydrophobic colloid depends on the relative magnitude of the potential energies of attraction and repulsion between a pair of particles approaching a collision with each other. Therefore, the electrokinetic or ζ potential has a direct bearing on the material of the preceding two chapters as far as the theory and practice of colloid stability are concerned.

Although the ζ potential is undoubtedly an important quantity in colloid chemistry, it is not totally free of ambiguity. The problem is this: It is not clear at what location within the double layer the potential is measured. The derivations of this chapter will show that the ζ potential is the double layer potential close to the surface, but the precise quantitative meaning of "close" cannot be defined.

11.2 COMPARISON OF SMALL IONS AND MACROIONS

The fact that positive ions migrate toward the cathode and negative ions migrate toward the anode is so well known as to be virtually self-evident. It seems equally evident, therefore, that positively and negatively charged colloidal particles should display similar migrations. Indeed, this is the case. Because we are relatively familiar with the conductivity of simple electrolytes, we shall start our discussion of electrokinetic phenomena with a comparison of the migrations of the particles in these two different size domains.

An isolated ion in an electric field experiences a force directed toward the oppositely charged electrode. This force is given by the product of the charge of the ion q times the electric field $\bar{E}$:

$$F_{el} = q\bar{E} \tag{1}$$

Since the charge of the ion may be given in either practical or esu–cgs units, the force in dynes may be evaluated by either of the following:

1. If q is in coulombs,

$$F_{el} = q\bar{E}\frac{\text{coulombs volt}}{\text{cm}} \times \frac{1 \text{ joule}}{\text{cm volt}} \times \frac{10^7 \text{ ergs}}{\text{joule}} \times \frac{1 \text{ dyne cm}}{\text{erg}} \tag{2}$$

2. If q is in esu units

$$F_{el} = q\bar{E}\frac{\text{esu volt}}{\text{cm}} \times \frac{1 \text{ statvolt}}{300 \text{ volts}} \times \frac{1 \text{ erg}}{1 \text{ esu statvolt}} \times \frac{1 \text{ dyne cm}}{\text{erg}} \tag{3}$$

These expressions are limited to situations in which the electric field at the ion is due to the applied potential gradient only, undisturbed by the effects of other ions in the solution (i.e., infinite dilution).

An ion in an electric field thus experiences an acceleration toward the oppositely charged electrode. However, its velocity does not increase without limit. An opposing force due to the viscous resistance of the medium increases as the particle velocity increases:

$$F_{vis} = fv \tag{4}$$

where f is the friction factor [see Eq. (3.2)]. A stationary state velocity is established quite rapidly in which these two forces are equal:

$$v = \frac{q\bar{E}}{f} \tag{5}$$

The situation is thus very much like the sedimentation velocity discussed in Chap. 3 in which the gravitational forces on a particle are opposed by viscous resistance.

As a further development, we may tentatively substitute the value for f given by Stokes' law [Eq. (3.11)] to obtain

$$v = \frac{q\bar{E}}{6\pi\eta R} \tag{6}$$

where R is the radius of the particle, assumed to be a sphere by this substitution. The charge of a simple ion can be written as the product of its valence z times the electron charge e:

$$q = ze \tag{7}$$

Substitution of this result into Eq. (6) yields

$$v = \frac{ze\bar{E}}{6\pi\eta R} \tag{8}$$

The velocity per unit field is defined to be the mobility of the ion u:

$$u = \frac{v}{\bar{E}} \tag{9}$$

For simple ions, mobilities are typically on the order of 10^{-4} cm s^{-1}/V cm^{-1} (cm^2 V^{-1} s^{-1}). It is shown in physical chemistry that the mobility of an ion is directly proportional to its equivalent conductance λ_{i0}:

$$u_i = F\lambda_{i0} \tag{10}$$

where F is the Faraday constant. We have stipulated the conductance at infinite dilution (subscript 0) as a reminder that these relationships all refer to isolated ions. When ion mobilities are analyzed by Eq. (9), quite reasonable values for the radii of the hydrated ions are obtained.

The success and relative simplicity of conductivity as a method of study for small ions prompt us to extend these ideas to particles in the colloidal size range. For certain colloids, the experimental aspects of this are simpler than for small ions, because of the possibility of measuring the velocity of high contrast particles by direct microscopic observation. If the velocity and the field responsible for the migration are known, the mobility of the colloid may be evaluated directly from Eq. (9). When the term is applied to colloidal particles, the mobility is known specifically as the electrophoretic mobility. In this case, the overall phenomenon is known as electrophoresis and the specific experimental technique of direct microscopic observation of the electrophoretic mobility is called microelectrophoresis. This and other electrophoretic techniques are described in more detail in Sec. 11.9.

Although the electrophoretic mobility is—at least in some cases—a readily measured quantity, its interpretation is considerably more difficult for colloidal particles than for small ions. First, we realize that the charge carried by a colloidal particle is not a constant known quantity as is the case for simple ions. This prevents us from using Eq. (8) to evaluate R, but suggests instead a method whereby the charge might be determined. Suppose, for example, we substitute Eq. (3.37) for f rather than use Stokes' law for this quantity. Then the electrophoretic mobility is given by

$$u = \frac{ze}{kT/D} = \frac{zeD}{kT} \tag{11}$$

It appears that the combination of electrophoresis and diffusion experiments would allow for the evaluation of the charge carried by the macroion. Again, the situation is reminiscent of the procedures described in Chap. 3 in which sedimentation and diffusion experiments were combined. However, this is only the beginning of the difficulty. The validity of Eq. (11) is limited to the situation in which a charged particle is considered in isolation from other ions. A charged colloid will be surrounded by an electrical double layer as we saw in Chap. 9. Thus the field at the particle is modified by the potential of the double layer. That is, the migrating unit is the charged colloidal particle *along with* its electrical double layer just as the same composite is the kinetic unit in flocculation, as we saw in Chap. 10.

Therefore, this strategy for determining the charge of a colloid from electrophoresis measurements is invalid except for the rather special case of determining the conditions of zero charge for the colloid. We shall return to a discussion of this point in Sec. 11.10.

In Chap. 9, we discussed the structure of the double layer in terms of the potential of the surface. This background plus the realization that the ion atmosphere also contributes to the electrophoretic mobility of a colloid suggests that potential rather than charge is the more useful parameter to pursue. This is the topic of the following section. In discussing the migration of charged colloidal particles through a solution containing small ions, it is convenient to begin by distinguishing between two extremes of particle size. We saw in Chap. 9 that the parameter κ^{-1} (see Table 9.2) is a convenient way to characterize the "thickness" of the ion atmosphere near a surface. Distances are regarded as large or small relative to this quantity. For simplicity, we shall restrict our consideration to spherical, nonconducting particles and shall begin by examining the two extremes of very small and very large particles. These designations acquire specific meaning when compared to κ^{-1}, taken as a standard length. Thus, the two cases we consider first are those in which R/κ^{-1} (or simply κR) is small and large.

Figure 11.1 shows schematically the shape of the flow streamlines around the particle in the two cases. The dashed line in the figure is displaced from the surface of the spherical particles by an amount κ^{-1}. In Fig. 11.1a, R is small (compared to κ^{-1}) and the streamlines undergo negligible displacement. In Fig. 11.1b, on the other hand, the streamlines follow the contours of the particle nearly tangentially. It should be noted that the electric field as well as the velocity obeys the equation of continuity [Eq. (2.8)]. It will be recalled that the equation of continuity reflects the

FIGURE 11.1 *Streamlines (which also represent the electric field) around spherical particles of radius R. The dashed lines are displaced from the surface of the spheres by the double layer thickness, κ^{-1}. In (a) κR is small; in (b) κR is large.*

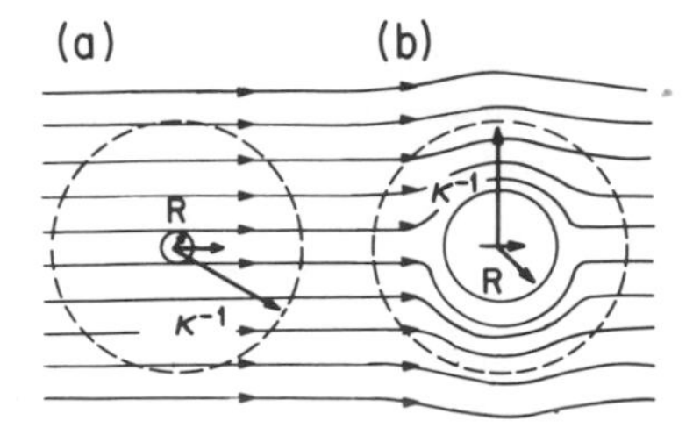

fact that matter is conserved. A streamline carrying matter into a volume element must also carry matter out of that volume element. The equation of continuity applied to the electric field implies that charge is conserved as well as matter. In an electrically neutral volume element, the same number of lines of force must enter and leave a volume element. Accordingly, the vectors in Fig. 11.1 may also be regarded as describing the electric field in the neighborhood of small and large particles.

11.3 THE ZETA POTENTIAL FOR SMALL VALUES OF κR

We know from Chap. 9 that the potential drops off gradually with distance from a charged surface, its range decreasing with increasing electrolyte content. Most of the expressions developed in Chap. 9, however, describe the potential situation adjacent to a planar wall. In the present context, we need to know how the potential varies with distance from the surface of a sphere. The Poisson equation (see Sec. 9.4) gives the fundamental differential equation for potential as a function of charge density. The Debye–Hückel approximation may be used to express the charge density as a function of potential as in Eq. (9.42) if the potential is low. Combining Eqs. (9.38) and (9.42) gives

$$\frac{1}{r^2}\frac{d}{dr}\left(r^2\frac{d\psi}{dr}\right)=\frac{4\pi}{\varepsilon}\frac{e^2}{kT}\left(\sum_i z_i^2 n_i\right)\psi=\kappa^2\psi \tag{12}$$

Equation (12) is the basic relationship of the Debye–Hückel theory and may be integrated as follows. The variable x is introduced with the following definition:

$$x=r\psi \tag{13}$$

Thus, Eq. (12) may be written

$$\frac{d}{dr}\left(r^2\frac{d\psi}{dr}\right)=\kappa^2 rx \tag{14}$$

Now let us consider the incorporation of Eq. (13) into the left-hand side of (14):

$$\frac{d\psi}{dr}=\frac{d(x/r)}{dr}=\frac{1}{r}\frac{dx}{dr}-\frac{x}{r^2} \tag{15}$$

and

$$\frac{d}{dr}\left(r^2\frac{d\psi}{dr}\right)=\frac{d}{dr}\left(r\frac{dx}{dr}-x\right)=r^2\frac{d^2x}{dr^2} \tag{16}$$

Combining Eqs. (14) and (16) gives

$$\frac{d^2x}{dr^2}=\kappa^2 x \tag{17}$$

for which a general solution is

$$x=A\exp(-\kappa r)+B\exp(\kappa r) \tag{18}$$

as may be readily verified by differentiation. Replacing x in this equation by its definition in (13) gives

$$\psi = \frac{A \exp(-\kappa r)}{r} + \frac{B \exp(\kappa r)}{r} \tag{19}$$

Since $\psi \to 0$ as $r \to \infty$, it is apparent that $B = 0$.

To evaluate A we proceed as follows. In the limit of infinite dilution—that is, as $\kappa \to 0$—the potential around the charged particle is given by the expression for the potential of an isolated charge. Elementary physics gives this as

$$\psi = \frac{q}{\varepsilon r} \tag{20}$$

a distance r from a charge q. As $\kappa \to 0$, Eqs. (19) and (20) must converge; therefore, A must equal q/ε. The general expression for potential around a spherical particle at low potential may be written

$$\psi = \frac{q}{\varepsilon r} \exp(-\kappa r) \tag{21}$$

Next, let us consider the application of this expression to a particle migrating in an electric field. We recall from Chap. 2 that the layer of liquid immediately adjacent to a particle moves with the same velocity as the surface. That is, whatever the relative velocity between the particle and the fluid may be some distance from the surface, it is zero at the surface. What is not clear is the actual distance from the surface at which the relative motion sets in between the immobilized layer and the mobile fluid. This boundary is known as the surface of shear. Although the precise location of the surface of shear is not known, it is presumably within a couple of molecular diameters of the actual particle surface for smooth particles. Ideas about adsorption from solution (e.g., Sec. 7.7) in general and about the Stern layer (Sec. 9.8) in particular give a molecular interpretation to the stationary layer and lend plausibility to the statement about its thickness. What is most important here is the realization that the surface of shear occurs well within the double layer, probably at a location roughly

FIGURE 11.2 *The relative magnitudes of various double layer potentials of interest.*

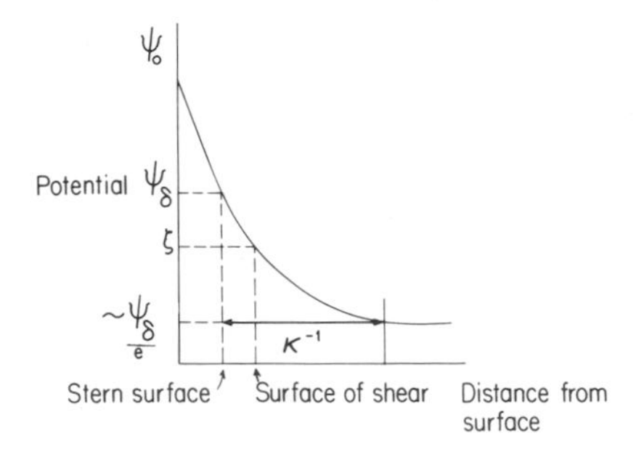

equivalent to the Stern surface. Rather than identify the Stern surface and the surface of shear, we define the potential at the surface of shear to be the zeta potential ζ. It is probably fairly close to the Stern potential ψ_δ in magnitude, and definitely less than the potential at the surface, ψ_0. The relative values of these different potentials are shown in Fig. 11.2.

Distances within the double layer are considered large or small depending on their magnitude relative to κ^{-1}. Thus in dilute solutions where κ^{-1} is large, the surface of shear—which is close to the particle surface even in absolute units—may be safely regarded as coinciding with the surface in units which are relative to the double layer thickness. Therefore, in the case where κ^{-1} is large (or κ small), Eq. (21) becomes

$$\zeta = \frac{q}{\varepsilon R} \exp(-\kappa R) \tag{22}$$

where R is the actual radius of the particle.

Since this result applies only when κ is small, the exponential may be expanded (Appendix A) to give

$$\zeta \simeq \frac{q}{\varepsilon R} \frac{1}{\exp \kappa R} \simeq \frac{q}{\varepsilon R}\left(\frac{1}{1+\kappa R}\right) \tag{23}$$

This result may also be written

$$\zeta = \frac{q}{\varepsilon R} \frac{\kappa^{-1}}{R+\kappa^{-1}} \tag{24}$$

which is the same as

$$\zeta = \frac{q}{\varepsilon R} - \frac{q}{\varepsilon (R+\kappa^{-1})} \tag{25}$$

This last result is interesting because it may be interpreted as the sum of two superimposed potentials: one arising from a charge q on a surface of radius R and a second arising from a charge $-q$ on a sphere of radius $R+\kappa^{-1}$. This is the net potential between two concentric spheres carrying equal but opposite charges and differing in radius by an amount κ^{-1}. Such a situation corresponds to a concentric sphere capacitor. As in Chap. 9, we again see the double layer behaving as if it were a capacitor with a characteristic spacing κ^{-1}.

Having explored the capacitor analogy, we no longer need to retain the second term in the series expansion of the exponential in Eq. (23). For our present purposes, it is sufficient to note that for small values of κR, Eq. (22) becomes

$$\zeta = \frac{q}{\varepsilon R} \tag{26}$$

Solving this result for q and substituting into Eq. (6), we obtain

$$u = \frac{\varepsilon \zeta}{6 \pi \eta} \tag{27}$$

The possible usefulness of this relationship—which is known as the Hückel equation—should not be overlooked. Throughout Chaps. 9 and 10 we were concerned with the potential surrounding a charged particle. Equation (9.2) provides a way of evaluating the potential at the surface, ψ_0, in terms of the concentration of potential-determining ions. Owing to ion adsorption in the Stern layer, this may may not be the appropriate value to use for the potential at the inner limit of the diffuse double layer. Although ζ is not necessarily identical to ψ_δ, it is nevertheless a quantity of considerable interest.

More elaborate theory shows that Eq. (27) is valid for spheres when κR is less than about 0.1. This imposes a rather severe restriction on the applicability of this result in aqueous systems since for $R = 10^{-6}$ cm the corresponding concentration is about 10^{-5} M for a 1 : 1 electrolyte. In nonaqueous media, however, ion concentrations may be very low and this result assumes increasing importance.

The next question to be considered is the relationship between u and ζ for the case where κR is not small.

11.4 THE ZETA POTENTIAL FOR LARGE VALUES OF κR

In this section, we consider the situation in which the thickness of the double layer is negligible compared to the radius of curvature of the surface. The derivation is not limited to any particle geometry, as long as the radius of curvature R is large compared to κ^{-1}. This situation may be brought about by making κ^{-1} small (i.e., κ large), which is equivalent to dealing with relatively high concentrations of electrolyte or to dealing with flat or slightly curved surfaces. For our purposes, it is convenient to consider a planar surface, but the results will apply equally to any case for which the product κR is large.

Suppose we consider a volume element of area A and thickness dx situated a distance x from a planar surface as shown in Fig. 11.3. The viscous force on the face nearest the surface is given by

$$F_x = \eta A \left(\frac{dv}{dx} \right)_x \tag{28}$$

FIGURE 11.3 *Location of a volume element of solution adjacent to a planar wall.*

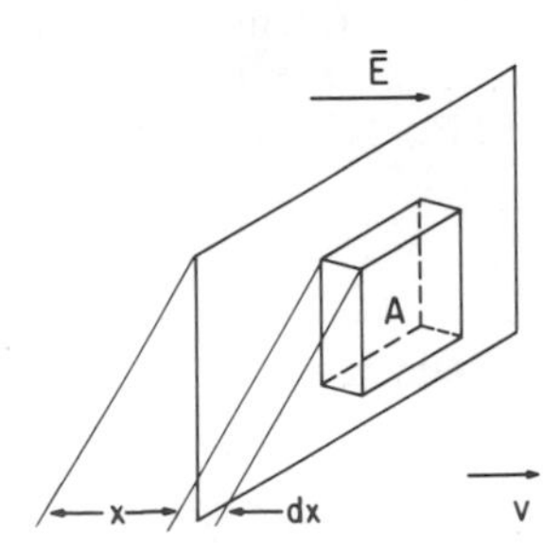

and the force exerted on the face farther from the surface is given by

$$F_{x+dx} = \eta A\left(\frac{dv}{dx}\right)_{x+dx} \tag{29}$$

In these equations, v is the relative velocity between the particle and the surrounding medium. The difference between Eqs. (28) and (29) therefore equals the net viscous force on the volume element:

$$F_{\text{vis}} = \eta A\left[\left(\frac{dv}{dx}\right)_{x+dx} - \left(\frac{dv}{dx}\right)_x\right] \tag{30}$$

Equation (2.29) can be used to relate $(dv/dx)_x$ to $(dv/dx)_{x+dx}$, so Eq. (30) becomes

$$F_{\text{vis}} = \eta A \frac{d^2v}{dx^2}\, dx \tag{31}$$

Under stationary state conditions, an equal and opposite force is exerted on the volume element by the electric field acting on the ions contained in the volume element. The force on the ions is given by the product of the field strength times the total charge. The latter equals the charge density ρ times the volume of the element; therefore,

$$F_{\text{el}} = \bar{E}\rho A\, dx \tag{32}$$

Poisson's equation [Eq. (9.40)] may now be used as a substitution for ρ to yield

$$\rho = -\frac{\varepsilon}{4\pi}\nabla^2\psi = -\frac{\varepsilon}{4\pi}\frac{d^2\psi}{dx^2} \tag{33}$$

where the second result applies specificially to the region adjacent to the planar surface and where ψ is the potential a distance x from the surface.

Setting Eqs. (31) and (32) equal to each other, substituting (33), and simplifying leads to the equation

$$\eta\frac{d^2v}{dx^2} = -\frac{\varepsilon}{4\pi}\bar{E}\frac{d^2\psi}{dx^2} \tag{34}$$

With certain assumptions, this result may be integrated twice to give the relation between v and ψ.

The integration of Eq. (34) is carried out by assuming that both η and ε are constants in the vicinity of the surface. We shall return to a discussion of this assumption in Sec. 11.8. Making this assumption, Eq. (34) can be written

$$\frac{d}{dx}\left(\eta\frac{dv}{dx}\right) = -\frac{\bar{E}}{4\pi}\frac{d}{dx}\left(\varepsilon\frac{d\psi}{dx}\right) \tag{35}$$

In this form the first integration is readily found to give

$$\eta\frac{dv}{dx} = -\frac{\varepsilon\bar{E}}{4\pi}\frac{d\psi}{dx} + C_1 \tag{36}$$

The constant of integration C_1 is evaluated by noting that both dv/dx and $d\psi/dx$ must equal zero at large distances from the surface; therefore, $C_1 = 0$.

The resulting expression is easily integrated again with the following limits: (a) at the surface of shear, $\psi = \zeta$ and $v = 0$, (b) at the outside edge of the double layer, $\psi = 0$ and v equals the observed velocity of particle migration. Therefore,

$$\eta \int_v^0 dv = -\frac{\varepsilon \bar{E}}{4\pi} \int_0^\zeta d\psi \tag{37}$$

or

$$\eta v = \frac{\varepsilon \bar{E} \zeta}{4\pi} \tag{38}$$

In terms of electrophoretic mobility, Eq. (38) can be written

$$u = \frac{v}{\bar{E}} = \frac{\varepsilon \zeta}{4\pi\eta} \tag{39}$$

Equation (39) is known as the Helmholtz–Smoluchowski equation. No assumptions are made in its derivation as to the actual structure of the double layer, only that the Poisson equation applies and that bulk values of η and ε apply within the double layer. It has been shown that this result is valid for values of κR larger than about 100.

We have now reached the position of having two expressions—Eq. (27) and (39)—to describe the relationship between the mobility of a particle (an experimental quantity) and the zeta potential (a quantity of considerable theoretical interest). The situation may be summarized by noting that both the Hückel and the Helmholtz–Smoluchowski equations may be written

$$u = C\frac{\varepsilon \zeta}{\eta} \tag{40}$$

where C is a constant the numerical value of which depends on the magnitude of κR. In the limit of both large and small values of κR, the value of C becomes independent of κR:

1. For $\kappa R < 0.1$:

$$C = \frac{1}{6\pi} \tag{41}$$

2. For $\kappa R > 100$:

$$C = \frac{1}{4\pi} \tag{42}$$

In view of the widely different pictures of the electric field surrounding the particles in the two extremes—as shown schematically in Fig. 11.1—it is not surprising that different results are obtained in the two limits.

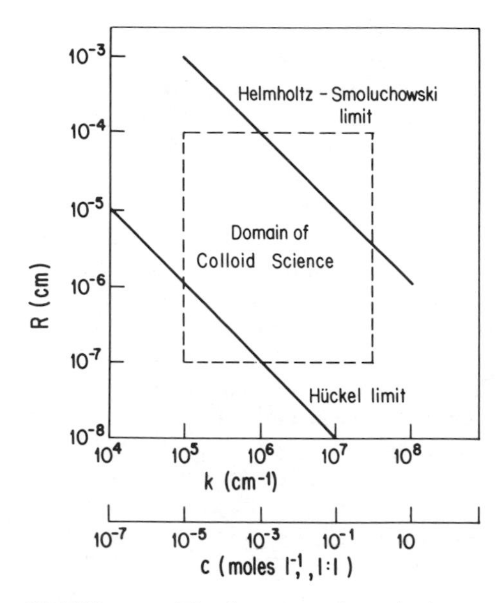

FIGURE 11.4 *The domain within which most investigations of aqueous colloidal systems lie in terms of particle radii and* 1 : 1 *electrolyte concentration. The diagonal lines indicate the limits of the Hückel and the Helmholtz–Smoluchowski equations. (From Overbeek* [5], *used with permission.)*

A major remaining problem is that many systems of interest in colloid chemistry do not correspond to either of these two limiting cases. The situation is summarized in Fig. 11.4 which maps the particle radii and 1 : 1 electrolyte concentrations which correspond to various κR values. Clearly, there is a significant domain of particle size and/or electrolyte concentration for which neither the Hückel nor the Helmholtz–Smoluchowski equations can be used to evaluate ζ from experimental mobility values. The relationship between ζ and u for intermediate values of κR is the topic of the following section.

11.5 THE ZETA POTENTIAL: GENERAL THEORY FOR SPHERICAL PARTICLES

It is apparent from the preceding sections that the understanding of electrophoretic mobility involves both the phenomena of fluid flow as discussed in Chap. 2 and the double layer potential as discussed in Chap. 9. In both places, we see that theoretical results are dependent on the geometry chosen to describe the boundary conditions of the system under consideration. This continues to be true in discussing electrophoresis where these two topics are combined. As was the case in Chaps. 2 and 9, solutions to the various differential equations which arise are possible only for rather simple geometries, of which the sphere is pre-eminent.

The generalized electrophoresis problem has been solved for spherical and rod-shaped particles, and more approximately, for random coils. In this section, we shall restrict our attention to spheres, although in the limit of large values of κR the Helmholtz–Smoluchowski equation is obtained, a result which is independent of particle shape. In the general theory, the conductivity of the particle is one of the parameters that must be considered. We shall discuss only the case of nonconducting spheres. It has been shown experimentally that mercury droplets for which κR is large follow Eq. (39) even though—as conductors—the full theory predicts they should show zero mobility. The explanation of this anomaly is that the surface of the metallic drops becomes sufficiently polarized to block the passage of current through the particle. Thus even a metallic particle may behave as an insulator, thereby justifying our choice of the nonconducting particle as the model for consideration.

In addition, we shall consider only the case in which the colloid is present in small concentration so that colloid–colloid interactions can be ignored. We shall assume that the diffuse part of the double layer is adequately described by the Gouy–Chapman theory. Since the surface of shear more or less coincides with the Stern surface, it is the diffuse part of the double layer and not the Stern layer (where specific adsorption occurs) in which we are interested. Specific adsorption in the Stern layer may have a large effect on the zeta potential itself, but should be unimportant when it comes to establishing the connection between u and ζ. The Gouy–Chapman theory ignores the actual discreteness of electrical charges and is also subject to the objections against the Poisson–Boltzmamn equation (see Sec. 9.5). An extensive body of research has been devoted either to circumventing these limitations or to estimating the approximation introduced by their use. Overbeek and Wiersma [8] have rightly noted that it is rather futile to introduce one or two corrections to the theory while neglecting other approximations that are probably of the same magnitude. A safer procedure, they note, is to use the simpler theory, keeping in mind the semiquantitative nature of the result.

By assuming that the external field—deformed by the presence of the colloidal particle—and the field of the double layer are additive, D. C. Henry derived the following expression for mobility:

$$u = \frac{\varepsilon}{4\pi\eta}\left(\zeta + 5R^5 \int_\infty^R \frac{\psi}{r^6}\,dr - 2R^3 \int_\infty^R \frac{\psi}{r^4}\,dr\right) \tag{43}$$

where r is the radial distance from the center of the particle. To go beyond Eq. (43), it is necessary to know ψ as a function of r. The resulting expressions are mathematically intractable unless a relatively simple expression is used for ψ. We may use the Debye–Hückel approximation given by Eq. (19) for this, but the constant in that equation is best evaluated somewhat differently before proceeding.

We return to the solution of the Poisson–Boltzmann equation for a spherical particle, Eq. (19), with $B = 0$:

$$\psi = \frac{A\exp(-\kappa r)}{r} \tag{44}$$

In the present development, we evaluate A by recalling that $\psi = \zeta$ when $r = R$.

Therefore,

$$A = R\zeta \exp(\kappa R) \tag{45}$$

and Eq. (44) becomes

$$\psi = \frac{R\zeta}{r} \exp[-\kappa(r-R)] \tag{46}$$

Combining Eqs. (43) and (46) and integrating leads to the result

$$u = \frac{\varepsilon\zeta}{6\pi\eta}\left\{1 + \tfrac{1}{16}(\kappa R)^2 - \tfrac{5}{48}(\kappa R)^3 - \tfrac{1}{96}(\kappa R)^4 + \tfrac{1}{96}(\kappa R)^5 \right.$$

$$\left. -[\tfrac{1}{8}(\kappa R)^4 - \tfrac{1}{96}(\kappa R)^6] \exp(\kappa R) \int_{\infty}^{\kappa R} \frac{e^{-t}\,dt}{t}\right\} \tag{47}$$

Equation (47) is called Henry's equation. Two specific assumptions underlying its derivation should be pointed out: (a) that the ion atmosphere is undistorted by the external field and (b) that the potential is low enough to justify writing $e\psi/kT < 1$, which is equivalent to requiring that $\psi < 25$ mV (Sec. 9.2). It should also be noted that in the limit of $\kappa R \to 0$, Eq. (47) reduces to the Hückel equation and in the limit of $\kappa R \to \infty$ it reduces to the Helmholtz–Smoluchowski equation. Thus the general theory confirms the idea introduced in connection with the discussion of Fig. 11.1 that the amount of distortion of the field surrounding the particles will be totally different in the case of large and small particles. The two values of C in Eq. (40) are a direct consequence of this difference. Figure 11.5a shows how the constant C varies with κR (shown on a logarithmic scale) according to Henry's equation.

We noted earlier that many systems of interest in colloid chemistry involve intermediate values of κR, so Henry's equation fills an important gap. At the same time, it explicitly introduces additional restrictions: low potentials and undistorted double layers. A topic of considerable importance is the *actual* distortion of the double layer which accompanies particle migration. The consequences of this distortion—known as the relaxation effect—are known to be important in the conductivity of simple electrolytes. A remaining development, therefore, is to consider the relaxation effect in colloidal systems.

Because the charged particle and its ion atmosphere move in opposite directions, the center of positive charge and the center of negative charge do not coincide. If the external field is removed, this asymmetry disappears over a period of time known as the relaxation time. Therefore, in addition to the fact that the colloid and its atmosphere move countercurrent with respect to one another (which is called the retardation effect), there is a second inhibiting effect on the migration which arises from the tug exerted on the particle by its *distorted* atmosphere. Retardation and relaxation both originate with the double layer, then, but describe two different consequences of the ion atmosphere. The theories we have discussed until now have all correctly incorporated retardation, but relaxation effects have not been included in any of the models considered so far.

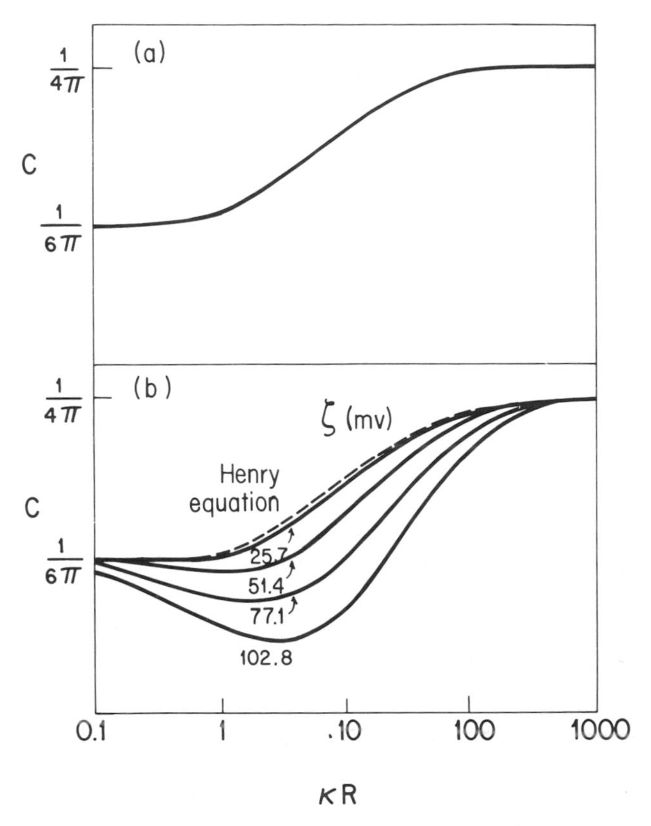

FIGURE 11.5 *Variation of the constant* C [*Eq.* (40)] *with* R (*log scale*). *In* (*a*), *at low potentials according to Henry's equation*; *in* (*b*), *for various potentials.* [*data from P. H. Wiersma, A. L. Loeb, and J. Th. G. Overbeek*, J. Colloid Interface Sci. **22**:78 (1966), *redrawn from Shaw* [9] *with permission.*]

A number of workers have tackled the problem of relaxation. The use of computers has greatly assisted this area of research because of the complexity of the mathematics involved. Loeb et al. [5], listed in Chap. 9, report the results of some numerical solutions to the mobility problem with relaxation specifically considered. Figure 11.5b summarizes some results from these studies for the case of a 1:1 electrolyte. The various curves correspond to values of zeta equaling 25.7, 51.4, 77.1, and 102.8 mV at 25°C. It will be noted that the restriction to low potentials no longer applies in the theory from which these curves were evaluated. It is evident from the figure that the relaxation effect is negligible when $\zeta < 25$ mV regardless of the value of κR and in the limit of both large and small values of κR regardless of the value of ζ. That is, intermediate values of κR and large potentials correspond to the condition of maximum resistance to flow arising from relaxation.

A family of curves qualitatively similar in appearance to those shown in Fig. 11.5b results when C is plotted versus κR at constant ζ with the valence of the electrolyte taken as the variable parameter. In that case, the relaxation effect is found to increase with the valence of the counterions. As the valence of those small ions which have the same charge as the macroion increases, the relaxation effect leads to a higher mobility (at constant ζ) than would be predicted from Henry's equation.

In this section, we have considered the relationship between u and ζ under conditions of intermediate κR values, a wide range of ζ values, and a number of ionic valence possibilities. The relationship is seen to be quite complex, except in the Hückel and Helmholtz–Smoluchowski limits. When the particle-size–electrolyte concentration conditions are such that one of these limits clearly applies, ζ can be evaluated unambiguously from experimental mobilities. The Helmholtz–Smoluchowski limit is independent of particle shape. The Hückel equation is equally free from ambiguity, although it does require spherical particles and—as already noted—the circumstances under which it holds are not especially useful for aqueous colloids. If a particle is of intermediate size with definite, known values of κ and R and with ζ known to be small, Henry's equation (or Fig. 11.5a) could be used to evaluate ζ from mobility measurements. As the complexity (i.e., higher potentials, mixed electrolyte valences) of the system increases, however, the feasibility of evaluating ζ from experimental mobilities becomes increasingly tenuous. In these circumstances, precise experimental results are best reported as mobilities with the corresponding value of ζ only an approximation.

11.6 ELECTRO-OSMOSIS

In all the sections of this chapter until now, we have focused attention on electrophoresis. We have seen that the potential at the surface of shear can be measured from electrophoretic mobility measurements, provided the system complies with the assumptions of a manageable model. One feature that has been conspicuously lacking from our discussions is any comparison between electrophoretically determined values of ζ and potential values determined by another method. The reasons for this are twofold:

1. Other techniques for measuring ζ are contingent on the same set of assumptions associated with electrophoresis and, therefore, do not constitute an independent determination.
2. Uncertainty as to the location within the double layer at which the shear surface is located makes it difficult to relate ζ to other double layer potentials, such as ψ_0 as determined from knowledge of the concentration of potential-determining ions [see Eq. (9.2)].

In this section, we shall describe electro-osmosis and in the following section, the streaming potential. These two electrokinetic techniques also permit the evaluation of ζ, but are subject to objection 1. In Sec. 11.8, we shall examine in greater detail the location of the surface of shear which is the essence of objection 2.

Earlier, we defined electrokinetic phenomena as arising from the relative motion of a charged surface and its associated double layer. In electrophoresis, it is the dispersed phase that moves, with the continuous phase remaining (more or less) stationary. It is apparent that the required relative motion between a surface and its double layer could also be brought about by causing the electrolyte solution to flow past a stationary charged wall. The complements of electrophoresis are electro-osmosis and streaming potential. The latter two measurements differ from each other as follows: (a) in electro-osmosis, it is an applied potential which induces the flow of solution; (b) in streaming potential, the solution is made to flow by applying a pressure and a potential is induced as a result. Cause and effect are thus interchanged in electro-osmosis and streaming potential.

The electro-osmosis apparatus shown in Fig. 11.6a consists of two capillaries in parallel attached at either end to reservoirs of electrolyte solution. One of the capillaries—the working capillary—is arranged with reversible electrodes at either end, while the measuring capillary contains an air bubble to indicate fluid displacement. It is the glass–solution interface in the working capillary at which the electro-osmotic phenomenon originates. Substances other than glass may also be investigated by this method, a particularly useful variation being the replacement of the capillary by a plug of powdered material which cannot be fabricated into a cylindrical tube. For the purpose of discussion, we shall continue to refer to the capillary. The conditions under which the same analysis applies to a plug will be clear from the following discussion.

When an electric field is applied across the working capillary, the double layer ions begin to migrate and soon reach the stationary state velocity. In the stationary state, electrical and viscous forces balance one another. The forces exerted on the ions by the medium are equal and opposite to the forces exerted on the medium by the ions; consequently, the liquid also attains a stationary state velocity. The tangential displacement of the fluid relative to the wall defines a surface of shear at which the potential equals ζ.

FIGURE 11.6 *Schematic illustrations of the apparatus used to measure (a) electro-osmosis and (b) streaming potential.*

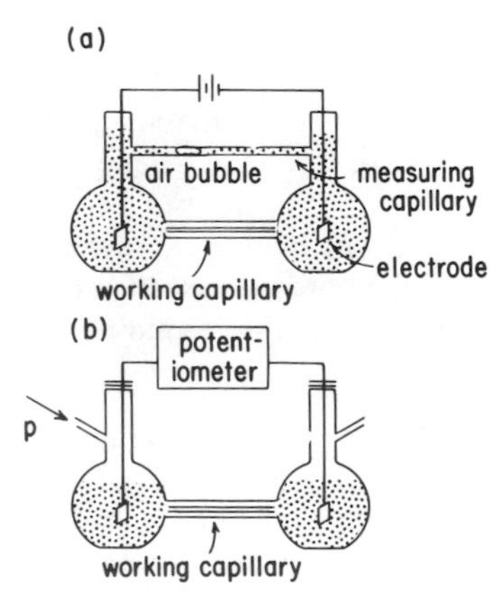

Although the Helmholtz–Smoluchowski equation was derived in reference to electrophoretic mobility, it clearly applies to electro-osmosis as well since the displacement of one part of the double layer relative to another part is common to both. Figure 11.3, for example, may be taken as an illustration of either electrophoresis or electro-osmosis. The condition of the Helmholtz–Smoluchowski equation—that R is large compared to κ^{-1}—is clearly applicable to capillaries of macroscopic dimensions. We noted earlier that the Helmholtz–Smoluchowski equation applies to the electrophoresis of nonspherical particles as long as κR is large; the same logic permits Eq. (39) to be applied to cylindrical capillaries as well as pores of irregular shape. It is this latter application that allows the replacement of a well-defined capillary by a porous plug of material in an apparatus such as that shown in Fig. 11.6a.

Equation (39) may therefore be used to describe the relationship between the potential at the capillary wall and at the velocity of electro-osmotic flow. The volume of liquid displaced per unit time V is given by multiplying both sides of Eq. (39) by the cross-sectional area of the capillary:

$$V = vA = \frac{\varepsilon\zeta\bar{E}A}{4\pi\eta} \tag{48}$$

Now suppose we apply Ohm's law to the capillary. The electric field is related to the current I and the conductivity k of the electrolyte solution as follows:

$$\bar{E} = \frac{I}{Ak} \tag{49}$$

This result may be substituted into Eq. (47) to yield

$$V = \frac{\varepsilon\zeta I}{4\pi\eta k} \tag{50}$$

This equation permits ζ to be evaluated from measurements of the rate of volume flow through the capillary; the latter are made by observing the rate of displacement of the air bubble in the measuring capillary of Fig. 11.6a.

The preceding equations are the first we have encountered in which conductivity plays a role. What is troublesome about this quantity is the fact that it is a property of bulk solutions and we are considering here an effect which arises precisely as a result of the uneven distribution of ions near a charged wall. It is essential, therefore, to examine the current carried by the ions in the double layer. Toward this end, current may be written as the sum of two contributions:

$$I = I_b + I_s \tag{51}$$

where the subscripts refer to bulk and surface contributions. Equation (49) may be used as a substitution for I_b with πR^2 as the area of a cylindrical capillary of radius R. An analogous expression may be written for the current carried by the surface layer. In this case, the bulk conductivity is replaced by surface conductivity and the cross-sectional area is replaced by the perimeter of the capillary. With these substitutions Eq. (51) becomes

$$I = \bar{E}(\pi R^2 k_b + 2\pi R k_s) = \bar{E}A\left(k_b + \frac{2k_s}{R}\right) \tag{52}$$

According to this relationship, the product $\bar{E}A$ in Eq. (48) should be replaced by

$$\bar{E}A = \frac{I}{k_b + 2k_s/R} \tag{53}$$

to give

$$V = \frac{\varepsilon \zeta I}{4\pi\eta(k_b + 2k_s/R)} \tag{54}$$

It will be noted that the importance of the correction for surface conductivity increases as R decreases and vanishes as $R \to \infty$. Equation (54) also suggests that the numerical evaluation of k_s may be accomplished by studying electro-osmosis in a set of capillaries identical in all respects except for variability in R. Finally, the expansion of Eq. (50) to (54) in correcting for surface conductivity explicitly assumes a cylindrical capillary. Experiments made with porous plugs cannot be corrected for surface conductivity by Eq. (54), but the qualitative conclusion that the effect of surface conductivity increases as the pore radius decreases is valid in this case also.

It has already been noted that there is a close similarity between electro-osmosis and streaming potential. Therefore, we shall consider this additional electrokinetic phenomenon next.

11.7 STREAMING POTENTIAL

Figure 11.6b is a sketch of an apparatus that may be used to measure streaming potential. As was the case with electro-osmosis, the capillary can be replaced by a plug of powdered material between perforated electrodes. An applied pressure difference p across the capillary causes the solution to flow through the capillary, thereby tangentially displacing the part of the double layer in the mobile phase from the stationary part.

The relationships developed in Chap. 2 for fluid flow through a capillary can be applied to this situation as follows:

1. The velocity of a cylindrical shell of radius r in a capillary of radius R and length l is given by Eq. (2.35):

$$V = \frac{p}{4\eta l}(R^2 - r^2) \tag{55}$$

2. The rate of volume flow from this cylindrical volume element is given by Eq. (2.36):

$$\frac{dV}{dt} = \frac{p}{4\eta l}(R^2 - r^2)2\pi r\, dr \tag{56}$$

3. The current associated with this volume element is

$$dI = \rho \frac{dV}{dt} = \frac{\rho p}{4\eta l}(R^2 - r^2)2\pi r\, dr \tag{57}$$

where ρ is the charge density.

4. Next, a change of variable is helpful. We replace r by a distance measured from the surface of shear, x, where

$$x = R - r \tag{58}$$

In terms of this substitution, Eq. (57) becomes

$$dI = -\frac{\rho p}{4\eta l}(2Rx - x^2)2\pi(R - x)\, dx \tag{59}$$

Our specific interest is in the region near the walls of the capillary where $x \ll R$. In this region, Eq. (59) may be approximated

$$dI \simeq -\frac{\rho \pi p}{\eta l} R^2 x\, dx \tag{60}$$

5. Substituting Eq. (9.40) for ρ yields

$$dI - \frac{\varepsilon p R^2}{4\eta l} \frac{d^2\psi}{dx^2} x\, dx \tag{61}$$

6. The total current carried by the capillary is obtained by integrating x over the radius of the cylinder. Integration by parts yields

$$I = \frac{\varepsilon p R^2}{4\eta l}\left(x \frac{d\psi}{dx} \Big|_0^R - \int_0^R \frac{d\psi}{dx} dx \right) = \frac{\varepsilon p R^2 \zeta}{4\eta l} \tag{62}$$

since $\psi = \zeta$ at $x = 0$ and $\psi = d\psi/dx = 0$ at $x = R$.

The quantity calculated by Eq. (62) is known as the streaming current. It is specifically due to the net displacement of the mobile part of the double layer relative to the stationary part of the double layer. The field associated with this current is given by combining Eqs. (52) and (62):

$$\bar{E} = \frac{\varepsilon \zeta}{4\pi\eta}\left(\frac{1}{k_b + 2k_s/R}\right)\frac{p}{l} \tag{63}$$

If both sides of Eq. (63) are multiplied by the length of the capillary l, the potential difference between the measuring electrodes—the streaming potential Φ—is obtained

$$\Phi = \frac{\varepsilon \zeta p}{4\pi\eta(k_b - 2k_s/R)} \tag{64}$$

The conditions under which Eq. (64) for streaming potential and Eq. (54) for electro-osmosis were derived are comparable inasmuch as each applies to the case of large κR. Comparison of Eqs. (54) and (64) in the limit of large R shows that

$$\frac{\Phi}{p}=\frac{V}{I}=\frac{\varepsilon\zeta}{4\pi\eta k} \tag{65}$$

The coupling of two different electrokinetic ratios (Φ/p and V/I) through Eq. (65) is an illustration of a very general law of reciprocity due to L. Onsager (Nobel Prize, 1968). The general theory of the Onsager relations, of which Eq. (65) is an example, is an important topic in nonequilibrium thermodynamics.

If the relationships shown in Eq. (65) are to be used in computations, it is essential that proper units be used. The easiest way of demonstrating the dimensional equivalency of each of the terms in Eq. (65) is to express all quantities in cgs units (see Table 9.1):

1. For V/I:

$$\frac{\text{cm}^3\ \text{s}^{-1}}{\text{esu s}^{-1}}=\text{cm}^3\ \text{esu}^{-1}$$

2. For Φ/p:

$$\frac{\text{statvolts}}{\text{dyne cm}^{-2}}\times\frac{1\ \text{dyne}}{\text{g cm s}^{-2}}\times\frac{1\ \text{erg esu}^{-1}}{\text{statvolt}}\times\frac{\text{g cm}^2\ \text{s}^{-1}}{\text{erg}}=\text{cm}^3\ \text{esu}^{-1}$$

3. For $\zeta/\eta k$:

$$\frac{\text{statvolts}}{(\text{dyne cm}^{-2}\ \text{s})(\text{ohm}^{-1}\ \text{cm}^{-1})}\times\frac{\text{dyne cm esu}^{-1}}{\text{statvolt}}\times\frac{\text{volts ampere}^{-1}}{\text{ohm}}$$

$$\times\frac{\text{ampere second}}{\text{coulomb}}\times\frac{\text{esu cm}^{-1}}{\text{statvolt}}\times\frac{1\ \text{statvolt}}{300\ \text{volts}}\times\frac{1\ \text{coulomb}}{3\times10^9\ \text{esu}}=\text{cm}^3\ \text{esu}^{-1}$$

In aqueous solutions at 25°C, Eq. (54) becomes

$$\frac{V}{I}=7.8\times10^{-9}\frac{\zeta}{k} \tag{66}$$

where ζ is measured in millivolts, I in milliamperes, k in reciprocal ohms per centimeter, and V in cubic centimeters per second. For water at 25°C, Eq. (64) becomes

$$\frac{\Phi}{p}=10^{-6}\frac{\zeta}{k} \tag{67}$$

where p is measured in millimeters of mercury and Φ and ζ are measured in the same units. Specific conductivity may also be written

$$k=\frac{c\Lambda}{1000} \tag{68}$$

where c is the normality of the electrolyte solution and Λ is the equivalent conductivity. Thus for 10^{-3} *M* NaCl for which Λ is about 126 $cm^2\, eq^{-1}\, ohm^{-1}$, $k = 1.26 \times 10^{-4}\, ohm^{-1}\, cm^{-1}$. Therefore, a surface with a ζ potential of 50 mV will displace about 11 $cm^3\, h^{-1}$ if a current of 1.0 mA flows through an electro-osmosis apparatus. With the same electrolyte and the same value of ζ, an applied pressure of 760 mm Hg would produce a streaming potential of about 300 mV.

In hydrocarbons, the specific conductivity may be lower than that of aqueous solutions by many orders of magnitude, so the streaming potentials generated by the high-pressure pumping of these materials may be quite spectacular. The danger of sparking at such voltages plus the flammability of these substances makes the petroleum industry an area in which streaming potential finds important applications. For example, gasoline (for which the specific conductivity would be as low as $10^{-14}\, ohm^{-1}\, cm^{-1}$ if untreated) pumping equipment must be grounded. In addition, a variety of organic-soluble electrolytes have been developed as antistatic additives for petroleum. Examples of two such compounds are tetraisoamylammonium picrate and calcium diisopropyl salicylate. Crude petroleum is less troublesome in this regard than refined products since the crude contains oxidation products, asphaltenes, and so on, which impart a natural conductivity to this material.

The objective of comparing values of ζ determined from electrophoresis with those determined by other electrokinetic methods was stated at the beginning of Sec. 11.6. Enough experiments have been conducted in which at least two of the electrokinetic methods we have discussed are compared to leave no doubt as to the self-consistency of ζ as determined by these different methods. There is no guarantee, however, that self-consistent ζ potentials are correct. Consistency means only that ζ has been extracted from experimental quantities by a self-consistent set of approximations. It should be emphasized, however, that the existence of a potential at the surface of shear—which is the common component in all the electrokinetic analyses we have discussed—is more than amply confirmed by these observations.

Two conditions must be met to justify comparisons between ζ values determined by different electrokinetic measurements: (a) the effects of relaxation and surface conductivity must be either negligible or taken into account, and (b) the surface of shear must divide comparable double layers in all cases being compared. This second limitation is really no problem when electro-osmosis and streaming potential are compared since, in principle, the same capillary can be used for both experiments. However, obtaining a capillary and a migrating particle with identical surfaces may not be as readily accomplished. One means by which particles and capillaries may be compared is to coat both with a layer of adsorbed protein. It is an experimental fact that this procedure levels off differences between substrates: The surface characteristics of each are totally determined by the adsorbed protein. This technique also permits the use of microelectrophoresis for proteins since adsorbed and dissolved proteins have been shown to have nearly identical mobilities.

11.8 THE SURFACE OF SHEAR

The surface of shear is the location within the electrical double layer at which the various electrokinetic phenomena measure the potential. We saw in Chap. 9 how the double layer extends outward from a charged wall. Its value at any particular distance from the wall can, in principle, be expressed in terms of the potential at the wall and the electrolyte content of the solution. In terms of electrokinetic phenomena, the question is how far from the interface is the surface of shear situated?

The very existence of a surface of shear implies some interesting behavior within the fluid phase of the system under consideration. In our discussion of all electrokinetic phenomena until now, we have assumed that the viscosity of the medium has its bulk value right up to the surface of shear. In addition, it has been implicitly assumed that the viscosity abruptly becomes infinite at the surface of shear.

At this point, it is convenient to recall Fig. 7.11 and the discussion thereof. In that context, we observed that there is generally a variation of properties in the vicinity of an interface from the values which characterize one of the adjoining phases to those which characterize the other. This variation occurs over a distance τ measured perpendicular to the interface. In the present discussion, viscosity is the property of interest and the surface of shear—rather than the interface per se—is the boundary of interest. The model we have considered until now has implied an infinite jump in viscosity, occurring so sharply that τ is essentially zero. From a molecular point of view, such an abrupt transition is highly unrealistic. A gradual variation in η over a distance comparable to molecular dimensions is a far more realistic model. With these ideas in mind, it is evident that we would do better to think of a *zone* of shear rather than a surface of shear. Although we shall continue to speak of the shear "surface," the term is not used in the mathematical sense of possessing zero thickness, but rather in the broader sense of Chap. 7.

How must the expressions derived in the earlier sections of this chapter be modified to take into account the finite distance over which η increases? The answer is that η—the viscosity within the double layer—must be written as a function of location. Our objective in discussing this variation is not to examine in detail the efforts that have been directed along these lines. Instead, it is to arrive at a better understanding of the relationship between ζ and the potential at the inner limit of the diffuse double layer and a better appreciation of the physical significance of the surface of shear.

Measurements of the viscosity of organic liquids in the presence of an electric field reveal that there is an increase in viscosity in high electric fields which is described by the expression

$$\frac{\eta_E - \eta_0}{\eta_0} = f\bar{E}^2 \tag{69}$$

where the subscripts indicate the presence (E) or absence (0) of a field. The factor f is called the viscoelectric constant and has a value of about $2 \times 10^{-12}\ \mathrm{V^{-2}\ cm^2}$ for several organic liquids. Thus a 10% increase in viscosity may be anticipated for a field strength of about $2 \times 10^5\ \mathrm{V\ cm^{-1}}$.

An expression such as Eq. (9.66) may be used to estimate $\bar{E}$ $(=d\psi/dx)$ in the double layer. Table 11.1 shows values of $\bar{E}$ evaluated by means of this equation for a variety of ψ_0 values and 1 : 1 electrolyte concentrations. It will be noted that for high values of ψ_0 and high ionic strengths, the field in the double layer may be large enough to produce a very significant viscoelectric effect.

TABLE 11.1 *Values of the electric field (in* $\mathrm{V\,cm^{-1}}$*) calculated in the double layer by Eq. (9.66) for various* ψ_0 *values and concentrations of* 1 : 1 *electrolyte*

	c (moles liter^{-1})		
ψ_0 (mV)	10^{-3}	10^{-2}	10^{-1}
50	6.36×10^4	2.01×10^5	6.36×10^5
100	1.98×10^5	6.24×10^5	1.98×10^6
150	5.49×10^5	1.74×10^6	5.49×10^6
200	1.51×10^6	4.77×10^6	1.51×10^7

Now suppose we re-examine the derivation of the Helmholtz–Smoluchowski equation as given in Sec. 11.4. Returning to Eq. (37), we note that the relationship between u and ζ is given by

$$u = \frac{\varepsilon}{4\pi}\int_0^\zeta \frac{d\psi}{\eta} \tag{70}$$

where η has been left inside the integral this time since its value is assumed to vary with ψ. We continue to assume that ε is a constant since the effect of the field is known to be less for this quantity than for η.

Now we substitute η_E from Eq. (69) for the viscosity in the double layer in (70) to obtain

$$u = \frac{\varepsilon}{4\pi\eta_0}\int_0^\zeta \frac{d\psi}{1+f(d\psi/dx)^2} \tag{71}$$

Finally, Eqs. (9.66) and (9.80) may be used to evaluate $d\psi/dx$ in the double layer:

$$u = \frac{\varepsilon}{4\pi\eta_0}\int_0^\zeta \frac{d\psi}{1+(f32\pi cRT)/1000\varepsilon\ \sinh^2(ze\psi/2kT)} \tag{72}$$

or

$$u = \frac{\varepsilon}{4\pi\eta_0}\int_0^\zeta \frac{d\psi}{1+A\sinh^2 B\psi} \tag{73}$$

where $A = 32\pi cRTf/1000\varepsilon$ and $B = ze/2kT$. If c (therefore A) and ψ are small, Eq. (73) becomes approximately

$$u \simeq \frac{\varepsilon}{4\pi\eta_0}\int_0^\zeta (1 - A\sinh^2 B\psi)\,d\psi \tag{74}$$

Under the same conditions, the hyperbolic sine function may be expanded (Appendix A) with only the leading term retained to obtain

$$u \simeq \frac{\varepsilon}{4\pi\eta_0}\int_0^\zeta [1 - A(B\psi)^2]\,d\psi \tag{75}$$

This equation is readily integrated to yield

$$u \simeq \frac{\varepsilon}{4\pi\eta_0}\left(\zeta - \frac{AB^2}{3}\zeta^3\right) = \frac{\varepsilon\zeta}{4\pi\eta_0}\left(1 + \frac{AB^2}{3}\zeta^2\right) \tag{76}$$

Under conditions in which the second term is negligibly small, Eq. (76) becomes identical to (39), the Helmholtz–Smoluchowski result. On the other hand, when the concentration and ζ increase, the value of ζ which would be associated with an observed mobility is larger than the Helmholtz–Smoluchowski equation would indicate.

Equation (72) may also be integrated analytically. Although we shall not consider the actual solutions, which are rather complex, Fig. 11.7 shows graphically the results of these integrations drawn for water at 25°C, assuming $f = 10^{-11}\ \mathrm{V}^{-2}\ \mathrm{cm}^2$. The abscissa shows values of ψ_0, the potential at the inner limit of the diffuse double layer, with $4\pi\eta u/\varepsilon$ plotted on the ordinate. It must be remembered that this latter quantity equals ζ according to Eq. (39)—which we shall designate ζ_{HS}—when the viscosity is assumed to be the bulk value throughout the double layer. The figure shows that $\zeta_{HS} = \psi_0$ at low values of the potential. As the potential increases, however, ζ_{HS} begins lagging behind ψ_0, the effect indicated by Eq. (76) in a limiting approximation. At still higher potentials, ζ_{HS} eventually reaches a constant value which is independent of the actual value of ψ_0. Note, further, that this leveling off occurs at progressively lower potentials as the concentration of electrolyte increases. Increasing both the potential and the electrolyte concentration tends to increase the field in the double layer (see Table 11.1) which, in turn, increases the viscosity of solvent in the double layer. As the effective viscosity of the medium increases, the surface of shear occurs progressively further from the surface. This accounts for the fact that ζ_{HS} falls behind ψ_0 as the latter increases. These conclusions are consistent with the experimental observation that ζ_{HS} for AgI becomes independent of the concentration of the potential-determining Ag^+ and I^- ions once the concentrations of these ions are well removed from the conditions at which the particles are uncharged.

The results shown in Fig. 11.7 illustrate quite clearly the relationship between ζ and ψ_0 and in this way reveal the dependence of the location of the surface of shear on the structure of the double layer. It might appear that one could consult curves such as those shown in Fig. 11.7 to read from the appropriate plot that value of ψ_0 which corresponds to a particular ζ, at least for values of ζ which are less than the limiting value. Although semiquantitative interpretations based on this figure may be trusted, some caution must be expressed about the numerous assumptions and approximations inherent in Fig. 11.7. In summary, the following may be cited as examples of such constraints:

1. The possible immobilization of solvent near the surface due to either chemical or mechanical (as opposed to viscoelectric) interaction with the solid phase has not been considered.
2. Use of the Gouy–Chapman theory [Eq. (72)] overlooks any specific effects arising from differences between ions, especially with regard to hydration.

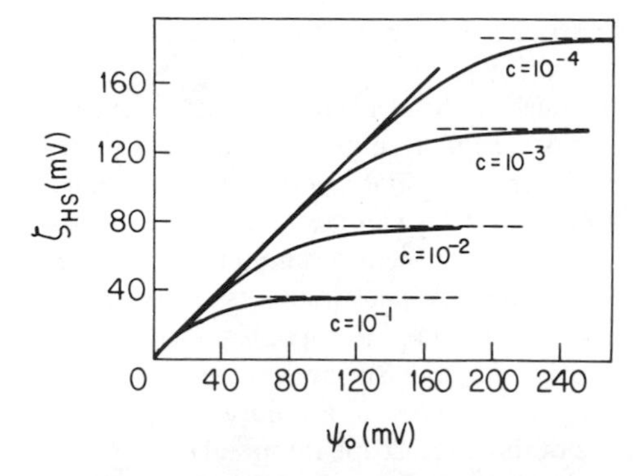

FIGURE 11.7 *Plot of* $4\pi\eta u/\varepsilon$ *versus* ψ_0 *or, in words, the zeta potential according to the Helmholtz–Smoluchowski equation* (39) *versus the potential at the inner limit of the diffuse part of the double layer. Curves drawn for various concentrations of* 1 : 1 *electrolyte with* $f = 10^{-11}\ V^{-2}\ cm^2$. [*From J. Lyklema and J. Th. G. Overbeek*, J. Colloid Sci., **16**:501 (1961), *used with permission.*]

3. The validity of Eq. (69) in electrolyte solutions, especially the dependence of f on concentration, has not been investigated as fully as might be desired.

11.9 EXPERIMENTAL ASPECTS OF ELECTROPHORESIS

Of the electrokinetic phenomena we have considered, electrophoresis is by far the most important. Until now, our discussion of experimental techniques of electrophoresis has been limited to a brief description of microelectrophoresis. The latter is easily visualized and has provided sufficient background for our considerations up to this point. Microelectrophoresis itself is subject to some complications that can be discussed now that we have some background in the general area of electrical transport phenomena. In addition, the methods of moving boundary electrophoresis and zone electrophoresis are sufficiently important to warrant at least brief summaries.

Microelectrophoresis depends on the visibility of the migrating particles under the microscope. As such, it is inapplicable to molecular colloids such as proteins. By adsorbing the protein molecules on suitable carrier particles, however, the range of utility for microelectrophoresis can be extended. The use of dark field illumination (see Sec. 1.5) can sometimes be used to advantage to extend microelectrophoresis observations to small, high-contrast particles.

The migrating particles are observed in a cell which may be either cylindrical or rectangular in shape. The walls must be optically uniform for observation and fewer optical corrections and thermostating difficulties are encountered if the walls are thin. The working part of the apparatus is thus fragile and auxiliary connecting rods are generally incorporated into the design to increase the mechanical strength of the cell. Figure 11.8 is a sketch of an electrophoresis apparatus with a rectangular working compartment.

The electric field in the cell is best established by means of reversible electrodes such as Ag–AgCl or $Cu–CuSO_4$. Care must be taken to prevent the electrolyte of the electrode from contaminating the dispersion. Platinized electrodes behave reversibly with low currents, but gas evolution causes troubles at higher currents.

The field strength is best obtained by including an accurate ammeter in the circuit to determine the current. Independent conductivity measurements in the cell with standard solutions permit the determination of the field through Eq. (49).

The rate of particle migration is determined by measuring with a stopwatch the time required for a particle to travel between the marks of a calibrated graticule in the microscope eyepiece. If the objective of the microscope is immersed during the electrophoresis measurement, the calibration of the graticule should be made with the same immersion liquid.

Electrophoretic migrations are always superimposed on other displacements which must either be eliminated or corrected to give accurate values for mobility. Examples of these other kinds of movement are Brownian motion, sedimentation, convection, and electro-osmotic flow. Brownian motion, being random, is eliminated by averaging a series of individual observations. Sedimentation and convection, on the other hand, are systematic effects. Corrections for the former may be made by observing a particle with and without the electric field, and the latter may be minimized by effective thermostating and working at low current densities.

Correcting for the superposition of electrophoresis and electro-osmosis is important because of the small inside dimensions of electrophoresis cells. Even in the absence of a colloid an electrolyte solution will display electro-osmotic flow at a rate given by Eq. (54). Since the system is closed in an electrophoresis cell, the flow of liquid along the walls is offset by a flow of liquid in the opposite direction down the middle of the cell. This effect is present regardless of the cross-sectional shape of the cell, but the details of the velocity profile depend on the shape.

FIGURE 11.8 *Schematic illustration of a microelectrophoresis cell with a rectangular working compartment.*

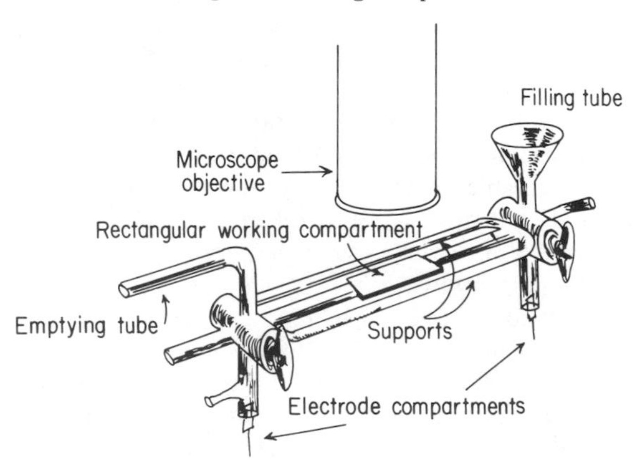

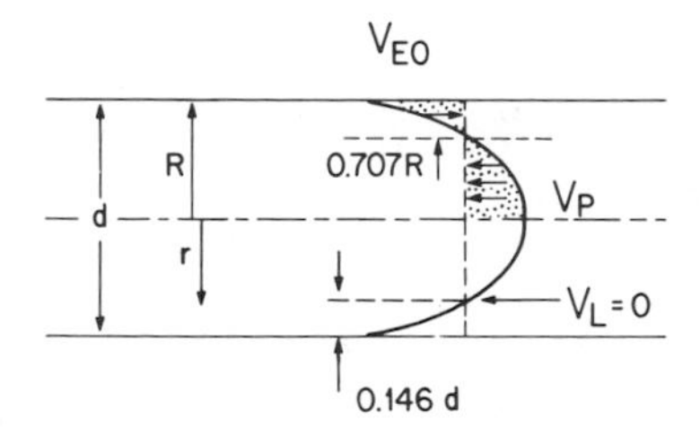

FIGURE 11.9 *Location of the surface of zero liquid velocity in a cylindrical capillary.*

The analysis of this effect in a closed cyclindrical cell is obtained by subtracting from the electro-osmotic velocity, v_{EO}, the velocity of flow through a capillary given by Poiseuille's equation [Eq. (2.35), v_P]:

$$v_L = v_{EO} - v_P = v_{EO} - C(r^2 - R^2) \tag{77}$$

where v_L is the velocity of the liquid and C is a constant. The requirement of no net displacement of liquid is incorporated by integrating Eq. (77) over the cross section of the cylinder and setting the result equal to zero:

$$\int_0^R v_L(2\pi r)\, dr = 0 \tag{78}$$

In this expression R is the radius of the capillary and r is the radial distance from the capillary axis as shown in Fig. 11.9. Substitution of Eq. (77) into (78) and integration gives

$$C = \frac{2v_{EO}}{R^2} \tag{79}$$

This result may be substituted back into Eq. (77) to evaluate that location in the cylinder where the net liquid displacement is zero:

$$v_L = 0 = v_{EO}\left[1 + \frac{2}{R^2}(r^2 - R^2)\right] = v_{EO}\left(\frac{2r^2}{R^2} - 1\right) \tag{80}$$

This result shows that electro-osmotic flow and backflow in the capillary cancel when the factor $(2r^2/R^2) - 1$ equals zero. This condition corresponds to $r/R = 0.707$. Thus, at 70.7% of the radial distance from the center of the capillary lies a circular surface of zero liquid flow. Any particle tracked at this position in the capillary will display its mobility uncomplicated by the effects of electro-osmosis. This location may also be described as lying 14.6% of the cell diameter inside the surface of the capillary. Experimentally, then, one establishes the inside diameter of the capillary and focuses the microscope 14.6% of this distance inside the walls of the capillary. Corrections for the effect of the refractive index must also be included. Additional details on this correction can be found in the book by Shaw [9].

The location of the surface of zero liquid flow in cells of rectangular cross section has also been worked out. For a cell in which the direction of migration is very long

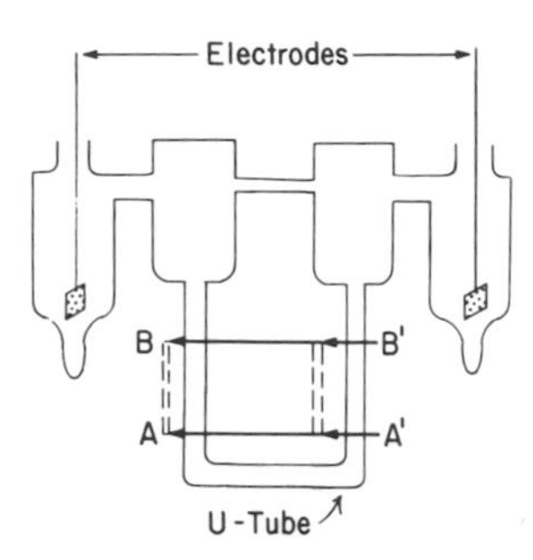

FIGURE 11.10 *Schematic illustration of a Tiselius-type moving boundary electrophoresis apparatus.*

compared to the width of the cell, the surface where $v = 0$ lies 21.1% of the cell depth above the bottom and below the top of the working compartment.

In addition to microelectrophoresis, another important method for the determination of mobility is the moving boundary method. In essence, this is no different from the moving boundary method as applied to simple ions. The apparatus most commonly used is that of A. Tiselius (Nobel Prize, 1948) which is illustrated schematically in Fig. 11.10. The Tiselius cell consists of a U-tube of rectangular cross section which is segmented in such a way that the sections between the lines AA' and BB' in the figure can be laterally displaced with respect to the rest of the apparatus. The offset segments of the U-tube are filled with the colloidal dispersion and, after thermal equilibration with the buffer solution contained in other parts of the apparatus, the various sections are aligned so that sharp boundaries are obtained. The location of the boundaries is usually observed by Schlieren optics which identify refractive index gradients (see Sec. 3.5). As the macroions migrate in the electric field, the Schlieren peak becomes displaced and the mobility of a colloidal component may be determined by measuring the rate of boundary movement per unit electric field. Relatively longer times are required for accurate mobility experiments than for microelectrophoresis since the particles must migrate over macroscopic distances rather than microscopic ones. To avoid contamination of the electrolyte in the U-tube with electrode products, the electrodes are generally located near the bottom of large reservoirs as shown in Fig. 11.10. Relatively concentrated salt solution is used to cover the electrodes, with the buffer solution layered on top.

Under optimum conditions, the dimensions of the cross section of the cell are such that the effects of electro-osmosis are minimal. The rectangular profile of the cross section allows for both good thermal equilibration (because one dimension is short) and good optical precision (because the other dimension is longer).

Moving boundary electrophoresis is most widely applied to protein mixtures. In such a case, each molecular species travels with a characteristic velocity. After sufficient time, the various components in a mixture become effectively separated, and the percentage of each may be determined by measuring the areas under the Schlieren peaks. Figure 11.11 shows a typical electrophoresis pattern for human

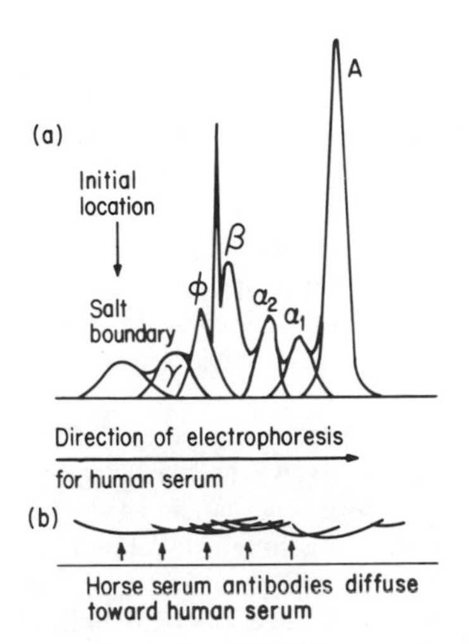

FIGURE 11.11 *Electrophoresis patterns for human serum. (a) Schematic of Schlieren profiles; (b) precipitates with horse serum antibodies. The latter is an illustration of immunoelectrophoresis.*

blood serum. In this figure, the protein albumin (A), α_1-, α_2-, β-, and γ-globulin, and fibrinogen (ϕ) are fairly clearly resolved. The remaining peak in the figure is the boundary between the original buffer and the colloid. This "false boundary" moves little in an electrophoresis experiment and is obviously not considered in determining the percentages of different proteins in a mixture.

When separation rather than determination of mobility is the primary objective of an electrophoresis experiment, a technique called zone electrophoresis is quite widely employed. In zone electrophoresis, a supporting medium such as moist filter paper or a gel such as polyacrylamide is the location of the particle migration. The method thus resembles solid–liquid chromatography, and many of the substrates and analytical methods of the latter are used in this electrophoretic procedure as well. As with chromatography, a spot or band of a mixture is applied to one end of the support medium. As the electrophoresis proceeds, spots or bands of the individual components appear at different locations along the axis of the voltage gradient. Sometimes the resolution is improved by following the electrophoresis by a chromatographic separation at right angles to the direction of the initial separation.

Zone electrophoresis is influenced by adsorption and capillarity as well as by electro-osmosis. Therefore, evaluation of mobility (and ζ) from this type of measurement is considerably more complex than from either microelectrophoresis or moving boundary electrophoresis. Nevertheless, zone electrophoresis is an important technique which is widely used in biochemistry and clinical chemistry. One particularly important area of application is the field of immunoelectrophoresis, which is described briefly in Sec. 11.11.

11.10 THE CHARGE OF PROTEIN MOLECULES

In the quantitative sections of this chapter, the primary emphasis has been on establishing the relationship between the electrophoretic properties of the system and the zeta potential. We saw in Chaps. 9 and 10 that potential is a particularly useful quantity for the characterization of lyophobic colloids. In this context, then, the ζ potential is a valuable property to measure for a lyophobic colloid. For lyophilic colloids such as proteins, on the other hand, the charge of the particle is a more useful way to describe the molecule. In this section, we shall consider briefly what information may be obtained about the charge of a particle from electrophoresis measurements.

We have lamented the fact that electrokinetic potentials cannot be evaluated independently to check the correctness of various theories. However, the charge of a protein can be evaluated from its titration curve. Therefore, if we can find a way of evaluating particle charge from electrokinetic data, the long-sought independent verification will be established. The net charge of a particle q is equal and opposite to the total charge in the double layer. The increment of charge in a spherical shell of radius r and thickness dr in the double layer is given by the area of the shell times its thickness times the charge density:

$$dq = 4\pi r^2 \rho \, dr \tag{81}$$

Integrating this expression over the entire double layer gives

$$q = -\int_R^\infty 4\pi r^2 \rho \, dr = \int_R^\infty \frac{\varepsilon}{4\pi} \nabla^2 \psi 4\pi r^2 \, dr = \varepsilon \int_R^\infty \frac{d}{dr}\left(r^2 \frac{d\psi}{dr}\right) dr \tag{82}$$

Where the Poisson equation (9.38) has been substituted for ρ. Integration yields

$$q = \varepsilon\left(r^2 \frac{d\psi}{dr}\right)\Bigg|_R^\infty = -\varepsilon R^2 \frac{d\psi}{dr}\Bigg|_R \tag{83}$$

where the derivative is evaluated at $r = R$.

Now Eq. (46) is used to evaluate the derivative in Eq. (83):

$$\frac{d\psi}{dr}\Bigg|_R = -\frac{\zeta}{R}(1 + \kappa R) \tag{84}$$

Substituting this result into Eq. (83) gives

$$q = \varepsilon \zeta R (1 + \kappa R) \tag{85}$$

for the charge enclosed by the surface of shear.

This discussion shows that the evaluation of charge from electrokinetic measurements involves all the complications inherent in the evaluation of ζ plus the additional restrictions of low potentials and spherical particles. Additional relationships have been developed which permit these restrictions to be relaxed, but we shall not discuss these here.

We conclude this section by comparing briefly the charge on protein molecules as determined by electrophoresis measurements through Eq. (85) and as determined by titration. Protein molecules carry acid and base functions in side groups along the macromolecule. In a strongly acidic solution, amine groups will be protonated and the protein will carry a positive charge. Addition of a known number of equivalents of strong base to a measured volume of protein solution results in a change of pH and a change in the state of charge of the protein. From the volume of the solution, the change of pH, and a knowledge of activity coefficients, the number of added equivalents of base which react with the protein may be determined. It should be noted that the added base may remove protons from either neutral groups or cationic groups. Thus in acid solution, a protein may have a charge corresponding to the binding of z H^+ ions: $+z$. After z OH^- ions have reacted with it, the molecule will have a *net* charge of zero. If the reactions consist exclusively in the removal of bound H^+ ions, the net charge (zero) would correspond to the *actual* charge of the particle. If all the reacting OH^- ions remove H^+ ions from neutral groups, on the other hand, the molecule would be twice as highly charged as the initial species: an equal number of positive and negative charges. In reality, both processes occur together so it cannot be inferred that the point of equivalency—called the isoionic point—corresponds to an uncharged state. All that can be said is that the *net* charge is zero at the isoionic point. Addition of more base beyond this point will increase (still by both processes) the negative charge of the molecule even further. The isoionic point corresponds to a point at which the polyelectrolyte changes sign. This discussion shows that the net charge relative to the initial condition of the colloid is readily determined from titration curves.

The electrophoretic mobility of a protein solution may also be measured as a function of pH. By this technique also it is observed that the colloid passes through a point of zero net charge where its mobility is zero. The point at which charge reversal is observed electrophoretically is called the isoelectric point.

Figure 11.12 shows the relationship between the charge of egg albumin as determined by titration and by electrophoresis. The points were determined electrophoretically, and the solid line was determined by titration. The titration curve has been shifted so that the isoionic point and the isoelectric point match. It will be observed that the two independent charge determinations lead to slightly different values. The charges determined electrophoretically are 60% of those determined analytically. If the titration results are multiplied by 0.60, the dashed line in Fig. 11.12 is obtained. This shows clearly that the two determinations are identical in pH dependence, but raises the question as to the origin of the constant percentage difference.

There are several minor corrections which tend to reduce the discrepancy between the two curves: for example, corrections for relaxation and finite ion size. It should also be remembered that electrophoresis measures the net charge inside the surface of shear. To the extent that this diverges from the "surface" of the molecule, the two techniques may very properly "see" different charges for the colloid. Additional studies in this area, therefore, might help to clarify the relationship between the actual surface and the surface of shear.

We noted earlier that proteins display essentially the same mobility both as free molecules and when adsorbed on carrier particles. Adsorption clearly increases the

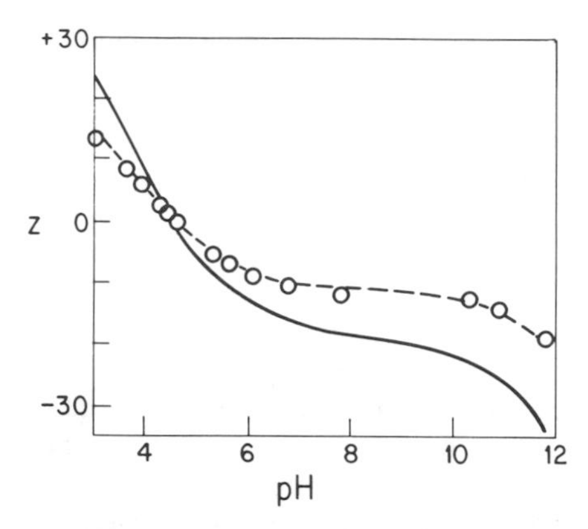

FIGURE 11.12 *Net charge of egg albumin versus pH. Points determined by electrophoresis; solid line, by titration; broken line,* 60% *of charge from titration.* [*Data from L. G. Longsworth,* Ann. N.Y. Acad. Sci. **41**:267 (1941), *redrawn from Overbeek* [5] *with permission.*]

radius of the kinetic unit appreciably, so this effect on mobility is unexpected. One way to rationalize this result is to assume that the protein adsorbs on the surface with very little alteration of the shape it has in free solution. Next, assume that it is the radius of these molecular protuberances rather than the overall radius of curvature of the carrier that governs the mobility.

11.11 APPLICATIONS OF ELECTROKINETIC PHENOMENA

Throughout most of this chapter, the emphasis has been on the evaluation of zeta potentials from electrokinetic measurements. This emphasis is entirely fitting in view of the important role played by the potential in the DLVO theory of colloidal stability. From a theoretical point of view, a fairly complete picture of colloidal stability can be built up from a knowledge of potential, electrolyte content, Hamaker constants, and particle geometry. This aspect of flocculation was stressed in Chaps. 9 and 10. From this perspective, the fundamental importance of the ζ potential is evident.

In addition to fundamental principles, however, there are many practical situations in which flocculation is a process of considerable importance. Often all that is desired in these cases is either to maximize or minimize flocculation in some experimental system. Systems of practical interest are frequently so complex that theoretical models apply to them only qualitatively at best. In this context, the concept of zeta potential emerges as a valuable practical parameter. If two systems of different ζ are compared—all other factors being equal—the one that has the higher ζ potential is expected to be more stable with respect to flocculation and the one with the lower potential, less stable. The second case is particularly important. At the isoelectric point, electrophoretic mobility is zero, ζ is zero, and the potential energy of repulsion between particles is minimum. Thus electrophoresis measurements can be used as an indicator for optimum conditions for flocculation. In this type of application, the technique is used as a null detector, hence it is independent of any model or equation for interpretation.

One important—if unattractive—example of this application is in sewage treatment. Industrial wastewater and domestic sewage contain an enormous assortment of hydrophilic and hydrophobic debris of technological and biological origin. The concentration of surface-active materials in sewage from household detergents alone is about 10 ppm. In addition, sewage abounds in amphipathic materials of natural and biological origin. These substances tend to adsorb on and impart a charge to the suspended solid and liquid particles in the polluted water. Negative zeta potentials in the range of 10 to 40 mV are fairly typical for the suspended particles in sewage.

A typical purification scheme consists of adding $NaHCO_3$ and $Al_2(SO_4)_3$ (alum) to water with agitation. The aluminum ion undergoes hydrolysis and precipitates as a gelatinous, polymeric hydrated oxide. Suspended material is enmeshed in this amorphous precipitate which produces flocs by bridging the particles together. The polymeric nature of the "$Al(OH)_3$" precipitate permits us to compare it with protein in its ability to coat particles and impart to the carrier particles its own characteristic potential. Like proteins, $Al(OH)_3$ is also capable of reacting with both H^+ and OH^- so that these ions determine the charge of the suspended units, whether these are flocs formed by the $Al(OH)_3$ network or individual particles with an adsorbed layer of $Al(OH)_3$. In either case, the charge is pH sensitive, the isoelectric point occurring near pH 6. It is under these pH conditions, then, that the flocculating effectiveness of the precipitate is optimum. In fact, the pH is often adjusted so that the hydrous aluminum oxide surface has a slightly positive value of ζ (about 5 mV). This promotes further interaction with slightly anionic polymeric materials which are also added to further build up and strengthen flocs. Once adequate flocculation has been accomplished, the dispersed particles are removed by sedimentation or filtration.

Numerous other applications could be listed in which electrokinetic characterization provides a convenient experimental way of judging the relative stability of a system to flocculation. Paints, printing inks, drilling muds, and soils are examples of additional systems whose properties are extensively studied and controlled by means of the ζ potential.

In addition to these applications in which ζ is used to monitor for optimum flocculation conditions, there are applications which explicitly depend on mobility or differences in mobility for their usefulness. We have already noted that zone electrophoresis is similar in many ways to chromatography. One important application of the ability of electrophoresis to segregate materials by mobility is immunoelectrophoresis. This technique uses known immunochemical reactions between antigen and antibody for the identification of proteins separated electrophoretically. Experimentally, an antigen mixture is subjected to electrophoresis on a suitable medium (usually agar gel). Next, the antibody mixture is introduced into a slit cut in the gel parallel to the axis of the separation. The antigen and antibody components then diffuse toward one another, producing an arc-shaped precipitate where the two fronts meet. Figure 11.11b shows a trace of the precipitated arcs which result from reacting separated human serum with horse serum antibodies. It is obvious from a comparison of parts a and b of Fig. 11.11 that each of the broad electrophoretic zones actually consists of many constituents of different immunochemical nature.

Tests of this sort are particularly useful for comparing either two antigen preparations (against a single antibody) or two antibody preparations (against a single antigen). In such a comparison, one of the samples serves as a control and differences between the two are revealed by an unpaired arc of precipate at a particular location along the path of separation.

Electrodeposition is another direct application of electrophoretic mobility. In this process, as in electroplating with metals, the substance to be coated is made into an electrode of opposite charge from the particles to be deposited. At one time, natural rubber latex was extensively fabricated in this way. Paint coatings which are quite dense and coherent with little tendency to sag or run can be prepared by electrodeposition. If the deposited layer has insulating properties, this technique is also self-regulating, producing a uniform thin covering of very good quality.

Although electrophoresis is the most important of the electrokinetic methods, it is not the only one with practical applications. We already noted in Sec. 11.7 that streaming potentials could be quite hazardous in low conducting, highly flammable substances such as purified hydrocarbons. In this case, our knowledge of the effect enables us to minimize it. Dewatering fine suspensions which are not amenable to filtration is an application of electro-osmosis. Peat, clay, and other minerals have been dewatered this way, and water may be removed from moist soil prior to excavation by electro-osmosis. Electrodes are driven into the ground, the cathode in the form of a perforated pipe. The surface of the soil particles carries a negative charge; therefore the diffuse part of the double layer is positive and the solution moves toward the cathode. The water which collects in the cathode is subsequently removed by pumping.

11.12 WRAP-UP: CHAPTERS 9 THROUGH 11

The common theme connecting Chaps. 9 through 11 is the stability of some dispersed lyophobic systems with respect to flocculation. To understand this stability, it is important to appreciate the electrical potential near a charged wall. This is the central consideration of Chap. 9. In Chap. 11, we examined what happens if one part of the electrical double layer moves with respect to the other. At first glance, this seems to have nothing to do with flocculation, although it clearly depends on the structure of the double layer. However, further analysis reveals that the potential in the double layer at some (unknown) location close to the surface is measured by electrokinetics. This measured potential is part of the total potential which is responsible for stability.

The discussion of stability would be purely academic if double layer repulsion were the only interaction between particles. This is not the case, and Chap. 10 is concerned with the attraction which operates among all particles. The origin of these ubiquitous attractions can be traced all the way back to the molecular level where they are seen to be the same forces responsible for gas nonideality, liquid miscibility, and so on.

van der Waals attractions are electrostatic in origin just as double layer interactions are. In this sense, Chaps. 9 to 11 have all been concerned with the electrical interactions among colloidal particles. In one way or another, the same statement can be made for all topics in chemistry, however, so there is little advantage in promoting this viewpoint to unify the material of these chapters.

The primary variables which determine the stability of an aqueous dispersion have been shown to be the potential at the surface or at the inner limit of the diffuse part of the double layer, the valence and concentration of the electrolyte in the continuous phase, and the Hamaker constant of water and the dispersed phase. These same principles undoubtedly apply in nonaqueous electrolyte solutions as well, but we have emphasized aqueous hydrophobic colloids in our discussion because they are relatively well understood and contain many systems of practical importance.

REFERENCES

1. H. A. Abramson, L. S. Moyer, and M. H. Gorin, *Electrophoresis of Proteins*, Hafner, New York, 1964.
2. C. C. Brinton, Jr., and M. A. Lauffer, "The Electrophoresis of Viruses, Bacteria and Cells, and the Microscope Method of Electrophoresis," in *Electrophoresis*, Vol. 1 (M. Bier, ed.), Academic Press, New York, 1959.
3. D. A. Haydon, "The Electrical Double Layer and Electrokinetic Phenomena," in *Recent Progress in Surface Science*, Vol. 1 (J. F. Danielli, K. G. A. Pankhurst, and A. C. Riddiford, eds.), Academic Press, New York, 1964.
4. L. G. Longsworth, "Moving Boundary Electrophoresis—Practice," in *Electrophoresis*, Vol. 1 (M. Bier, ed.), Academic Press, New York, 1959.
5. J. Th. G. Overbeek, "Quantitative Interpretation of the Electrophoretic Velocity of Colloids," in *Advances in Colloid Science*, Vol. 3 (H. Mark and E. J. W. Verwey, eds.), Wiley-Interscience, New York, 1950.
6. J. Th. G. Overbeek, "Electrokinetic Phenomena," in *Colloid Science*, Vol. 1 (H. R. Kruyt, ed.), Elsevier, Amsterdam, 1952.
7. J. Th. G. Overbeek and L. Lyklema, "Electric Potentials in Colloidal Systems," in *Electrophoresis*, Vol. 1 (M. Bier, ed.), Academic Press, New York, 1959.
8. J. Th. G. Overbeek and P. H. Wiersma, "The Interpretation of Electrophoretic Mobilities," in *Electrophoresis*, Vol. 2 (M. Bier, ed.), Academic Press, New York, 1967.
9. D. J. Shaw, *Electrophoresis*, Academic Press, New York, 1969.

PROBLEMS

1. Particles of Fe_2O_3 with an average diameter of 1 μm were dispersed in xylene containing 5×10^{-3} moles liter^{-1} of copper(I) oleate. These showed an electrophoretic mobility of 0.110 $\mu s^{-1}\ V^{-1}\ cm^{-1}$. The conductivity of the solution

was 4.7×10^{-10} ohm^{-1} cm^{-1}, indicating an ion concentration about 10^{-11} *M*.* Calculate κ^{-1} for this concentration. Which limiting form of Eq. (40) is most applicable in this system? Would the same conclusion be true for a 5×10^{-3} *M* aqueous solution of a 1:1 electrolyte? What is ζ for these particles? For xylene, $\varepsilon = 2.3$ and $\eta = 0.0065$ P.

2. Criticize or defend the following proposition: "Zeta potentials for three different polystyrene latex preparations were calculated by the Helmholtz–Smoluchowski equation from electrophoresis measurements made in different concentrations of KCl.†

Latex designation	$R \times 10^8$ (cm)	ζ (mV)		
		10^{-1} *M* KCl	10^{-2} *M* KCl	10^{-3} *M* KCl
L	475	21	29	40
M	610	29	39	53
N	665	34	47	64

These zeta potentials are inaccurate because the range of κR values exceeds the range of validity for the Helmholtz–Smoluchowski equation. The nature of the error is such as to make the estimated values of ζ too low."

3. The electrophoretic mobility of sodium dodecyl sulfate micelles was determined by the moving boundary method after the micelles were made visible by solubilizing dye in them. This quantity was measured at the cmc in the presence of varying concentrations of NaCl. The radius of the micelles was determined by light scattering.‡

moles NaCl liter^{-1}	$u \times 10^4$ (cm^2 s^{-1} V^{-1})	κR
0	4.55	0.61
0.05	3.63	1.69

Estimate from Fig. 11.5a the appropriate value of C to be used in Eq. (40) according to Henry's equation. Calculate ζ using these estimated C values. Figure 11.5b shows that Henry's equation overestimates C (and, therefore, underestimates ζ). Estimate C from Fig. 11.5b using the curve for the ζ value which is nearest—on the high side—to the values obtained by using Henry's equation. Re-evaluate ζ on the basis of these "constants." For this system, $\varepsilon = 78.5$ and $\eta = 0.0089$ P.

4. The accompanying mobility data for colloidal SiO_2 at a constant ionic strength of 10^{-3} *M* reveal the superpositioning of specific chemical effects on general electrostatic phenomena. Adjustment of pH was made by addition of HNO_3 or KOH, maintaining the ionic strength. The following results were

* H. Koelmans and J. Th. G. Overbeek, *Discuss. Faraday Soc.* **18**:52 (1954).
† A. Kitahara and H. Ushiyama, *J. Colloid Interface Sci.* **43**:73 (1973).
‡ D. Stigter and K. J. Mysels, *J. Phys. Chem.* **59**:45 (1955).

obtained*:

	$u \times 10^4$ ($cm^2\ s^{-1}\ V^{-1}$)								
pH of solution:	2.0	3.0	4.0	5.0	6.0	7.0	8.0	9.0	10.0
SiO_2	0	−1.4	−1.7	−2.0	−2.3	−2.5	−2.6	−2.8	−3.0
$SiO_2 + 10^{-4}\,M$ $La(NO_3)_3$	0	−1.1	−1.2	−1.2	−1.1	−0.1	+2.2	+0.5	−1.2

Criticize or defend the following propositions: "H^+ and OH^- are potential determining for SiO_2—in the absence of a hydrolyzable cation—with an isoelectric point of 2.0. For solid $La(OH)_3$ the zero point of charge is known (by independent studies) to be 10.4. The solid surface apparently becomes coated by $La(OH)_3$ at higher pH levels and goes through a transition from one character to another at intermediate pH values."

5. In their study of the effects of hydrolyzable cations on electrokinetic phenomena (see Problem 4), James and Healy compared the electrophoretic behavior of colloidal silica with the streaming potential through a silica capillary. In both sets of experiments, the solution was $10^{-3}\,M$ KNO_3 and $10^{-4}\,M$ $Co(NO_3)_2$. The following results were obtained:

pH	6.0	7.0	7.5	8.0	9.0	10.0
ζ (mV) from streaming potential	−65	−55	−30	+10	+25	+20
$u \times 10^4$ ($cm^2\ s^{-1}\ V^{-1}$)	−2.5	−2.5	−2.2	−1.8	+0.5	−0.3

The silica surface area-to-solution volume ratio was $2 \times 10^{-3}\ m^2\ liter^{-1}$ for the streaming potential experiment and $1.0\ m^2\ liter^{-1}$ for the electrophoresis experiment. Calculate ζ_{HS} at each pH from the electrophoresis data ($\eta = 0.00894$ P, $\varepsilon = 78.5$). Propose an explanation for the charge reversal behavior of the silica. Discuss the origin of the difference between ζ_{HS} and $\zeta_{St\ Pot}$ in terms of this model.

6. P. Somasundaran and R. D. Kulkarni† measured the streaming potential of $10^{-3}\,N$ KNO_3 against quartz at 25°C, obtaining the following results:

Φ (mV)	−9.0	−18.0	−26.0	−35.0
p (mm Hg)	50	100	150	200

Use these data to evaluate ζ/k. What would be the value of the ratio V/I for this system? What would be the rate of volume displacement if a current of 1.0 mA flowed through the apparatus? Evaluate ζ for the quartz–solution interface, assuming $\Lambda \simeq 145\ cm^2\ eq^{-1}\ ohm^{-1}$ for $10^{-3}\,N$ KNO_3.

* R. O. James and T. W. Healy, *J. Colloid Interface Sci.* **40**:42 (1972).
† P. Somasundaran and R. D. Kulkarni, *J. Colloid Interface Sci.* **45**:591 (1973).

7. It has been estimated* that a specific conductivity of 10^3 picomho m^{-1} would provide an ample margin of safety against electrokinetic explosions for the handling of refined petroleum products. These authors also measured the concentrations of various additives needed to reach this level of conductivity:

Solvent	Additive	Concentration (kmol m^{-3})
Benzene	Tetraisoamylammonium picrate	1×10^{-4}
Benzene	Calcium diisopropylsalicylate	5×10^{-3}
Gasoline	Ca salt of di-(2-ethylhexyl)sulfosuccinic acid	1×10^{-3}
Gasoline	Cr salt of mono- and dialkyl (C_{14}–C_{18}) salicylic acid	2.5×10^{-6}

Calculate the apparent value of the equivalent conductance Λ for each of these electrolytes in the conventional units cm^2 eq^{-1} ohm^{-1}. How do the Λ values of these compounds compare with Λ_0 for simple electrolytes in aqueous solutions?

8. The pH variation of the electrophoretic mobility of solid $Th(OH)_4$ in 10^{-2} *M* HNO_3–KOH electrolyte is as follows†:

pH	7.6	8.0	9.0	9.6	10.0	10.3	10.6	11.3
$u\times10^4$ (cm^2 s^{-1} V^{-1})	+2.4	+2.2	+1.3	+1.0	−0.1	−1.1	−1.5	−1.8

Use these data to evaluate the isoelectric point for $Th(OH)_4$. Since H^+ and OH^- appear to be potential determining, ψ may be estimated at various pH levels according to Eq. (9.2) if we identify the isoelectric point with the true point of zero charge. Compare these values with values of ζ calculated by means of the Helmholtz–Smoluchowski equation ($\eta=0.0089$ P, $\varepsilon=78.5$). Are the results qualitatively (quantitatively?) consistent with Fig. 11.7?

9. The aggregation number n and radius of sodium dodecyl sulfate micelles (by light scattering) and the zeta potential (from electrophoresis, by an accurate formula) were determined in the presence of varying concentrations of NaCl‡:

moles NaCl $liter^{-1}$	ζ (mV)	$R\times10^8$ (cm)	κR	n
0.01	92.3	22.1	0.86	89
0.03	80.9	23.0	1.32	100
0.10	68.3	24.0	2.40	112

Use Eq. (85) to estimate the charge of the micelles. What approximation(s) in the derivation of Eq. (85) prevents this expression from applying exactly to this

* A. Klinkenberg and B. V. Poulston, *J. Inst. Pet.* **44**:379 (1958).
† R. O. James and T. W. Healy, *J. Colloid Interface Sci.* **40**:42 (1972).
‡ D. Stigter and K. J. Mysels, *J. Phys. Chem.* **59**:45 (1955).

system? On the basis of the charges evaluated by Eq. (85), calculate the ratio of charge to aggregation number, the effective degree of dissociation, of these micelles. How do these results compare with the numbers given in Table 7.3?

10. An electrophoretic technique which is especially interesting for the study of proteins is called "isoelectric focusing." In this method, electrophoresis is carried out across a medium which supports a pH gradient. The pH gradient and cell polarity are such that the cathode end of the column is relatively basic. Thus a positively charged protein gradually loses its charge as it migrates, finally coming to rest at a pH corresponding to its isoelectric point. A. Carlstrom and D. Vesterberg* used this method to study the heterogeneity of peroxidase from cow's milk. After focusing was achieved, the column was drained and the pH and absorbance (at 280 nm) of successive fractions of eluent were measured:

Fraction No.	Absorbance	pH	Fraction No.	Absorbance	pH
12	0.9	9.83	28	1.4	9.49
14	3.0	9.80	30	0.9	9.45
16	2.1	9.75	32	0.6	9.38
18	1.2	9.70	34	0.8	9.31
20	2.7	9.69	36	0.5	9.30
21	2.2	9.685	38	0.3	9.28
22	2.8	9.68	40	0.4	9.23
24	1.6	9.60	41	0.5	9.16
26	1.2	9.55	42	0.4	9.10

How many components does this sample apparently contain? What are the values of the isoelectric points for each?

11. Criticize or defend the following proposition as a forensic technique for "fingerprinting" bloodstains: "Antigens are eluted from dried bloodstains into a buffer solution. Blood serum antibodies are prepared in two rabbits, one by injection with the antigens from the bloodstain, the other by injection with serum from the suspect. Immunoelectrophoresis of the antigen against the two antibody preparations should allow both qualitative and quantitative measures of the individuality of bloodstains."†

* A. Carlstrom and D. Vesterberg, *Acta Chem. Scand.* **21**:271 (1967).
† G. H. Sweet and J. W. Elvins, *Science* **192**:1012 (1976).

Appendix A

Examples of expansions encountered in this book.

(i) $1/(1-x) = 1 + x + x^2 + x^3 + \cdots$

(ii) $\ln(1+x) = x - \frac{1}{2}x^2 + \frac{1}{3}x^3 - \cdots \quad (-1 < x < +1)$

(iii) Binomial: $(1 \pm x)^n = 1 \pm nx \pm [n(n-1)/2!]x^2$

$$\pm [n(n-1)(n-2)/3!]x^3 \pm \cdots \quad (x^2 < 1)$$

(iv) Taylor: $f(x) = f(x_0) + (x - x_0)f'(x_0) + (x - x_0)^2/2! f''(x_0) + \cdots$

(v) $\sin x = x - x^3/3! + x^5/5! - \cdots$

(vi) $\sinh x = x + x^3/3! + x^5/5! + \cdots$

(vii) $\cosh x = 1 + x^2/2! + x^4/4! + \cdots$

Appendix B

Examples of gamma functions encountered in this book.

$$\int_0^\infty x^n e^{-ax^2}\, dx =$$

$$\tfrac{1}{2}\sqrt{\pi/a} \quad \text{if} \quad n = 0 \qquad\qquad \frac{1}{2a} \quad \text{if} \quad n = 1$$

$$\frac{1}{4a}\sqrt{\pi/a} \quad \text{if} \quad n = 2 \qquad\qquad \frac{1}{2a^2} \quad \text{if} \quad n = 3$$

$$\frac{3}{8a^2}\sqrt{\pi/a} \quad \text{if} \quad n = 4 \qquad\qquad \frac{1}{a^3} \quad \text{if} \quad n = 5$$

Appendix C

Units: CGS–SI Interconversions

From time to time, probably all science students find themselves entangled in a problem of units. For those who have advanced through physical chemistry to the level of this book, these problems have obviously not been insurmountable. It is likely, however, that—along with feelings of frustration—they have left us with wishes that everyone used the same units, specifically those units with which we are most comfortable. In response to the recognized need for uniformity, IUPAC recommends the use of *Système International* (SI) units which are essentially standardized mks units. In the United States, most chemistry students are more comfortable with cgs units (think, for example, of the familiar constants of Boltzmann and Planck) and such common, nonmetric units as the calorie and the atmosphere. As noted in the Preface, this book uses the more familiar cgs/common units on the premise that learning new chemistry is troublesome enough without compounding the difficulty by learning new units as well. SI units are appearing with increasing frequency in textbooks and, especially, journals, so it seems appropriate to include a few remarks about the connections between these "new" units and the ones used in this book. Perhaps by the time this book is ready for a second edition, the use of SI units in the United States will have become sufficiently well established that SI units may be used throughout, so an appendix such as this will only be an aid to those consulting the "old" literature.

The SI system is based on mutually consistent units assigned to the nine physical quantities listed in Table C.1. In addition to the SI units for these nine quantities, the table also lists cgs or other commonly encountered units as well as the conversion factors between the two. In this table, the headings at the top of the table indicate how the conversion factors are to be used in going from SI to cgs/common units, whereas the bottom headings indicate the use of these factors for calculations in the reverse direction.

From these nine basic quantities numerous other SI units may be derived. Table C.2 lists a number of these derived units, particularly those which are relevant to colloid and surface chemistry. The table is arranged alphabetically according to the name of the physical quantity involved. Note that instructions for the use of the conversion factors—depending on the direction of the conversion—are given in the top and bottom headings of the columns. Table C.2 is by no means an exhaustive list of the various derived SI units; Hopkins [1] reports on many additional conversions, as do most handbooks and numerous other references.

TABLE C.1 *Basic* SI *units and their relation to* cgs *or other common units. Note that different column headings are given at the top and bottom of the table to facilitate conversions from* SI *to* cgs *and from* cgs *to* SI, *respectively*

Physical quantity	SI unit		× Conversion factor →	cgs/common unit	
	Name	Symbol		Name	Symbol
Length	meter	m	10^2	centimeter	cm
Mass	kilogram	kg	10^3	gram	g
Time	second	s	1	second	s
Electric current	ampere	A	2.998×10^9	statampere	statamp
Thermodynamic temperature	kelvin	K	1	kelvin	K
Luminous intensity	candela	cd	π	Lambert	(cm^2)
Amount of substance	mole	mole	1	mole	mole
Plane angle	radian	rad	$180/\pi$	degree (angle)	°
Solid angle	steradian	sr	1	steradian	sr
	SI unit		← Conversion factor ÷	cgs/common unit	

TABLE C.2 *Derived SI units and their relation to cgs or other common units. Note that different column headings are given at the top and bottom of the table to facilitate conversions from SI to cgs and from cgs to SI, respectively*

Physical quantity	SI unit ×	Conversion factor →	cgs/common unit
Acceleration	$m\ s^{-2}$	10^2	$cm\ s^{-2}$
Acceleration, angular	$rad\ s^{-1}$	1	$rad\ s^{-1}$
Area	m^2	10^4	cm^2
		10^{20}	$Å^2$
Capacitance (farad)	$F = m^{-2}\ kg^{-1}\ s^4\ A^2 = C\ V^{-1}$	8.99×10^{11}	statfarad
Charge (Coulomb)	$C = A\ s = J\ V^{-1}$	3.00×10^9	statcoulomb (esu)
Charge density, surface	$C\ m^{-2}$	3.00×10^5	statcoul cm^{-2}
Charge density, volume	$C\ m^{-3}$	3.00×10^3	statcoul cm^{-3}
Conductance (siemens)	$S = m^{-2}\ kg^{-1}\ s^3\ A^2 = ohm^{-1}$	8.99×10^{11}	statmho ($statohm^{-1}$)
Conductivity	$ohm^{-1}\ m^{-1}$	8.99×10^9	statmho cm^{-1}
		10^{-2}	mho cm^{-1} ($ohm^{-1}\ cm^{-1}$)
Density	$kg\ m^{-3}$	10^{-3}	$g\ cm^{-3}$
Diffusion coefficient	$m^2\ s^{-1}$	10^4	$cm^2\ s^{-1}$
Dipole moment	$C\ m$	3.00×10^{11}	statcoul cm
Electric field	$V\ m^{-1}$	3.34×10^{-5}	statvolt cm^{-1}
		10^{-2}	$V\ cm^{-1}$
Electric potential (volt)	$V = m^2\ kg\ s^{-3}\ A^{-1} = J\ C^{-1}$	3.34×10^{-3}	statvolt
Energy (joule)	$J = m^2\ kg\ s^{-2} = N\ m$	10^7	erg
		0.2390	calorie
Entropy	$J\ K^{-1}$	0.2390	cal K^{-1}
Force (newton)	$N = m\ kg\ s^{-2}$	10^5	dyne
Frequency (hertz)	$Hz = s^{-1}$	1	s^{-1}
Friction factor	$kg\ s^{-1}$	10^3	$g\ s^{-1}$
Heat capacity	$J\ K^{-1}$	0.2390	cal K^{-1}
Molarity	moles dm^{-3}	1	moles $liter^{-1}$

Quantity	SI unit	Conversion factor	cgs/common unit
Moment, dipole	C m	3.00×10^{11}	statcoul cm
Moment, force	N m	10^7	dyne cm
Moment, inertia	kg m^2	10^7	g cm^2
Momentum	N s	10^5	dyne s
Momentum, angular	J s	10^7	erg s
Period	s	1	s
Permittivity	F m^{-1}	8.99×10^9	statfarad cm^{-1}
Polarization, electric	C m^{-2}	3.00×10^5	statcoul cm^{-2}
Potential (volt)	V	3.34×10^{-3}	statvolt
Power (watt)	$W = m^2\ kg\ s^{-3} = J\ s^{-1}$	10^7	ergs s^{-1}
Pressure (pascal)	$Pa = m^{-1}\ kg\ s^{-2} = N\ m^{-2}$	10	dynes cm^{-2}
		9.87×10^{-6}	atm
Radius of gyration	m	10^2	cm
Resistance (ohm)	$ohm = m^2\ kg\ s^{-3}\ A^{-2} = VA^{-1}$	1.11×10^{-12}	statohm
Specific heat capacity	J kg^{-1} K^{-1}	2.39×10^{-4}	cal g^{-1} K^{-1}
Stress	N m^{-2}	10	dynes cm^{-2}
Surface energy	J m^{-2}	10^3	ergs cm^{-2}
Surface tension	N m^{-1}	10^3	dynes cm^{-1}
Torque	N m	10^7	dyne cm
Velocity	m s^{-1}	10^2	cm s^{-1}
Velocity, angular	rad s^{-1}	1	rad s^{-1}
Viscosity	N s m^{-2}	10	dyne s cm^{-2} (P)
Volume	m^3	10^6	cm^3
		10^3	dm^3 (liter)
Wave number	m^{-1}	10^{-2}	cm^{-1}
Weight	N	10^5	dyne
	SI unit ←	Conversion factor ÷	cgs/common unit

TABLE C.3 *Multiples of units, their names, and symbols*

Multiple	Prefix	Symbol
10^{12}	tera	T
10^{9}	giga	G
10^{6}	mega	M
10^{3}	kilo	k
10^{2}	hecto	h
10	deca	da
10^{-1}	deci	d
10^{-2}	centi	c
10^{-3}	milli	m
10^{-6}	micro	μ
10^{-9}	nano	n
10^{-12}	pico	p
10^{-15}	femto	f
10^{-18}	atto	a

One reason for the great diversity of units in existence is the fact that quantities of such diverse magnitude are measured. A general rule is that the unit should be appropriate in magnitude to the quantity being measured. To obtain a dimension of convenient size in SI units, the SI unit is multiplied by a power of 10 and the prefixes listed in Table C.3 are affixed to the unit. Accordingly, a surface tension of 50 dynes cm^{-1} becomes 50 mN m^{-1}, a viscosity of 0.01 P becomes 1.0 mN s m^{-2}, and a dimension of 100 Å becomes 10 nm.

REFERENCES

1. R. A. Hopkins, *The International (SI) Metric System and How It Works*, Polymetric Services, Inc., Reseda, California, 1973.
2. M. L. McGlashan, *Pure Appl. Chem.* **21**:577 (1970).
3. C. H. Page and P. Vigoureux (eds.), *The International System of Units (SI)*, National Bureau of Standards, Special Publication 330, 1974.
4. M. A. Paul, *J. Chem. Educ.* **48**:569 (1971).

Index

The notation F or T which accompanies certain entries refers to figures or tables, respectively, appearing on the indicated pages.

A

B

C

D

E

F

I

K

L

M

Q

R

T

U

V

W

X

Y

Z